Symmetry in Mechanical Engineering

Symmetry in Mechanical Engineering

Special Issue Editors

Adam Glowacz
Grzegorz Królczyk
Jose A. Antonino-Daviu

MDPI • Basel • Beijing • Wuhan • Barcelona • Belgrade • Manchester • Tokyo • Cluj • Tianjin

Special Issue Editors

Adam Glowacz
AGH University of Science and Technology
Poland

Grzegorz Królczyk
Opole University of Technology
Poland

Jose A. Antonino-Daviu
Universitat Politecnica de València
Spain

Editorial Office
MDPI
St. Alban-Anlage 66
4052 Basel, Switzerland

This is a reprint of articles from the Special Issue published online in the open access journal *Symmetry* (ISSN 2073-8994) (available at: https://www.mdpi.com/journal/symmetry/special_issues/Symmetry_Mechanical_Engineering).

For citation purposes, cite each article independently as indicated on the article page online and as indicated below:

LastName, A.A.; LastName, B.B.; LastName, C.C. Article Title. *Journal Name* **Year**, *Article Number*, Page Range.

ISBN 978-3-03936-214-1 (Pbk)
ISBN 978-3-03936-215-8 (PDF)

Contents

About the Special Issue Editors

Adam Glowacz received his Ph.D. in Computer Science from the AGH University of Science and Technology, Cracow, Poland, in 2013. Adam Glowacz is the author/coauthor of 106 scientific papers (58 papers indexed by Web of Science) that correspond to a h-index of 21 and 1026 citations in Web of Science and a h-index of 23 and 1407 citations in Google Scholar. He has supervised 30 B.Sc. and 12 M.Sc. theses. Adam Glowacz is an Associate Editor of Symmetry, Electronics, Measurement, and Advances in Mechanical Engineering, and has also authored 300 scientific reviews.

Grzegorz Krolczyk is Professor and Vice-Rector for Research and Development at Opole University of Technology and author and coauthor of 200 scientific publications (110 JCR papers), as well as around 30 studies and industrial applications. His main scientific activities are in the analysis and improvement of manufacturing processes, surface metrology, and surface engineering. His research focuses on sustainable manufacturing as a tool for the practical implementation of the concept of social responsibility in the area of machining. Grzegorz Krolczykis is a member of several scientific organizations, including an expert in the Section of Technology of the Committee on Machine Building of the Polish Academy of Sciences. In addition, he is a member of several editorial committees of scientific journals. He has participated in advisory and opinion-forming bodies, including the advisory team of the Minister of Science and Higher Education. The coauthor of two patent applications, Grzegorz Krolczyk has been awarded on numerous occasions for his scientific activities in Poland and around the world.

Jose A. Antonino-Daviu received his M.Sc. and Ph.D. degrees in Electrical Engineering, both from the Universitat Polit'ecnica de Val'encia, Valencia, Spain, in 2000 and 2006, respectively. He has worked for IBM, where he was involved in several international projects. He is currently Full Professor in the Department of Electrical Engineering, Universitat Polit'ecnica de Val'encia. He was an Invited Professor at Helsinki University of Technology, Finland, in 2005 and 2007; Michigan State University, USA, in 2010; Korea University, South Korea, in 2014; Université Claude Bernard Lyon 1, France; and Coventry University, U.K., in 2016. He is a coauthor of more than 200 papers published in technical journals and conference proceedings and one international patent. Dr. Antonino-Daviu is Associate Editor of IEEE Transactions on Industrial Informatics, IEEE Industrial Electronics Magazine, and IEEE Journal of Emerging and Selected Topics in Industrial Electronics. He received the IEEE Second Prize Paper Award from the Electric Machines Committee of the IEEE Industry Applications Society (2013). He also received the Best Paper Award in the conferences IEEE ICEM 2012, IEEE SDEMPED 2011, and IEEE SDEMPED 2019 and "Highly Commended Recognition" of the IET Innovation Awards in 2014 and in 2016. He was the General Co-Chair of SDEMPED 2013 and is a member of the Steering Committee of IEEE SDEMPED. In 2016, he received the Medal of the Spanish Royal Academy of Engineering (Madrid, Spain) for his contributions in new techniques for predictive maintenance of electric motors. In 2018, he was awarded the prestigious "Nagamori Award" from the Nagamori Foundation (Kyoto, Japan). In 2019, he received the SDEMPED Diagnostic Achievement Award (Toulouse, France) for his contributions to the advanced diagnosis of electric motors.

Editorial

Introduction to Special Issue on Symmetry in Mechanical Engineering

Grzegorz Krolczyk [1,*], Stanislaw Legutko [2], Zhixiong Li [3] and
Jose Alfonso Antonino Daviu [4]

1   Faculty of Mechanical Engineering, Opole University of Technology, 76 Proszkowska St.,
    45-758 Opole, Poland
2   Faculty of Mechanical Engineering and Management, Poznan University of Technology, 3 Piotrowo Street,
    60-965 Poznan, Poland; stanislaw.legutko@put.poznan.pl
3   School of Mechanical, Materials, Mechatronic and Biomedical Engineering, University of Wollongong,
    Wollongong, NSW 2522, Australia; zhixiong.li@ieee.org
4   Instituto Tecnológico de la Energía, Universitat Politècnica de València (UPV), Camino de Vera s/n,
    46022 Valencia, Spain; joanda@die.upv.es
*   Correspondence: g.krolczyk@po.opole.pl

Received: 22 January 2020; Accepted: 22 January 2020; Published: 5 February 2020

1. Introduction

Recent advancements in mechanical engineering are an essential topic for discussion. The topics relating to mechanical engineering include the following: measurements of signals of shafts, springs, belts, bearings, gears, rotors, machine elements, vibration analysis, acoustic analysis, fault diagnosis, construction, analysis of machine operation, analysis of smart-material systems, integrated systems, stresses, analysis of deformations, analysis of mechanical properties, signal processing of mechanical systems, and rotor dynamics. Mechanical engineering deals with solid and fluid mechanics, rotation, movements, materials, and thermodynamics.

2. The Content

This Special Issue, with 15 published articles, presents the topic "Symmetry in Mechanical Engineering". The presented topic is interesting. It is categorized into eight different sections:

- deformation;
- stresses;
- mechanical properties;
- tribology;
- thermodynamic;
- measurement;
- fault diagnosis;
- machine;

The authors of the first paper analysed the self-excited vibration of a thin spur gear caused by the initial transverse vibration [1]. The article [2] described a new technique to identify sectional deformation modes of the doubly symmetric thin-walled cross-section. The matching model of the dual mass flywheel and the power transmission by integration of the sensitivity analysis method was presented in the paper [3]. In the paper [4], the authors presented an approach for the active control of structural vibration. The authors of the paper [5] presented an approach to the correction of optical measurement results of fertilizer particles. Fault diagnosis of the rolling bearing using vibration signals was presented in [6]. A study of the effect of medium viscosity on breakage parameters for wet grinding was presented in [7]. The monitoring method for the wear state of a tool using a convolutional bidirectional LSTM model was shown in the article [8]. The use of structural symmetries of a U12 engine using vibration analysis was presented in [9].

The Special Issue contains other interesting papers about mechanical engineering. The presented solutions, methods, and approaches can be improved and used in the future. Moreover, mechanical engineering is essential for fault diagnosis of machines [10–22] and the analysis of temperature [23–25]. The mechanical properties of materials are also investigated in the literature [26–28]. Acoustic analysis is also profitable for the analysis of the power transformer and detection of defects in on-load tap-changers [29,30]. Acoustically induced cavitation bubbles in insulating oil are also presented in the literature [31].

3. Summary

The development of techniques and methods related to mechanical engineering is growing every month. The described articles have contribution to mechanical engineering. The proposed research can find applications in factories, oil refineries, and mines. It is essential to develop new improved methods, techniques and devices related to mechanical engineering.

Author Contributions: All the authors contributed equally to the conception of the idea, implementing and analyzing the experimental results, and writing the manuscript. All authors have read and agreed to the published version of the manuscript.

Acknowledgments: The Guest Editors would like to thank all authors, reviewers and the editorial board of the MDPI Symmetry journal for their valuable contributions to this Special Issue.

Conflicts of Interest: The author declares no conflict of interest.

References

1. Wang, Y.; Ye, H.; Yang, L.; Tian, A. On the Existence of Self-Excited Vibration in Thin Spur Gears: A Theoretical Model for the Estimation of Damping by the Energy Method. *Symmetry* **2018**, *10*, 664. [CrossRef]
2. Zhang, L.; Ji, A.; Zhu, W.; Peng, L. On the Identification of Sectional Deformation Modes of Thin-Walled Structures with Doubly Symmetric Cross-Sections Based on the Shell-Like Deformation. *Symmetry* **2018**, *10*, 759. [CrossRef]
3. Chen, L.; Zhang, X.; Yan, Z.; Zeng, R. Matching Model of Dual Mass Flywheel and Power Transmission Based on the Structural Sensitivity Analysis Method. *Symmetry* **2019**, *11*, 187. [CrossRef]
4. Bai, L.; Feng, Y.-W.; Li, N.; Xue, X.-F.; Cao, Y. Data-Driven Adaptive Iterative Learning Method for Active Vibration Control Based on Imprecise Probability. *Symmetry* **2019**, *11*, 746. [CrossRef]
5. Laucka, A.; Adaskeviciute, V.; Andriukaitis, D. Research of the Equipment Self-Calibration Methods for Different Shape Fertilizers Particles Distribution by Size Using Image Processing Measurement Method. *Symmetry* **2019**, *11*, 838. [CrossRef]
6. Lu, L.; Yuan, Y.; Wang, H.; Zhao, X.; Zheng, J. A New Second-Order Tristable Stochastic Resonance Method for Fault Diagnosis. *Symmetry* **2019**, *11*, 965. [CrossRef]
7. Osorio, A.M.; Bustamante, M.O.; Restrepo, G.M.; Lopez, M.M.M.; Menendez-Aguado, J.M. A Study of the Effect of Medium Viscosity on Breakage Parameters for Wet Grinding. *Symmetry* **2019**, *11*, 1202. [CrossRef]
8. Chen, Q.; Xie, Q.; Yuan, Q.; Huang, H.; Li, Y. Research on a Real-Time Monitoring Method for the Wear State of a Tool Based on a Convolutional Bidirectional LSTM Model. *Symmetry* **2019**, *11*, 1233. [CrossRef]
9. Mihalcica, M.; Vlase, S.; Paun, M. The Use of Structural Symmetries of a U12 Engine in the Vibration Analysis of a Transmission. *Symmetry* **2019**, *11*, 1296. [CrossRef]
10. Irfan, M. A Novel Non-intrusive Method to Diagnose Bearings Surface Roughness Faults in Induction Motors. *J. Fail. Anal. Prev.* **2018**, *18*, 145–152. [CrossRef]
11. Caesarendra, W.; Tjahjowidodo, T.; Kosasih, B.; Tieu, A.K. Integrated Condition Monitoring and Prognosis Method for Incipient Defect Detection and Remaining Life Prediction of Low Speed Slew Bearings. *Machines* **2017**, *5*, 11. [CrossRef]
12. Glowacz, A. Recognition of acoustic signals of induction motor using FFT, SMOFS-10 and LSVM. *Eksploat. I Niezawodn. Maint. Reliab.* **2015**, *17*, 569–574. [CrossRef]

13. Glowacz, A. Recognition of Acoustic Signals of Loaded Synchronous Motor Using FFT, MSAF-5 and LSVM. *Arch. Acoust.* **2015**, *40*, 197–203. [CrossRef]

14. Sikora, M.; Szczyrba, K.; Wrobel, L.; Michalak, M. Monitoring and maintenance of a gantry based on a wireless system for measurement and analysis of the vibration level. *Eksploat. I Niezawodn. Maint. Reliab.* **2019**, *21*, 341–350. [CrossRef]

15. Glowacz, A.; Glowacz, Z. Recognition of rotor damages in a DC motor using acoustic signals. *Bull. Pol. Acad. Sci. Tech. Sci.* **2017**, *65*, 187–194. [CrossRef]

16. Caesarendra, W.; Wijaya, T.; Tjahjowidodo, T.; Pappachan, B.K.; Wee, A.; Roslan, M.I. Adaptive neuro-fuzzy inference system for deburring stage classification and prediction for indirect quality monitoring. *Appl. Soft Comput.* **2018**, *72*, 565–578. [CrossRef]

17. Irfan, M.; Saad, N.; Ibrahim, R.; Asirvadam, V.S.; Alwadie, A. Analysis of distributed faults in inner and outer race of bearing via Park vector analysis method. *Neural Comput. Appl.* **2019**, *31*, 683–691. [CrossRef]

18. Glowacz, A.; Glowacz, W.; Kozik, J.; Piech, K.; Gutten, M.; Caesarendra, W.; Liu, H.; Brumercik, F.; Irfan, M.; Khan, Z.F. Detection of Deterioration of Three-phase Induction Motor using Vibration Signals. *Meas. Sci. Rev.* **2019**, *19*, 241–249. [CrossRef]

19. Glowacz, A. Acoustic fault analysis of three commutator motors. *Mech. Syst. Signal Process.* **2019**, *133*, 106226. [CrossRef]

20. Stief, A.; Ottewill, J.R.; Baranowski, J.; Orkisz, M. A PCA and Two-Stage Bayesian Sensor Fusion Approach for Diagnosing Electrical and Mechanical Faults in Induction Motors. *IEEE Trans. Ind. Electron.* **2019**, *66*, 9510–9520. [CrossRef]

21. Xi, W.K.; Li, Z.X.; Tian, Z.; Duan, Z.H. A feature extraction and visualization method for fault detection of marine diesel engines. *Measurement* **2018**, *116*, 429–437. [CrossRef]

22. Li, Z.X.; Wu, D.Z.; Hu, C.; Terpenny, J. An ensemble learning-based prognostic approach with degradation-dependent weights for remaining useful life prediction. *Reliab. Eng. Syst. Saf.* **2019**, *184*, 110–122. [CrossRef]

23. Chen, J.L.; Su, J.; Kochan, O.; Levkiv, M. Metrological Software Test for Simulating the Method of Determining the Thermocouple Error in Situ during Operation. *Meas. Sci. Rev.* **2018**, *18*, 52–58. [CrossRef]

24. Wang, J.F.; Kochan, O.; Przystupa, K.; Su, J. Information-measuring System to Study the Thermocouple with Controlled Temperature Field. *Meas. Sci. Rev.* **2019**, *19*, 161–169. [CrossRef]

25. Maruda, R.W.; Feldshtein, E.; Legutko, S.; Krolczyk, G.M. Analysis of Contact Phenomena and Heat Exchange in the Cutting Zone Under Minimum Quantity Cooling Lubrication conditions. *Arab. J. Sci. Eng.* **2016**, *41*, 661–668. [CrossRef]

26. Krolczyk, G.; Legutko, S.; Stoic, A. Influence of cutting parameters and conditions onto surface hardness of Duplex Stainless Steel after turning process. *Teh. Vjesn. Tech. Gaz.* **2013**, *20*, 1077–1080.

27. Kumar, R.; Chattopadhyaya, S.; Hloch, S.; Krolczyk, G.; Legutko, S. Wear characteristics and defects analysis of friction stir welded joint of aluminium alloy 6061-t6. *Eksploat. I Niezawodn. Maint. Reliab.* **2016**, *18*, 128–135. [CrossRef]

28. Krolczyk, J.B.; Krolczyk, G.M.; Legutko, S.; Napiorkowski, J.; Hloch, S.; Foltys, J.; Tama, E. Material flow optimization—A case study in automotive industry. *Teh. Vjesn. Tech. Gaz.* **2015**, *22*, 1447–1456.

29. Borucki, S.; Cichon, A.; Boczar, T.; Fracz, P. The Analysis of the Impact Point of the Power Transformer Core of Torsional Load on the Measured Parameters of the Vibroacoustics Signals. In Proceedings of the 2012 IEEE International Symposium on Electrical Insulation (ISEI), San Juan, PR, USA, 10–13 June 2012; pp. 175–178.

30. Cichon, A.; Fracz, P.; Boczar, T.; Zmarzly, D. Detection of Defects in On-Load Tap-Changers Using Acoustic Emission Method. In Proceedings of the 2012 IEEE International Symposium on Electrical Insulation (ISEI), San Juan, PR, USA, 10–13 June 2012; pp. 184–188.

31. Szmechta, M.; Zmarzly, D.; Boczar, T.; Lorenc, M. Acoustic Spectra of Ultrasound Induced Cavitations in Insulating Oils. *Acta Phys. Pol. A* **2008**, *114*, A231–A238. [CrossRef]

Article

A Novel Surface Inset Permanent Magnet Synchronous Motor for Electric Vehicles

Baojun Qu [1,2], Qingxin Yang [1], Yongjian Li [1,*], Miguel Angel Sotelo [3], Shilun Ma [4,*] and Zhixiong Li [5,6]

1   State Key Laboratory of Reliability and Intelligence of Electrical Equipment, Hebei University of Technology, Tianjin 300130, China; qbj22@sina.com (B.Q.); qxyang@tjpu.edu.cn (Q.Y.)
2   School of Mechanical Engineering, Shandong University of Technology, Zibo 255049, China
3   Department of Computer Engineering, University of Alcalá, 28801 Alcalá de Henares, Madrid, Spain; miguel.sotelo@uah.es
4   School of Transportation and Vehicle Engineering, Shandong University of Technology, Zibo 255049, China
5   Suzhou Automotive Research Institute, Tsinghua University, Suzhou 215134 China; zhixiong_li@uow.edu.au
6   School of Mechanical, Materials, Mechatronic and Biomedical Engineering, University of Wollongong, Wollongong, NSW 2522, Australia
*   Correspondence: liyongjian@hebut.edu.cn (Y.L.); msl@sdut.edu.cn (S.M.); Tel.: +86-022-6043-5928 (Y.L.)

Received: 17 December 2019; Accepted: 14 January 2020; Published: 19 January 2020

Abstract: Aiming to successfully meet the requirements of a large output torque and a wide range of flux weakening speed expansion in permanent magnet synchronous motors (PMSM) for electric vehicles, a novel surface insert permanent magnet synchronous motor (SIPMSM) is developed. The method of notching auxiliary slots between the magnetic poles in the rotor and unequal thickness magnetic poles is proposed to improve the performance of the motor. By analyzing the magnetic circuit characteristics of the novel SIPMSM, the notching auxiliary slots between the adjacent magnetic poles can affect the q-axis inductance, and the shape of magnetic pole effects the d-axis inductance of the motor. The combined action of the two factors not only weakens the cogging torque, but also improves the flux weakening capability of the motor. In this paper, the response surface methodology (RSM) is used to establish a mathematical model of the relationship between the structural parameters of the motor and the optimization objectives, and the optimal design of the motor is completed by solving the mathematical model. Experimental validation has been conducted to show the correctness and effectiveness of the proposed SIPMSM.

Keywords: electric vehicles; PMSM; auxiliary slot; response surface methodology

---

1. Introduction

With the rapid development of the automobile industry, the two global problems of environmental pollution and energy shortage are becoming more and more serious. Under such a severe situation, many countries have begun to formulate plans to ban the sale of fuel vehicles. Electric vehicles are powered by electricity and have the advantages of zero emission, low noise and energy saving. Therefore, the development and promotion of electric vehicles is highly valued by governments all over the world [1].

As the core component of electric vehicle, the performance of the driving motor directly affects the performance of electric vehicles. The research and development of high-performance electric vehicle drive motors has become one of the important factors restricting the development of electric vehicles [2]. The main types of drive motors for electric vehicles are brushless DC motors, induction motors, switched reluctance motors and permanent magnet synchronous motors (PMSM). The PMSM has a series of advantages, such as a simple structure, high efficiency and excellent performance of flux-weakening speed expansion—making it more and more widely used as an electric vehicle drive motor in recent years [3]. However, for PMSM, the magnetic energy generated by the permanent magnet will interact with the stator slot, which will produce the slot effect, increase the harmonic content in the air gap, and reduce the control accuracy of the drive system. Therefore, reducing the cogging torque and improving the performance of flux-weakening speed expansion are the important research contents of PMSM for electric vehicles.

The cogging torque is reduced by changing the size and shape of the magnetic barrier of PMSM [4]. The auxiliary slot in the stator of the surface mounted PMSM is designed, and the mathematical model of the size of the auxiliary slot is established [5,6]. Through the analytical mathematical model, the cogging torque of the motor is optimized. A method of the axial combination of different permanent magnets in a rotor is proposed to reduce harmonic content in airgap and torque ripple [7]. Stator tooth modification is used to reduce the harmonic content of the teeth, thereby reducing the eddy current loss and vibration noise, and improving the efficiency of the motor [8–10]. In summary, most of the structure optimization methods of PMSM are based on a single parameter or index. The optimization values are determined by the parameters, then the other structure parameters are optimized one by one, and, finally, the optimized parameters are combined.

However, PMSM is difficult to obtain the optimal results by optimizing a single index or parameter. The Taguchi method is used to optimize the shape of permanent magnet, which improves the efficiency and reduces the torque ripple of the motor [11]. However, the optimal value obtained by this method can only be a combination of the levels used in the experiment, and the optimal result has certain limitations. A genetic algorithm is used to optimize the structural parameters of the interior asymmetric V type magnetic pole, which reduces the torque ripple of the motor [12]. However, the genetic algorithm can easily fall into the extreme point near the optimal solution in the later stage of calculation, so the results obtained by this method tend to approach the optimal solution rather than the optimal value. A particle swarm optimization algorithm is used to optimize the piecewise width and pole arc coefficient of the permanent magnet of the surface mounted PMSM, which improves the output characteristics of the motor [13]. However, the disadvantage of the particle swarm optimization algorithm is that the optimal results can easily fall into the problem of local optimum [14]. The structure optimization of flexible rotor of hollow traveling wave ultrasonic motor is carried out based on response surface methodology (RSM), and the experiment results verify the correctness of the optimization method. The correctness of the optimization method is verified by the experiment. RSM is used to optimize the design of slotless permanent magnet linear synchronous motor to increase the average reasoning and reduce the torque ripple [15]. The accuracy of RSM design method is verified by experiments. The emergence of RSM is the result of the close connection of statistics, mathematics and computer science. Owing to this optimization method takes many factors into account and it establishes complex multi-dimensional surface which is closer to the actual situation than other optimization methods, response surface method is widely used, so RSM is widely used in practical engineering.

In this paper, a novel surface insert permanent magnet synchronous motor (SIPMSM) is proposed, and the primary design parameters of the motor are determined by empirical formulas. By establishing the mathematical model of the cogging torque and inductance of the magnetic circuit of the SIPMSM, the influence factors of the cogging torque and flux-weakening speed expansion of the developed SIPMSM are deduced. Then the multi-objective mathematical model of the relationship between the structural parameters and the influencing factors of the motor is established by RSM. Optimal structural parameters are obtained by solving the mathematical model. Lastly, the traditional SIPMSM and the novel SIPMSM are trial-manufactured and compared.

The reminders of this study are organized as follows. Section 2 describes the mathematical model of the proposed SIPMSM. Section 3 performs the parameter optimization for the SIPMSM. Experimental validation is carried out in Section 4 and conclusions are drawn in Section 5.

2. The Proposed SIPMSM

2.1. Structure Design

The shape of magnetic pole effects the output characteristics of PMSM directly. The permanent magnet of typical SIPMSM is a tile shape with inner and outer arc centers at the same point—as shown in Figure 1. A PMSM with this kind of magnetic pole usually has the disadvantages of large cogging torque, large leakage and poor flux weakening capability [16]. Therefore, a novel SIPMSM is developed, as shown in Figure 2. The permanent magnet in the novel SIPMSM is an unequal thickness magnetic pole with different inner and outer radians, which results in the uneven distribution of the radial air-gap flux density and remarkable magnetic congregate effect. In order to reduce the leakage flux and the high harmonic content in the air-gap, an auxiliary slot is notched in the rotor, as shown in Figure 3.

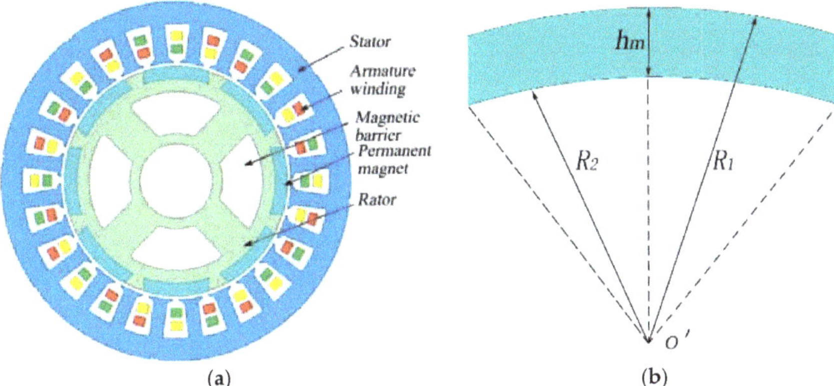

Figure 1. Schematic diagram of typical SIPMSM and tile shape magnetic poles. (**a**) Schematic diagram of traditional SIPMSM; (**b**) Schematic diagram of tile shape magnetic poles.

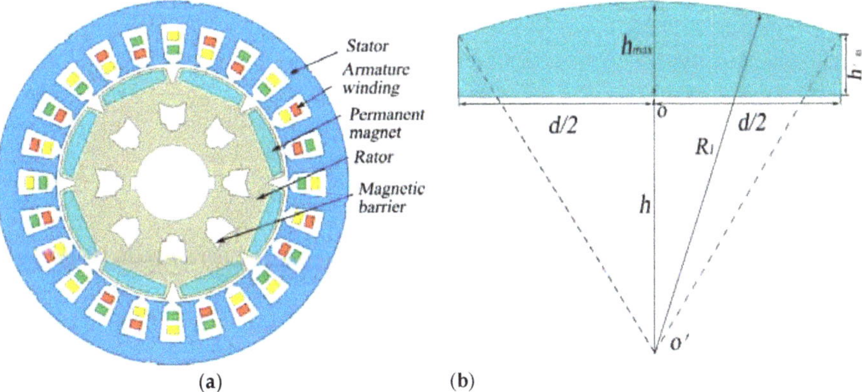

Figure 2. Schematic diagram of the novel SIPMSM and unequal thickness magnetic poles. (**a**) Schematic diagram of the novel SIPMSM; (**b**) Schematic diagram of unequal thickness magnetic poles.

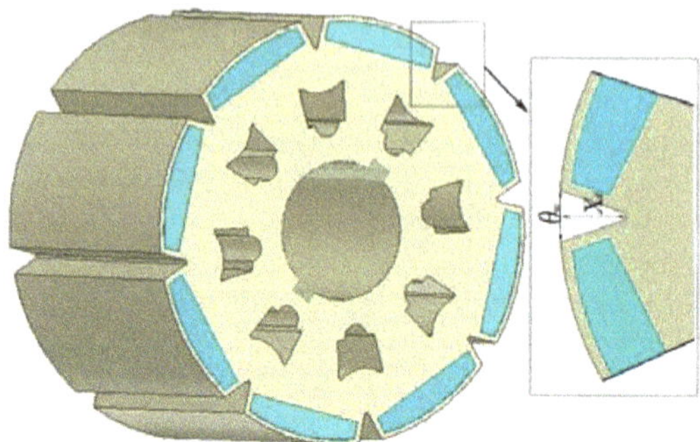

Figure 3. Schematic diagram of auxiliary slot in rotor.

According to the performance requirements of PMSM for electric vehicles, the structure parameters of the novel SIPMSM are determined by using the empirical formula (see Table 1).

Table 1. Initial Design Parameters of the novel SIPMSM.

Parameters	Numberical Value	Parameters	Numberical Value
Rated voltage (V)	60	Rated speed (r/min)	3000
Rated power (kW)	3	Rated torque (N·m)	89
Number of pole pairs	4	Rotor outer diameter (mm)	30
Number of slots	24	Rator inner diameter (mm)	70
Stator inner diameter (mm)	90	Number of turns per slot	12
Stator outer diameter (mm)	145	Magnet width	29
Maximun magnet thickness (mm)	6	Slot width of rotor/mm	7
Minimum magnet thickness(mm)	4	Slot depth of rotor/mm	5

2.2. Infulence on Cogging Torque

Compared with the traditional SIPMSM, the shape of the auxiliary slot and the unequal thickness of the permanent magnet between magnetic poles changes the harmonic content in the air gap flux density, which inevitably affects the cogging torque. In this paper, a mathematical model of the cogging torque of the novel SIPMSM is established based on the energy method, and the influence of auxiliary slot in the rotor on the cogging torque is analyzed.

The cogging torque is defined as the negative derivative of the magnetic field energy, relative to the position angle when the armature winding does not turn on the current [17]. The cogging torque of the permanent magnet motor can be expressed as

$$T_{cog} = -\frac{\partial E}{\partial \alpha} \tag{1}$$

where *E* is the energy of the magnetic field in the air-gap; *B* is the air-gap flux density; *V* is the volume of air-gap between stator and rotor; α is the relative position angle between stator and rotor.

The energy of the air-gap magnetic field can be expressed as

$$E = \frac{1}{2\mu_0} \int_V B^2(\theta, \alpha) dV \tag{2}$$

where θ is the angle between the air-gap magnetic density and the central line of the magnetic pole; μ_0 is vacuum permeability. The mathematical model of air-gap density distribution along the rotor surface with the unequal thickness magnetic pole is expressed as

$$B(\theta, \alpha) = B_r(\theta) \frac{h_m'(\theta)}{h_m'(\theta) + \delta(\theta, \alpha)} \tag{3}$$

where, $B_r(\theta)$ is the distribution of the remanence of permanent magnet along the circumferential direction; $\delta(\theta, \alpha)$ is the distribution of effective air-gap length along the circumferential direction; $h_m'(\theta)$ is the distribution of the direction of magnetization along the circumferential direction at the minimum thickness of permanent magnet.

Substituting (2) into the E, we obtain

$$E = \frac{1}{2\mu_0} \int_V B_r^2(\theta) \left[\frac{h_m'(\theta)}{h_m'(\theta) + \delta(\theta, \alpha)} \right]^2 dV \tag{4}$$

Perform Fourier decomposition of $B_r^2(\theta)$ without considering the influence of the relative position of stator and rotor.

$$B_r^2(\theta) = B_{r0} + \sum_{n=1}^{\infty} B_{rn} \cos 2np\theta = \alpha_p B_r^2 + \sum_{n=1}^{\infty} \frac{2}{n\pi} B_r^2 \sin n\alpha_p \pi \tag{5}$$

where, p is the number of pole pairs and α_p is polar arc coefficient.

By Fourier transform formula $\left[\frac{h'_m(\theta)}{h'_m(\theta) + \delta(\theta,\alpha)} \right]$, we can get

$$\left[\frac{h_m'(\theta)}{h_m'(\theta) + \delta(\theta, \alpha)} \right] = G_0 + \sum_{n=1}^{\infty} G_n \cos nz\theta \tag{6}$$

Substituted (2), (3), (4) and (5) into (1) and the expression of cogging torque is obtained by integrating the trigonometric functions within $[0, 2\pi]$.

$$T_{cog}(\alpha) = \frac{\pi L_a}{4\mu_0} (R_1^2 - h^2) \sum_{n=1}^{\infty} n G_n B_{r\frac{nz}{2p}} \sin nz. \tag{7}$$

where L_a is the axial length of the motor; R_1 is the outer arc radius of unequal thickness magnetic poles; h is the vertical distance from the center of the outer arc to the permanent magnet; n is an integer that makes $(nz/2p)$ an integer.

According to the triangle similarity theorem, one can get

$$h^2 = R_1^2 - \frac{h_{max}^2 d^2}{4h'^2_m} \tag{8}$$

where h_{max} is the maximum thickness of permanent magnet; h'_m is the minimum thickness of permanent magnet; d is the width of permanent magnet.

Substitute (8) into (7), then the cogging torque expression of the SIPMSM with unequal thickness magnetic poles can be described as

$$T_{cog}(\alpha) = \frac{\pi L_a h_{max}^2 d^2}{16\mu_0 h'^2_m} \sum_{n=1}^{\infty} n G_n B_{r\frac{nz}{2p}} \sin nz. \tag{9}$$

When the auxiliary slot is notched between the magnetic poles, the back EMF and magnetic field distribution of the motor will be greatly affected, and the high-order harmonic content in the air-gap flux density will be reduced, so the cogging torque of the motor will be weakened [18].

When the number of auxiliary slots is k, the Fourier decomposition coefficients of $\left[\frac{h'_m}{h'_m + \delta(\theta,\alpha)}\right]$ in the $\left[-\frac{\pi}{z}, \frac{\pi}{z}\right]$ can be expressed as:

$$G_n = \frac{2}{n\pi}\left(\frac{h'_m}{h'_m + X_d}\right)^2 [2\cos\frac{n\pi}{2}\sin(\frac{n\pi}{2} - \frac{nz\theta_{s0}}{2}) - 2\sin\frac{nz\theta_{s0}}{2}\sum_{i=1}\cos\frac{2in\pi}{k+1}] \tag{10}$$

where θ_{s0} is the width of auxiliary slot in the rotor, X_d is auxiliary slot in rotor, and $k = 1$.

By substituting (10) into (9), the expression of cogging torque with slots between magnetic poles of unequal thickness can be obtained as

$$T_{cog}(\alpha) = \frac{\pi L_a h_{max}^2 d^2}{4\mu_0(h'_m + X_d)^2}\sum_{n=1}^{\infty}[\cos\frac{n\pi}{2}\sin(\frac{n\pi}{2} - \frac{nz\theta_{s0}}{2}) - \sin\frac{nz\theta_{s0}}{2}\sum_{i=1}\cos in\pi]B_{r\frac{nz}{2p}}\sin nz \tag{11}$$

From (11), it can be seen that the structural parameters affecting the cogging torque include the axial length of the motor, the number of pole pairs, the maximum thickness of permanent magnet, the minimum thickness of permanent magnet, the width of permanent magnet, the width of slotting, the depth of the auxiliary slot and the number of stator slots. This paper mainly studies the influence of the notching auxiliary slot between adjacent magnetic poles.

2.3. Influence of Notching Auxiliary Slots

The auxiliary slots will increase the magnetic reluctance of the q axis magnetic circuit of SIPMSM. On the one hand, the difference of inductance between the d and q axis will produce reluctance torque. On the other hand, increasing the d axis inductance or reducing the q axis inductance can improve the flux weakening capability of the motor [19,20]. This section mainly studies the influence of the shape of unequal thickness magnetic poles and size of auxiliary slots on the flux weakening capability of the motor.

When the speed of PMSM exceeds the base speed, the phase current and phase voltage will reach the maximum value. In order to ensure that the limiting voltage does not exceed the limit voltage of the controller, the flux weakening control of PMSM is needed [21,22]. When the motor reaches the maximum speed, the stator current is used to weaken the magnetic field. The voltage amplitude is equal to the voltage limit of the controller. It can be seen that the d and q axis voltage in the rotating coordinate system can be expressed as:

$$\begin{cases} u_d = R_s i_d + \frac{d\psi_d}{d_t} - \omega\psi_q \\ u_q = R_s i_q + \frac{d\psi_q}{d_t} + \omega\psi_d \end{cases} \tag{12}$$

where u_d is the voltage of d axis; u_q is the voltage of q axis; R_s is the resistance of armature winding; i_d is the current of d axis; i_q is the current of q axis; Ψ_d is the flux linkage of d axis; Ψ_q is the flux linkage of q axis; ω is the angular speed of rotor rotation.

The change of current and flux linkage is zero when the motor is in steady state, so the steady state d and q axis voltage equation of the motor is obtained as:

$$\begin{cases} u_d = R_s i_d - \omega L_q i_q \\ u_q = R_s i_q + \omega(L_d i_d + \psi_{PM}) \end{cases} \tag{13}$$

When the motor is controlled by flux weakening control, the base speed of the motor can be expressed as:

$$\omega_0 = \frac{u_{\lim}}{p\sqrt{(\psi_{PM} + L_d i_d)^2 + (L_q i_q)^2}} \tag{14}$$

where $u_{\lim}$ is the limit voltage; ψ_{PM} is the flux linkage of permanent magnet; L_d is the d axis inductance of motor; L_q is the q axis inductance of motor.

When the output current of the inverter is all d axis demagnetizing current, that is $i_q = 0$. The ideal maximum speed of the motor can be expressed as

$$\omega_{max} = \frac{u_{\lim}}{\left|\psi_{PM} - L_d i_{\lim}\right|} \tag{15}$$

According to (14), the expansion of the weakening speed of the magnetic flux can be improved by reducing the permanent magnetic flux chain and increasing the d-axis inductance or reducing the q-axis inductance. According to (15), the expansion of the weakening speed of the magnetic flux can also be improved by increasing the limiting current and limiting voltage. Therefore, this paper mainly changes the inductance of the d and q-axis magnetic circuits by optimizing the design of unequal thickness magnetic poles and the slot size of the rotor. Changing these inductances can improve the performance of magnetic flux weakening speed. The ratio of the maximum speed to the base speed is the flux weakening expansion rate [23]. The expression of the flux weakening speed expansion can be expressed as:

$$\rho = \frac{\omega_0}{\omega_{max}} = \frac{p\sqrt{(\psi_{PM} + L_d i_d)^2 + (L_q i_q)^2}}{\left|\psi_{PM} - L_d i_{\lim}\right|} \tag{16}$$

Inductance is a physical quantity to measure the ability of a coil to produce electromagnetic induction, and it is the flux linkage produced by the unit current [24]. According to the definition, we can get the inductance of SIPMSM

$$L = \frac{\psi_{PM}}{i} = \frac{N\left(\frac{F_c}{R_M}\right)}{i} = \frac{N^2}{R_M} \tag{17}$$

where F_c is the magnetomotive of permanent magnet; R_M is the magnetic resistance of magnetic circuit through which the self-induction flux; N is the number of turns of conductor.

The expressions for calculating the d axis inductance of the novel SIPMSM as:

$$L_d = \frac{N^2}{R_r + R_z + R_a + R_s + R_y} \tag{18}$$

where, R_r is the magnetic resistance at the maximum thickness of permanent magnet steel; R_z is the rotor core magnetic resistance between the permanent magnet steel and the air-gap; R_a is the magnetic resistance of the air-gap; R_s is the magnetic resistance of stator; R_y is the magnetic resistance of the stator yoke.

The expressions for calculating the q axis inductance of the novel SIPMSM as:

$$L_q = \frac{N^2}{R_x + R_a + R_d + R_y} \tag{19}$$

where, R_d is the magnetic resistance of auxiliary slot.

The developed SIPMSM adopts the asymmetric magnetic circuit; the d axis flux linkage passes through unequal thickness magnetic poles, rotor core and air-gap, stator teeth and stator yoke; the q

axis flux linkage passes through the slot of slot, air-gap, stator boots and stator yoke. Due to large magnetic resistance of permanent magnet, the thickness of magnetic poles is approximately equal to the air-gap. When the auxiliary slot is notched between the magnetic poles in the rotor, the inductance in the q axis decreases with the increase in the depth of the auxiliary slot.

In summary, the depth of the auxiliary slot and maximum thickness of the permanent magnet influence the cogging torque and flux weakening speed expansion. This paper will optimize these two parameters.

3. Optimization Model

In the proposed SIPMSM, the complicated rotor topology and different saturation degree of the iron core make it difficult to solve the relationship between the object and the independent variables by analytic method. Because of the flexibility of the real response surface and the interaction between factors, the second order or higher order model is usually used to approximate the response objective function. The relationship between the maximum thickness of permanent magnet h_{max}, the depth of auxiliary slot X_d, the cogging torque T_{cog} and the flux weakening expansion rate ρ can be obtained by the combination of RSM and finite element method. The experimental arrangements and finite elements results are shown in Table 2, where X_1 and X_2 are the factor coded values, Y_1 is the cogging torque, and Y_2 is the flux weakening expansion rate. The second-order model can be expressed as follows:

$$y = \beta_0 + \sum_{i=1}^{k} \beta_i x_i + \sum_{i=1}^{k} \beta_{ii} x_i^2 + \sum_{i<j} \beta_{ij} x_i x_j + \varepsilon \tag{20}$$

where x_i is the coded variables; β_i is the linear effect of x_i; β_{ij} is the interaction between x_i and x_j; β_{ii} is the quadratic effect of x_i.

Table 2. The design matrix and finite elements results.

No.	Experimental Factor		Code Conversion		T_{cog}	ρ
	h_{max}	X_d	x_1	x_2	Y_1	Y_2
1	5	5	−1	−1	1.82	2.07
2	7	5	0	−1	0.75	1.95
3	5	9	−1	1	2.04	1.98
4	7	9	1	1	1.32	2.06
5	5	7	−1	0	2.05	1.87
6	7	7	1	0	0.73	2.03
7	6	5	0	−1	1.51	1.91
8	6	9	0	1	2.19	2.02
9	6	7	0	0	1.90	1.93

Because of the space limitation of rotor, the parameter level is limited in a certain range. Therefore, the central composite surface design (CCF) is adopted in this paper. All test points do not exceed the requirements of the cube and meet the design requirements [25–27].

The regression models of Y_1 and Y_2 are analyzed using Design-Expert software. According to the regression model equation, the response surface and contour map of the interaction factors to peak value of cogging torque and flux magnetic speed expansion can be established, as shown in Figures 4 and 5.

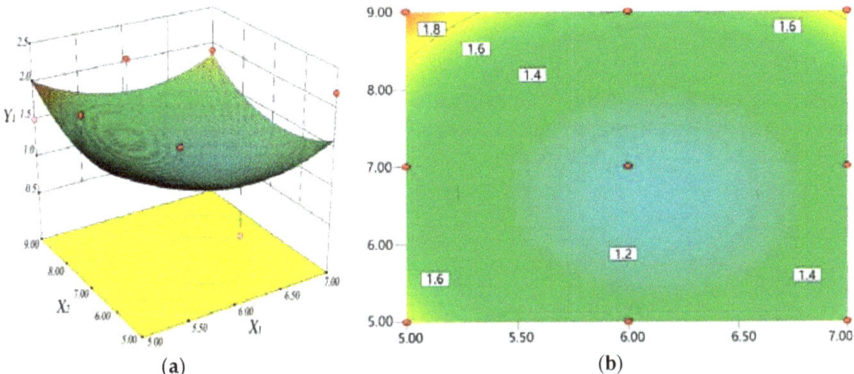

(a)   (b)

Figure 4. Effect of interaction factors on peak value of cogging torque Y_1: (**a**) 3D and (**b**) 2D contours.

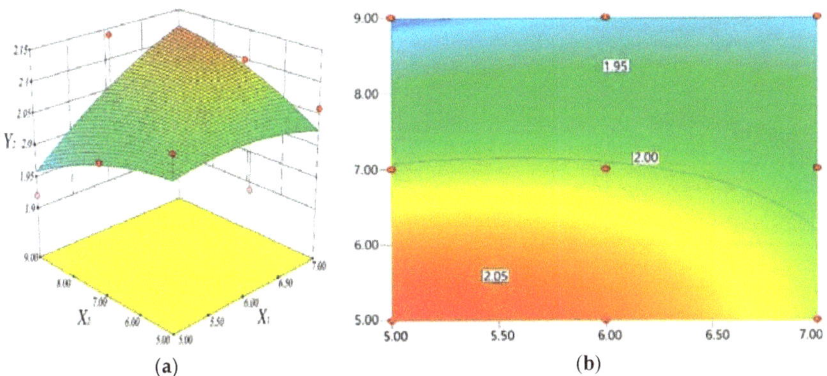

(a)   (b)

Figure 5. Effect of interaction factors on peak value of cogging torque Y_2: (**a**) 3D and (**b**) 2D contours.

It can be seen from Figure 4, when x_1 varies from 5 to 7 mm, the peak value of the cogging torque decreases first and then increases. If x_1 is fixed at a certain level, the peak value of the cogging torque decreases first and then increases. Therefore, according to the analysis of partial regression equation and contour plot, the effect of the depth of auxiliary slot in rotor on the flux magnetic speed expansion is greater than the maximum thickness of the permanent magnet.

It can be seen from Figure 5, when x_1 changes from 5 to 7 mm, the value of flux weakening expansion ratio of the motor increases first and then decreases. If x_1 is fixed at a certain level, with the increase in x_2, the peak value of cogging torque trends to decrease. Therefore, according to the analysis of partial regression equation and the contour plot, the effect of the maximum thickness of the permanent magnet is greater than the depth of auxiliary slot in the rotor.

In the motor design, the cogging torque is always expected to be small, while the flux weakening speed expansion is always expected to be large. So, a quasi-function is constructed using RSM. The relationship between independent variables and response variables in the motor and the nonlinearity is solved by solving the mathematical model for multi-objective optimization. Thus, the global solution of optimization calculation is obtained: $(h_{max}, X_d) = (6.40, 7)$. The optimal response is $(T_{cog}, \rho) = (1.16, 2.08)$.

4. Experimental Validation

In order to verify the correctness of the optimization design and the superiority of the novel SIPMSM. In this paper, two SIPMSMs with the same-rated parameters were trial-produced, as shown in Figures 6 and 7. The optimum design structures of traditional SIPMSM and the novel SIPMSM were tested and verified, and the performance of the test results was compared. The structure of the stator, winding layout, outer diameter of rotor, volume of permanent magnet and pole arc coefficient of the two motors are the same—the main difference between the two motors is that the maximum thickness of the rotor pole and slotting structure are different.

Figure 6. The rotor structure of traditional SIPMSM.

Figure 7. The rotor structure of the novel SIPMSM.

The two prototypes were tested on the eddy current dynamometer, as shown in Figure 8. With the increase in the motor speed, the load increases continuously, so that the motor current reaches the limit value. Then the sensor detects and collects signals. Finally, the output characteristic curve and efficiency map of the two SIPMSMs are shown in Figures 9 and 10.

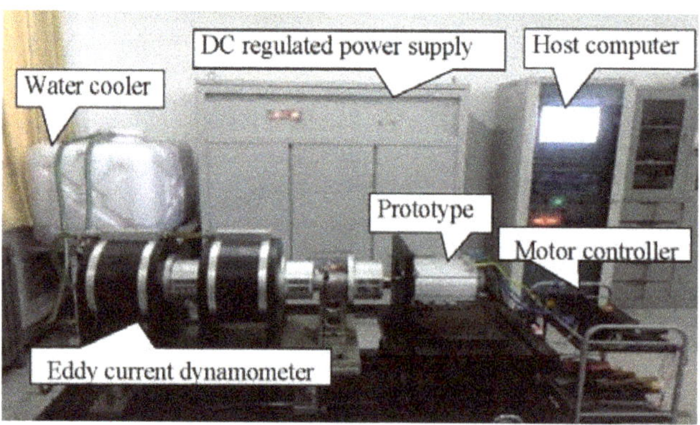

Figure 8. Test platform of prototype.

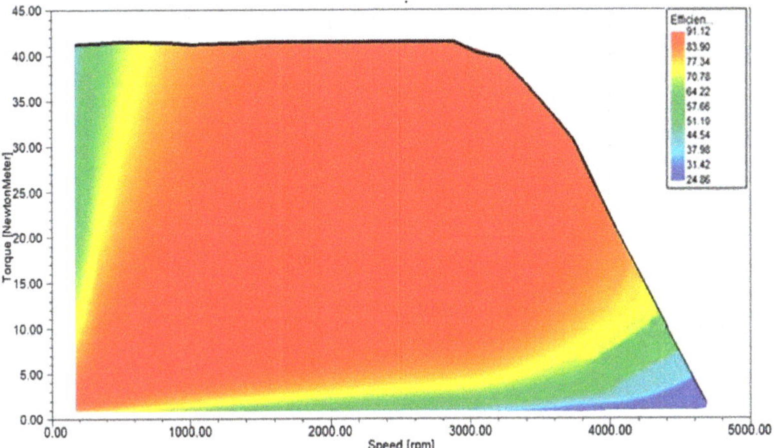

Figure 9. Efficiency map and characteristic curve of traditional SIPMSM.

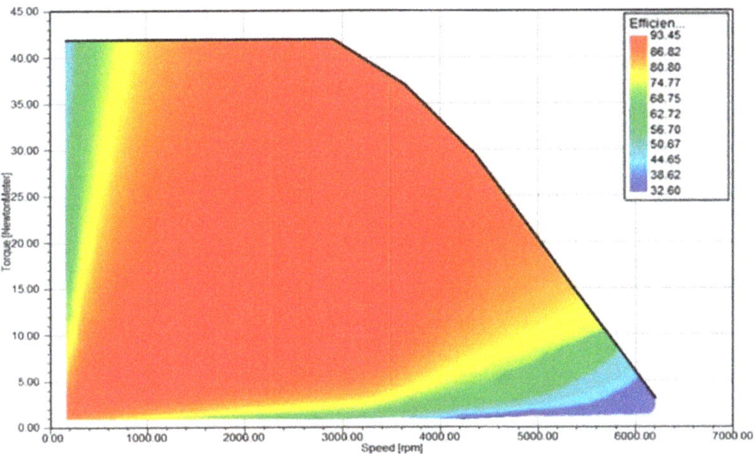

Figure 10. Efficiency map and characteristic curve of the novel SIPMSM.

Compared with Figures 9 and 10, it can be seen that the maximum efficiency of the novel SIPMSM is 91.12%, and the maximum speed that can be achieved is only 4750 rpm, while the maximum efficiency of traditional SIPMSM is 93.45%, and the maximum speed is 6250 rpm. Moreover, the novel SIPMSM has a wider range of high efficiency working area, which shows that the novel SIPMSM has a better performance in terms of flux magnetic speed expansion, thus verifying the veracity of theoretical analysis.

At the same time, PMSM for electric vehicles needs a variable speed drive and wide speed range, because electric vehicles need to start and stop frequently at low speeds under traffic congestion conditions, and good acceleration performance is required when traffic is smooth or running on highways [28,29]. Obviously, the novel SIPMSM has this output characteristic.

Finally, the experiment on the torque ripple was tested on the test platform from startup to related load, and the test results are shown in Figure 11.

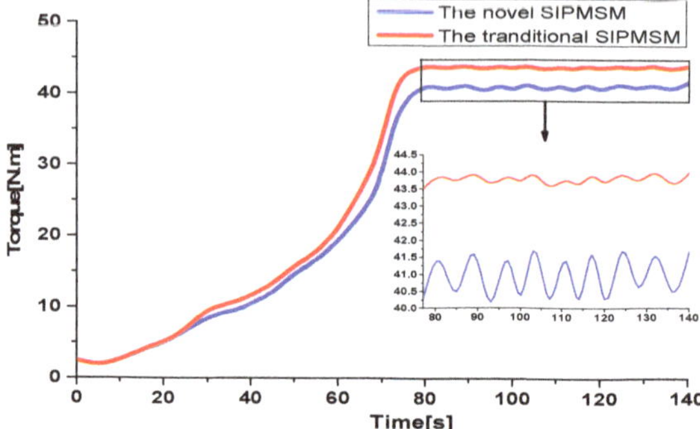

Figure 11. Torque ripple comparison between motor startup and rated load.

From Figure 11, it can be seen that the output torque of the two SIPMSMs reaches a stable state at 77 s, owing to the same control mode and rated parameters. However, the output torque of the novel SIPMSM is clearly higher than the traditional SIPMSM. This is because the reasonable design of the permanent magnets with unequal thickness makes the air-gap magnetic field have a significant magnetic congregate effect and enhances the output torque. It is not difficult to see that the torque ripple of the novel SIPMSM is lower than the traditional SIPMSM when it reaches steady state.

Torque ripple of PMSM is composed of cogging torque and ripple torque [30]. However, ripple torque is mainly caused by the motor control mode. In order to control a single variable, this paper uses the same controller for two motors. In this case, torque ripple is only determined by the cogging torque. Compared with Figures 10 and 11, it can be seen that the torque ripple of the novel SIPMSM is significantly lower than the traditional SIPMSM. Therefore, notching auxiliary slots in the rotor reduces the high order harmonic content in the air-gap flux density, weakens the cogging torque, and verifies the correctness of the design. The veracity of the theoretical analysis and optimization design is confirmed.

5. Conclusions

A novel SIPMSM by the method of notching auxiliary slots between the magnetic poles in the rotor and unequal thickness magnetic poles is developed. Compared with the traditional SIPMSM, it has the characteristics of large output torque, a wide range of speed regulation and a low cogging torque. The findings of the experimental results are described as follows:

(1) The mathematical model of cogging torque of the novel SIPMSM is established, and the influence factors of the cogging torque and flux weakening speed expansion of the novel SIPMSM were deduced. Combining the RSM and finite element method, the multi-objective optimization design of the influencing factors was carried out. The optimization results show that when the maximum thickness of magnetic poles is 6.4mm and the depth of rotor slot is 7 mm, the peak value of cogging torque is 1.16 nm and the flux weakening speed rate is 2.08.

(2) The prototype test shows that—compared with the traditional SIPMSM—the new SIPMSM not only enhances the output torque and reduces the torque ripple, but also improves the performance of flux weakening speed expansion. At the same time, the high efficiency range of the constant power operation is widened, and more in line with the performance requirements of PMSM for electric vehicles. Therefore, the novel SIPMSM is more suitable for electric vehicles.

Author Contributions: S.M. and Y.L. conceived and designed the experiments; B.Q. and S.M. performed the experiments and wrote the paper; M.A.S. and Z.L. revised the paper; S.M., B.Q., M.A.S. and Z.L. analyzed the data. All authors have read and agreed to the published version of the manuscript.

Funding: The research was funded by the National Natural Science Foundation of China (No. 51875327, 51777055 and 51690181), the Natural Science Foundation of Shandong Province (No. ZR2018LE010 and ZR 2017MF045) and Australia ARC DECRA (No. DE190100931). And the APC was funded by 51777055.

Acknowledgments: The research is supported by the National Natural Science Foundation of China (No. 51875327, 51777055 and 51690181), the Natural Science Foundation of Shandong Province (No. ZR2018LE010 and ZR 2017MF045) and Australia ARC DECRA (No. DE190100931).

Conflicts of Interest: The authors declare no conflict of interest.

References

1. Zhu, Z.Q.; Howe, D. Electrical Machines and Drives for Electric Hybrid and Fuel Cell Vehicles. *IEEE Proc.* **2007**, *95*, 746–765. [CrossRef]
2. Zhang, X.; Du, Q.; Ma, S.; Geng, H.; Hu, W.; Li, Z.; Liu, G. Permeance Analysis and Calculation of the Double-Radial Rare-Earth Permanent Magnet Voltage-Stabilizing Generation Device. *IEEE Access* **2018**, *6*, 23939–23947. [CrossRef]
3. Kim, K.; Lee, J.; Kim, H.; Koo, D. Multiobjective Optimal Design for Interior Permanent Magnet Synchronous Motor. *IEEE Trans. Magn.* **2009**, *45*, 1780–1783.
4. Ping, J.; Shuhua, F.; Ho, S.-L. Distribution Characteristic and Combined Optimization of Maximum Cogging Torque of Surface-Mounted Permanent-Magnet Machines. *IEEE Trans. Magn.* **2018**, *54*, 1–5. [CrossRef]
5. Xia, C.; Chen, Z.; Shi, T.; Wang, H. Cogging Torque Modeling and Analyzing for Surface-Mounted Permanent Magnet Machines with Auxiliary Slots. *IEEE Trans. Magn.* **2013**, *49*, 5112–5123. [CrossRef]
6. Abbaszadeh, K.; Alam, F.R.; Teshnehlab, M. Slot opening optimization of surface mounted permanent magnet motor for cogging torque reduction. *Energy Convers. Manag.* **2012**, *55*, 108–115. [CrossRef]
7. Bao, X.; Wu, C.; Fang, J. Cogging torque reduction in surface-mounted permanent magnet synchronous motor by combining different permanent magnets in axial direction. *J. Electr. Eng.* **2018**, *33*, 4231–4238.
8. Hwang, S.M.; Eom, J.B.; Hwang, G.B.; Jeong, W.B.; Jung, Y. Cogging torque and acoustic noise reduction in permanent magnet motors by teeth pairing. *IEEE Trans. Magn.* **2000**, *36*, 3144–3146. [CrossRef]
9. Kim, T.H.; Won, S.H.; Bong, K.; Lee, J. Reduction of cogging torque in flux-reversal machine by rotor teeth pairing. *IEEE Trans. Magn.* **2005**, *41*, 3964–3966.
10. Song, J.; Dong, F.; Zhao, J.; Lu, S.; Dou, S.; Wang, H. Optimal Design of Permanent Magnet Linear Synchronous Motors Based on Taguchi Method. *IET Electr. Power Appl.* **2016**, *11*, 41–48. [CrossRef]
11. Wang, H.; Qu, Z.; Tang, S.; Pang, M.; Zhang, M. Analysis and optimization of hybrid excitation permanent magnet synchronous generator for stand-alone power system. *J. Magn. Magn. Mater.* **2017**, *436*, 117–125. [CrossRef]
12. Ren, W.; Xu, Q.; Li, Q. Asymmetrical V-Shape Rotor Configuration of an Interior Permanent Magnet Machine for Improving Torque Characteristics. *IEEE Trans. Magn.* **2015**, *51*, 1–4. [CrossRef]
13. Niu, Z.; Sun, Z.; Chen, C.; Shi, Y. Optimization of the rotor structure of a hollow traveling wave ultrasonic motor based on response surface methodology and self-adaptive genetic algorithm. *Proc. Chin. Soc. Electrical. Eng.* **2014**, *34*, 5378–5385.
14. Ashabani, M.; Mohamed, Y.A.-R.I. Multiobjective Shape Optimization of Segmented Pole Permanent-Magnet Synchronous Machines with Improved Torque Characteristics. *IEEE Trans. Magn.* **2011**, *47*, 795–804. [CrossRef]
15. Kim, S.I.; Hong, J.P.; Kim, Y.K.; Nam, H.; Cho, H.I. Optimal design of slotless-type PMLSM considering multiple responses by response surface methodology. *IEEE Trans. Magn.* **2006**, *42*, 1219–1222.
16. Si, M.; Yang, X.Y.; Zhao, S.W.; Gong, S. Design and analysis of a novel spoke-type permanent magnet synchronous motor. *IET Electr. Power Appl.* **2016**, *10*, 571–580. [CrossRef]
17. Zhou, Y.; Li, H.; Meng, G.; Zhou, S.; Cao, Q. Analytical calculation of magnetic field and cogging torque in surface-mounted permanent magnet machines accounting for any eccentric rotor shape. *IEEE Trans. Ind. Electron.* **2015**, *62*, 3438–3447. [CrossRef]
18. Soong, W.; Ertugrul, N. Field-weakening performance of interior permanent-magnet motors. *IEEE Trans. Ind. Appl.* **2002**, *38*, 1251–1258. [CrossRef]
19. Ji, P.; Wang, X.; Wang, D.; Yang, Y. Study of Cogging Torque in Surface-Mouted Permanent Magnet Motor. *Proc. Chin. Soc. Electrical. Eng.* **2004**, *24*, 188–191.

20. Kim, K.-C. A Novel Magnetic Flux Weakening Method of Permanent Magnet Synchronous Motor for Electric Vehicles. *IEEE Trans. Magn.* **2012**, *48*, 4042–4045. [CrossRef]
21. Yin, S.; Wang, W. Study on the flux-weakening capability of permanent magnet synchronous motor for electric vehicle. *Mechatronics* **2016**, *38*, 115–120. [CrossRef]
22. Pan, C.T.; Liaw, J.H. A robust field-weakening control strategy for surface-mounted permanent-magnet motor drives. *IEEE Trans. Energy. Conver.* **2005**, *20*, 701–709. [CrossRef]
23. Xia, C.; Guo, L.; Zhang, Z.; Shi, T.; Wang, H. Optimal Designing of Permanent Magnet Cavity to Reduce Iron Loss of Interior Permanent Magnet Machine. *IEEE Trans. Magn.* **2015**, *51*, 1–9. [CrossRef]
24. Cheng, S.K.; Yu, Y.J.; Chai, F.; Cho, H. Analysis of the inductance of interior permanent magnetsynchronous motor. *Proc. Chin. Soc. Electrical. Eng.* **2009**, *29*, 94–99.
25. Kim, S.I.; Lee, J.Y.; Kim, Y.; Hong, J. Optimization for reduction of torque ripple in interior permanent magnet motor by using the Taguchi method. *IEEE Trans. Magn.* **2005**, *41*, 1796–1799.
26. Abbaszadeh, K.; Alam, F.R.; Saied, S. Cogging torque optimization in surface-mounted permanent-magnet motors by using design of experiment. *Energy Convers. Manag.* **2011**, *52*, 3075–3082. [CrossRef]
27. Shin, P.S.; Woo, S.H.; Koh, C.S. An optimal design of large scale permanent magnet pole shape using adaptive response surface method with Latin hypercube sampling strategy. *IEEE Trans. Magn.* **2009**, *45*, 1214–1217. [CrossRef]
28. Trancho, E.; Ibarra, E.; Arias, A.; Kortabarria, I.; Jurgens, J.; Marengo, L.; Fricasse, A.; Gragger, J. PM-Assisted Synchronous Reluctance Machine Flux Weakening Control for EV and HEV Applications. *IEEE Trans. Ind. Electron.* **2018**, *65*, 2986–2995. [CrossRef]
29. Zhao, W.; Zhao, F.; Lipo, T.A.; Kwon, B.-I. Optimal Design of a Novel V-Type Interior Permanent Magnet Motor with Assisted Barriers for the Improvement of Torque Characteristics. *IEEE Trans. Magn.* **2014**, *50*, 1–4. [CrossRef]
30. Liu, X.; Chen, H.; Zhao, J.; Belahcen, A. Research on the Performances and Parameters of Interior PMSM Used for Electric Vehicles. *IEEE Trans. Ind. Electron.* **2016**, *63*, 3533–3545. [CrossRef]

Article

Modelling of Material Removal in Abrasive Belt Grinding Process: A Regression Approach

Vigneashwara Pandiyan [1,2,3], Wahyu Caesarendra [4,*], Adam Glowacz [5] and Tegoeh Tjahjowidodo [1,6]

[1] School of Mechanical and Aerospace Engineering, Nanyang Technological University, Singapore 639815, Singapore; vigneash002@e.ntu.edu.sg (V.P.); tegoeh.tjahjowidodo@kuleuven.be (T.T.)

[2] Smart Manufacturing Group, Advanced Remanufacturing and Technology Centre, A*STAR, Singapore 637143, Singapore; vigneashwara_pandiyan@artc.a-star.edu.sg

[3] Empa, Swiss Federal Laboratories for Materials Science & Technology, Laboratory for Advanced Materials Processing, Feuerwerkerstrasse 39, 3602 Thun, Switzerland; vigneashwara.solairajapandiyan@empa.ch

[4] Faculty of Integrated Technologies, Universiti Brunei Darussalam, Jalan Tungku Link BE1410, Brunei

[5] Department of Automatic Control and Robotics, AGH University of Science and Technology, Al. A. Mickiewicza 30, 30-059 Kraków, Poland; adglow@agh.edu.pl

[6] Department of Mechanical Engineering, KU Leuven, Technology campus De Nayer, Jan De Nayerlaan 5, 2860 St.-Katelijne-Waver, Belgium

* Correspondence: wahyu.caesarendra@ubd.edu.bn; Tel.: +673-734-5623

Received: 25 November 2019; Accepted: 28 December 2019; Published: 5 January 2020

Abstract: This article explores the effects of parameters such as cutting speed, force, polymer wheel hardness, feed, and grit size in the abrasive belt grinding process to model material removal. The process has high uncertainty during the interaction between the abrasives and the underneath surface, therefore the theoretical material removal models developed in belt grinding involve assumptions. A conclusive material removal model can be developed in such a dynamic process involving multiple parameters using statistical regression techniques. Six different regression modelling methodologies, namely multiple linear regression, stepwise regression, artificial neural network (ANN), adaptive neuro-fuzzy inference system (ANFIS), support vector regression (SVR) and random forests (RF) have been applied to the experimental data determined using the Taguchi design of experiments (DoE). The results obtained by the six models have been assessed and compared. All five models, except multiple linear regression, demonstrated a relatively low prediction error. Regarding the influence of the examined belt grinding parameters on the material removal, inference from some statistical models shows that the grit size has the most substantial effect. The proposed regression models can likely be applied for achieving desired material removal by defining process parameter levels without the need to conduct physical belt grinding experiments.

Keywords: abrasive belt grinding; predictive model; regression; material removal

---

1. Introduction

A compliant belt grinding resembles an elastic grinding in its operating principle, and it offers some potentials like milling, grinding and polishing applications [1]. The abrasive belt grinding process essentially is a two-body abrasive compliant grinding processes wherein the abrasive belt is forced against the components to remove undesired topographies, such as burrs and weld seams, to achieve the required material removal and surface finish [2]. The belt grinding process is widespread in aerospace industries owing to its compliant nature, to machine components of intricate geometry such as fan blades for achieving uniform material removal. Analogous to other abrasive machining processes, many process parameters in the belt grinding impact the material removal performance, which include cutting speed, loading belt tension, the force imparted, infeed rate, workpiece topographies, polymer wheel hardness, wheel geometry and belt topography features, e.g., backing material, grain composition, and grit size [3]. Changing any of these parameters will result in different belt grinding performance. However, the effect will vary from parameter to parameter [4]. Abrasive belt grinding process is highly nonlinear due to the complexity of the underlying compliance mechanism, where some of them remain unknown. The presence of multiple parameters working in different regimes creates a dynamic condition which is not entirely realised well within the workforce. Most belt grinding processes in industry are still primarily based on empirical rules and operator experience. The amount of removed material relies heavily on the distinct local contact conditions, which is completely influenced by the state of the grinding parameters. Material removal in the belt grinding process is determined by force distribution in the contact area between the workpiece and the elastic contact [5]. Zhang et al. [5,6] formed a local grinding model based on support vector regression (SVR) and artificial neural network (ANN) to obtain the force distribution in the contact area between the workpiece and the elastic contact wheel, which offer faster analysis compared to that with the conventional finite element method (FEM). A three-dimensional numerical model to determine the distribution of pressure and real contact and give a better understanding of the material removal mechanism has been presented by Jourani et al. [7]. Ren et al. [3,8] simulated a local process model based to calculate the material removal before machining by calculating the acting force from the information on the local geometry of the workpiece. The result of the simulation offers a methodology to optimise the tool path. Table 1. lists the research outcomes in the literature on abrasive belt grinding to predict material removal and to model the contact conditions.

Hamann [9] had proposed a simple linear mathematical model which involves C_A (grinding process constant), K_A (constant of resistance of the workpiece with grinding ability of the belt), k_t (belt wear factor), V_b (grinding rate), V_w (feed-in rate), L_w (machining width), and F_A (normal force). The model states that the overall material removal rate (MRR) r is either proportional or inversely proportional to belt grinding parameters as shown in Equation (1). However, this model does not take into account the interaction between belt grinding parameters.

$$r = C_A.K_A.k_t.\frac{V_b}{V_w.L_w}.F_A \tag{1}$$

Preston's fundamental polishing equation as shown in Equation (2) states that MRR, $\partial R/\partial t$, of a belt grinding process has a direct relationship with relative velocity, R_v, and polishing pressure, P [10]. The constant, C, is established to denote other influential parameters and it is determined experimentally for each polishing system.

$$\frac{\partial R}{\partial t} = CPR_v$$
$\frac{\partial R}{\partial t}$ is the material removal (R) with time (t)
$$\tag{2}$$

Archard's wear based on an equation as shown in Equation (3) predicts wear volume, V_w, to be a function of normal load, Fn, sliding distance, S, and hardness of the softest contacting surface, H, while K is a dimensionless constant [11].

$$V_w = \frac{KFnS}{H} \tag{3}$$

Though the equations from Preston, Archard, and Hamann give a holistic view on the relationship between MRR and a few process parameters, dimensionless constants in each equation (C_A, K and C) need to be determined after many exhaustive physical experiments. Developing analytical models for such nonlinear processes with a large number of parameters and assumptions may introduce biases and will not be a viable option to model the process. Though the correlation of individual parameters on material removal is understood using ANOVA and statistical techniques, their combined consequence on effective material removal is not well established [12]. The adaptive neuro-fuzzy inference system (ANFIS) model with sigmoidal membership function has been used to determine material removal in a belt grinding process [12].

Table 1. Research efforts so far in modelling of belt grinding process.

Investigators	Contribution
Y. Wang et al. [13]	Developed a controllable material removal strategy to control the acting force and grinding dwell time by modelling the global and local material removal process of belt grinding. A finite element method (FEM) has been adopted to calculate the local force and global grinding model based on the Hertz contact theory.
X. Ren et al. [14]	Established a simulation system using Surfel to visualise the material removal process interactively and to optimise the tool path planning.
S. Wu et al. [15]	Presented a comprehensive platform to simulate the belt grinding system incorporating a kinematic model of the robot for tool path planning, dynamic model of the robot joint, along with the material removal model of the grinding process.
S. Mezgahani et al. [16]	Performed a comparative study of contact pressure and abrasive grit size to material removal keeping parameters such as speed of workpiece, tool hardness, cycle time, coolant, abrasive feed, and tool wear constant. The study showed that a decrease in grain size results in more ploughing action rather than cutting action.
J. Shibata et al. [17]	Offered a metal removal model incorporating the belt wear factor to explain the belt grinding characteristics quantitatively.
A. Khellouki et al. [18]	Theoretically modelled contact conditions between abrasive film and the surface and investigated the effect of average contact pressure, contact duration and the number of active grains in the contact.
V. T. Thien et al. [19] and Y. Sun et al. [20]	Demonstrated that pressure distribution obtained from pressure films can be correlated with a Hertzian model under different loads and hardness of the polymer wheel.
H. Lv et al. [21]	Presented a material removal modelling technique for free-form surface using an echo state network.
W. Wang et al. [22]	Proposed a grinding depth predicting frame working using a local stress model and a local material removal model taking into account the contact wheel deformation.
Y. Sun et al. [11] and V. Pandiyan et al. [23]	Proposed a novel methodology using a dynamic pressure sensor to predict material removal considering belt grinding parameters such as force, workpiece geometry and different types of contact wheel geometry.
Y. J. Wang et al. [24]	Demonstrated that the nonlinear material model performs better than the linear material removal model.

Most of the developed local material removal models have concentrated on simulating the contact condition of the tool and surface. These local models neglect the granularity parameter of the belt tool and have made assumptions during the development of the material removal model. A more systematic model taking into account the granularity parameter has not been reported yet. Incorporating actual values of parameters will help in developing a conclusive material removal model. A conclusive material removal model with other regression techniques in such a dynamic process has not yet been studied and compared in detail. This paper presents a systematic approach to model material removal using statistical regression techniques. The paper is organised as follows. A brief outline of the abrasive belt grinding process and the problem statement is presented in Section 1, followed by a brief theoretical basis on the belt grinding process and the regression modelling techniques in Section 2. The grinding conditions, belt grinding setup, process parameters and the Taguchi orthogonal array experimental data are listed in Section 3. The results of the six different regression models are discussed in Section 4. Finally, the findings of this paper are reviewed in Section 5.

2. Theoretical Basis

2.1. Abrasive Belt Grinding

An abrasive belt grinding process consists of a coated abrasive belt that is fixed firmly around, at least, two rotating polymer contact wheels (see Figure 1). The polymer wheel enables the grinding process to appropriately manufacture free-form surfaces due to its capability to adjust to the grinding surface [5,18].

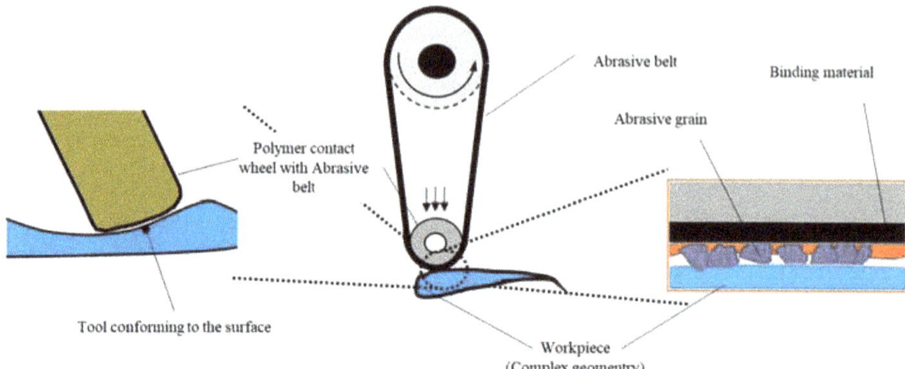

Figure 1. Principle of the belt grinding process.

The significant benefit of the belt grinding process is that there is no requirement of a coolant system as the grinding belts with the typical length of 2 to 5 m can cool down during the return strokes on the process [25]. Similar to traditional grinding processes, many machining parameters, e.g., the grinding belt topography features and belt grinding parameters, have an impact on the grinding result. Although super finishing by belt grinding is a straightforward and inexpensive process, essential aspects of material removal phenomenon are not well grasped. Hence the industry relies heavily on operators' knowledge.

2.2. Multiple Linear and Stepwise Regression

Multiple linear regression is a promising supervised learning algorithm and the most common form of linear regression analysis. As a predictive analytical tool, multiple linear regression is used to associate one continuous dependent variable to two (or more) independent variables by finding the best suitable fitted line [26]. The best fitted line is a line with total minimum error to all the points. Multiple linear regression analysis helps us to comprehend the rate at which the dependent variable changes when changes are made in the independent variables. Regression models describe the relationship between a dependent variable, y_i also referred to as the response variable and independent variables, X_{in}, or predictor variables, by fitting a linear equation as shown in the following equation:

$$y_i = \beta_0 + \beta_1 X_{i1} + \beta_2 X_{i2} + \beta_3 X_{i3} + \ldots + \beta_n X_{in} + \varepsilon_i \tag{4}$$

where the constant β_0 represents the intercept in the model and β_n ($n \neq 0$) refers to a coefficient. Since the response values for y_i vary about their means, the multiple regression model includes a residual term, ε_i, representing the variation. The goal of the multilinear regression is to create a linear model in hyperplane that minimises the sum of the square of the residuals concerning all predictor variables as shown in Figure 2. Multilinear regression identifies the hyperplane, in terms of the slope β_n and intercept β_0, through the sample data with the minimum sum of the squared errors thereby predicting the regression line.

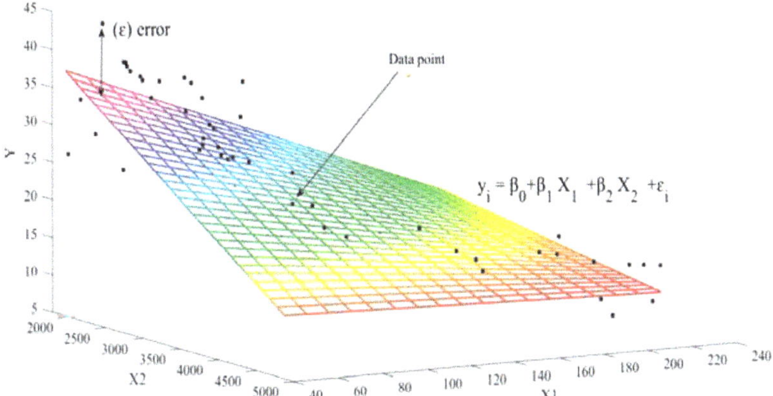

Figure 2. Surface that corresponds to the simplest multiple regression models.

Interaction effects of the predictors through linear regression can be achieved by using the stepwise regression. Stepwise regression essentially does multiple regression a number of times, each time removing the weakest correlated variable based solely on the t-statistics of their estimated coefficients. In the end, the variables that give the best distribution will remain. A linear model containing only the linear terms is used as a starting model whereas a quadratic model containing an intercept, linear terms, interactions, and squared terms are used as a terminating model. Predictor variables are added one at a time as the regression model progresses. At each step, predictor variable or their interaction that increases R^2 the most are considered significant whereas others are removed.

2.3. Artificial Neural Networks

Artificial neural networks (ANNs) is a method of developing logical systems by imitating the biological architecture of the human brain [27]. Establishing the networks between neuronal nodes determines the structure of the network. Neurons in ANNs are organised in a layered order. Each ANN structure comprises three different neurons namely input, hidden and output neurons. The input values are conveyed to the network through the input neurons, and these nodes pass the information to the next neurons in the hidden layer. Experiments determine the size of hidden layers and the number of neurons that are present according to the problem at hand [28]. The output neurons are where the output values of the network are generated. An ANN processes the information acquired from the input neurons by means of a linear/non-linear activation function and communicates to the subsequent neuron connected to it as illustrated in Figure 3.

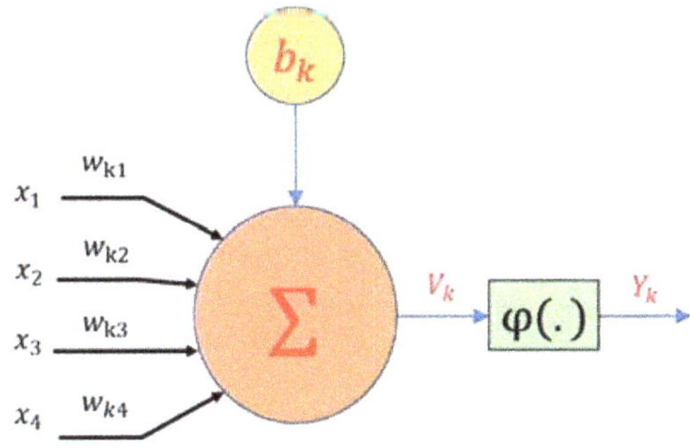

Figure 3. The mathematical model of a neuron.

Each connection between neurons is articulated in terms of a weight value, and these connections are governed based on the training of the network. Expected nodes in the input layer and inputs in every other node are the sum of the weighted outputs of the previous layer. Each node is brought as an active case depending on the input of one node, the activation function and the threshold value of the node. In mathematical terms, a neuron u_k can be described with weights w_{ji} and input x_i by Equation (5)

$$u_k = \sum_{j=1}^{m} w_{ji} x_i \tag{5}$$

and

$$v_k = u_k + b_k \tag{6}$$

where u_k is the linear combination output due to input signals and b_k is the bias. The output of the neuron is represented as y_k

$$y_k = \varphi(u_k + b_k) = \varphi(v_k) \tag{7}$$

A neuron in the network produces its output by processing its net input v_k through an activation function otherwise called a transfer function. Every neuron needs a (non-linear) activation function to cater for the non-linear property inside the network. There are several types of activation functions used in ANN, where the three forms of sigmoidal activation function are most the utilised functions Equation (8):

$$f(x) = \frac{1}{1 + e^{-x}} \text{ range } (0, 1) \text{ or } \frac{2}{1 + e^{-x}} - 1 \text{ range } (-1, 1) \text{ or } \frac{e^x - e^{-x}}{e^x + e^{-x}} \text{ range } (-1, 1) \tag{8}$$

The methodology to train connections to achieve anticipated results determines the learning algorithm of the network. There are several learning algorithms out of which the backpropagation learning algorithm is most commonly used. The backpropagation (BP) algorithm is based on the main principle of minimisation of errors in a neural network output and modification of network values according to the minimised values. The error of a neural network is described as the difference between the desired output D_0 and the calculated output C_0 of the network as shown in Equation (9).

$$E_p = \frac{1}{2} \sum_{k=1}^{K} (D_0 - C_0)^2 = \frac{1}{2} \sum_{k=1}^{K} \sum_{p=1}^{p} (D_0 - C_0)^2 \tag{9}$$

where p indicates the total number of instances and K denotes the number of neurons in the output of the network. By comparing the output value of the network with the desired value, the error of the network is determined which is further minimised by adjusting the weights of the networks.

Backpropagation is an algorithm to minimise the error in the network output based on a gradient descent minimisation method. The process of altering the weights starts at the output neuron and propagates backwards to the hidden layer. The altered weight w_{ji}^{new} is given by the η learning parameter

$$w_{ji}^{new} = w_{ji}^{old} \pm \Delta w_{ji} \tag{10}$$

$$\Delta w_{ji} = \eta \frac{\partial E_p^2}{\partial w_{ji}} \tag{11}$$

once the weights of all the links of the network are decided, the decision mechanism is then developed.

2.4. Adaptive Neuro-Fuzzy Inference System

Adaptive neuro-fuzzy inference system (ANFIS) overcomes the fundamental problem in fuzzy if-then rules by exploiting the learning competence of ANN for automated optimisation of fuzzy if-then rules during training. This results in an automatic modification of the fuzzy system based on input-output space [29], which lead to optimized membership function of parameters. In other words, ANFIS architecture is a superimposition of FIS on ANN architecture, which will empower FIS to self-tune its rule base based on the output. The ANFIS forms a FIS initially, and the membership function is altered by the backpropagation algorithm and least square method available with ANN [30]. Recent ANFIS application in the manufacturing field to predict the surface finishing quality is presented in Reference [31]. Figure 4 illustrates ANFIS with five layers of neurons with two inputs x and y, which form two fuzzy if-then rules based on a first-order Sugeno fuzzy model [32].

$$\text{Rule 1: If } x \text{ is A1 and } y \text{ is B1, then } f1 = p1x + q1y + r1;$$

$$\text{Rule 2: If } x \text{ is A2 and } y \text{ is B2, then } f2 = p2x + q2y + r2;$$

where $p1, p2, q1, q2, r1,$ and $r2$ are linear parameters and A1, A2, B1, and B2 are nonlinear parameters. The output of the ith node in the membership function layer is denoted as $O_{l,i}$.

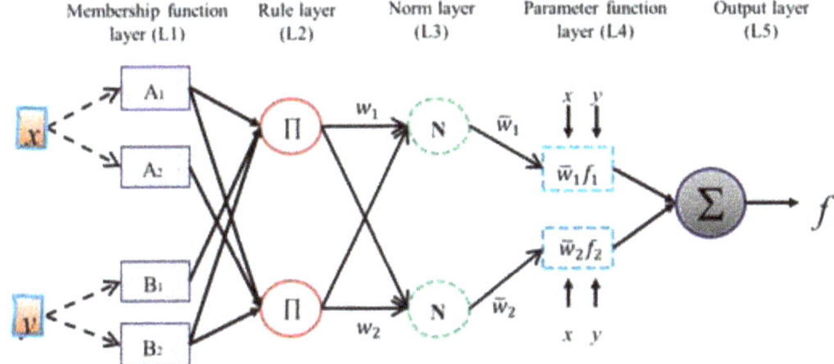

Figure 4. Adaptive neuro-fuzzy inference system structure.

Membership function layer (L_1): Every adaptive node i in the L_1 has a node function.

$$O_{l,i} = \mu_{A_i}(x) \text{ for } i = 1, 2 \text{ or } O_{l,i} = \mu_{B_{i-2}}(y) \text{ for } i = 3, 4 \tag{12}$$

where x (or y) is the input to nodes i and A_i (or B_{i-2}) generating a linguistic label coupled with the node as given by Equation (12). The membership function for A (or B) can be any, such as a sigmoidal membership function given by Equation (13).

$$\mu_{A_i}(x) = \frac{1}{1 + e^{-a(x-c)}} \tag{13}$$

where (c_i, a_i) is the parameter set. These are called premise parameters. As the values of the parameters change, the shape of the membership function varies.

Rule layer (L_2): Every node in L_2 is a fixed node labelled $\prod$. Each node calculates the firing strength of each rule, which is the output using the simple product operator. The rule premises result is evaluated as the product of all of the incoming signals and given by the Equation (14)

$$O_{2,i} = w_i = \mu_{A_i}(x) \times \mu_{B_i}(y) \text{ for } i = 1,2 \tag{14}$$

Norm layer (L_3): The ratio of the ith rule's firing strength to the sum of all of the rule's firing strengths is calculated by Equation (15) in L_3. The output of L_3 is called normalised firing strengths.

$$O_{3,i} = \overline{w}_i = \frac{w_i}{\sum_i^2 w_i} = \frac{w_i}{w_1 + w_2} \quad i = 1,2 \tag{15}$$

Parameter function layer (L_4): Every node i in L_4 is an adaptive node with a node function. The nodes compute a parameter function on the L_4 output. Parameters in this L_4 are referred to as consequent parameters.

$$O_{4,i} = \overline{w}_i f_i = \overline{w}_i(p_i x + q_i y + r_i) \tag{16}$$

where w_i is a normalised firing strength from L_3 and (p_i, q_i, r_i) is the parameter set for the node.

Output layer (L_5): L_5 has a single fixed node labelled Σ, which computes the overall output as the summation of all of the incoming signals, as shown in Equation (17). The Σ gives the overall output of the constructed adaptive network, having the same functionality as the Sugeno fuzzy model.

$$O_{5,i} = \sum_i \overline{w}_i f_i = \frac{\Sigma_i w_i f_i}{\Sigma_i w_i} \tag{17}$$

2.5. Support Vector Regression

A support vector regression (SVR) is a regression version of the support vector machine (SVM) that offers a powerful technique to solve regression problems. The regression problem is solved by introducing an alternative loss function. The SVR is centred on the idea of mapping the data x into a high-dimensional feature space to solve the regression problem [33]. Consider a data set $\{(x_i, y_i) \mid i = 1, 2, \ldots l\}$ where x_i is a D-dimensional input vector, y_i is a scalar output or target, and l is the number of points. In SVR, the goal is to find a function $f(x)$ that has at most ε deviation from the obtained targets y_i for all the training data. The nonlinear relationship between input and output is then described by a regression function $f(x)$ with weight factor $\omega = \sum_{i=1}^{n} a_i x_i$ and bias b. The SVR algorithm can be extended to nonlinear cases by simply preprocessing the training patterns x_i, by a map $\phi : X \to \mathcal{F}$, into some feature space $\mathcal{F}$ and then applying the standard SV regression algorithm. In the feature space $\mathcal{F}$, the regression function takes shape using a nonlinear mapping function $\phi(x)$:

$$f(x) = \omega^T \phi(x) + b = \sum_{i=1}^{n} \alpha_i \phi(x_i)^T \phi(x_j) + b \tag{18}$$

Nonlinear SVRs are trained by exchanging the inner products with the corresponding kernel $K(x_i, x_j) = \phi(x_i)^T \phi(x_j)$ and the resulting non-linear SVR is then represented as a kernel function:

$$f(x) = \sum_{i=1}^{n} \alpha_i K(x_i, x_j) + b \tag{19}$$

For the SVR model to have good generalisation performance, ω needs to be as flat as possible. This means that the norm ($\|.\|$) of the ω vector needs to be minimised for every data $i = 1, 2, \ldots l$

$$\text{minimise } \frac{1}{2}\|\omega\|^2 \tag{20}$$

$$\text{subject to } \begin{cases} y_i - \langle \omega, x_i \rangle - b \leq \varepsilon \\ \langle \omega, x_i \rangle + b - y_i \leq \varepsilon \end{cases} \tag{21}$$

Figure 5 shows the regression line, the upper and lower boundary lines and shows the radius of the $\pm\varepsilon$ insensitive loss function. The insensitive zone ($+\varepsilon$ to $-\varepsilon$) controls a number of support vectors. A point is well estimated if the point is within the zone, otherwise, it contributes to the training error loss. If the radius of the insensitive zone increases, numbers in the support vector group reduces, and robustness of the model diminishes.

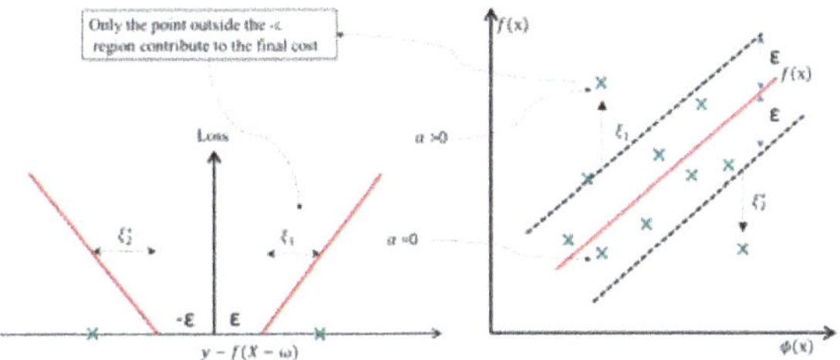

Figure 5. The regression line of support vector regression (SVR) is shown with the loss function and slack variables.

A penalisation is calculated by introducing C for every point that falls outside the insensitive zone ($+\varepsilon$ to $-\varepsilon$) which also defines the flatness and complexity of the model. A higher value of C increases the chance of overfitting, while a smaller value may increase the training errors. An optimum choice of this two-regularisation parameter is necessary for an ideal SVR model. To obtain the optimal hyperplane, the positive slack variable ξ_i is introduced to solve the following optimisation problem as illustrated in Figure 5. The constraint problem can be reformulated as shown in Equation (22).

$$\text{minimise } \frac{1}{2}||\omega||^2 + C\sum_{i=1}^{l}\left(\xi_i + \xi_i^*\right) \tag{22}$$

$$|\xi|_\varepsilon = \begin{cases} 0, & if \ |\xi| < \varepsilon \\ |\xi| - \varepsilon & \text{otherwise,} \end{cases} \tag{23}$$

$$\text{subject to } \begin{cases} y_i - \langle \omega, x_i \rangle - b \leq \left(\xi_i + \xi_i^*\right) \\ \langle \omega, x_i \rangle + b - y_i \leq \left(\xi_i + \xi_i^*\right) \\ \left(\xi_i, \xi_i^* \geq 0\right) \end{cases} \tag{24}$$

The regression function can be obtained by solving the following optimisation problem using the standard dualization principle utilising the Lagrange multiplier which will give the estimates of $\omega = \sum_{i=1}^{n}\left(\alpha_i^* - \alpha_i\right)\phi(x_i)$ and b. After solving the dual form, the regression function becomes:

$$f(x) = \sum_{i=1}^{n}\left(\alpha_i^* - \alpha_i\right)K\left(x_i, x_j\right) + b \qquad \text{where } \alpha_i^*, \alpha_i = \text{Lagrange multipliers} \tag{25}$$

Gaussian radial basis function (RBF) with γ standard deviation is commonly used due to its better potential to handle higher dimensional input space. For accurate model fitting, three internal parameters of SVM, namely the regularisation parameter (C), the radius of loss of insensitive zone (ε), and standard deviation (γ) of kernel function are to be correctly set [34]. A Bayesian optimisation approach is used to identify the optimised regularisation parameter.

2.6. Random Forest

One disadvantage of using a single decision tree (DT) is attributed to the risk of overfitting in the training data. A random forest (RF) model is a cumulative model that makes predictions by augmenting decisions from a group of decision trees as base learners [35]. The RF models can be written as:

$$g(x) = f0(x) + f1(x) + f2(x) + \ldots + fN(x) \tag{26}$$

where the final model g is the sum of simple base models f_i which basically is a regression decision tree. Unlike linear models, RF can capture the non-linear interaction between the features and the target.

One of the most significant advantages of RF over DT is that the algorithm works on bootstrapping [35]. The RF creates a lot of individual DT by re-sampling the data many times with replacement and makes the final prediction at a new point by averaging the predictions from all the individual binary regression trees on this point. Averaging over all the decision trees results in a reduction of variance thereby enhancing the accuracy of the prediction. The accuracy of the RF can be estimated from observations that are not used for individual trees otherwise called "out of the bag data" (OOB) as shown in Equation (27)

$$\text{OOB} \sim \text{MSE} = \frac{1}{n} \sum_{i=1}^{n} \left(y_i - \bar{\hat{y}}_{i\,OOB} \right)^2 \tag{27}$$

where $\bar{\hat{y}}_{i\,OOB}$ denotes the average prediction for the ith observation from all trees for which this observation has been OOB. The elements not embraced in the bootstrap sample are denoted to as OOB. At each bootstrap iteration, the response value for data not included in the bootstrap sample (OOB data) is predicted and averaged over all trees t_n as shown in Equation (28)

$$OOBMSE_t = \frac{1}{n_{OOB,\,t}} \sum_{\substack{i=1: \\ i \in OOB_t}}^{n} \left(y_i - \bar{\hat{y}}_{i,t} \right)^2 \tag{28}$$

where $\bar{\hat{y}}_i$, y_i, $n_{OOB,\,t}$ denote average prediction, observed output and the number of OOB observations in tree t respectively. The importance of each predictor is considered by computing the increase in mean squared error (MSE). Then, the variable importance assessments are manipulated to rank the predictors in terms of the strength of their relationship to the response or the process outcome.

3. Experimental Setup

3.1. Methodology

A hand-held electrically powered abrasive belt grinder was customised with a fixture to be used as a belt grinding tool. A contact wheel made of ester polyurethane polymer of different Shore A hardness was used as the grinding end of the custom built setup, as shown in Figure 6. The experiments were conducted by mounting the variable speed electric belt grinder to an ABB 6660-205-193 multi-axis robot as shown in Figure 6. The robot arm was mainly used to deliver designated toolpath trajectories. An ATI Industrial automation force sensor (Omega 160) interfaced the belt sander and the robot arm end effector to ensure that the applied force was maintained constant according to the desired level. Control force was used explicitly to obtain a uniform material removal along the grinding path. In addition, before the experiment, it was also ensured that the surface condition of the machined Aluminium 6061 coupons was uniform for all experimental trials.

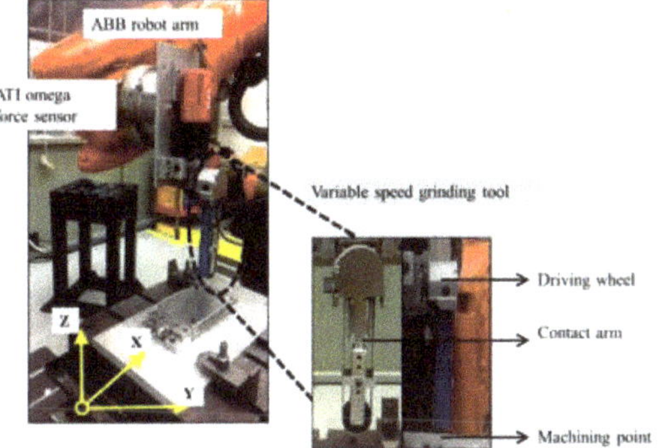

Figure 6. Compliant abrasive belt grinding experimental setup [12].

3.2. Taguchi Design of Experiments (DoE) and Data Collection

The experiments were performed based on Taguchi's L_{27} orthogonal array (five-factor, three-level) model. The familiarity of the probable interactions of the belt grinding parameters on material removal was not known therefore the Taguchi-based design of experiments (DoE) methodology was chosen Table 2 shows a list of belt grinding parameters (factors) and their levels used during the belt grinding trials. Material removal was quantified based on the depth of cut, which was measured using a Mitutoyo stylus profilometer with a tip radius of five microns with the accuracy of two decimal places used measure to the depth of cut. The depth of cut was calculated as the distance from the deepest point in the ground path from the unground surface of the workpiece as shown in Figure 7.

Table 2. Belt grinding parameters and their levels.

Parameter	Unit	Levels		
		L1	L2	L3
RPM	(m/min)	250	500	700
Feed	(mm/s)	10	20	30
Force	(N)	10	20	30
Rubber hardness	(Shore A)	30	60	90
Grit Size	-	60	120	220

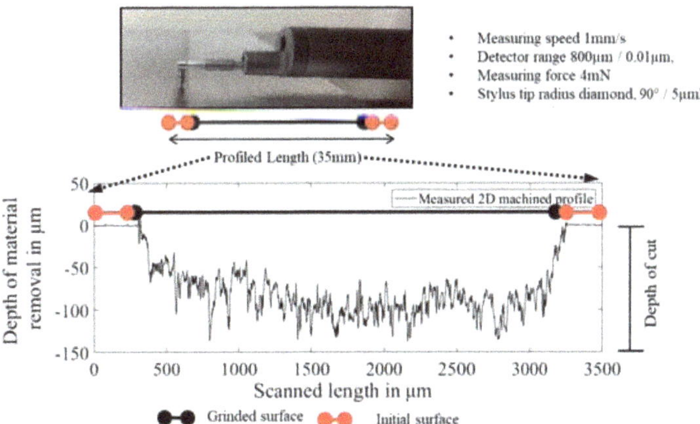

Figure 7. Mitutoyo contact profilometer used to measure the depth of cut across the grinded path.

Based on the parameter combinations from the Taguchi method, 27 combinations were obtained as presented in Table 3. A database of 243 depth of cut readings was created based on 27 Taguchi parametric combinations with nine readings for each combination. The dataset from Table 3. corresponds to only one reading value which was reused from previously published work by the authors [12]. The database with 243 rows of parameter level and depth of cut was used to model the material removal. Six different regression modelling methodologies, namely multiple linear regression, stepwise regression, ANN, ANFIS, SVR and RF were applied, and results obtained by the models are discussed in Section 4.

Table 3. Taguchi design of experiments (DoE) using the L_{27} orthogonal array and the corresponding depth of cut [12].

Trial No.	Factors					MRR
	RPM	Feed	Force	Hardness	Grit	Depth of Cut
	(m/min)	(mm/s)	(N)	(Shore A)		(μm)
1	250	10	10	30	60	65.60076
2	250	10	10	30	120	25.87109
3	250	10	10	30	220	13.34471
4	250	20	20	60	60	86.10453
5	250	20	20	60	120	44.20156
6	250	20	20	60	220	23.53456
7	250	30	30	90	60	93.8753
8	250	30	30	90	120	54.33391
9	250	30	30	90	220	23.55062
10	500	10	20	90	60	142.9324
11	500	10	20	90	120	86.37583
12	500	10	20	90	220	59.38035
13	500	20	30	30	60	120.6638
14	500	20	30	30	120	57.50747
15	500	20	30	30	220	45.55799
16	500	30	10	60	60	77.47286
17	500	30	10	60	120	26.08495
18	500	30	10	60	220	13.54166
19	700	10	30	60	60	134.8952
20	700	10	30	60	120	76.88529
21	700	10	30	60	220	58.97687
22	700	20	10	90	60	103.8255
23	700	20	10	90	120	56.9663
24	700	20	10	90	220	35.31606
25	700	30	20	30	60	114.009
26	700	30	20	30	120	56.65924
27	700	30	20	30	220	44.31528

4. Regression Modelling for Abrasive Belt Grinding Process

4.1. Multiple Linear Regression Based Modelling

Multilinear regression was performed to determine the relationships between the five grinding parameters Rotation per minute (RPM), force, rubber hardness, grit size, and feed rate to the depth of cut. The schematic model is illustrated in Figure 8.

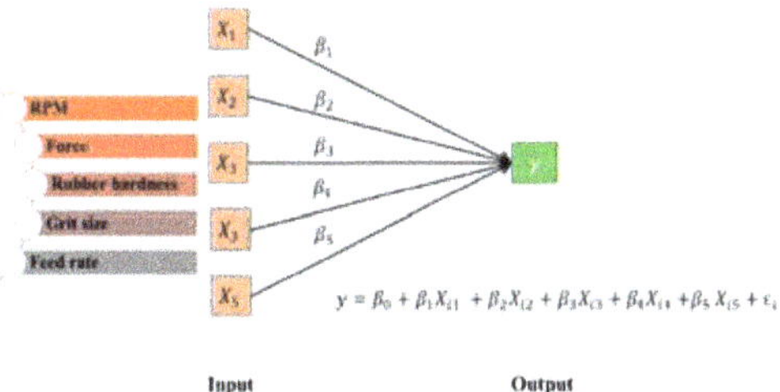

Figure 8. Schematic illustration of the multilinear regression model for prediction of the material removal.

The linear correlation model was developed using MATLAB. The dataset ($n = 243$) consisting of the five predictors and the response variable was randomly split into 70% training set and 30% testing set. The multilinear model fit the data to a hyperplane upon least square error minimisation. Figure 9 shows the comparison between observed and predicted material removal values on the testing dataset using the developed multilinear regression model. The figure also shows the deviation the predicted values against actual values with multilinear model highlighting a higher root mean squared error (RMSE) of 16.12 which was used as a measure to predict the performance of the model. The developed multilinear model was deficient at predicting the material removal, i.e., depth of cut. The calculated R-squared (R^2) gave a value of 0.886, which implied that the fitness of the linear model was inadequate.

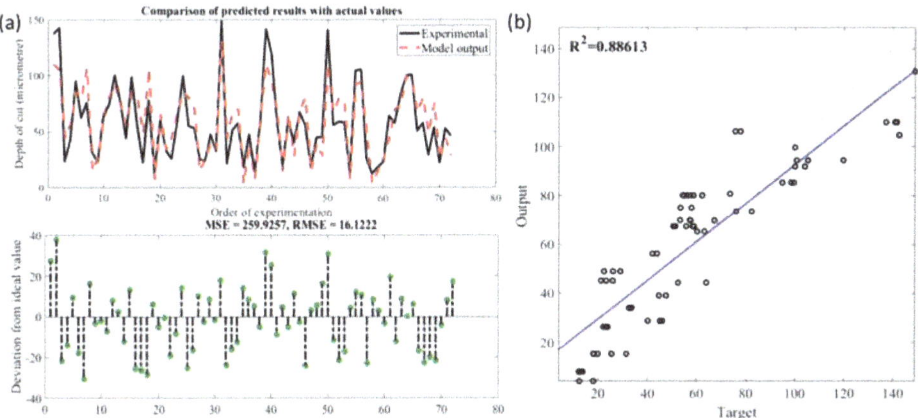

Figure 9. (**a**). Comparison of the observed and predicted depth of cut using multilinear regression; (**b**). Statistical analysis fit of the multilinear regression model.

4.2. Stepwise Regression-Based Modelling

The simple multilinear regression based on the least square method as discussed in the previous Section 4.1 indicated its incapability to accurately predict the depth of cut that was attributed to the nonlinear nature of the relationship in grinding process. The deviation from predicted to the original values may occur as the interaction effects between the parameters were ignored which could be analysed using stepwise regression. The training parameters used in the stepwise regression are listed in Table 4. The output of the developed stepwise model containing 13 terms which included five significant interactions and few higher-order polynomials are shown in Table 5.

Table 4. Stepwise multilinear regression training parameters.

Parameter	Value
Training set	70%
Testing set	30%
Method	Forward stepwise regression
Starting model	Linear
Upper limit	Quadratic

Figure 10 shows the comparison between the observed and the predicted material removal values from the 30% testing dataset. In addition, the figure shows the deviation of the predicted values from the actual ones highlighting a reduction in RMSE of 7.77 as compared to least square based multilinear regression. The proposed model was capable but insufficiently robust at predicting material removal. The R-squared (R^2) of value 0.975 increased when the regression took in a quadratic form which also took into account the interaction effect between the belt grinding parameters. The order in which predictor variables were removed or added could provide valuable information about the quality of the predictor variables. Referring to the t-stat value of the developed regression model, especially between the predictor interaction from Table 5, it was apparent that grit parameter played a dominant role. The estimated coefficients of the regression model were significant when the interaction happened with the grit size predictor.

Figure 11 depicts the comparison of the residuals from the multilinear model and the stepwise multilinear model. It explicitly indicated that the latter fits the data better. Also, comparing the R^2 statistical metric values from both approaches, it was clear that the latter performed better than the former as it incorporated a quadratic form, which addressed the influence of the interaction between the grinding parameters and the nonlinear behaviour of the belt grinding process. The proposed methodologies on the multilinear regression and the stepwise multilinear regression provided a useful tool to predict material removal depending on the grinding parameters. The multilinear model may not have been the best choice according to the prediction accuracy and R^2, but still, we could use it to find the nature of the relationship between the two variables. Interpreting the performance of the model, it was apparent that the data were intrinsically nonlinear and a straight-line relationship should never be assumed in the belt grinding process. The use of a quadratic form in the stepwise regression helped to tune the model to have a better fit.

Table 5. Estimated coefficients from stepwise regression.

S. No	Predictors	Estimate	Std Error	t-Stat	*p*-Value
1	(Intercept)	86.755	7.956	10.904	5.392×10^{-21}
2	RPM	0.049856	0.0084825	5.8776	2.3937×10^{-8}
3	Feed	2.0067	0.51282	3.9131	13.509×10^{-5}
4	Force	−2.7861	1.2896	−2.1605	03.2243×10^{-4}
5	Rubber hardness	1.7358	0.18569	9.3481	8.2594×10^{-17}
6	Grit size	−1.3237	0.062244	−21.266	3.2042×10^{-48}
7	RPM: grit size	−0.00014814	4.1438×10^{-5}	−3.575	4.6498×10^{-4}
8	Feed: force	0.062239	0.011008	5.654	7.1536×10^{-8}
9	Feed: rubber hardness	−0.069002	0.0091514	−7.5401	3.4527×10^{-12}
10	Force: grit size	−0.0035275	0.00096854	−3.642	3.6624×10^{-4}
11	Rubber hardness: grit size	−0.0010501	0.00031645	−3.3184	1.1237×10^{-3}
12	Force^2	0.084463	0.029849	2.8297	5.265×10^{-3}
13	Grit size^2	0.0039407	0.00018283	21.554	6.5639×10^{-49}

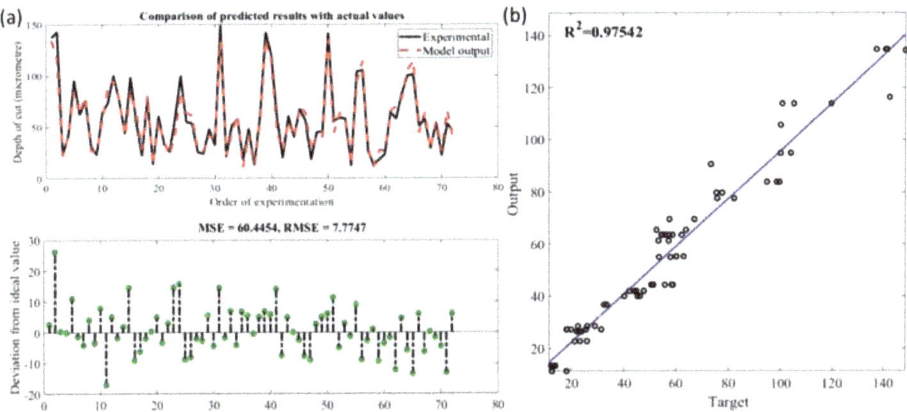

Figure 10. (**a**). Comparison of observed and predicted depth of cut using stepwise regression; (**b**) Statistical analysis fit of the stepwise regression model.

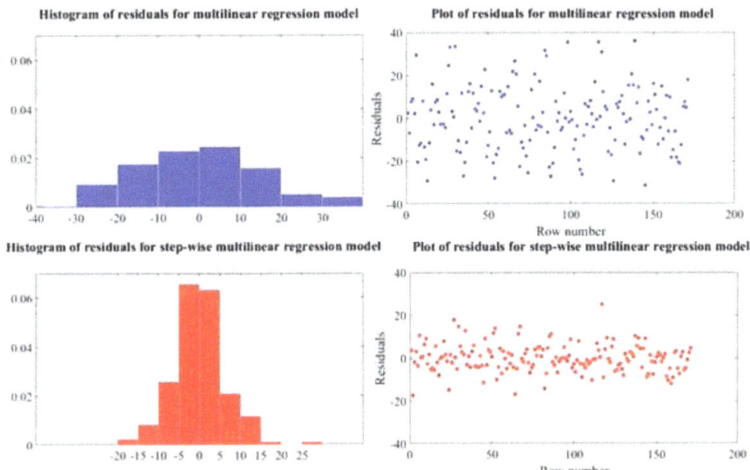

Figure 11. Residual plot observed between the multilinear regression model and stepwise multilinear regression model.

4.3. Artificial Neural Networks Based Modelling

The architecture of the ANN that estimated the material removal in abrasive belt grinding is shown in Figure 12. The accuracy of an ANN model depends on its architecture and on how it is trained. The input layer of the proposed network had five neurons that represented the prediction input, while the output layer had only one neuron to predict the depth of cut. In this study, the backpropagation neural network was used to construct non-linear functions between several inputs and the output. The weights of the networks were adjusted in first place at the output neuron and worked backwards to the hidden layer during the learning, i.e., training procedure until the expected error was achieved. With a hyperbolic tangent sigmoid transfer function, the learning rate of 0.01, and momentum rate of 0.05, the modelling result and the MSE value of material removal were generated by using the ANN toolbox in the MATLAB software.

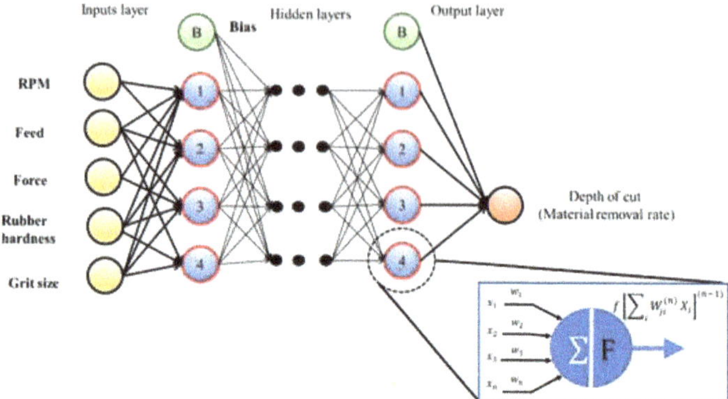

Figure 12. Schematic illustration of an artificial neural network (ANN) model for prediction of the material removal.

The configuration parameters used to determine the best network structure of the ANN prediction model is presented in Table 6. The ANN algorithm used 70% of the input data for model training with the backpropagation method as Bayesian regularisation and used the remainder for model validation. The performance of ANNs depends on the number of hidden layers in the ANN network structure. It is also to be noted that the increase in the number of hidden layers has a direct impact on the cost and the modelling time. The determination of the number of layers and nodes in the hidden layers in the training using the trial-and-error method is based on a minimum number of iterations required to reach the necessary performance goal. Training stops when the error is minimised to the performance goal. Three ANNs with a different number of hidden layer configuration, i.e., 10, 20, and 30 were tested on the same training dataset. The performance of the three configurations was evaluated on minimum epochs required to achieve minimum MSE. From Table 7 it is clear that the epochs rate decreased with the increase of hidden layers. The relationship between the number of iteration and the sum-squared error for an ANN structure with 30 hidden layers is shown in Figure 13.

Table 6. ANN training algorithm configuration parameters.

Parameter	Value
Maximum number of epochs to train	200
Performance goal	0
Backpropagation method	Bayesian regularisation
Initial μ	0.005
Hidden layers	10, 20, 30
Training	70% of the dataset
Testing	30% of the dataset
Weight function	Dot product
Activation function	tansig
Predictors	5
Response	1

Table 7. Comparison of hidden layers against epochs to reach the performance goal.

Hidden Layers	Epochs
10	151
20	62
30	40

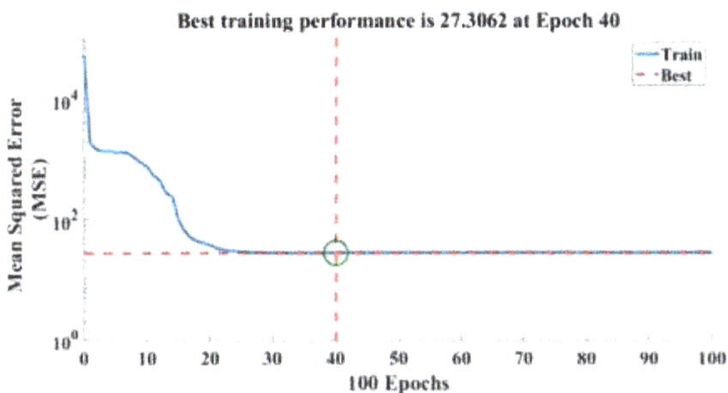

Figure 13. The reduction of the error with an increase in the number of iterations.

It could be noted that the MSE decreased when the iteration numbers using 30 hidden layers is increased. There was a sudden reduction error after 25 iterations, and this reduction went further until it saturated after 40 iterations. Also, while the prediction accuracy increased as the number of hidden layers increased, the performance was not significantly improved by adding more than 30 hidden layers. Upon the completion of the training stage, the network was tested with the validation set. Figure 14 shows the comparison of the actual material removal value with the results of the neural network analysis for all validation data set (71 points). The results indicate the low deviation of the predicted values against actual values with ANNs architecture with 30 hidden layers as represented by an RMSE of 6.64 as shown in Figure 14. The trained ANN model based on backpropagation algorithm was tested with 81 data from the experimental results with the RMSE of 6.649. The predicted values were found to be close to the measured values. The calculated R^2 of 0.981 suggested a satisfactory fitness model. The results lead to the conclusion that the proposed models could be used to predict the depth of cut in the belt grinding process effectively. Though this network correlates the parameters on its own, as ANN essentially functions as a "black box", it is trivial to evaluate the association between each independent variable and the dependent variable inside a neural network. Moreover, there is no specific rule for determining the structure of the ANN as the network is configured based on trial and error.

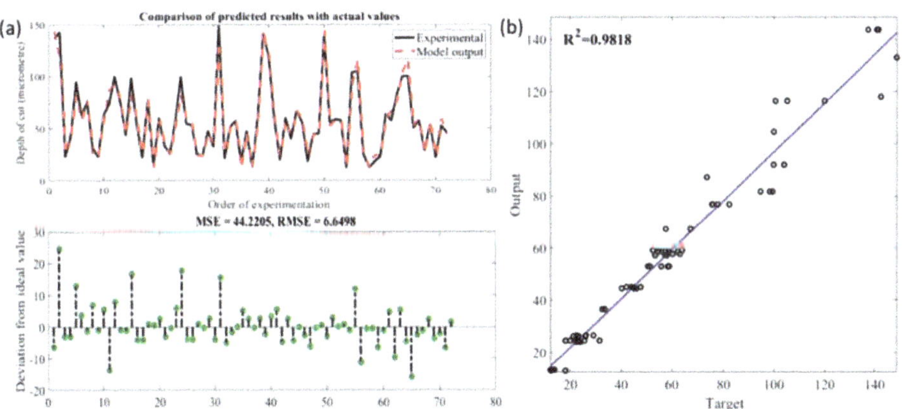

Figure 14. (**a**). Comparison of observed and predicted depth of cut using neural network regression; (**b**) Statistical analysis fit of the neural network regression model.

4.4. Adaptive Neuro-Fuzzy Inference System Based Modelling

The ANFIS model developed had multiple inputs, and a single output hence it resembled a multiple inputs–single-output (MISO) system. The architecture of MISO based ANFIS architecture for prediction of material is illustrated in Figure 15. The figure indicates that the input membership functions were non-linear, whereas the output membership functions were linear. For the training set, 70% of the data was chosen stochastically and the remaining 30% used for testing, while the step size for parameter adaptation was 0.1. The ANFIS training parameters for our case are listed in Table 8. The proposed ANFIS contained 243 rules, where each membership function was assigned to each input grinding variable. The MISO model was executed under the MATLAB platform using a Sugeno-type fuzzy inference system with the help of the fuzzy logic toolbox. ANFIS learning used the backpropagation technique for revising membership function shape.

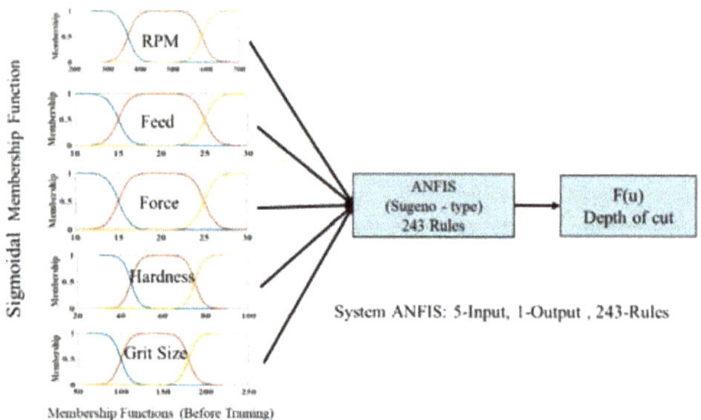

Figure 15. Adaptive Neuro-Fuzzy Inference System (ANFIS) model for belt grinding showing inputs and output [12].

Table 8. The ANFIS training parameters with four different membership functions.

Parameter	Value
andMethod	Prod
orMethod	Max
defuzzMethod	Wtaver (Weighted average)
impMethod	Prod
aggMethod	Max
Membership function	/Sigmoidal/Gaussian bell/Gaussian/Bell-shaped
Learning rules	Gradient descent algorithm

A fuzzy membership function assigns grades of membership between zero and one. The lower the grade, the less significant the membership to the fuzzy set Figure 16 presents the initial and final membership functions of the five belt grinding input parameters derived by using Gaussian membership function training. Comparing the shape of initial and final membership functions of the belt grinding parameters, it could be concluded that the grit size had the most significant impact on the material removal followed by RPM and force. Four types of the membership function, i.e., sigmoidal, Gaussian bell, Gaussian, and bell-shaped, were used and compared to model the belt grind process. Out of all the four types of membership function, we could conclude that sigmoidal membership gave better accuracy, as shown in Table 9.

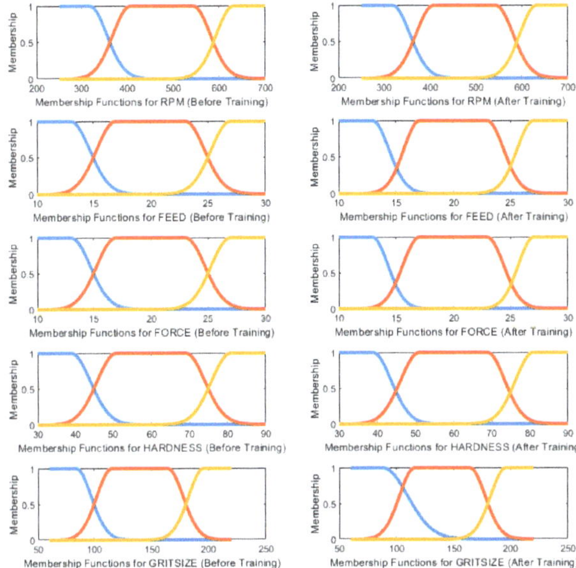

Figure 16. Change in shape of the Gaussian bell membership function for each input before and after training.

Table 9. Comparison of prediction accuracy and membership function.

Membership Function	RMSE
Sigmoidal membership [12]	5.648
Gaussian bell membership	6.7941
Gaussian membership	7.100
Bell-shaped membership	7.0341

Figure 17 gives the comparison of the predicted depth of cut using the established fuzzy model and the data taken from the belt grinding experiments from the validation set. The developed ANFIS system gave an overall RMSE of 6.79 on using Gaussian membership function. The deviation of the outputs generated by the fuzzy model with the experimental data is illustrated in Figure 17. A strong correlation between the simulated results and the experimental results obtained at the same machining conditions is demonstrated in the figure. The R^2 calculated based on the fitted regression line was 0.980 which also indicated a satisfactory fit of the model. The prediction results showed that by employing a hybrid learning algorithm such as ANFIS, the quality of generated relevant fuzzy if-then rules could be tuned to model the material removal behaviour.

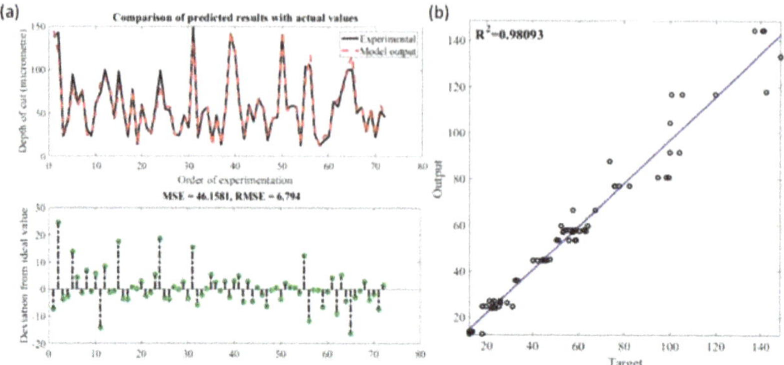

Figure 17. (**a**). Comparison of observed and predicted depth of cut using ANFIS with Gaussian bell membership function; (**b**) Statistical analysis fit of the ANFIS model with Gaussian bell membership.

In short, the relationship between the input parameters and the output behaviour of the nonlinear belt grinding process, which was not possible to be modelled using a black-box approach, such as ANN, could be represented satisfactorily using ANFIS. Understanding the input-output correlation from ANFIS architecture helped to characterise the material removal behaviour for which there is no concrete analytical description to date.

4.5. Support Vector Regression Based Modelling

Figure 18 shows the architecture of a regression machine for the belt grinding process. The construction of the material removal model using SVR can be summarised as follows:

1. Define the data, predictors, and response for learning and testing.
2. Decide a fitting kernel function (linear, Gaussian, RBF, polynomial, etc.).
3. Select an ideal model for training on the input data with predictors and response. The training is done by using a Bayesian optimisation to model material in MATLAB.
4. Validation of the testing data set.

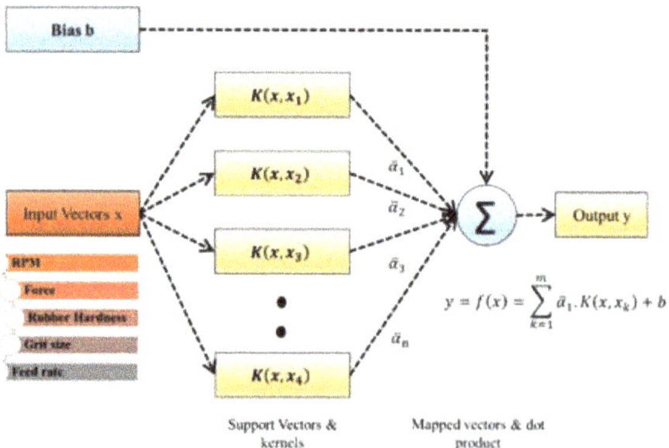

Figure 18. The architecture of a regression machine constructed by the SVR algorithm.

Similar to the previous analyses or the development of the model 70% of the recorded data was selected stochastically for training and 30% for testing. The RBF was used as the kernel, where the kernel parameter (γ) was determined by Bayesian optimisation. The training was performed under the MATLAB platform. From our initial analysis, it was found that regularisation parameter (C = 989.65), the radius of loss insensitive zone ($\varepsilon = 0.12457$), and standard deviation ($\gamma = 0.10894$) gave the minimal value of the MSE based on the Bayesian optimisation algorithm as depicted in Figure 19. There was a sudden reduction error after five evaluation and this reduction goes further before it saturates after 15 iterations.

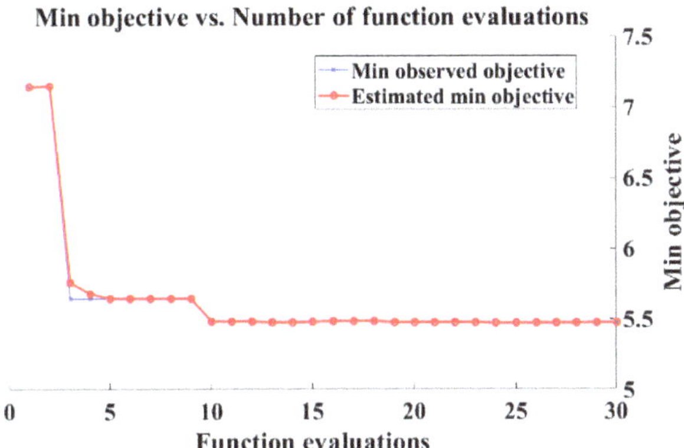

Figure 19. Optimisation of SVR parameters with respect to the number of iterations.

A higher value of C in the feature space signified that the material removal model was highly complex with large random noises which also indicated the nonlinear behaviour of the belt grinding process. The lower values of ε for material removal indicated that the SVR model could capture the intricacy of the belt grinding process effectively. The tuned parameters used in the SVR model during the learning process is listed in Table 10. Once the model completed the training stage, it was tested with the validation data. The effectiveness of the training could be concluded from its ability to forecast material removal from unseen data. The R^2 calculated based on the fitted regression line was 0.980 which also showed the good fit of the model.

Table 10. Optimised SVR training parameters.

Parameter	Value
Epsilon (ε)	0.12457
Box Constraint (C)	989.65
Optimisation Method	Bayesian optimisation
Kernel	Gaussian RBF
Kernel scale (γ)	0.10894

Figure 20 shows the deviation of the predicted values against actual values using the SVR model, highlighting an RMSE of 6.9989. The trained SVR model with optimum parameter settings C, ε, and γ obtained using Bayesian optimisation could efficiently predict material removal responses. Though the testing result of developed SVR algorithm favoured the practical use of the model in the chosen range, the main limitation of the support vector approach was attributed to the kernel selection, which was mostly based on trial and error.

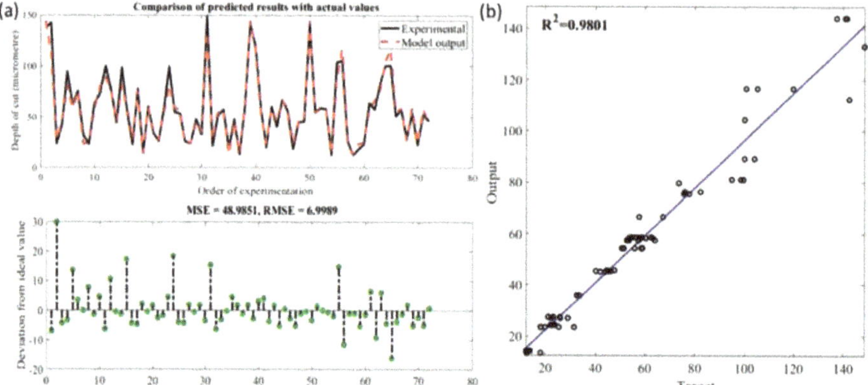

Figure 20. (**a**) Comparison of observed and predicted depth of cut using SVR; (**b**) Statistical analysis fit of the SVR model.

4.6. Random Forest Based Modelling

The framework for predicting material removal using a random forest (RF) is illustrated in Figure 21. The algorithm for material removal regression can be summarised as follows:

1. A bootstrap sample $(x_1, x_2, x_3, \ldots x_n)$ of size, N is to be drawn from the training data consisting of the five predictors and the response variable.
2. For each bootstrap sample x_i, a regression tree model is constructed by optimising the parameters, such as the number of trees t_n and leaf size based on MSE error.
3. Prediction at a new point Z is achieved by aggregating the predictions of the regression tree models.
4. The accuracy of the RF model is calculated based on the deviation of the predicted value, x, from the ideal value in the validation data set.

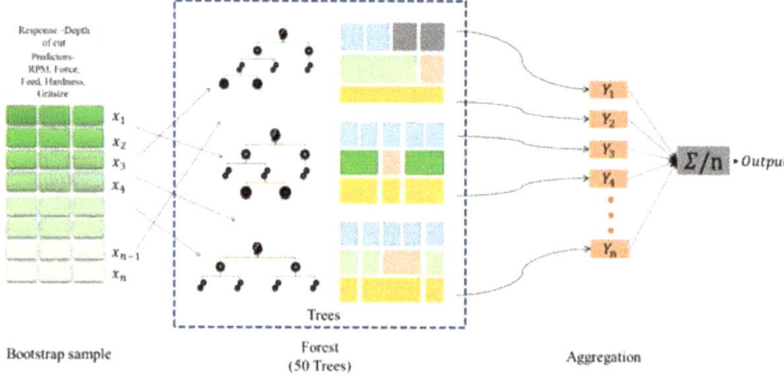

Figure 21. Material removal prediction using a random forest.

The development of the relationship between the belt grinding parameters and material removal RF model was carried out using MATLAB. Some parameters such as the minimal size of the terminal nodes of the trees, i.e., leaf size and a number of regression trees grown based on a bootstrap sample were optimised in the random forest model upon the minimisation of the MSE. The number of regression trees in a random forest model defines the strength of each tree in the forest and its correlation with other trees. A number of regression trees were optimised based on the MSE by testing with five different tree values (5, 10, 20, 50, and 100) on the training dataset. Figure 22 indicates how the number of regression trees grown in RF affected the prediction error. Inferring from the figure, it is evident that the MSE saturated after 35 trees and did not improve further as the number of trees increased.

The minimum leaf size parameter specifies the smallest number of observations a node can have. Letting the trees split down to leaf nodes with a minimum observation results in the most accurate random forest models. However, it is noted that smaller leaf size results in deeper trees with higher accuracy but also increases the cost of computation time and memory. Figure 23. indicates that the effect of the minimum leaf size and it was quite clear that with minimum leaf size, i.e., deeper trees the prediction ability of the model was higher. The splitting process stopped at the child node if the number of observations in a node was less than five which was used as a stopping criterion in the regression model developed. Based on results from fine-tuning discussed in the section above, an RF was constructed using 50 regression trees. A bootstrap sample of size 40 (0.25% of the training data) was drawn from the training dataset for every iteration. After 50 regression trees were constructed, a prediction at a new point could be made by aggregating the predictions from all the individual binary regression trees on this point. The tuning parameter used for developing the RF regression model is listed in Table 11.

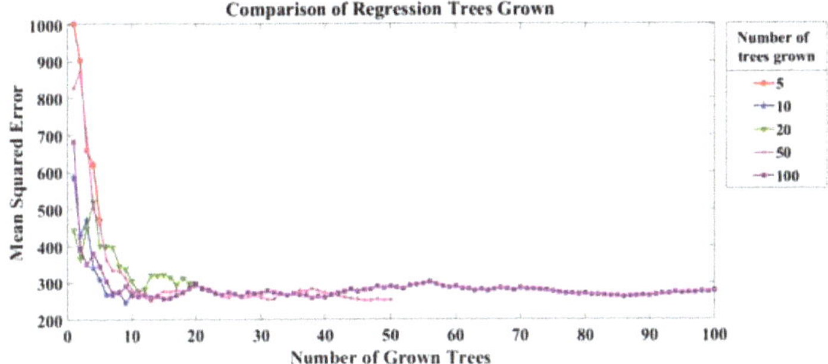

Figure 22. Optimisation of random forest (RF) parameters (number of grown trees) using mean square error (MSE).

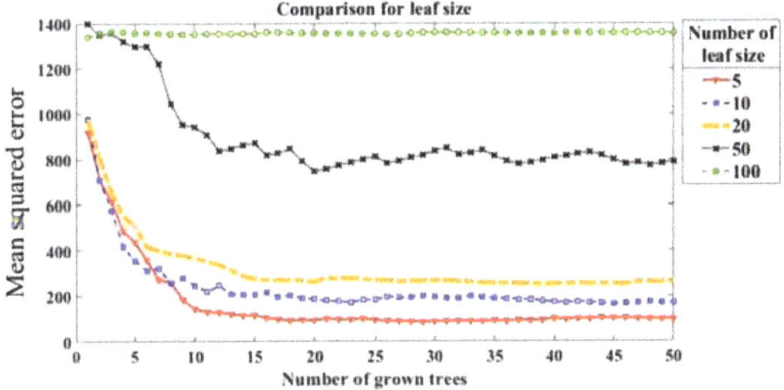

Figure 23. Optimisation of RF parameters (number of leaf size) using MSE.

Table 11. Random forest training parameters.

Parameter	Value
In bag fraction	0.25% of the training dataset
Method used by trees	Regression
Min leaf size	5
Number of trees	50

The RF model can rank the predictors (RPM, hardness, force, grit size and feed rate in our case) based on their importance by estimating out-of-bag (OOB) error. Figure 24 shows the predictors importance measured in terms of the increase in OOB error which states that grit size had a higher correlation with material removal than others. The developed random forest model robustness was evaluated by identifying the deviation of the observed value in the validation dataset and value predicted by the model. The RF model predicted a new point by aggregating the predictions from all the individual 50 regression trees ensembled inside the model.

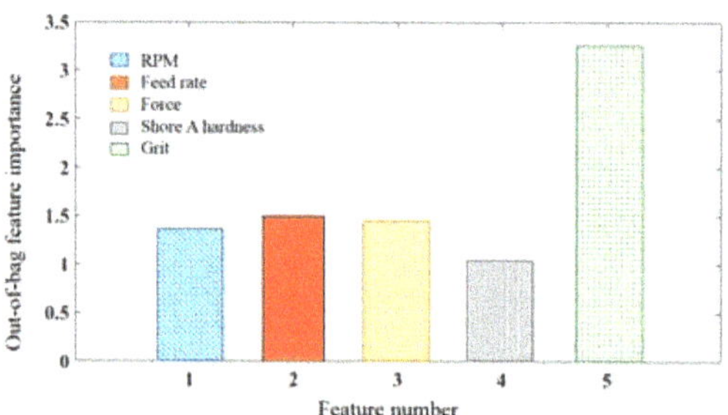

Figure 24. Variables' importance in predicting material removal using RF.

Figure 25 shows the one to one relationship between observed material removal values as predicted with the test dataset using the RF regression model. It appeared that the proposed model was good at predicting the material removal, i.e., depth of cut. Figure 25 shows the deviation of the predicted values against actual values with RFs highlighting the RMSE of 8.94. The R^2 calculated based on the fitted regression line was of value 0.975 which also showed that goodness of fit of the RF model was good. Although RF seems more of a "black box" approach compared to regression trees since individual trees cannot be assessed separately, it still provides means for interpretation in giving measures of variable importance.

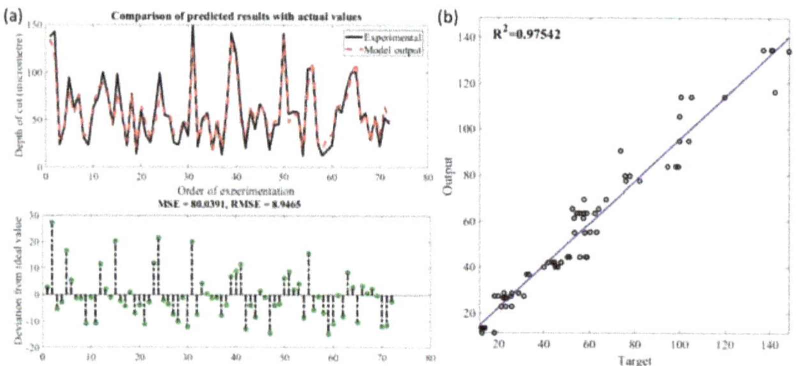

Figure 25. (**a**). Comparison of observed and predicted depth of cut using RF regression; (**b**) Statistical analysis fit of the RF regression model.

5. Conclusions

This paper illustrated the application of different statistical regression techniques coupled with the Taguchi design of experiments for the belt grinding process. The outcome demonstrated the practicality of the methodologies in developing a material removal model for the abrasive belt grinding process. Based on the regression models developed the following generalised conclusions were drawn:

- Observing the performance of the multilinear and stepwise regression models, it was seen that belt grinding parameters were intrinsically nonlinear and a straight-line relationship assumption could not satisfy the material removal.
- Although predicted values using ANN networks were close to the measured values, it functioned as a black box model correlating the parameters on its own, and the structure determination was based on trial and error.
- SVR modelling implemented using a Gaussian kernel function showed good accuracy on material removal. However, the drawback of the model was attributed to the selection of the kernel function that was based on trial and error.
- The ANFIS model developed in this research work had acceptable deviations between the predicted and the real results and was viable to predict the depth of cut in the abrasive belt grinding process compared to other regression techniques. The performance of the six algorithms in terms of RMSE is summarised in Figure 26.
- In addition, ANFIS could interpret the relationship between the input parameters towards the output behaviour which was not possible using other modelling techniques.
- The random forest model which was based on the frequency table was not able to predict at higher accuracy compared to other complex predictive models even though it performed better than the multilinear regression.
- Analysis of the sigmoidal membership function after training in ANFIS, interpretation on the variable importance results from RF, and the coefficient from stepwise regression techniques indicated that the grit size factor in the belt grinding parameters was the most significant factor on the material removal process.

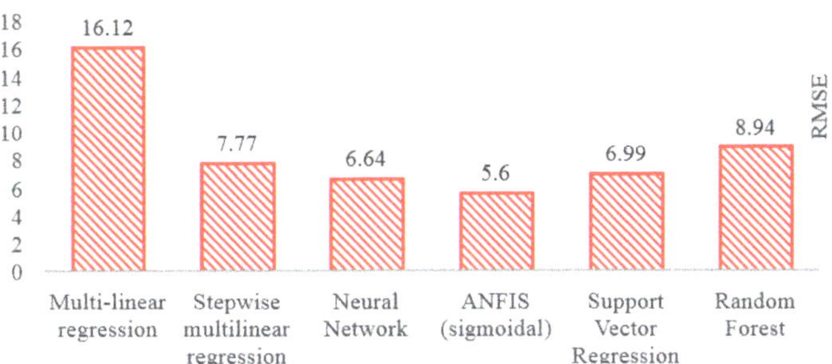

Figure 26. Predictive performance of the regression models.

Overall results from the paper suggest that statistical techniques that have the adaptability to capture nonlinear space performed well and is viable to model the material removal in the abrasive belt grinding process. The proposed regression models can be convenient for defining parameter levels for realising preferred material removal to the operator.

Author Contributions: V.P. conducted data curation, investigation, data processing, formal analysis and original draft preparation. W.C. conducted formal analysis, validation, review and editing. T.T. provided funding acquisition, conceptualization, resources, formal analysis, review and editing. A.G. conducted provided funding acquisition, review, and editing. All authors have read and agreed to the published version of the manuscript.

Funding: This research received no external funding.

Acknowledgments: This work was conducted within the Rolls-Royce@NTU Corporate Lab with support from the National Research Foundation (NRF) Singapore under the Corp Lab@University Scheme.

Conflicts of Interest: The authors declare no conflict of interest.

References

1. Wang, S.; Li, C. Application and development of high-efficiency abrasive process. *Int. J. Adv. Sci. Technol.* **2012**, *47*, 51–64.
2. Pandiyan, V.; Tjahjowidodo, T. In-process endpoint detection of weld seam removal in robotic abrasive belt grinding process. *Int. J. Adv. Manuf. Technol.* **2017**, *93*, 1699–1714. [CrossRef]
3. Ren, X.; Cabaravdic, M.; Zhang, X.; Kuhlenkötter, B. A local process model for simulation of robotic belt grinding. *Int. J. Mach. Tools Manuf.* **2007**, *47*, 962–970. [CrossRef]
4. Malkin, S.; Guo, C. *Grinding Technology: Theory and Applications of Machining with Abrasives*, 2nd ed.; Industrial Press: New York, NY, USA, 2008.
5. Zhang, X.; Kuhlenkötter, B.; Kneupner, K. An efficient method for solving the Signorini problem in the simulation of free-form surfaces produced by belt grinding. *Int. J. Mach. Tools Manuf.* **2005**, *45*, 641–648. [CrossRef]
6. Zhang, X.; Kneupner, K.; Kuhlenkötter, B. A new force distribution calculation model for high-quality production processes. *Int. J. Adv. Manuf. Technol.* **2006**, *27*, 726–732. [CrossRef]
7. Jourani, A.; Dursapt, M.; Hamdi, H.; Rech, J.; Zahouani, H. Effect of the belt grinding on the surface texture: Modeling of the contact and abrasive wear. *Wear* **2005**, *259*, 1137–1143. [CrossRef]
8. Ren, X.; Kuhlenkötter, B.; Müller, H. Simulation and verification of belt grinding with industrial robots. *Int. J. Mach. Tools Manuf.* **2006**, *46*, 708–716. [CrossRef]
9. Hamann, G. *Modellierung des abtragsverhaltens Elastischer Robotergefuehrter Schleifwerkzeuge*; University of Stuttgart: Stuttgart, Germany, 1998.
10. Wang, Y.Q.; Hou, B.; Liu, H.B.; Wang, F.B. Propeller material belt grinding parameters optimisation using Taguchi technique. *Int. J. Ind. Syst. Eng.* **2017**, *25*, 1–13. [CrossRef]
11. Sun, Y.; Vu, T.T.; Halil, Z.; Yeo, S.H.; Wee, A. Material removal prediction for contact wheels based on a dynamic pressure sensor. *Int. J. Adv. Manuf. Technol.* **2017**, *93*, 945–951. [CrossRef]
12. Pandiyan, V.; Caesarendra, W.; Tjahjowidodo, T.; Praveen, G. Predictive Modelling and Analysis of Process Parameters on Material Removal Characteristics in Abrasive Belt Grinding Process. *Appl. Sci.* **2017**, *7*, 363. [CrossRef]
13. Wang, Y.; Hou, B.; Wang, F.; Ji, Z. A controllable material removal strategy considering force-geometry model of belt grinding processes. *Int. J. Adv. Manuf. Technol.* **2017**, *93*, 241–251. [CrossRef]
14. Ren, X.; Kuhlenkötter, B. Real-time simulation and visualization of robotic belt grinding processes. *Int. J. Adv. Manuf. Technol.* **2008**, *35*, 1090–1099. [CrossRef]
15. Wu, S.; Kazerounian, K.; Gan, Z.; Sun, Y. A simulation platform for optimal selection of robotic belt grinding system parameters. *Int. J. Adv. Manuf. Technol.* **2013**, *64*, 447–458. [CrossRef]
16. Mezghani, S.; Mansori, M.E.; Zahouani, H. New criterion of grain size choice for optimal surface texture and tolerance in belt finishing production. *Wear* **2009**, *266*, 578–580. [CrossRef]
17. Shibata, J.; Inasaki, I.; Yonetsu, S. The relation between the wear of grain cutting edges and their metal removal ability in coated abrasive belt grinding. *Wear* **1979**, *55*, 331–344. [CrossRef]
18. Khellouki, A.; Rech, J.; Zahouani, H. The effect of abrasive grain's wear and contact conditions on surface texture in belt finishing. *Wear* **2007**, *263*, 81–87. [CrossRef]
19. Thien, V.T.; Halil, Z.B.A.; Yajuan, S.; Hock, Y.S.; Wee, A. Surface Finishing: Experimental Study of Pressure Distribution by Compliant Contact Wheels. In Proceedings of the 3rd International Conference on Mechatronics and Robotics Engineering, Paris, France, 8–12 February 2017; pp. 88–94.
20. Sun, Y.; Vu, T.T.; Yeo, S.H. Study of Pressure Distribution in Compliant Contact Wheels for Robotic Surface Finishing. *MATEC Web Conf.* **2016**, *42*, 03007. [CrossRef]
21. Lv, H.; Song, Y.; Jia, P.; Gan, Z.; Qi, L. An adaptive modeling approach based on ESN for robotic belt grinding. In Proceedings of the 2010 IEEE International Conference on Information and Automation, Harbin, China, 20–23 June 2010; pp. 787–792.
22. Wang, W.; Liu, F.; Liu, Z.; Yun, C. Prediction of depth of cut for robotic belt grinding. *Int. J. Adv. Manuf. Technol.* **2017**, *91*, 699–708. [CrossRef]

23. Pandiyan, V.; Tjahjowidodo, T.; Caesarendra, W.; Praveen, G.; Wijaya, T.; Pappachan, B.K. Analysis of Contact Conditions Based on Process Parameters in Robotic Abrasive Belt Grinding Using Dynamic Pressure Sensor. In Proceedings of the 2018 Joint 10th International Conference on Soft Computing and Intelligent Systems (SCIS) and 19th International Symposium on Advanced Intelligent Systems (ISIS), Toyama, Japan, 5–8 December 2018; pp. 1217–1221.

24. Wang, Y.J.; Huang, Y.; Chen, Y.X.; Yang, Z.S. Model of an abrasive belt grinding surface removal contour and its application. *Int. J. Adv. Manuf. Technol.* **2016**, *82*, 2113–2122. [CrossRef]

25. Tschätsch, H.; Reichelt, A. Abrasive belt grinding. In *Applied Machining Technology*; Tschätsch, H., Ed.; Springer: Berlin/Heidelberg, Germany, 2009; pp. 297–299. [CrossRef]

26. Fox, J. *Applied Regression Analysis and Generalized Linear Models*; SAGE Publications: Southend Oaks, CA, USA, 2008.

27. Hagan, M.T.; Demuth, H.B.; Beale, M.H. *Neural Network Design*; PWS Publishing: Boston, MA, USA, 1997.

28. Jain, A.K.; Jianchang, M.; Mohiuddin, K.M. Artificial neural networks: A tutorial. *Computer* **1996**, *29*, 31–44. [CrossRef]

29. Jang, J.S.R. ANFIS: Adaptive-network-based fuzzy inference system. *IEEE Trans. Syst. Man Cybern.* **1993**, *23*, 665–685. [CrossRef]

30. Ying, L.-C.; Pan, M.-C. Using adaptive network based fuzzy inference system to forecast regional electricity loads. *Energy Convers. Manag.* **2008**, *49*, 205–211. [CrossRef]

31. Caesarendra, W.; Wijaya, T.; Tjahjowidodo, T.; Pappachan, B.K.; Wee, A.; Roslan, M.I. Adaptive neuro-fuzzy inference system for deburring stage classification and prediction for indirect quality monitoring. *Appl. Soft Comput.* **2018**. [CrossRef]

32. Zadeh, L.A. Fuzzy sets. *Inf. Control* **1965**, *8*, 338–353. [CrossRef]

33. Smola, A.J.; Schölkopf, B. A tutorial on support vector regression. *Stat. Comput.* **2004**, *14*, 199–222. [CrossRef]

34. Drucker, H.; Burges, C.J.C.; Kaufman, L.; Smola, A.; Vapnik, V. Support vector regression machines. In Proceedings of the 9th International Conference on Neural Information Processing Systems, Denver, CO, USA, 3–5 December 1996; pp. 155–161.

35. Breiman, L. Random Forests. *Mach. Learn.* **2001**, *45*, 5–32. [CrossRef]

Article

A Hybrid Mechanism for Helicopters

Kevin Kuan-Shun Chiu [1],*, Jeou-Long Lee [2], Ming-Lang Tseng [3] , Rosslyn Hsiu-Ling Hsu [2] and Yen-Jen Chen [4]

[1]   Department of Business Administration, Lunghwa University of Science and Technology, Taoyuan city 33306, Taiwan
[2]   Department of Chemical & Materials Engineering, Lunghwa University of Science and Technology, Taoyuan city 33306, Taiwan; semxrdedx@yahoo.com.tw (J.-L.L.); rosslyn@mail.lhu.edu.tw (R.H.-L.H.)
[3]   Department of Business Administration, Asia University, Taichung city 41354, Taiwan; tsengminglang@gmail.com
[4]   Department of Multimedia and Game Science, Lunghwa University of Science and Technology, Taoyuan city 33306, Taiwan; clive_chen@mail.lhu.edu.tw
*   Correspondence: kevinchiu@msn.com; Tel.: +886-9-5858-3311

Received: 16 November 2019; Accepted: 19 December 2019; Published: 22 December 2019

Abstract: This study successfully provides the empirical practicability of a hybrid mechanism for helicopters. A turbine engine and a set of electricity power systems can operate simultaneously and/or independently as a symmetric structure. The latter power source works as an immediate supplementary device if the former one has malfunction. We look forward to promoting this experimental evidence in the helicopter industry. The ultimate purpose of this manuscript is to decrease the incidents of crashes and save people's lives.

Keywords: hybrid mechanism; helicopter

1. Introduction

Riding in a helicopter is a great experience with entertainment, excitement, and convenience. Surprisingly, "the helicopter accident rate was 7.5 per 100,000 hours of flying, whereas the airplane accident rate was approximately 0.175 per 100,000 flying hours" [1]. In other words, the risk of flying in a helicopter is 42.86 times greater than in an airplane. Further, shockingly, based on a statistical analysis of helicopter accidents between 2005 and 2015 [2], every crash caused an average of 2.3 fatalities.

The three most common reasons for helicopter crashes are personnel mistakes (i.e., pilots and tower staff), mechanical failures, and lack of fuel [3]. The latter cause leads to an engine's immediate flameout. As such, the helicopter's rotor head speed will slow down right away, and then lose the power of lift. In such an emergency, the pilot has only one chance to maneuver auto rotation for safe landing. However, this unusual happy ending needs an excellent combination of sufficient altitude, nice weather, skillful piloting, and a perfect landing spot.

As mentioned above, since lack of fuel is unavoidable, is it possible to build a backup system to solve this crisis? However, to the best knowledge of the authors, only a scarcity of empirical research has focused on an auxiliary device in the helicopter industry. The current work attempts to choose the best possible solution for overcoming a helicopter engine's flameout. To fill this gap, the major focus of this study is to propose a hybrid mechanism (turbine engine + electricity power system) to provide instant power right after the engine's flameout so as to maintain the helicopter's rotor head speed for safe landing. Currently, industry and academia are undergoing an evolution in developing the next generation of drone applications [4]. Due to financial budget constraint, the present experiment attempted to design and build a remote control helicopter for testing this concept.

2. Problem Formulation and Methods

- Lack of Fuel

Several reasons cause lack of fuel: fuel starvation (i.e., fuel does not reach the engine which refers to mechanical failures), fuel exhaustion (i.e., inadequate fuel management exists for intended flight/s), or fuel gauge needle problems (i.e., instrument malfunction) [3,5]. Lack of fuel seems to be inevitable. No matter what the real reason is, the engine's misfire takes place at the same time. At least 2–3 minutes is needed to restart the engine. However, it takes only a few seconds to crash the helicopter [6].

- Electricity Power System

An electricity power system seems to be the best method for solving the above threat. "Getting electrons from a battery to an electric motor is much faster than getting fuel from a gas tank to a piston. Electrons travel much faster along a wire than fuel does along a fuel line, and the electrons basically go straight to the place where they are needed" [7]. Therefore, we truly believe that an electrical power system offers the fastest response for keeping the helicopter's rotor head speed right after any flameout.

- Hybrid Operation

Hybrid operation has received significant attention recently in several industries, such as wind farming [8] and automobile manufacturing [9]. Such a device generally uses two or more diverse forms of power sources in various situations, shown as Figure 1.

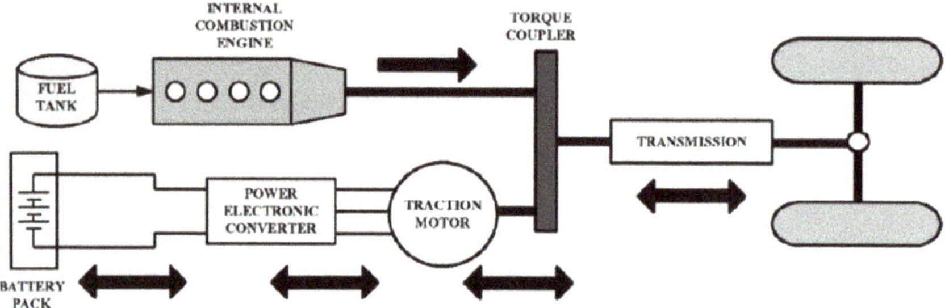

Figure 1. Parallel hybrid electric vehicle mechanism [10].

The basic principle is that the different motors work better at different speeds; the electric motor is more efficient at producing torque, or turning power, and the combustion engine is better for maintaining a high speed (better than a typical electric motor) [11]. Changing power sources with appropriate timing and in the right circumstances facilitates energy efficacy and fuel efficiency. Abundant hybrid models (i.e., the Boeing Fuel Cell Demonstrator Airplane, Hybrid FanWings) have been presented in order to take advantage of each and handle uncertainties [11]. This hybrid methodology is proposed, taking benefits of both the conventional power source (i.e., a turbine engine's consistent output) and an electricity power system (i.e., fastest response for utility) [12] so as to enhance safety if engine failure happens.

3. Conceptual Framework and Design for Experiment

This drone application proposal [4] is shown as Figure 2.

Figure 2 explains the sequence for the operation procedure of the proposed hybrid mechanism. Generally speaking, (1) A turbine engine offers the power for rotor head speed; (2) if the turbine engine suffers from lack of fuel/engine failure, (3) the electricity power system offers auxiliary dominance so as to (4) keep the full rotor head speed for a helicopter.

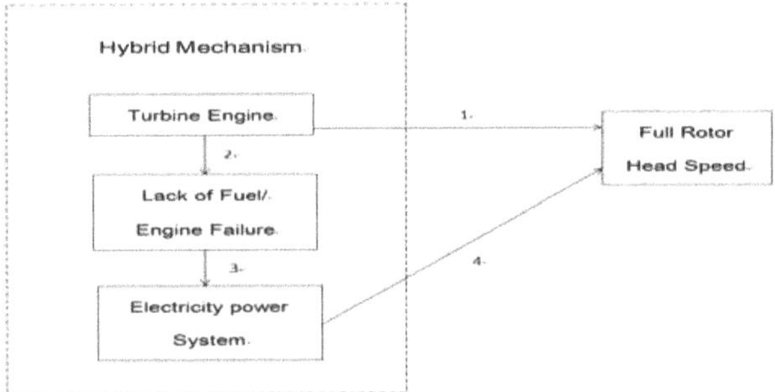

Figure 2. Proposed theoretical foundation of the hybrid mechanism.

Moreover, this experiment contains the following sections:

- 3.1. MD-8 remote control helicopter: We decided to choose the MD-8 as the model for our test because this helicopter has great reputation in the R/C industry due to its agility, rigidity, and simplicity for maintenance. The dimensions for this remote control helicopter are 230 (L) × 30 (W) × 55 cm (H). Further, we also needed to design and build a conversion kit which contained new reinforced helicopter body frames so as to accommodate both a complete turbine engine device and an electricity power system. This conversion kit was made by composite materials of carbon-fiberglass and aluminum-alloy poles with the dimensions of 50 (L) × 15 (W) × 30 cm (H). The turbine engine was installed in this kit via a set of custom-made engine mount and gear box mount.

- 3.2. Kingtech K-45 turbo-prop airplane turbine engine: This merchandise is a miniature version of real aircraft turbine engines with the dimensions of 33.5 (L) × 15 (W) × 8 cm (H). Not only the theory, but the operations are the same as those of authentic aircraft turbine engines. This engine weighs 1.8 kg, offers the output of 5.2 kW (6.97 hp), and consumes 5%-lubricant jet A1 fuel at 180 g/min with engine's rpm 170,000 (maximum output shaft, 9000 rpm). Most importantly, Kingtech is a world-famous but domestic company in Taiwan. We are proud to add to the impression of products made in Taiwan.

- 3.3. An electricity power system: This category includes an electric motor (with dimensions of 5.77 (L) × 2.8 cm (R)) which weighs 526 g (with the maximum rated output of 4.4Kw/6.03hp at the maximum rated current of 100 Amps); an electricity amperage supplier—160 Amps, (i.e., also known as motor speed controller); and 44.4 V (3250 mAh) batteries.

- 3.4. Electronic control devices: This group has servos, gyro, transmitter, receiver, and 7.4 V batteries.

- 3.5. Supplementary parts: 3 × 810 mm main and 3 × 120 mm tail blades, and ground support accessories. Most important is the next point.

- 3.6. The hybrid mechanism. Since we have to operate both Kingtech K-45 turbo-prop airplane turbine engine and the electronic motor consecutively, we had to design and make a brand new, innovative hybrid mechanism for testing our concept. This hybrid mechanism had to meet the following requirements. First of all, based on the conventional design, the electronic motor's driver shaft needed to vertically attach to the main driving gear. There is no other space for the turbine engine to engage onto the main driving gear. Therefore, we had to add a 2nd set of driving gears for the turbine engine. Moreover, due to a helicopter's restricted accommodation for keeping vertical as well as horizontal balance, we decided to create a set of bevel gears so as to cope with the turbine engine's output and rpm. Most important, both sets of driving gears have an individual one-way bearing module so as to independently transmit either the turbine engine's performance or the electronic motor's output onto the extended main shaft for spinning the rotor head system. The proposed hybrid mechanism is shown as Figure 3.

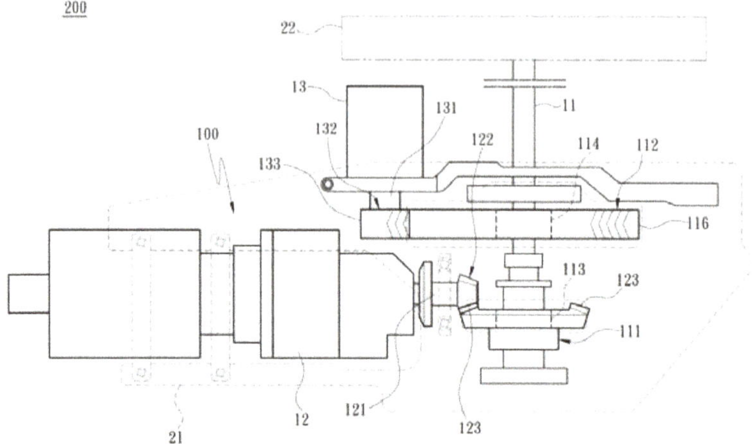

Figure 3. The hybrid mechanism proposal.

In more specific terms, number in Figure 3 denote the following:

100—hybrid mechanism;
11—driver shaft for propellers;
111—first transmission gear;
112—second transmission gear;
113,114—one-way bearing module;
116—first v-shaped teeth gear;
12—turbine engine;
121—first output shaft;
122—first bevel driving gear;
123—first bevel gear;
13—electronic motor;
131—second output shaft;
132—second driving shaft;
133—second v-shaped teeth gear;
200—drone application/remote control helicopter;
21—carbon-fiberglass frame of conversion kit;
 22—propellers.

Operationally, once we start the turbine engine (12), this engine rotates the first output shaft (121) and the first bevel driving gear (122) so as to drive first bevel gear (123). Consequently, through the first transmission gear (111), the one-way bearing module (113) is driven to turn the driver shaft for the propellers (11). Finally, the propellers (22) revolve at a specified speed so as to provide sufficient thrust for lifting the remote control helicopter. In more specific terms, we set the turbine engine's (12) output shaft rpm to 9000; the gear ratio between the first bevel driving gear (122) and the first bevel gear (123) was 1:4. Accordingly, the rpm of driver shaft for propellers (11) was 2250. However, the realistic rpm reaches only 70–80% (1575–1800) of the calculated result due to aerodynamic drag [13,14].

On the other hand, if accidentally, the turbine engine suffers from lack of fuel/engine failure, the electronic motor (13) works right away to drive second output shaft (131), second driving shaft (132), and second v-shaped teeth gear (133) so as to rotate first v-shaped teeth gear (116). Consequently, the second transmission gear (112) and one-way bearing module (114) are driven to turn driver shaft for the propellers (11). Finally, propellers (22) revolve at specified speed so as to provide sufficient thrust for lifting the remote control helicopter. In more specific terms, the electronic motor's (13) maximum rated output shaft rpm/v is 520 KV. The gear ratio between the second v-shaped teeth gear (133) and the first v-shaped teeth gear (116) is 1:11. Accordingly, the rpm of driver shaft for propellers (11) is 520 KV × 44.4 V/11 = 2099. However, the realistic rpm reaches only 70–80% (1469–1679) of the calculated result due to aerodynamic drag [13,14].

Figure 4 represents a transparent model for Figure 2.

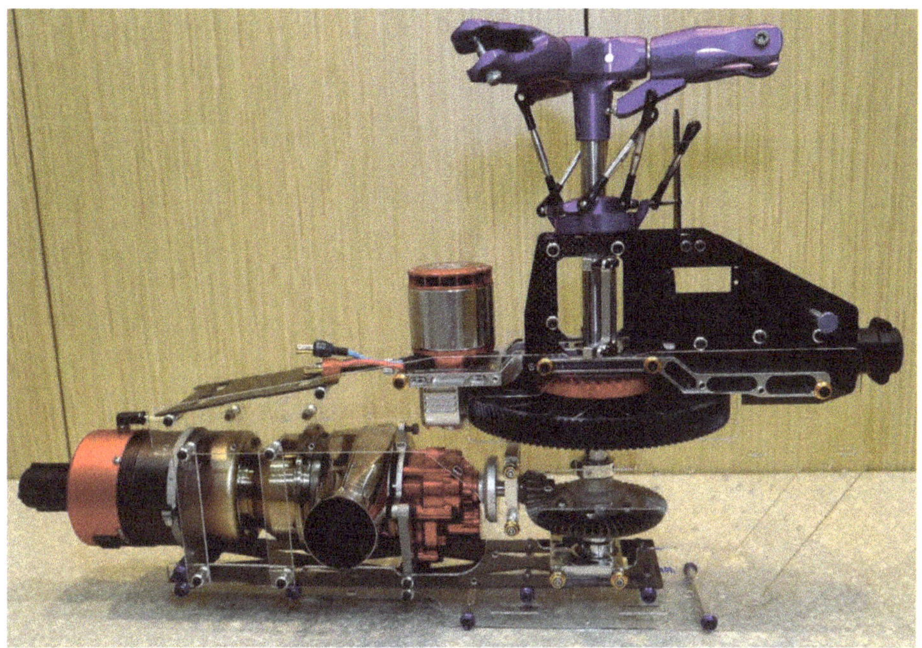

Figure 4. Transparent model.

Figure 5 represents the complete hybrid mechanism proposal in the conversion kit.

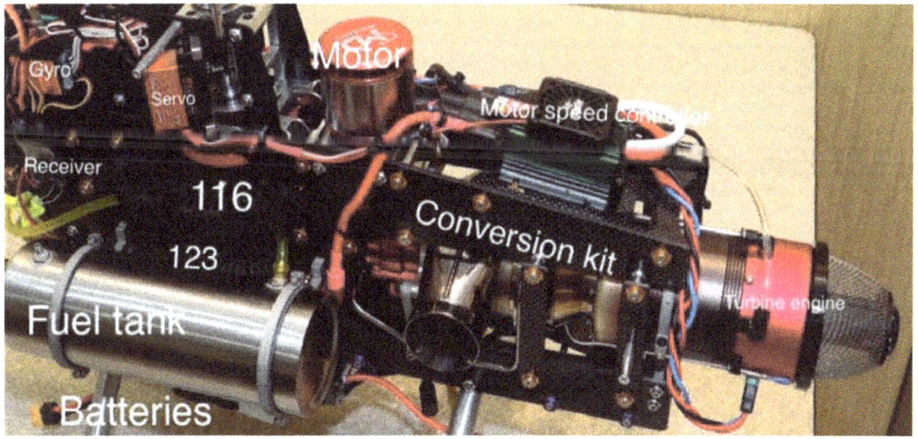

Figure 5. Complete hybrid mechanism. (116- first v-shaped teeth gear, 123- first bevel gear).

4. Discussion

We successfully test ran and flew this newly designed R/C helicopter, as shown in Figure 6.

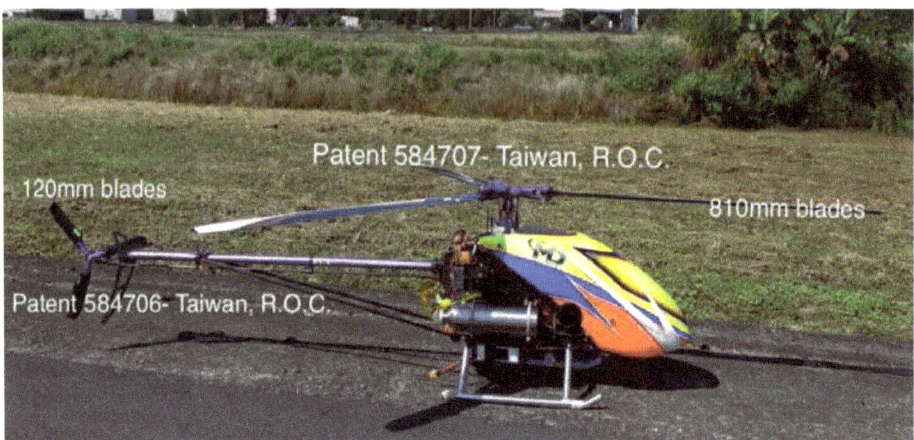

Figure 6. The new design of an R/C helicopter.

Operationally, we used a radio control transmitter to maintain the main blades' rpm at 1500 for both turbine engine and electronic motor power sources, respectively. Generally speaking, the maximum speed for the tip of the helicopter's blades is approximately 180–220 m/s. In other words, the speed is around 0.55–0.66 Mach at standard atmospheric pressure (i.e., the environmental temperature might have moderate influence on the speed) [15]. So as such, 1500 rpm $\times$ 0.9 m $\times$ 2 $\times$ 3.14 = 141.3 m/s which is roughly 70% of the maximum speed mentioned above for producing much less vicious variations in altitude or velocity of the helicopter [16]. The pitch administered to the main blades was between –5 and 10 degrees. This new design R/C helicopter started to lift at the pitch of 2° and hovered at that of 4.5°. The propellers are of the controllable-pitch type so that they could have a low pitch when taking off and a higher pitch for high speed, horizontal flight [16].

Empirically validated, within 1 second, the electricity power system offers supplementary motive force as soon as the turbine engine shuts down (and/or has a flameout). Furthermore, the electricity power system offers sufficient dynamism for the above helicopter (9.5 kg) not only hover at a specific altitude but also ascend to the sky. The capacity of those batteries is adequate for this model helicopter to fly for 5 minutes. In other words, the 5-minute airtime is a very safe and satisfactory period of time for a harmless landing. Through this hybrid mechanism, two individual power sources are interconnected to make either turbine engine or electronic motor flight possible [16].

This paper contains both theoretical and experimental pioneering research, presenting a model for the solution to an engineering problem: the fastest way to solve the crisis of unavoidable/unpredicted engine failure for helicopters. In the near future, we hope to apply this practical evidence in the helicopter industry for decreasing human injuries and/or fatalities. The ultimate purpose of this manuscript is to save people's lives.

For any specific helicopter, the hybrid mechanism needs to be redesigned for adaptation. The impending electricity power system (i.e., kW output of an electric motor, capacity of an electricity amperage supplier, and measurements of batteries) needs to be redefined. In other words, the overall weight-to-power ratio must be re-calculated.

The findings of this study solve only the lack of fuel/engine failure jeopardy. We might need to conduct some other experiments for straightening out personnel mistakes and/or mechanical failures.

5. Patents and Recognitions

In Figure 5, the complete hybrid mechanism is in the process of patent application on the date of submission for publication. The conversion kit which contains new, reinforced helicopter body frames (Patent pending) won a Gold Medal in the 2016 Kaoshiung International Invention and Design Expo. In Figure 6, both Triple Blades' Flybarless Main Rotor Head Complete System (Patent M584707, Taiwan, R.O.C.) and Triple Blades' Tail Rotor Assembly (Patent M584706, Taiwan, R.O.C.) won Gold Medals in the 2016 Kaoshiung International Invention and Design Expo. Additionally, the latter innovation obtained The Award of the President of the Jury in the same exhibition [17].

Author Contributions: Conceptualization, methodology, and experimental project administration (construction of this experimental R/C helicopter, and test flight), K.K.-S.C.; writing—original draft preparation, K.K.-S.C.; writing—review and editing, M.-L.T. and Y.-J.C.; confirmation of the rigidity of the materials in the construction of this experimental R/C helicopter, J.-L.L. and R.H.-L.H. All the authors have read and approved the final version for submission. All authors have read and agreed to the published version of the manuscript.

Funding: This research received no external funding.

Acknowledgments: This work has been supported by the KingTech Turbines and MingDa Technology Corp.

Conflicts of Interest: The authors declare no conflict of interest.

References

1. Crouse Law Offices. What Causes Helicopter Accidents? Available online: http://www.helicopterlawyers.com/causes_helicopter_accidents.html (accessed on 18 August 2018).
2. IHST-CIS. *Helicopter accidents: Statistics, trends and causes*; IHST Regional Partners Panel: Louisville, KY, USA, 2 March 2016; Available online: http://www.ihst.org/portals/54/symposium/2016/Presentation%20IHST-CIS_2016.pdf (accessed on 18 August 2018).
3. Byrne, G. Why helicopters crash: Fuel, pilot error and mechanical failure. *The Irish Times*. 15 March 2017. Available online: https://www.irishtimes.com/news/ireland/irish-news/why-helicopters-crash-fuel-pilot-error-and-mechanical-failure-1.3010682 (accessed on 18 August 2018).
4. Lynskey, J.; Thar, K.; Thant, Z.O.; Choong, S.H. Facility location problem approach for distributed drones. *Symmetry* **2019**, *11*, 118–129. [CrossRef]
5. Robb, G.C. Helicopter crashes 101, 2014. Robb & Robb Web site. Available online: https://www.robbrobb.com/helicopter-crashes-101 (accessed on 18 August 2018).
6. Pilot tells police what caused New York City helicopter crash that killed 5. *CBS News*. 13 March 2018. Available online: https://www.cbsnews.com/news/pilot-tells-police-what-caused-new-york-city-helicopter-crash-that-killed-5/ (accessed on 18 August 2018).
7. Barnard, M. Why are Teslas quicker than gas cars? *CleanTechnica*. 19 December 2015. Available online: https://cleantechnica.com/2015/12/19/teslas-quicker-gas-cars/ (accessed on 18 August 2018).
8. Dhiman, H.S.; Deb, D.; Muresan, V.; Unguresan, M.-L. Multi-criteria decision making approach, for hybrid operation of wind farms. *Symmetry* **2019**, *11*, 675–692. [CrossRef]
9. Wikipedia. Hybrid Vehicle. Available online: https://en.wikipedia.org/wiki/Hybrid_vehicle (accessed on 9 November 2019).
10. Hussein, M.M. An Argument for the Adoption of Hybrid Electric Vehicles. December 2014. Available online: https://www.researchgate.net/publication/274371964_An_Argument_for_the_Adoption_of_Hybrid_Electric_Vehicles (accessed on 14 December 2019).
11. Liu, Y.; Martinez, L.; Qin, K. A comparative study of some soft rough sets. *Symmetry* **2017**, *9*, 252–272. [CrossRef]
12. Pawar, P.; Nielsen, R.; Prasad, N.; Ohmori, S.; Prasad, R. Hybrid mechanisms: Towards an efficient wireless sensor network medium access control. In Proceedings of the 2011 14th International Symposium on Wireless Personal Multimedia Communications (WPMC), Brest, France, 3–7 October 2011.
13. Hoerner, S.F. *Fluid-Dynamic Drag: Theoretical, Experimental and Statistical Information*; Hoerner Fluid Dynamics: Vancouver, BC, Canada, 1965.
14. Hoerner, S.F. *Fluid Dynamic Drag: Practical Information on Aerodynamic Drag and Hydrodynamic Resistance*; Hoerner Fluid Dynamics: Vancouver, BC, Canada, 1965.

15. Hoerner, S.F. *Fluid-Dynamic Lift: Practical Information on Aerodynamic and Hydrodynamic Lift*; Hoerner Fluid Dynamics: Vancouver, BC, Canada, 1992.

16. Piancastelli, L.; Gatti, A.; Frizziero, L.; Ragazzi, L.; Cremonini, M. CFD analysis of the Zimmerman's V173 stol aircraft. *J. Eng. Appl. Sci.* **2015**, *10*, 8063–8070.

17. Kaohsiung KIDE international invention and design exhibition remote control jet helicopter shines. *RC Tech Magazine*. 13 December 2016. Available online: http://www.rctech.com.tw/2016/12/kide.html (accessed on 30 October 2018).

 *symmetry*

Article

An Analysis of the Dynamical Behaviour of Systems with Fractional Damping for Mechanical Engineering Applications

Ondiz Zarraga *[iD], Imanol Sarría [iD], Jon García-Barruetabeña [iD] and Fernando Cortés [iD]

Department of Mechanics, Design and Industrial Management, University of Deusto,
Avenida de las Universidades 24, 48007 Bilbao, Spain; isarria@deusto.es (I.S.);
jgarcia.barruetabena@deusto.es (J.G.-B.); fernando.cortes@deusto.es (F.C.)
* Correspondence: ondiz.zarraga@deusto.es

Received: 21 November 2019; Accepted: 8 December 2019; Published: 11 December 2019

Abstract: Fractional derivative models are widely used to easily characterise more complex damping behaviour than the viscous one, although the underlying properties are not trivial. Several studies about the mathematical properties can be found, but are usually far from the most daily applications. Thus, this paper studies the properties of structural systems whose damping is represented by a fractional model from the point of view of a mechanical engineer. First, a single-degree-of-freedom system with fractional damping is analysed. Specifically, the distribution of the poles and the dynamic response to several excitations is studied for different model parameter values highlighting dissimilarities from systems with conventional viscous damping. In fact, thanks to fractional models, particular behaviours are observed that cannot be reproduced by classical ones. Finally, the dynamics of a machine shaft supported by two bearings presenting fractional damping is analysed. The study is carried out by the Finite Element method, deriving in a system with symmetric matrices. Eigenvalues and eigenvectors are obtained by means of an iterative method, and the effect of damping is visualised on the mode shapes. In addition, the response to a perturbation is computed, revealing the influence of the model parameters on the resulting vibration.

Keywords: fractional damping; vibration; dynamic behaviour

---

1. Introduction

In the context of dynamics, fractional models allow us to easily represent the vibratory behaviour of elements that would otherwise require complex formulations, such as multielement or hereditary models, because they are able to reproduce correctly the damping mechanisms that come into play using a low number of variables [1,2]. Fractional models are specially advantageous for polymeric materials that show some level of dependence to frequency and arise naturally, for example, from certain motions of Newtonian fluids [3] or the molecular theories that predict the behaviour of certain type of polymeric materials [4]. In fact, fractional models are used to capture with more ease the viscoelastic nature of materials such as rubber or concrete [5] whose behaviour was previously modelled with a power law [6,7]; fractional operators appear in the non-linear stress-strain relation of metals [8]; and viscoelastic constitutive models based on fractional derivatives have been proposed to reproduce the time dependent behaviour of real materials [9–14].

Several works treating the behaviour of dynamic systems following fractional models from a mathematical perspective can be found in the literature. For instance, in [15], a fractionally damped single-degree-of-freedom (1 DOF) oscillator was analysed using the Laplace transform. It was found that a system of this kind exhibits nine distinct behaviour cases in opposition to the three shown by a traditional oscillator. Also, in the [16–18] series, the behaviour of a fractional oscillator whose derivative order was between one and two was studied, first when subjected to initial conditions and, after, when applying a sinusoidal force to the system. The main conclusion of the series was that a fractional oscillator like the one analysed presented damping even if not explicitly stated.

Computing methods have been widely proposed as well, mostly numerical [19] due to the difficulty to obtain closed-form solutions. Apart from the Laplace transform, which has been already mentioned, the Fourier transform [4], an eigenvector expansion in the state space [20], modal synthesis [21] or finite elements [22,23] have been used to obtain the response of systems presenting fractional damping.

In general, these types of systems are not analysed from the engineering point of view, even if fractional models have been used in the actual engineering practice, for instance, to linearise equations or to reduce the number of parameters needed to represent elastoplastic [24] or viscoelastic behaviours [11]. For this reason, the objective of this work is to shed light on the dynamic behaviour of a system whose damping is modelled using a fractional model. To this aim, first the theoretical background is presented with the aid of a 1 DOF system, for which the distribution of poles is studied and whose dynamic response to different excitations is computed. Instead of focusing on obtaining the analytical response for certain values, the behaviour of the system is analysed numerically in a range of values in order to identify trends. The goal is to identify the differences between a system whose damping is modelled as fractional and one following a traditional viscous damping formulation. This understanding is used afterwards to represent the dynamics of a machine shaft supported by two bearings with fractional damping.

2. Theoretical Background: A 1 DOF System with Fractional Damping

A single-degree-of-freedom mechanical system that presents fractional damping follows the equation of motion

$$m\ddot{x} + c\mathrm{D}^{\alpha}x + kx = F(t) \ / \ \alpha \in [0,1], \tag{1}$$

where m, c and k are respectively the mass, damping and stiffness of the system, $F(t)$ is an external force and α is the order of the fractional derivative that, in order to describe a viscoelastic damper, is considered to be between zero and one—if the order of the derivative were zero, the damper would behave as a spring and would be considered "elastic"; if one, the damping would be viscous.

The fractional derivative of a function $f(t)$ is computed following the Caputo definition using the gamma function Γ as

$$\mathrm{D}^{\alpha}f(t) = \frac{d^{\alpha}f(t)}{dt^{\alpha}} = \frac{1}{\Gamma(n-\alpha)}\frac{d}{dt}\int_{0}^{t}\frac{f^{(n)}(u)}{(t-u)^{\alpha+1-n}}du, \tag{2}$$

where $n - 1 \leq \alpha < n$. In this work, the Caputo definition is preferred to the Riemman-Liouville because the constant terms produced by the Laplace transform of the former have direct physical meaning [15] whereas the interpretation of such terms in the latter is more complex [12].

In the case under study, as the order of the derivative α is between zero and one and, in consequence, $n = 1$, the fractional derivative in (2) becomes

$$\mathrm{D}^{\alpha}f(t) = \frac{d^{\alpha}f(t)}{dt^{\alpha}} = \frac{1}{\Gamma(1-\alpha)}\frac{d}{dt}\int_{0}^{t}\frac{f(u)}{(t-u)^{\alpha}}du. \tag{3}$$

For more detail about the definition or computation of fractional derivatives, any of the classical texts on the subject [2,25–27] is recommended.

2.1. Poles of the System

One of the peculiarities of a system with fractional damping is the nature of its poles. In order to compute them, (1) is expressed in the Laplace domain, where adopts the form

$$\left(ms^2 + cs^\alpha + k\right) X(s) = s^{\alpha-1}x(0), \tag{4}$$

if $F(t) = 0$ is considered, $X(s) = \mathcal{L}\{x(t)\}$ being the Laplace transform of the displacement $x(t)$. Then, the characteristic equation

$$ms^2 + cs^\alpha + k = 0 \tag{5}$$

is solved. Supposing that α is a rational number and can be therefore expressed as the quotient of two integers $\alpha = p/q$, (5) becomes the polynomial of order $2q$

$$mr^{2q} + cr^p + k = 0 \ / \ r = s^{\frac{1}{q}} \tag{6}$$

from which the poles of the system can be obtained. Resorting to the state space is a more elegant way to determine the poles of the system [8,20,28,29], but the presented process allows us to focus on the qualitative properties of the fractionally damped oscillator keeping the mathematical complexity to a minimum.

From (6), one could think that a fractionally damped system has $2q$ poles but, in reality, only two of them—a complex conjugate pair [15], actually—are solutions to the original system (5), the rest are extraneous solutions due to the solving process.

The fact that (5) has only two complex conjugate poles can be explained from the physics of the real system it models: a 1 DOF system can only vibrate at a single frequency. Also, the fact that the poles of the system are complex conjugates implies that the response is always oscillatory and that there are no overdamped or critically damped cases as defined for a traditional 1 DOF system [20]. From a mathematical perspective, this phenomenon is related to the fact that the solutions of fractional order differential equations are Mittag-Leffler functions, whose properties explain the oscillatory dynamics of fractionally damped systems [16,29]. This feature will be further discussed when studying its dynamic response in Section 2.2.

In order to better understand the behaviour of the system, the variation in the position of the poles with the order of the derivative α is studied. Without loss of generality, for the numerical computations the mass m, damping c and stiffness k parameters are considered unitary (in any coherent system of units) while the value of the order of the derivative α ranges between zero and one, the completely elastic and the completely viscoelastic behaviours respectively.

If the poles of the system for different values of α are plotted in the complex plane, the shape shown in Figure 1 is obtained. In it, the solutions obtained from the variable change in (6) are shown in black while the ones that satisfy (5) are highlighted in red.

For every value of α, the centroid of each of the two distributions shown in Figure 1 computed as

$$C = \frac{1}{q}\sum_{j-1}^{q} s_j, \tag{7}$$

s_j being the jth pole of the system and q the amount of solutions of (6) for each distribution, the extraneous ones included, fall in $\pm w_0 = \pm\sqrt{k/m}$.

Another conclusion that can be drawn from the variation of the position of the poles with the value α is that, contrary to what could be expected, the number of decimal positions used to express α and, thus, the p and q values used to approximate it as a rational number, do not affect the position of the poles. This means that considering α or a slightly perturbed value $\alpha + \delta$ provides the same results both in terms of poles and, in consequence, of time response.

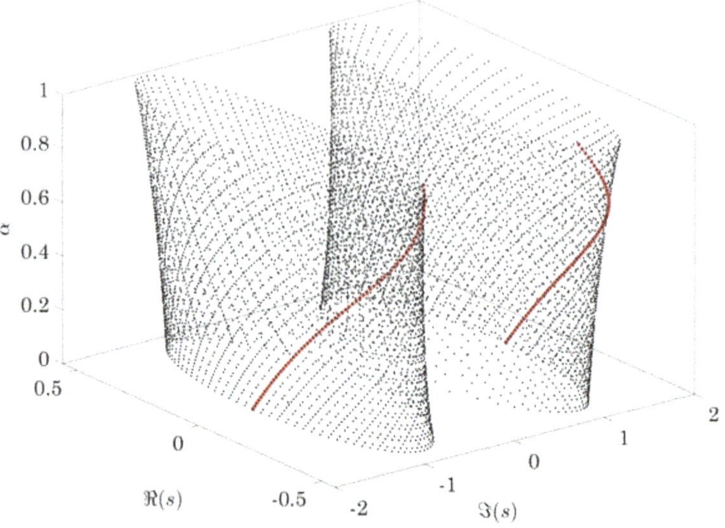

Figure 1. Distribution of poles for different values of α. The solutions that comply with (5) are highlighted in red.

Then, the relationship of the damped frequency of the system, represented by the imaginary part of the poles, with the order of the derivative is studied. As shown in Figure 2, the damped frequency of the system ranges from

$$\omega_\mathrm{d} = \sqrt{\frac{k+c}{m}} \tag{8}$$

when $\alpha = 0$ (it should be noted that in this case, as the damper behaves like a spring, the system will be *undamped*) to the well known formula for the viscous case when $\alpha = 1$

$$\omega_\mathrm{d} = \omega_0 \sqrt{1 - \xi^2}, \tag{9}$$

where $\xi = c/(2\sqrt{km})$ and $\omega_0 = \sqrt{k/m}$. The damped frequency decays with the order of the derivative α because, due to the viscoelastic nature of the model, the lower this order, a bigger part of the damping behaves as stiffness [30].

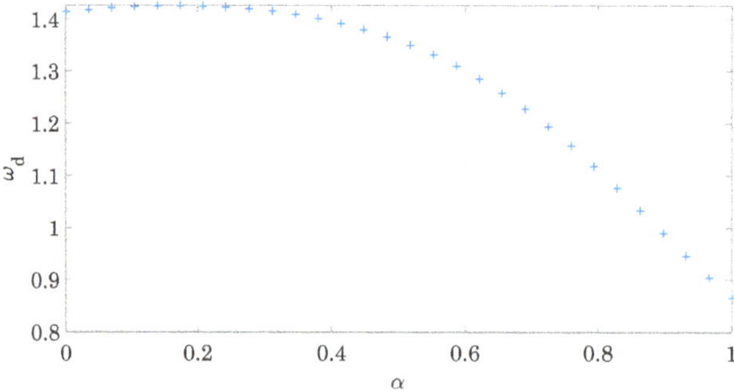

Figure 2. Evolution of damped frequency ω_d with the order of the derivative α. The mass, stiffness and damping parameters are unitary. The damped frequency ranges from $\omega_\mathrm{d} = \sqrt{\frac{k+c}{m}} = \sqrt{2}$ to $\omega_\mathrm{d} = \omega_0 \sqrt{1 - \xi^2} = 0.866$.

Then, the value of the damping parameter c is changed keeping the order of the fractional derivative constant, in order to study the evolution of the damped frequency (Figure 3). In contrast to what happens in systems with viscous damping, increasing the damping parameter c in a fractionally damped system increases the vibration frequency. This can be explained again by the viscoelastic nature of fractional damping: while in a system with viscous damping a change in the value of the damping parameter c does not affect the stiffness; if the damping is fractional, both magnitudes are interrelated and an increase in the damping parameter c also stiffens the system. It is worth noting that, even if the tendency is the same for different values of the order of the derivative α—increasing the damping parameter c produces an increase in the damped frequency—the relationship between the two magnitudes is not linear and varies depending on the value of the order of the derivative.

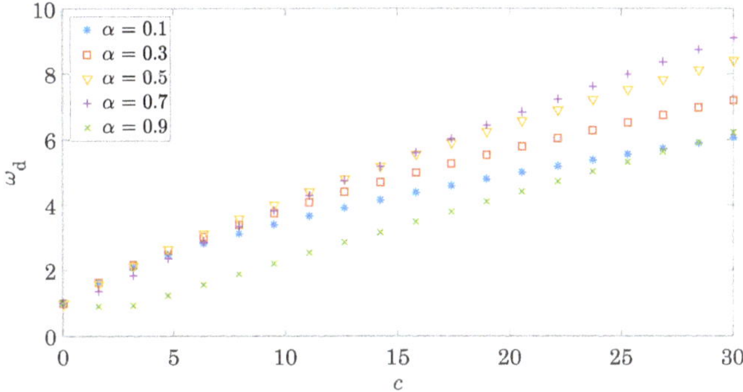

Figure 3. Evolution of the damped frequency ω_d with the damping parameter c for different values of the order of the derivative α.

2.2. Dynamic Response

Another aspect in which a system with fractional damping differs from one showing viscous damping is the dynamic response. These differences are analysed in the 1 DOF system, first, when it is subjected to initial conditions and, after, when a impulse or step force is applied.

In order to make the analysis extensible to systems with several degrees of freedom or subjected to any type of excitation for which analytical solutions are not available, the process to solve the differential equation (1) numerically is presented. The analytical response of a 1 DOF system to the cited excitations—initial conditions or impulse and step forces—can be found, for instance, in [2,20,29,31].

That said, the Grünwald-Letnikov method is used to compute the values of the fractional derivative for a given time t. Its objective is to take advantage of the definition of the derivative and approximate the gamma functions with some recursively computed coefficients. The derivative of order α is, thus, expressed as

$$D^{\alpha} f(t) = \lim_{\Delta t \to 0} \left(\frac{1}{(\Delta t)^{\alpha}} \sum_{j=0}^{S-1} \frac{\Gamma(j-\alpha)}{\Gamma(-\alpha)\Gamma(j+1)} f(t - j\Delta t) \right) \tag{10}$$

where $S = t/\Delta t$ is the number of samples. Considering

$$A_{j+1} = \frac{\Gamma(j-\alpha)}{\Gamma(-\alpha)\Gamma(j+1)}, \tag{11}$$

the A_j coefficients can be computed recursively by

$$A_1 = 1, \tag{12}$$

$$A_{j+1} = \frac{j - \alpha - 1}{j} A_j \tag{13}$$

and the derivative of order α takes the simplified form

$$\mathrm{D}^\alpha f(t) \approx \frac{1}{(\Delta t)^\alpha} \sum_{j=0}^{S-1} A_{j+1} f(t - j\Delta t), \tag{14}$$

which can be easily introduced in a numerical method, a forward Euler in this case.

It should be noted that fractional operators are convolution integrals and, as such, they have memory, making it is necessary to store the previous history so as to be able to compute the response in a certain step [1]. If needed, techniques as the one described in [30] could be used to reduce the storing requirements.

The presented methodology is followed in the next sections to compute the response of the fractionally damped 1 DOF system under study to initial conditions and to a impulse and step force, two of the most typical excitations.

2.2.1. Response to Initial Conditions

The response of a fractionally damping system whose m, c and k parameters are unitary and subjected to the initial conditions $\{x_0, v_0\} = \{1, 0\}$ is shown in Figure 4. As predicted from the poles of the system, the response is oscillatory and decays in time.

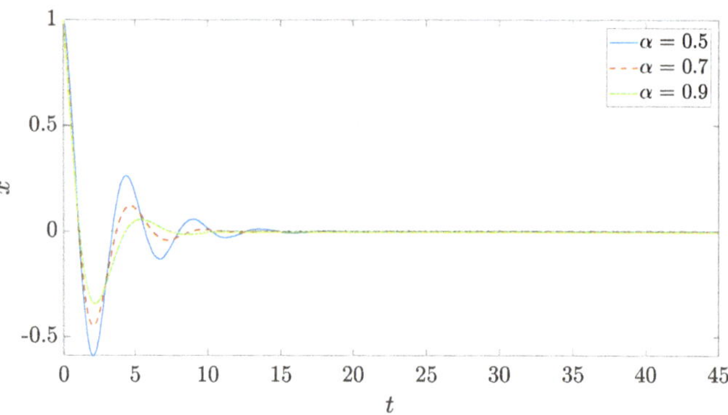

Figure 4. Time response of a fractionally damped system for different values of the order of the derivative.

The main difference that a system with fractional damping presents in comparison to one that shows viscous damping regarding its response to initial conditions is the appearance of a non oscillatory term that decays in time [19]. This effect can be seen in Figure 5, where the response in the frequency domain, obtained by the Fast Fourier Transform of the time response, is represented for different values of the order of derivative α. During the initial part of the response (Figure 5a), only a peak at the vibration frequency can be noticed; in the final part of the decay (Figure 5b), however, it is possible to identify also a peak in the spectrum at zero frequency, that is related to the non oscillatory term, together with the peak at the vibration frequency.

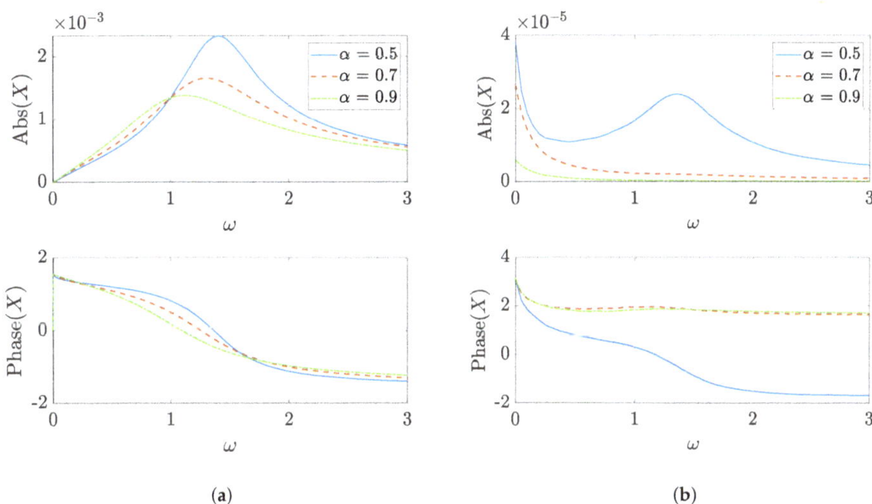

(a)   (b)

Figure 5. Response of a fractionally damped system in the frequency domain: (**a**) from $t = 0$ to $t = 20$, (**b**) for $t > 20$.

From a physical point of view, the appearance of a non oscillatory term has two implications. First, it means that the equilibrium position of the system is not constant but it varies with time. Taking into account that the equilibrium position is defined by the mass and the stiffness of the system and, in this case, the mass is constant, the single explanation to this phenomenon is that the stiffness of the system varies with time. Secondly, it implies that the system does not oscillate with decreasing amplitude until it finally stops, but that it reaches a point in which it approaches the original equilibrium position without oscillating.

Another aspect in which a system with fractional damping differs from one with viscous damping is in the fact that a critical damping ratio in which the system does not oscillate around the equilibrium position cannot be defined [20,30]. The response always crosses the origin of coordinates at least once due to the nature of the poles of the system, that are always complex conjugates. For this reason, fractional damping allows us to model dampers whose response only crosses the equilibrium position a single time, a case that cannot be represented with accuracy using a traditional model. For example, the 1 DOF system under study presents this behaviour when the damping parameter $c = 2$ and the order of the derivative $\alpha = 0.9$ (Figure 6).

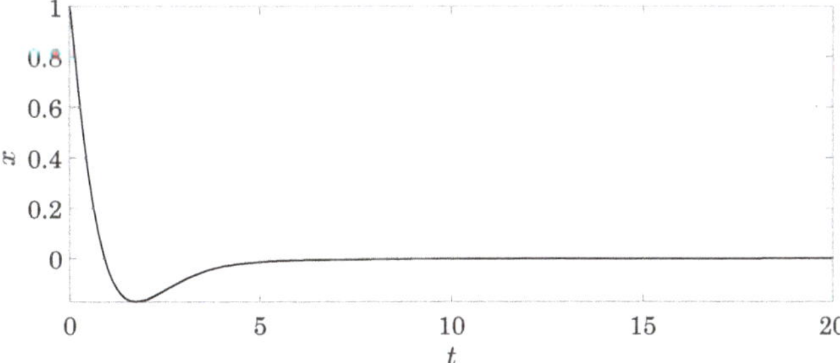

Figure 6. Response of a single-degree-of-freedom (1 DOF) fractionally damped system with unitary mass and stiffness parameters, the order of the derivative being $\alpha = 0.9$ and the damping parameter $c = 2$ when subjected to initial conditions $\{x_0, v_0\} = \{1, 0\}$.

2.2.2. Response to Impulse

The response to an impulse is of interest because it provides a way to define critical damping. According to [20], critical damping can be defined for systems with fractional damping as the value for which the response is tangent to the axis of zero displacement when an impulse force

$$F(t) = \begin{cases} 1/\Delta t \,, \ 0 < t < \Delta t \\ 0 \,, \ \text{otherwise} \end{cases} \tag{15}$$

is applied to the system. This means that for values of damping greater than this the response does not change sign, even if it still oscillates, around a position that varies with time in this case (Figure 7).

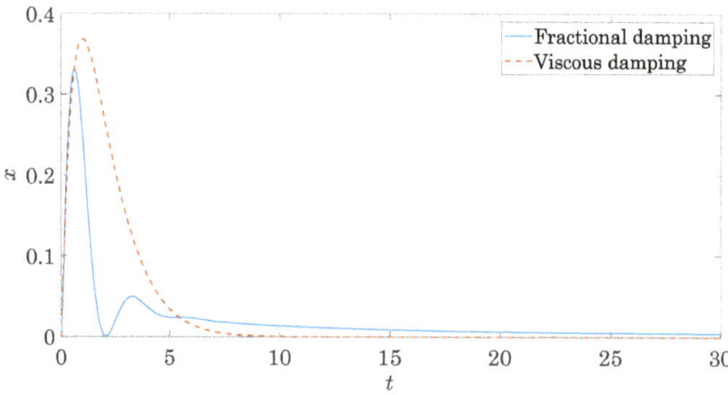

Figure 7. Difference between an overdamped systems with viscous or fractional damping.

Following this reasoning, the value of critical damping can be estimated for different values of the order of the derivative α. For example, in Figure 8, the response of the 1 DOF system with unitary mass and stiffness parameters and critical damping when subjected to a impulse is presented. For low values of α the response always changes sign no matter how high the added damping: increasing the damping parameter makes the system stiffer and the effect of the increase in the natural frequency is greater than that of the dissipation.

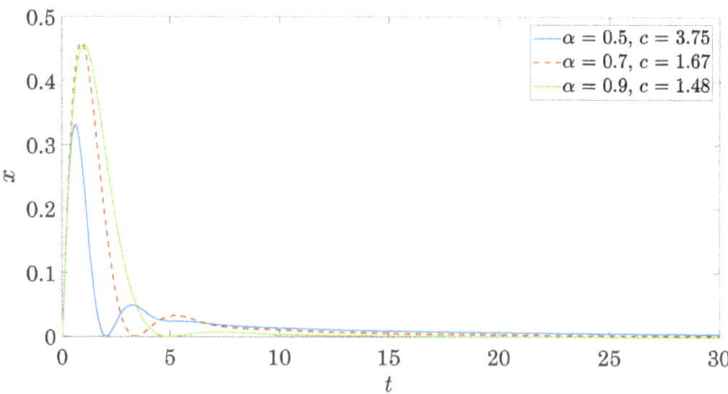

Figure 8. Response to impulse of a system with critical damping for different values of the order of the derivative α.

2.2.3. Response to Step

The response of the system with fractional damping to a step force

$$F(t) = \begin{cases} 0 \,, \, t < 0 \\ 1 \,, \, 0 \leq t \end{cases} \tag{16}$$

has the peculiarity that it stays under the response of a system with viscous damping. As an example, the response of the 1 DOF of freedom system with unitary mass, damping and stiffness constants under study is presented in Figure 9 depending of the order of the derivative α for a unit step excitation. It can be noted that for systems with fractional damping the overshoot decreases but the settling time is much longer, specially when decreasing the value of the order of the derivative.

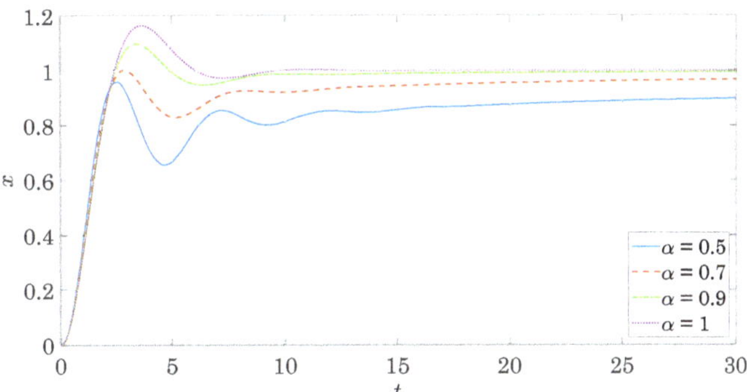

Figure 9. Response to step for different values of α.

3. Application: Bearing Support

In order to illustrate the properties of the fractional models presented in the previous sections in a real life application, a shaft supported by one bearing in each end is considered. The shaft is modelled by means of beam finite elements; both supports, following the typical model of a bearing [32], as a combination of a spring and fractional damper (Figure 10).

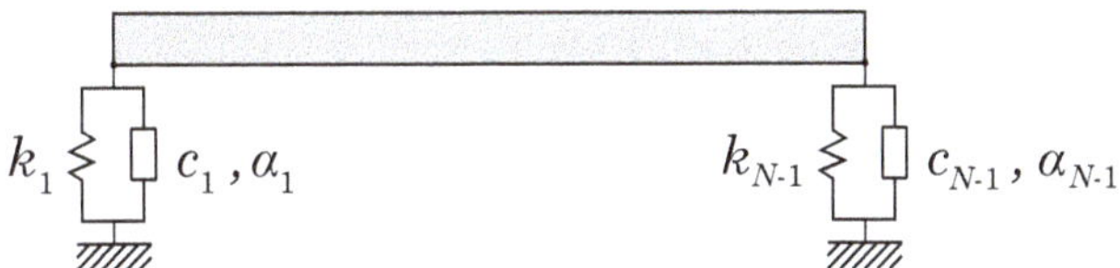

Figure 10. Shaft supported by a bearing in each end, where N stands for the number of degrees of freedom of the finite element model of the shaft. The bearing support affects the $(N-1)$th degree of freedom, as the Nth is related to the rotation of the end.

For the numerical application, a shaft of length $L = 1$ m and circular section of diameter $d = 0.4$ m made of steel with a Young's modulus of $E = 210$ GPa and density $\rho = 7900$ kg/m^3 is considered. The natural frequencies, mode shapes and response of the shaft when changing the parameters of the bearings are studied, the two bearings being identical ($k_1 = k_{N-1} = k$, $c_1 = c_{N-1} = c$ and $\alpha_1 = \alpha_{N-1} = \alpha$).

3.1. Natural Frequencies and Mode Shapes

The eigenpairs of a the shaft supported by bearings that present fractional damping satisfy the relationship

$$\left(-\lambda_r^* \mathbf{M} + \mathbf{K}_{\text{shaft}} + \mathbf{K}_{\text{supp}}^*(\omega_r) \right) \boldsymbol{\phi}_r^* = \mathbf{0}, \tag{17}$$

where λ_r^* and $\boldsymbol{\phi}_r^*$ are respectively the eigenvalue and the eigenvector of the rth mode, $\mathbf{M}$ the mass matrix of the system, that corresponds to the mass matrix of the shaft as the supports are considered massless, $\mathbf{K}_{\text{shaft}}$ and $\mathbf{K}_{\text{supp}}^*(\omega)$ the stiffness matrices of the shaft and the bearing supports. This last complex matrix $\mathbf{K}_{\text{supp}}^*(\omega)$ is frequency dependent and is defined as

$$\mathbf{K}_{\text{supp}}^*(\omega) = \sum_{j \in B}(k_j + c_j(i\omega)^\alpha)\mathbf{e}_j\mathbf{e}_j^\mathsf{T} \ / \ B = \{1, N-1\}, \tag{18}$$

where B holds the positions of the bearing supports, in this case the first and penultimate (N-1) degrees of freedoms, the ones that represent the vertical displacement of the first and last nodes, respectively and $\mathbf{e}_j$ stands for the unit vector that corresponds to the jth degree of freedom. As the matrix $\mathbf{K}_{\text{supp}}$ is frequency dependent, an iterative method is needed to obtain the eigenpairs of the system: the reader can find the details in [33,34].

With the aim of studying the evolution of the vibration frequencies of the system with the order of the derivative α, the values of the stiffness and damping parameters of the bearing supports are set to $k = 10^4$ N/m and $c = 50$ N s$^\alpha$/m respectively. When the shaft is discretised with 60 beam elements in its length the results shown in Figure 11 are obtained. It can be observed that, as deduced for the 1 DOF system, changing the order of the derivative affects the stiffness of the supports and, thus, the vibration frequencies.

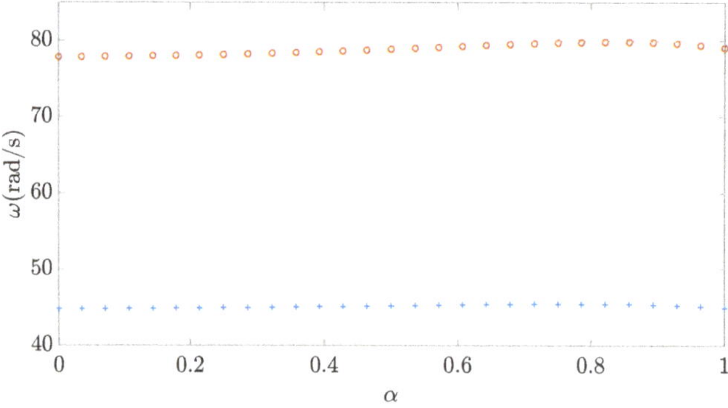

Figure 11. Evolution of the frequency of the first two modes with the order of the derivative α.

Another aspect that should be taken into account regarding the eigenpairs is that the modes are complex due to the non-proportionality of the damping. When the value of the damping parameter c is low this effect is barely noticeable so, for the sake of clarity, its value is increased to $c = 1000$ N s$^\alpha$/m in the computation of the mode shapes presented in Figure 12. The phase of the mode shape in each point of the shaft is represented with the aim of bringing to light the complex nature of the modes: if the mode were normal, the phase would change abruptly from $-\pi$ to π (or vice versa) in the nodal points, but it is not what occurs.

It should be noted also that the mode shapes are not affected by the order of the derivative α but the complex nature of the modes is more acute for high values of α as the effect of the damping is higher.

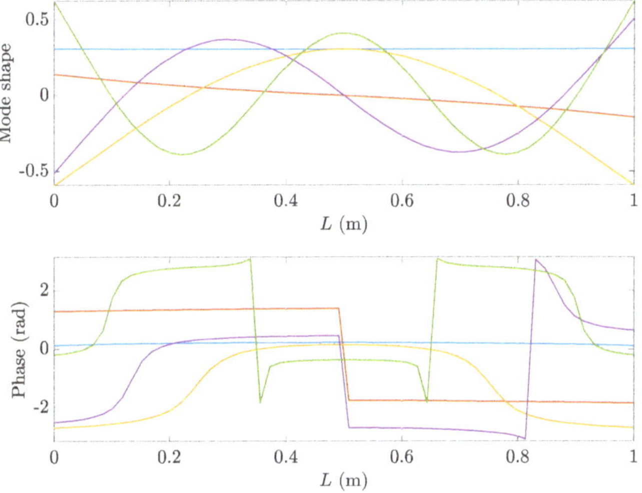

Figure 12. First five mode shapes of the supported shaft: real part and phase. The vibration frequencies of the mode shapes are the following: $\omega_1 = 116.20$ rad/s, $\omega_2 = 320.24$ rad/s, $\omega_3 = 1201.0$ rad/s, $\omega_4 = 3144.6$ rad/s and $\omega_5 = 6194.3$ rad/s.

3.2. Dynamic Response

Next, the time response of the shaft to a perturbation from the base is computed following the procedure in [35] by means of a **M-C-K** Newmark method. The details of the implementation of such method can be found in [22], but, in short, it consists of expressing the equation of motion of the system in the instant $n + 1$ as

$$\mathbf{M}\ddot{\mathbf{u}}_{n+1} + \mathbf{C}\left(D^\alpha \mathbf{u}\right)_{n+1} + \mathbf{K}\mathbf{u}_{n+1} = \mathbf{F}_{n+1}, \tag{19}$$

and introducing the notation

$$\mathbf{M}\ddot{\mathbf{u}}_{n+1} + \bar{\mathbf{C}}\dot{\mathbf{u}}_{n+1} + \mathbf{K}\mathbf{u}_{n+1} = \bar{\mathbf{F}}_{n+1}, \tag{20}$$

where

$$\bar{\mathbf{C}} = (\Delta t)^{1-\alpha}\mathbf{C} \tag{21}$$

and

$$\bar{\mathbf{F}}_{n+1} = \mathbf{F}_{n+1} - \frac{1}{(\Delta t)^\alpha}\mathbf{C}\left[\mathbf{u}_n + \sum_{j=1}^{n} A_{j+1}\mathbf{u}_{n+1-j}\right] \tag{22}$$

in order to use the Grünwald-Letnikov definition of the fractional derivative in the computations. In the case under study the mass matrix $\mathbf{M}$ is that of the shaft; the stiffness matrix $\mathbf{K}$ is the sum of the one of the shaft and the one of the supports

$$\mathbf{K} = \mathbf{K}_{\text{shaft}} + \mathbf{K}_{\text{supp}} \tag{23}$$

where

$$\mathbf{K}_{\text{supp}} = \sum_{j \in B} k_j \mathbf{e}_j \mathbf{e}_j^{\mathsf{T}} \ / \ B = \{1, N-1\}, \tag{24}$$

and the damping matrix $\mathbf{C}$ is given by the damping of the bearing supports as

$$\mathbf{C} = \sum_{j \in B} c_j \mathbf{e}_j \mathbf{e}_j^{\mathsf{T}} \ / \ B = \{1, N-1\}. \tag{25}$$

The time response of the supported shaft to a seismic excitation is computed as follows: first, the static response to a displacement of the rightmost end is computed; then, it is introduced into the presented Newmark scheme as an initial deformation $\mathbf{u}_0$. Four distinct cases are studied:

- **Case 1**: a viscous case ($\alpha = 1$) with low stiffness ($k_{\text{ref}} = 10^4$ N/m) and low damping ($c_{\text{ref}} = 50$ N s/m) that is used for reference, in which the rigid body movement prevails.
- **Case 2**: a viscous case ($\alpha = 1$) with low stiffness ($k = k_{\text{ref}}$) and supercritical damping ($c = 10c_{\text{ref}}$), in which the shaft returns to the equilibrium position without oscillating.
- **Case 3**: a fractionally damped case ($\alpha = 0.6$) with low stiffness ($k = k_{\text{ref}}$) and high damping ($c = 100c_{\text{ref}}$) so that the system returns to its original position oscillating around a variable equilibrium position.
- **Case 4**: a fractionally damped case ($\alpha = 0.6$) with high stiffness ($k = 10k_{\text{ref}}$) and low damping ($c = c_{\text{ref}}$), in which the movement is a combination of the rigid body motion and the first modes of the system.

The four cases are simulated for a time $T = 1$ s using 1000 samples and discretising the shaft with 60 beam elements. In order to analyse the difference between the cases, the displacement of the right end, to which a vertical displacement of 3 mm is imposed in the first instant, is plotted over time (Figure 13).

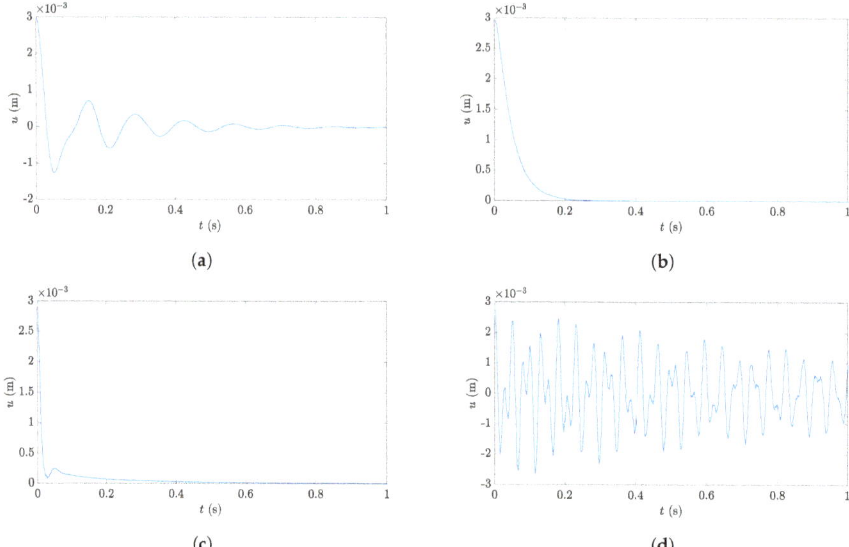

Figure 13. Displacement of the right end of the shaft: (**a**) case 1, (**b**) case 2, (**c**) case 3, (**d**) case 4.

The first case represents the classical vibration damped in time: this behaviour can be modelled both with a viscous and a fractional damper. The motion of the second case, instead, is characteristic of a overdamped system and only achievable if the damping is viscous: a fractionally damped system will always oscillate, even if the response does not have to necessarily change sign, as the third case makes clear. Finally, the fourth case represents a more general situation in which the response consists of a combination of several modes and the rigid body motion. In this case, by tuning the stiffness and damping parameters and the order of the derivative, different behaviours can be reproduced.

The evolution of the shaft in time can be seen in the four videos provided as Supplementary Materials with this article.

4. Conclusions

This work discusses the behaviour of a 1 DOF system with fractional damping from the point of view of dynamics, specially focusing in the divergence of such a model of damping from a classical viscous one.

The first difference between a system with viscous and a fractional damping is that the poles of the latter are always complex conjugates, which implies that the response is always oscillatory and that, therefore, the concept of critical damping as defined for viscous systems is not applicable. Also, due to the viscoelastic nature of fractional damping, the damped frequency of the system decreases with the order of the derivative, as the system resembles more a pure viscous system, but increases with the damping parameter, as the contribution to the stiffness of the system increases too.

Several particularities can be observed regarding the time response as well. First, the response to initial conditions is characterised by the appearance of a non oscillatory term, which can be understood as a change in the position of the equilibrium position in time. Secondly, the response to an impulse allows us to define critical damping for this kind of systems as the amount of damping needed for the response not to change its sign. Finally, the response to a step function is characterised by its reduced overshoot but longer settling time is comparison to a system with viscous damping.

Despite their peculiarities, fractional models are useful to ease the modelling process and replicate behaviours that would otherwise require complex formulations. This is the case, for example, of the bearing support represented in this work with a spring and fractional damper for which behaviours that range from a traditional viscous case to an oscillating case with variable equilibrium position can be represented with a single model.

As a final remark, we would like to warn against using fractional models as mathematical artifacts that allow us to simplify the modelling process. It is advisable to check if the behaviour of the mechanical system under study follows the same trends as the assumed model, or at least, that the effects attributable to the model do not interfere in the computations.

Supplementary Materials: The following are available online at http://www.mdpi.com/2073-8994/11/12/1499/s1.

Author Contributions: Conceptualization, O.Z., I.S., J.G.-B. and F.C.; methodology, O.Z., I.S., J.G.-B. and F.C.; software, O.Z.; formal analysis, O.Z and I.S.; investigation, O.Z.; resources, J.G.-B.; writing—original draft preparation, O.Z.; writing—review and editing, I.S., J.G.-B. and F.C.; visualization, O.Z.; supervision, F.C.; project administration, F.C.

Funding: This research received no external funding.

Conflicts of Interest: The authors declare no conflict of interest.

References

1. Adolfsson, K.; Enelund, M.; Olsson, P. On the fractional order model of viscoelasticity. *Mech. Time Depend. Mater.* **2005**, *9*, 15–34. [CrossRef]
2. Podlubny, I. *Fractional Differential Equations: An Introduction to Fractional Derivatives, Fractional Differential Equations, to Methods of Their Solution and Some of Their Spplications*; Mathematics in Science and Engineering; Academic Press: Cambridge, MA, USA, 1999.
3. Torvik, P.J.; Bagley, R.L. On the Appearance of the Fractional Derivative in the Behavior of Real Materials. *J. Appl. Mech.* **1984**, *51*, 294–298. [CrossRef]

4. Bagley, R.L.; Torvik, P.J. A theoretical basis for the application of fractional calculus to viscoelasticity. *J. Rheol.* **1983**, *27*, 201–210. [CrossRef]

5. Di Paola, M.; Pirrotta, A.; Valenza, A. Visco-elastic behavior through fractional calculus: An easier method for best fitting experimental results. *Mech. Mater.* **2011**, *43*, 799–806. [CrossRef]

6. Nutting, P.G. A new general law of deformation. *J. Frankl. Inst.* **1921**, *191*, 679–685. [CrossRef]

7. Gemant, A. A method of analyzing experimental results obtained from elasto-viscous bodies. *Physics* **1936**, *7*, 311–317. [CrossRef]

8. Pinnola, F.P.; Zavarise, G.; Prete, A.D.; Franchi, R. On the appearance of fractional operators in non-linear stress–strain relation of metals. *Int. J. Non-Linear Mech.* **2018**, *105*, 1–8. [CrossRef]

9. Makris, N. Three-dimensional constitutive viscoelastic laws with fractional order time derivatives. *J. Rheol.* **1997**, *41*, 1007–1020. [CrossRef]

10. Alotta, G.; Barrera, O.; Cocks, A.C.; Di Paola, M. On the behavior of a three-dimensional fractional viscoelastic constitutive model. *Meccanica* **2017**, *52*, 2127–2142. [CrossRef]

11. Alotta, G.; Barrera, O.; Cocks, A.; Paola, M.D. The finite element implementation of 3D fractional viscoelastic constitutive models. *Finite Elem. Anal. Des.* **2018**, *146*, 28–41. [CrossRef]

12. Heymans, N.; Podlubny, I. Physical interpretation of initial conditions for fractional differential equations with Riemann-Liouville fractional derivatives. *Rheol. Acta* **2006**, *45*, 765–771. [CrossRef]

13. Di Paola, M.; Pinnola, F.P.; Zingales, M. A discrete mechanical model of fractional hereditary materials. *Meccanica* **2013**, *48*, 1573–1586. [CrossRef]

14. Schiessel, H.; Metzler, R.; Blumen, A.; Nonnenmacher, T.F. Generalized viscoelastic models: their fractional equations with solutions. *J. Phys. Math. Gen.* **1995**, *28*, 6567. [CrossRef]

15. Naber, M. Linear fractionally damped oscillator. *Int. J. Differ. Equations* **2010**, *2010*. [CrossRef]

16. Achar, B.N.N.; Hanneken, J.W.; Enck, T.; Clarke, T. Dynamics of the fractional oscillator. *Phys. Stat. Mech. Appl.* **2001**, *297*, 361–367. [CrossRef]

17. Achar, B.N.N.; Hanneken, J.W.; Clarke, T. Response characteristics of a fractional oscillator. *Phys. Stat. Mech. Appl.* **2002**, *309*, 275–288. [CrossRef]

18. Achar, B.N.N.; Hanneken, J.W.; Clarke, T. Damping characteristics of a fractional oscillator. *Phys. Stat. Mech. Appl.* **2004**, *339*, 311–319. [CrossRef]

19. Shokooh, A.; Suárez, L. A comparison of numerical methods applied to a fractional model of damping materials. *J. Vib. Control.* **1999**, *5*, 331–354. [CrossRef]

20. Suarez, L.E.; Shokooh, A. An eigenvector expansion method for the solution of motion containing fractional derivatives. *J. Appl. Mech.* **1997**, *64*, 629–635. [CrossRef]

21. Fenander, A. Modal synthesis when modeling damping by use of fractional derivatives. *AIAA J.* **1996**, *34*, 1051–1058. [CrossRef]

22. Cortés, F.; Elejabarrieta, M.J. Finite element formulations for transient dynamic analysis in structural systems with viscoelastic treatments containing fractional derivative models. *Int. J. Numer. Methods Eng.* **2007**, *69*, 2173–2195. [CrossRef]

23. Cortés, F.; Elejabarrieta, M.J. Homogenised finite element for transient dynamic analysis of unconstrained layer damping beams involving fractional derivative models. *Comput. Mech.* **2007**, *40*, 313–324. [CrossRef]

24. Mendiguren, J.; Cortés, F.; Galdos, L. A generalised fractional derivative model to represent elastoplastic behaviour of metals. *Int. J. Mech. Sci.* **2012**, *65*, 12–17. [CrossRef]

25. Oldham, K.; Spanier, J. *The Fractional Calculus Theory and Applications of Differentiation and Integration to Arbitrary Order*; Elsevier: Amsterdam, The Netherlands, 1974.

26. Hilfer, R. *Applications of Fractional Calculus in Physics*; World Scientific: Singapore, 2000.

27. Miller, K.S.; Ross, B. *An Introduction to the Fractional Calculus and Fractional Differential Equations*; Wiley-Interscience: Hoboken, NJ, USA, 1993.

28. Bagley, R.L.; Calico, R.A. Fractional order state equations for the control of viscoelasticallydamped structures. *J. Guid. Control. Dyn.* **1991**, *14*, 304–311. [CrossRef]

29. Kilbas, A.A.A.; Srivastava, H.M.; Trujillo, J.J. *Theory and Applications of Fractional Differential Equations*; Elsevier Science Limited: Amsterdam, The Netherlands, 2006; Volume 204.

30. Yuan, L.; Agrawal, O.P. A numerical scheme for dynamic systems containing fractional derivatives. *Trans. Am. Soc. Mech. Eng. J. Vib. Acoust.* **2002**, *124*, 321–324. [CrossRef]

31. Pinnola, F.P. Statistical correlation of fractional oscillator response by complex spectral moments and state variable expansion. *Commun. Nonlinear Sci. Numer. Simul.* **2016**, *39*, 343–359. [CrossRef]

32. Matsubara, M.; Rahnejat, H.; Gohar, R. Computational modelling of precision spindles supported by ball bearings. *Int. J. Mach. Tools Manuf.* **1988**, *28*, 429–442. [CrossRef]

33. Cortés, F.; Elejabarrieta, M.J. Computational methods for complex eigenproblems in finite element analysis of structural systems with viscoelastic damping treatments. *Comput. Methods Appl. Mech. Eng.* **2006**, *195*, 6448–6462. [CrossRef]

34. Cortés, F.; Elejabarrieta, M.J. An approximate numerical method for the complex eigenproblem in systems characterised by a structural damping matrix. *J. Sound Vib.* **2006**, *296*, 166–182. [CrossRef]

35. Cortés, F.; Jesús Elejabarrieta, M. Finite element analysis of the seismic response of damped structural systems including fractional derivative models. *J. Vib. Acoust.* **2014**, *136*, 050901. [CrossRef]

 *symmetry*

Article

Complete Geometric Analysis Using the Study SE(3) Parameters for a Novel, Minimally Invasive Robot Used in Liver Cancer Treatment

Iosif Birlescu [1], Manfred Husty [2], Calin Vaida [1], Nicolae Plitea [1], Abhilash Nayak [3] and Doina Pisla [1,*]

[1] Research Center for Industrial Robots Simulation and Testing, Technical University of Cluj-Napoca, 400114 Cluj-Napoca, Romania; ibirlescu@mail.utcluj.ro (I.B.); calin.vaida@mep.utcluj.ro (C.V.); plitea.nicolae@mep.utcluj.ro (N.P.)
[2] Unit Geometry and CAD, University of Innsbruck, 6020 Innsbruck, Austria; Manfred.Husty@uibk.ac.at
[3] Laboratoire des Sciences du Numerique de Nantes, Université de Nantes, 44035 Nantes, France; Abhilash.Nayak@ls2n.fr
* Correspondence: doina.pisla@mep.utcluj.ro; Tel.: +40-264-401-684

Received: 8 November 2019; Accepted: 5 December 2019; Published: 7 December 2019

Abstract: The paper presents a complete geometric analysis of a novel parallel medical robotic system designed for minimally invasive treatment of hepatic tumors using brachytherapy, ablation or targeted chemotherapy. An algebraic method based on the study parameters of the special Euclidean transformation Lie group SE(3) was used to determine the mechanism kinematics singularities and workspace. Moreover, two particular medical tool manipulations for the minimally invasive medical procedures are defined in terms of the Study parameters. The first manipulation of the medical tool refers to the linear insertion (of e.g., needles) and the second one is the remote center of motion manipulation of specific medical instruments (e.g., ultrasound probes). The constraint equations of the robotic system are derived and then, the operational workspace is illustrated for the novel parallel robotic system. Lastly, a numerical simulation is presented showing the behavior of the robotic system manipulating the ultrasound probe constrained by the remote center of motion. The geometric analysis of the operational workspace and the numerical simulation show promising results that validate the novel robotic system (safe-wise) for the medical procedure.

Keywords: parallel robot; minimally invasive procedures; algebraic modeling; Study parameters

---

1. Introduction

In robotics and more specifically in mechanism analysis and synthesis there exists various mathematical approaches to describe the mechanism kinematics, singularities and workspace, which have both advantages and disadvantages [1]. Usually, the mathematical formulation used in describing a mechanism is chosen to benefit from the specific advantages that the method provides and depends strictly on the type (and even complexity) of the mechanism and sometimes on the task of the robot. One algebraic method based on the Study parameters (or dual quaternion) of SE(3) which is presented in [2–4], has the advantages (over some vector-based methods) that it describes the global kinematics of the mechanisms (showing all the working modes and being a powerful tool to find the mechanism singularities) and it is free of parameterization singularities. The Study parameters method was used in the analysis of various robotic systems such as the Stewart-Gough platform [5], the 3-RPS parallel manipulator [3], in the analysis of medical robots [4,6,7] and it was even tailored to analyze rehabilitation robots such as the one found in [8].

In this paper, the authors exploited the advantages of the algebraic method based on the Study parameters to achieve two main objectives: (i) a complete geometric analysis of the Pro-Hep-LCT robotic system (which is designed for the liver cancer treatment through brachytherapy, ablation or targeted chemotherapy under ultrasound imaging), and (ii) the analysis of the behavior of the robotic system while it performs the medical task. The motivation of using the Study parameters for the first objective is simply due to the fact that the Study parameters offer a description of the global kinematics of the mechanism which in turn shows all the kinematic solutions, all the singularities and all the working modes of the robotic system. All this information, especially the singularities of the mechanism (since they define configurations where the control of the robotic system may be lost), should be valuable for understanding the robotic system capabilities to perform the medical task from the safety point of view. The motivation of the second objective was to provide the general Study parameter relations for two fundamental ways in which the medical instruments are manipulated in minimally invasive procedures (the linear insertion and the Remote Center of Motion (RCM) manipulation [9]) and use these general relations (in combination with the constraint equations of the Pro-Hep-LCT robotic system) to assess the robotic system behavior during the medical task (while the instruments are inserted inside the patient body) to determine the geometry of the operational workspace of the robot). The constraint equations (derived from the Study kinematic model) together with the general formulations for the medical insertion and RCM manipulation will enable the development of a robust fail-safe control algorithm. As explained in Ref. [10], interpolation may be used in combination to inverse kinematics of the robotic system to determine accurate relations between the medical tool coordinates (constrained by the RCM) and the active joints of the robot (this method was also demonstrated in Ref. [11] for serial robots). Moreover, in Ref. [12] the authors used a so called RCM-constrained Jacobian for the control of 7 DOFs and 9 DOFs MIS robotic systems. In Ref. [13] the authors described a collaborative framework for a 7 DOFs serial robot for minimally invasive surgery, showing also numerical simulations which describe the operating workspace of the robot. Line symmetric motion generators (which are sub manifold on the Study quadric) were presented in Ref. [14], showing their application in applications that require RCM.

The paper is structured as follows: Section 2 presents the mathematical framework of the Study parameters and it defines the mathematical relations for the linear insertion/retraction of a medical instrument and for the RCM manipulation; Section 3 presents the Pro-Hep-LCT robotic system and shows the kinematics, the singularity analysis and the workspace of the robotic system; Section 4 presents some numerical simulations for the Pro-Hep-LCT robotic system based on the previous obtained results; Section 5 presents the conclusions and future work.

2. Mathematical Framework

In the fields of multi-body systems kinematics (e.g., mechanism analysis), the special Euclidean group SE(3) describes every Euclidean displacement (with 6 DOFs) by means of homogeneous transformation matrices which encode both translations and orientation information. A kinematic mapping k, as found extensively in the scientific literature (see for example [2–4]), is defined as:

$$k : SE(3) \rightarrow Q \in P^7$$
$$D(x_i, y_i) \rightarrow [x_0 : x_1 : x_2 : x_3 : y_0 : y_1 : y_2 : y_3]^T \neq [0 : 0 : 0 : 0 : 0 : 0 : 0 : 0]^T \tag{1}$$

and maps every Euclidean displacement D into a point Q from the projective space P^7. The coordinates of the point Q are based on a dual quaternion and are also called the Study parameters in the scientific literature. The x_i parameters show the orientation of a mobile coordinate frame (relative to a fixed one) whereas the parameters y_i represent the translation between the origins of the two coordinate frames. The trivial solution (where all Study parameters are 0) is excluded since it has no use in kinematics. Moreover, the Study parameters must fulfill two conditions [2–4]:

$$x_0^2 + x_1^2 + x_2^2 + x_3^2 = 1 \tag{2}$$

$$x_0 y_0 + x_1 y_1 + x_2 y_2 + x_3 y_3 = 0 \tag{3}$$

where the first condition Equation (2) is the normalizing condition and the second one Equation (3) is called the Studys' quadric [15]. The Study parameters may be obtained directly from a homogeneous transformation matrix using the following ratios [2–7]:

$$
\begin{aligned}
x_0 : x_1 : x_2 : x_3 \;\; &= 1 + a_{11} + a_{22} + a_{33} : a_{32} - a_{23} : a_{13} - a_{31} : a_{21} - a_{12} \\
&= a_{32} - a_{23} : 1 + a_{11} - a_{22} - a_{33} : a_{12} + a_{21} : a_{31} + a_{13} \\
&= a_{13} - a_{31} : a_{12} + a_{21} : 1 - a_{11} + a_{22} - a_{33} : a_{23} + a_{32} \\
&= a_{21} - a_{12} : a_{31} + a_{13} : a_{23} + a_{32} : 1 - a_{11} - a_{22} + a_{33}
\end{aligned}
\tag{4}
$$

$$
\begin{aligned}
y_0 &= -\tfrac{1}{2}(t_z x_3 + t_y x_2 + t_x x_1) y_1 = -\tfrac{1}{2}(t_z x_2 - t_y x_3 + t_x x_0) \\
y_2 &= -\tfrac{1}{2}(-t_z x_1 - t_y x_0 + t_x x_3) y_3 = -\tfrac{1}{2}(-t_z x_0 + t_y x_1 - t_x x_2)
\end{aligned}
\tag{5}
$$

where the terms a_{ij} are the entries of a 3×3 rotation matrix within a homogeneous 4×4 matrix and $[t_x, t_y, t_z]^{\mathrm{T}}$ is the translation vector from the homogeneous matrix. As seen in Equation (4) there are four ways to write the Study parameters and it is guaranteed that at least one way yields a nonsingular representation of the Euclidean displacement (i.e., the Study parameters are free of parametric singularities). As an example, as long as the Study parameters do not represent a rotation by a value of π, the first ratio from Equation (4) may be used [2].

In the context of mechanism analysis, the relative position between a mobile coordinate frame, which is located on the robot mobile platform (or end-effector) and a fixed coordinate frame (which is at the base of robot) is studied. Homogeneous transformation matrices may be used in this context (which are also called the Denavit-Hartenberg convention [16]) and the algorithm of applying this method is straightforward: the first step is to define each transformation matrix for each joint and link; the second step is to multiply the transformation matrices starting from the base coordinate frame towards the mobile coordinate frame (for robots that contain multiple kinematic chains this procedure is done for every chain). The Study parameters may be obtained thereafter from the transformation matrix using Equations (4) and (5). The trigonometric functions (sines and cosines) may be substituted with the tangent of half angle formulae resulting in a rational form of the Study parameters which may be simplified further (by factoring the denominator and dividing through the greatest common divisor as in [7]) to obtain polynomials (achieving thus an algebraic description of the constraint equations). A more direct method to obtain the constraint equations of a mechanism is to write directly the Study parameters for each element of the kinematic chain (joints and links) and then to use quaternion multiplication to obtain the constraints (this is analog to multiplying the homogeneous matrices). As mentioned before, the Study parameters are based on a dual quaternion which has a general form:

$$
Q = x_0 + x_1 i + x_2 j + x_3 k + \frac{1}{2}\varepsilon(y_0 + y_1 i + y_2 j + y_3 k)
\tag{6}
$$

where i, j, k are the imaginary parts and the following multiplication rules apply: $i^2 = j^2 = k^2 = -1$, $ij = k$, $jk = i$, $ki = j$, $ji = -k$, $kj = -i$, $ik = -j$, $\varepsilon \neq 0$, $\varepsilon^2 = 0$. The conjugate of the dual quaternion (which represent the inverse transformation) is defined as [2,9]:

$$
\overline{Q} = x_0 - x_1 i - x_2 j - x_3 k + \frac{1}{2}\varepsilon(y_0 - y_1 i - y_2 j - y_3 k); \qquad Q \cdot \overline{Q} = [1:0:0:0:0:0:0:0]
\tag{7}
$$

The Study Parameters for MIS Procedures

For the MIS procedures the two most fundamental instrument manipulations are: (i) the *linear insertion/retraction* of a medical tool (e.g., needle); (ii) the *Remote Center of Motion* (RCM) manipulation which is a combination of the insertion of a medical instrument with the orientation of the instrument where the entry-point remains unchanged. Consequently the insertion/retraction procedure has 1 DOF, whereas the RCM has a maximum of 4 DOF (3 orientations and one translation). The Study parameters equations which describe these two procedures are based on dual quaternion multiplication and are presented further.

Let $P(x_0{:}x_1{:}x_2{:}x_3{:}y_0{:}y_1{:}y_2{:}y_3)$ be a point on the Study quadric $x_0y_0 + x_1y_1 + x_2y_2 + x_3y_3 = 0$ which describe the position and orientation of the mobile platform (the end-effector) under the constraint of a robotic device:

(1) The *linear insertion/retraction* of a medical tool is achieved by the translation on a given axis and is defined by:

$$P_{ins} = P \cdot P_l \tag{8}$$

$$P = P_{ins} \cdot \overline{P}_l \tag{9}$$

where the P_{ins} represents the coordinates of the target point (encoding also the orientation of the medical tool) P represents the insertion point and P_l represents the insertion length; $[1{:}0{:}0{:}0{:}0{:}{-}l/2{:}0{:}0]$ for the insertion along the X' axis (of the mobile frame), $[1{:}0{:}0{:}0{:}0{:}0{:}{-}l/2{:}0]$ for the insertion along the Y' axis and $[1{:}0{:}0{:}0{:}0{:}0{:}0{:}{-}l/2]$ for the insertion along the Z' axis. Equation (8) represents the linear insertion whereas Equation (9) represents the retraction (i.e., the inverse operation) and it is given by multiplying with the conjugate of the dual quaternion.

(2) The *RCM manipulation* between two configurations of the robot is given by:

$$P_{RCM} = P \cdot P_o \cdot P_l \tag{10}$$

$$P = P_{RCM} \cdot \overline{P}_l \cdot \overline{P}_o \tag{11}$$

where P_{RCM} represents the Study parameters for a general point that is manipulated using the RCM concept, P_o are the Study parameters which describe a change in orientation and P_l are the Study parameters describing the insertion of the medical instrument. Assuming that u is the tangent of the half angle of a rotation angle, the Study parameters $[1{:}u{:}0{:}0{:}0{:}0{:}0{:}0]$, $[1{:}0{:}u{:}0{:}0{:}0{:}0{:}0]$ and $[1{:}0{:}0{:}u{:}0{:}0{:}0{:}0]$ represent rotations (by a value of u) about the X', Y' and Z' axes respectively. Equation (10) describes the manipulation using RCM whereas Equation (11) describes the inverse operation which is achieved by multiplying with the quaternion conjugate. To describe a general way in which the medical tool is displaced using RCM both Equations (10) and (11) are used:

$$P_{RCM} = P \cdot P_{o,1} \cdot P_{l,1} \cdot \overline{P}_{l,1} \cdot \overline{P}_{o,1} \cdot P_{o,2} \cdot P_{l,2} \cdot \overline{P}_{l,2} \cdot \overline{P}_{o,2} \cdot \ldots \cdot P_{o,n} \cdot P_{l,n} \tag{12}$$

Although the quaternions offer a nice analytic method to operate the constraint equations, the Euclidean representation of specific points (e.g., the insertion point) is also necessary to create an intuitive description of the robotic assisted medical act (i.e., the Cartesian coordinates are used for an unambiguous robot interface targeting medical experts, whereas the quaternions are used for the computation). The relations between the Study parameters and points in 3D Cartesian space may be obtained from Equation (5) together with the strong normalizing condition in Equation (1) and are [2–4]:

$$\begin{aligned}
t_x &= -2(x_0y_1 - x_1y_0 + x_2y_3 - x_3y_2) \\
t_y &= -2(x_0y_2 - x_1y_3 - x_2y_0 + x_3y_1) \\
t_z &= -2(x_0y_3 + x_1y_2 - x_2y_1 - x_3y_0)
\end{aligned} \tag{13}$$

where again the $t = [t_x, t_y, t_z]^{\mathrm{T}}$ is the translation vector which describes the position of mobile frame origin relative to the fixed frame origin.

3. Pro-Hep-LCT Parallel Robotic System

The Pro-Hep-LCT parallel robotic system is designed for the minimally invasive procedure of the palliative treatment of the unresectable liver tumors using brachytherapy, chemotherapeutic agent delivery or ablation. The procedure requires the accurate insertion of brachytherapy needles on linear trajectories (and releasing them afterwards since the brachytherapy treatment is delivered in specialized rooms), under the real live imaging given by an intra-operatory ultrasound probe. Consequently, two independent modules are needed, one for the linear insertion of the needles, and the second one for the RCM manipulation of the intra-operatory ultrasound probe [17]. In previous work, the authors developed medical instruments with redundant DOFs for the needle insertion procedures (such as the ones in [18,19]). The use of such instruments increases the precision of insertion, and reduces the risk of the medical procedure. Such medical instruments will also be integrated on the Pro-Hep-LCT robotic system, namely: one brachytherapy instrument which will insert and release the brachytherapy needle with a redundant DOF; one US probe instrument which will insert/retract the probe (with a redundant DOF).

3.1. Pro-Hep-LCT Parallel Robotic System Description

The two guiding modules of the Pro-Hep-LCT robotic system are identical and operate "in mirror" relative to each other (as illustrated in Figure 1). This allows more access possibilities for the medical tools since both of the guiding modules may guide either the needle instrument or the ultrasound probe instrument.

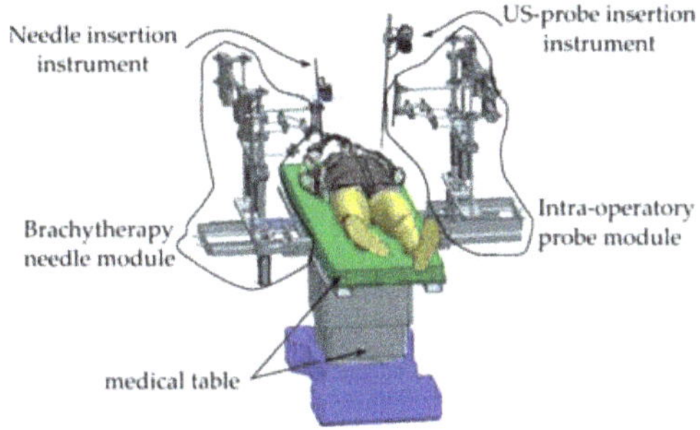

Figure 1. The CAD model for the Pro-Hep-LCT robotic system.

The kinematic scheme of the intra-operatory US probe guiding module is presented in Figure 2. The robotic module contains two planar modules which guide the robot end effector through a pair of universal joints. Since the two planar mechanisms are displaced by a constant value, the end effector must have a free prismatic joint to allow its orientation. To facilitate the computation, the ultrasound probe guiding module is decoupled into 3 mechanisms and the relation among them is the following:

1. The *upper planar mechanism* guides a point P_u (X_u, Y_u) in a plane parallel to the XOY plane (of the fixed coordinate system and displaced by D_z). The point P_u is located at the intersection of the rotation axes of the Universal joint U_1. The upper chain type is PRR-PP (prismatic-revolute-revolute coupled with prismatic-prismatic) and has 2 DOF actuated by the q_4 and q_5 active joints (see Figure 2);

2. The *lower planar mechanism* guides a point P_d (X_d, Y_d) in the plane XOY and it is located on a platform with constant orientation (hence it has only two translational DOFs actuated by the two active rotatory joints q_2 and q_3). The point P_d is chosen to be at the middle point of the link d. The lower planar chain type is RRR-RRR (revolute-revolute-revolute coupled with revolute-revolute-revolute) which is similar to a six bar planar mechanism but has a platform with constant orientation due to the two concatenated parallelogram mechanisms Par_1 and Par_2;

3. The ***mobile platform*** (or the end-effector) of the robot is guided by the two planar mechanisms through a pair of Universal joints. Consequently the platform is of type U-P-U (universal-prismatic-universal) and it relates the points P_u and P_d with the mobile coordinate system $X'Y'Z'$ which has the origin at the tip of the medical instrument (see Figure 2);

4. The last active DOF is actuated by the active joint q_1 and it displaces the entire mechanism in the Z direction of the fixed coordinate system by moving the mechanical frame on which the two planar mechanisms are assembled. This DOF will be used only to guide the medical tool at the patient entry point and the manipulation of the medical instrument thereafter is done by the robot end-effector using redundant DOFs.

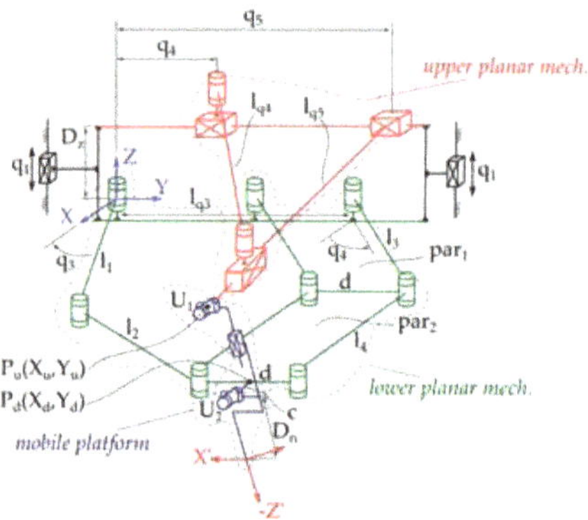

Figure 2. The kinematic scheme of the intra-operatory US probe guiding module.

3.2. Pro-Hep-LCT Kinematics

The algebraic method used in the paper, for the Pro-Hep-LCT analysis has the potential to offer a generalized solution for the kinematics (both forward and inverse) and shows a complete description of the workspace (together with the singularities). Although at times the constraint equations become complicated the method exploits the fact that the constraints are algebraic varieties and Groebner bases may be computed to obtain "fundamental" polynomials which describe the constraint varieties. The constraint equations for the Pro-Hep-LCT parallel robotic system are obtained for all three component mechanisms and the forward and inverse kinematics are obtained using the following stepwise procedures:

For the Forward Kinematics:	*For the Inverse Kinematics:*
1. Substitute the numerical values for the geometric parameters and the active joints (t_1, w_1, q_4, q_5) into the constraint equations;	1. Substitute the numerical values for the geometric parameters and the Study parameters into the constraint equations;
2. Solve the constraints of the **upper planar mechanism** for X_u Y_u (Figure 2);	2. Solve the constraints of the **mobile platform** yielding X_u, Y_u, X_d, Y_d (Figure 2);
3. Solve the constraints of the *lower planar mechanism* for X_d, Y_d (Figure 2);	3. Substitute X_u, Y_u, into the **upper planar mechanism** constraints and solve for q_4, q_5 (the inputs);
4. Substitute X_u, Y_u, X_d, Y_d into the constraints of the **mobile platform** and solve for the Study parameters (the outputs).	4. Substitute X_d, Y_d, into the lower planar mechanism and solve for t_1, w_1 (the inputs).

Moreover, since the robotic system uses end-effectors for the insertion of the medical tools, the mathematical description for the medical procedure using the quaternion viewpoint is straightforward and also easy to implement in the robot control. Assuming that the robotic system is in a configuration where the medical tool tip is at the patient insertion point, the Study parameters of the mobile platform are known. The manipulation of the medial tool constrained by RCM is achieved using Equation (12) and the active joint parameters are obtained using the *inverse kinematics* algorithm.

Observation 1: Although the Study parameters show computational advantages, there is a fundamental problem that must be overcome. For the displacement of the ultrasound probe from an initial configuration to the next configuration by using Equation (12) the Cartesian coordinates of the entry point will not change. However, in-between the two configurations, there is no guarantee that the entry point will not move. A numerical solution for this drawback is to use smaller displacements for the tool manipulations and interpolation to determine the tool path.

3.2.1. The Upper Planar Mechanism Kinematics

The *upper planar mechanism* (Figure 2) is described by the constraints of the point P_u (located at the intersection axes of U_1). Using dual quaternion multiplication (described by Equation (6) and the rules of the imaginary parts multiplication) the constraint equations for the point P_u are:

$$P_u = P_{u1} \cdot P_{u2} \cdot P_{u3} \cdot P_{u4}$$

$$P_u(x_i, y_i) \rightarrow [x_0 : x_1 : x_2 : x_3 : y_0 : y_1 : y_2 : y_3]^T = [1 : 0 : 0 : 0 : 0 : -\tfrac{1}{2}(l_{q5} + l_{q4}) : -\tfrac{1}{2}q_5 : 0]^T \tag{14}$$

$$l_{q4} = \sqrt{l_{d4}^2 - (q_5 - q_4)^2}$$

where the dual quaternions used in obtaining the P_u map are detailed in Table 1.

Table 1. The quaternions used in defining the constraints of the upper planar chain.

Symbol	Details	Description
P_{u1}	$[1{:}0{:}0{:}0{:}0{:}0{:}0{:}{-}1/2\,D_z]$	Geometric parameter; translation along OZ axis
P_{u2}	$[1{:}0{:}0{:}0{:}0{:}\ {-}1/2\,q_5{:}0]$	Active motion parameter; translation along OY axis
P_{u3}	$[1{:}0{:}0{:}0{:}0{:}\ {-}1/2\,l_{q5}{:}0{:}0]$	Geometric parameter; translation along OX axis
P_{u4}	$[1{:}0{:}0{:}0{:}0{:}\ {-}1/2\,l_{q4}{:}0{:}0]$	Active motion parameter; translation along OX axis

3.2.2. The Lower Planar Mechanism Kinematics

For the *lower planar mechanism* (Figure 2) the Study parameters are written for the two kinematic chains (both $\underline{R}RR$) which are coupled at the point P_d (located at the midpoint of the link d). Therefore the point P_d must simultaneously satisfy two equations:

$$\begin{cases} P_d = P_{d1} \cdot P_{d2} \cdot P_{d3} \cdot P_{d4} \cdot P_{d5} \cdot P_{d6} \\ P_d = P_{d7} \cdot P_{d8} \cdot P_{d9} \cdot P_{d10} \cdot P_{d11} \cdot P_{d12} \end{cases}$$

$$P_d(x_i, y_i) \rightarrow [x_0 : x_1 : x_2 : x_3 : y_0 : y_1 : y_2 : y_3]^T = [x_0 : 0 : 0 : x_3 : 0 : y_1 : y_2 : 0]^T$$

$$x_0 = 4(t_1 t_2 + t_1 t_3 + t_2 t_3 - 1) \quad x_0 = 4(w_1 w_2 + w_1 w_3 + w_2 w_3 - 1)$$

$$x_3 = 4(t_1 t_2 t_3 + t_1 + t_2 - t_3) \quad x_3 = 4(w_1 w_2 w_3 + w_1 + w_2 - w_3)$$

$$y_1 = t_1 t_2(dt_3 + 2l_1 - 2l_2) + t_1 t_3(2l_1 + 2l_2) + t_2 t_3(-2l_1 + 2l_2) - d(t_1 + t_2 + t_3) + 2(l_1 + l_2) \tag{15}$$

$$y_1 = w_1 w_2 w_3(d + 2l_{q3}) + 2w_1 w_2(l_3 - l_4) + 2w_1 w_3(l_3 + l_4) + 2w_2 w_3(-l_3 + l_4) - d(w_1 + w_2 + w_3) +$$
$$2l_{q4}(w_1 + w_2 + w_3) + 2(l_3 + l_4)$$

$$y_2 = 2t_1 t_2 t_3(l_2 - l_1) - d(t_1 t_2 + t_2 t_3 + t_1 t_3) + 2l_1(t_1 - t_2 - t_3) + 2l_2(t_1 + t_2 - t_3) + d$$

$$y_1 = w_1 w_2 w_3(2l_4 - 2l_3) - w_1 w_2(d + 2l_{q4}) - w_1 w_3(d + 2l_{q4}) - 2w_2 w_3(d + 2l_{q4}) + l_3(w_1 + w_2 - w_3) +$$
$$2l_4(w_1 + w_2 - w_3) + d + 2_{q4}$$

The twelve quaternions used in obtaining the map P_d are detailed in Table 2. It is already visible from the zeroes of the Study parameters in Equation (15) that the point P_d is constrained to move on a plane. However, Equation (15) describe the constraints of a six bar mechanism and a change in orientation (of the link d) is still possible. The solution to this is to simply set $x_0 = 1$ and $x_3 = 0$, which will also satisfy the normalization condition defined in Equation (2).

Table 2. The quaternions used in defining the constraints of the lower planar chain.

Symbol	Details	Description
P_{d1}	$[1{:}0{:}0{:}t_1{:}0{:}0{:}0{:}0]$ [1]	Active motion parameter; rotation around OZ axis
P_{d2}	$[1{:}0{:}0{:}0{:}0{:}-l_1/2{:}0{:}0]$	Geometric parameter; translation along OX axis
P_{d3}	$[1{:}0{:}0{:}t_2{:}0{:}0{:}0{:}0]$ [1]	Free motion parameter; rotation around OZ axis
P_{d4}	$[1{:}0{:}0{:}0{:}0{:}-l_2/2{:}0{:}0]$	Geometric parameter; translation along OX axis
P_{d5}	$[1{:}0{:}0{:}t_3{:}0{:}0{:}0{:}0]$ [1]	Free motion parameter; rotation around OZ axis
P_{d6}	$[1{:}0{:}0{:}0{:}0{:}0{:}-d/4{:}0]$	Geometric parameter; translation along OY axis
P_{d7}	$[1{:}0{:}0{:}0{:}0{:}0{:}-l_{q4}/2{:}0]$	Geometric parameter; translation along OY axis
P_{d8}	$[1{:}0{:}0{:}w_1{:}0{:}0{:}0{:}0]$ [1]	Active motion parameter; rotation around OZ axis
P_{d9}	$[1{:}0{:}0{:}w_2{:}0{:}0{:}0{:}0]$ [1]	Free motion parameter; rotation around OZ axis
P_{d10}	$[1{:}0{:}0{:}0{:}0{:}-l_3/2{:}0{:}0]$	Geometric parameter; translation along OX axis
P_{d11}	$[1{:}0{:}0{:}w_3{:}0{:}0{:}0{:}0]$[1]	Free motion parameter; rotation around OZ axis
P_{d12}	$[1{:}0{:}0{:}0{:}0{:}0{:}d/4{:}0]$	Geometric parameter; translation along OY axis

[1] The t_i and w_i represent the tangent of half angles parameters.

The fact that the Study parameters Equation (15) describe varieties in the Study parameter space, allows the elimination of the t_2, t_3, w_2, w_3 parameters (which are the free motion parameters in the planar mechanisms) by computing elimination ideals with Groebner bases (this technique is detailed in [20] and was used in the analysis of various robotic systems such as [4,7]); it is worth mentioning that the parameter elimination can be also achieved with the Linear Implicitization Algorithm (LIA) [21] which was used in [6]. The resulting constraints in this case will be the union of the equations found in the two Groebner bases which were computed to eliminate the free motion parameters and are:

$$
\begin{aligned}
G_d = {} & < 16(t_1^2 + 1)y_1^2 - 16l_1(t_1^2 - 1)y_1 + 16(t_1^2 + 1)y_2^2 + 8(dt_1^2 + 4l_1t_1 + d)y_2 + (d^2 + 4l_1^2 - 4l_2^2)t_1^2 + 8dl_1t_1 \\
& + d^2 + 4(l_1^2 - l_2^2); 16(w_1^2 + 1)y_1^2 - 16l_3(w_1^2 - 1)y_1 + 16(w_1^2 + 1)y_2^2 + 8((2l_{q4} + d)w_1^2 + 4l_3w_1 + 2l_{q4} + d)y_2 \\
& + (d^2 + 4(dl_{q4} + l_3^2 - l_4^2 + l_{q4}^2))w_1^2 + 8l_3(d + 2l_{q4})w_1 + d^2 + 4(dl_{q4} + l_3^2 - l_4^2 + l_{q4}^2) >
\end{aligned}
\tag{16}
$$

3.2.3. The Mobile Platform Mechanism Kinematics

For the *mobile platform* (Figure 2) the forward kinematics is written using the Study parameters with the observation that the inputs are the two points $P_u(X_u, Y_u)$ and $P_d(X_d, Y_d)$ (were the eight dual quaternions used in the P map are presented in Table 3):

$$
\begin{cases}
P = P_1 \cdot P_2 \cdot P_3 \cdot P_4 \\
P = P_5 \cdot P_6 \cdot P_7 \cdot P_8
\end{cases}
$$

$$
\begin{aligned}
P_u(x_i, y_i) \to [x_0 : x_1 : x_2 : x_3 : y_0 : y_1 : y_2 : y_3]^T \quad & = [2 : 2u_1 : -2u_2 : -2u_1u_2 : y_0 : y_1 : y_2 : y_3]^T \\
& = [2 : 2v_1 : -2v_2 : -2v_1v_2 : y_0 : y_1 : y_2 : y_3]^T \\
y_0 = -D_n u_1 u_2 + X_d u_1 - Y_d u_2 + cu_1 \quad & y_0 = v_1v_2(D_n - D_z + p) + X_u v_1 - Y_u v_2 \\
y_1 = Y_d u_1 u_2 + D_n u_2 - X_d - c \quad & y_1 = v_2(Y_u v_1 - D_n - D_z - p) - X_u \\
y_2 = -(X_d + c)u_1 u_2 + D_n u_1 - Y_d \quad & y_2 = v_1(X_u v_2 - D_n - D_z - p) - Y_u \\
y_3 = (X_d + c)u_2 + Y_d u_1 - D_n \quad & y_3 = X_u v_2 + Y_u v_1 + D_n - D_z + p
\end{aligned}
\tag{17}
$$

Table 3. The quaternions used in defining the constraints of the robot mobile platform.

Symbol	Details	Description
P_1	$[1{:}0{:}0{:}0{:}0{:}(X_d{+}c)/2{:}\ Y_d/2{:}0]$	Active motion parameter; translation on the XY plane
P_2	$[1{:}0{:}u_1{:}0{:}0{:}0{:}0{:}0]\ ^1$	Free motion parameter; rotation around OY axis
P_3	$[1{:}0{:}u_2{:}0{:}0{:}0{:}0{:}0]\ ^1$	Free motion parameter; rotation around OZ axis
P_4	$[1{:}0{:}0{:}0{:}0{:}0{:}0{:}\ D_n/2]$	Geometric parameter; translation along OZ axis
P_5	$[1{:}0{:}0{:}0{:}0{:}\ {-}X_u/2{:}\ {-}Y_u/2{:}\ {-}D_z/2]$	Active motion parameter; translation on a plane parallel to the XY plane
P_6	$[1{:}0{:}v_1{:}0{:}0{:}0{:}0{:}0]\ ^1$	Free motion parameter; rotation around OY axis
P_7	$[1{:}0{:}v_2{:}0{:}0{:}0{:}0{:}0]\ ^1$	Free motion parameter; rotation around OZ axis
P_8	$[1{:}0{:}0{:}0{:}0{:}0{:}0{:}p/2]$	Active motion parameter; rotation around OZ axis
P_9	$[1{:}0{:}0{:}0{:}0{:}0{:}0{:}D_n/2]$	Active motion parameter; rotation around OZ axis

^{1} The u_i and v_i represent the tangent of half angles parameters.

The Study parameters equations found in Equation (17) represent a variety in Study space. Like in the previous case the free motion parameters are eliminated from the equations (the rotation parameters u_1, u_2, v_1, v_2 and the translation parameter p) by computing elimination ideals using Groebner bases. Consequently, the resulting input/output equations relate the terms X_d, X_u, Y_d, Y_u with the Study parameters (the mobile platform coordinates and orientation). The constraints of the mobile platform are given by the union of the Study parameter equations of the two elimination ideals (describing P—see Equation (17)) which are four linear equations (in the Study parameters), nine quadratic equations, one cubic equation, the Study quadric and the normalizing condition (due to the complexity of the equations they are shown only in general form—e.g., g_1 is a linear equation in the unknowns x_3, y_0, y_1, y_2, g_5 is an quadratic equation in the unknowns y_0, y_1, y_2, y_3 and so on):

$$G =<\ g_1(x_3, y_0, y_1, y_2), g_2(x_2, y_0, y_1, y_3), g_3(x_1, y_2, y_0, y_3), g_4(x_0, y_1, y_2, y_3), g_5^2(y_0, y_1, y_2, y_3),$$
$$g_{6..10}^2(x_0, x_1, x_2, x_3, y_0, y_1, y_2, y_3), g_{11..12}^2(x_0, x_1, x_2, x_3), g_{13}^2(x_0, x_1, x_2, x_3, y_1), x_0^2 + x_1^2 + x_2^2 + x_3^2 - 1, \tag{18}$$
$$x_0 y_0 + x_1 y_1 + x_2 y_2 + x_3 y_3, g_{16}^3(x_0, x_1, x_2, x_3, y_0, y_1, y_2, y_3) >$$

At this point it can be checked that the basis G has the Hilbert dimension 0 which corresponds to the forward kinematic solution being points (not necessarily unique since for one set of input parameters the mechanism may pose in different ways). Moreover, the basis may be solved for both the Study parameters (when the active joints are inputs) achieving the forward kinematics and for the active joints (when the Study parameters are inputs) achieving the inverse kinematics.

3.3. Pro-Hep-LCT Singularities

This section describes the Pro-Hep-LCT robotic system singularities which are determined from the constraint equations obtained in the previous section. Consequently, the singularities are analyzed for each component mechanism individually. This is possible due to the fact that the robot is a hybrid (and the end effector is guided by two planar mechanisms). The inputs of the lower planar mechanism (t_1 and w_1 being the tangents of the half angles of q_2 and q_3 respectively) do not influence the point P_u (of the upper mechanism) and the inputs of the upper chain (q_4 and q_5) do not influence the point P_d. For the input singularities, while the robot is rigid, an infinitesimal change in the inputs will produce no change in the outputs, whereas for the output singularities, an infinitesimal change in the output is possible even though there is no change in the input. The singularities for the robotic system are determined using the rank of the input and output Jacobians for the three presented mechanisms.

If the Jacobians of any of the mechanism are rank deficient a singularity occurs. The input and output Jacobians were determined for each component mechanism and are:

$$J_{i,1} \cdot \dot{q}_u + J_{o,1} \cdot \dot{P}_u = 0 \quad J_{i,1} = \left[\frac{\partial(G_u[1],G_u[2])}{\partial(q_4,q_5)} \right]; \quad J_{o,1} = \left[\frac{\partial(G_u[1],G_u[2])}{\partial(y_1,y_2)} \right]; \dot{q}_u = [\dot{q}_4, \dot{q}_5]^T;$$
$$\dot{P}_u = [\dot{y}_1, \dot{y}_2]^T \tag{19}$$

$$J_{i,2} \cdot \dot{q}_d + J_{o,2} \cdot \dot{P}_d = 0 \quad J_{i,2} = \left[\frac{\partial(G_u[1],G_u[2])}{\partial(q_4,q_5)} \right]; \quad J_{o,2} = \left[\frac{\partial(G_d[1],G_d[2])}{\partial(y_1,y_2)} \right]; \dot{q}_d = [\dot{t}_1, \dot{w}_1]^T$$
$$\dot{P}_d = [\dot{y}_1, \dot{y}_2]^T \tag{20}$$

$$J_{i,3} \cdot \dot{q} + J_{o,3} \cdot \dot{P} = 0 \quad J_{i,3} = \left[\frac{\partial(G[1]...G[16])}{\partial(q_4,q_5)} \right]; \quad J_{o,3} = \left[\frac{\partial(G^*[1]...G^*[8])}{\partial(x_i,y_i)} \right]; \dot{q} = [\dot{X}_u, \dot{Y}_u, \dot{X}_d, \dot{Y}_d]^T$$
$$\dot{P}_d = [\dot{x}_0 ... \dot{x}_3, \dot{y}_0 .. \dot{y}_3]^T \tag{21}$$

The G^* basis represents a pure lexicographic Groebner base computed from the G base and has the advantage that it has 8 equations and the Jacobian $J_{o,3}$ remains a square matrix. This is not true for the $J_{i,3}$, therefore, all the determinants of the square minors must be computed for the singularity analysis.

Analyzing the rank deficiency conditions of the Jacobians, the output singularities are given by the following 12 polynomial equations (Equations (24)–(27) and (32)–(36) are only presented in the general form since they are lengthy):

$$sg_1 : w_1^2 + 1 = 0 \tag{22}$$

$$sg_2 : t_1^2 + 1 = 0 \tag{23}$$

$$sg_3(t_1, w_1, l_1, l_3) = 0 \tag{24}$$

$$sg_4(t_1, w_1, l_1, l_3, l_{q4}, d) = 0 \tag{25}$$

$$sg_5(t_1, w_1, l_1, l_2, l_3, l_{q4}, d) = 0 \tag{26}$$

$$sg_6(t_1, w_1, l_1, l_2, l_3, l_{q4}, d) = 0 \tag{27}$$

$$sg_7 : Yd - Yu = 0 \tag{28}$$

$$sg_8 : Xd - Xu + c = 0 \tag{29}$$

$$sg_9 : D_z = 0 \tag{30}$$

$$sg_{10} : D_z^2 + Y_d^2 - 2Y_dY_u + Y_u^2 = 0 \tag{31}$$

$$sg_{11}(X_u, Y_u, X_d, Y_d, D_z, c) = 0 \tag{32}$$

$$sg_{12}(X_u, Y_u, X_d, Y_d, D_z, c) = 0 \tag{33}$$

and the input singularities are given by three polynomials and an irrational function (due to the substitution found in the map defined in Equation (14)):

$$sg_{13}(y_1, y_2, l_3, l_4, l_{q4}, d) = 0 \tag{34}$$

$$sg_{14}(y_1, y_2, l_3, l_4, l_{q4}, d) = 0 \tag{35}$$

$$sg_{15}(y_1, y_2, l_3, l_4, l_{q4}, d) = 0 \tag{36}$$

$$sg_{16} : \frac{q_4 - q_5}{\sqrt{(q_4 - q_5 + l_{d4})(-q_4 + q_5 + l_{d4})}} = 0 \tag{37}$$

Equations (22)–(37) represent the general singularities of the Pro-Hep-LCT robotic system. The simple equations are straight forward to analyze: Equations (22), (23) and (31) have no real solutions and are not of interest in kinematics; Equation (30) cannot be true since it implies that the two planar mechanisms operate in the same plane (which is not the case); Equation (37) represents a singularity when active joints q_4 and q_5 are overlapped or are at a distance l_{q4} apart (both conditions being physically impossible due to mechanism design considerations). Although the conditions from Equations (24), (28) and (29) make the Jacobians rank to drop, they are not singularities of the constraint varieties. Further investigations showed that the algorithm used in computing the Groebner bases G_d and G divided through $(t_1 \pm w_1)$, $(X_d - X_u + c)$ and $(Y_d - Y_u)$ respectively. Consequently the bases are not correct when these conditions are met since the division by 0 is not allowed. In a sense, Equations (24), (28) and (29) represent singularities of the computed Groebner bases and are not singularities of the constraint varieties. To check that this is true, the conditions were substituted into the kinematic model before the computation of the Groebner bases (G_d and G) and no singularity was found.

For the lengthy polynomial equations which describe singularities of the Pro-Hep-LCT the numerical values for the geometric parameters resulted from the robot design (the links lengths as shown in Table 4) were used to evaluate the equations to allow a geometric interpretation of the singularities.

Table 4. Numerical values for the lengths of the Pro-Hep-LCT robotic system.

Component Mechanism	Link Notation/Length [mm]
Upper planar mechanism	$l_{q5} = 160$; $l_{q4} = 284$.
Lower planar mechanism	$l_1 = 168$; $l_2 = 168$; $l_3 = 168$; $l_4 = 168$; $l_{q3} = 400$; $d = 150$; $c = 61$.
Mobile platform	$D_z = 108$; $D_n = 110$; $p = [108, 125]$

After the numerical evaluation of the remaining equations, Equations (25) and (27) showed only complex solutions and are not of interest. Equation (26) showed a configuration where the link d is aligned with either of the links l_2 or l_4 (see Figure 3a). This singularity represents the separation of two working modes of the lower planar mechanism and it is part of the boundary of the operational workspace (as illustrated in following sections). Equations (32) and (33) showed a configuration which was already described in Equation (30), namely the D_z parameter being zero. However, this condition is not allowed since the mechanism in not constructed in such way.

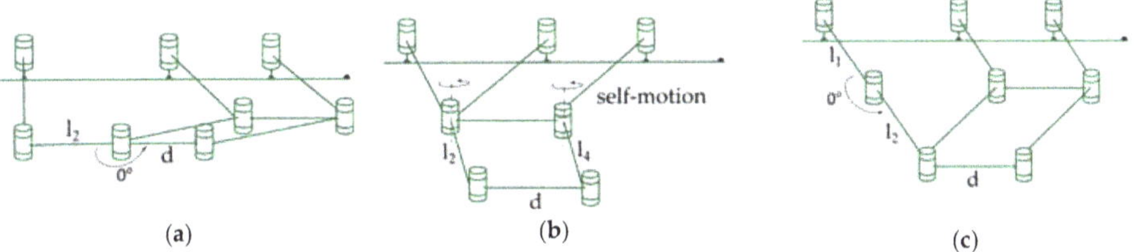

Figure 3. Singularities of the Pro-Hep-LCT robotic system; (**a**) configuration where the links l_2 and d are aligned; (**b**) self-motion when the links l_2 and l_4 are parallel; (**c**) configuration were the links l_1 and l_2 are aligned.

For the input singularities the equations were also evaluated with the numerical values detailed in Table 4. Through a close analysis it was found that Equation (34) describes a self-motion of the link d which is present when the links l_2 and l_4 are parallel (Figure 3b). This configuration must be avoided using the robot control since it may cause harm to the patient during the medical procedure. Equation (35) describes a singularity when $t_1 = 1$ and $w_1 = -1$ which is physically impossible since in the robot design it implies that various mechanical links (l_3, l_4) will overlap. Equation (36) describes configurations where the link l_1 is aligned with l_2 (Figure 3c) or the link l_3 is aligned with l_4. This actually represents the boundary of the workspace of the lower planar mechanism.

3.4. Pro-Hep-LCT Workspace

Analyzing the constraint equations for the Pro-Hep-LCT robotic system (Equations (14), (16) and (18)) it can be shown that the lower planar mechanism has four working modes, the upper planar mechanism has two working modes and even the end effector has four working modes (at least theoretically). For this reason, the geometry of the workspace is complex and the focus hereafter switches towards the operational workspace (the set of configurations intended for achieving the medical task).

The operational workspace was analyzed in two ways: at first the operational workspace describing the tip of the medical tool was determined (which for the medical task at hand represents the skin entry-point); and second, the redundant DOF for the medical tool was introduced to geometrically illustrate the Pro-Hep-LCT robotic system capabilities to achieve the medical task. Figure 4a shows the boundary of the operational workspace in which the end-effector of the Pro-Hep-LCT robotic system has no orientation (i.e., the medical tool points vertically downwards) and Figure 4b shows a cross-section of the space. Within this cross-section of the workspace, there exists a subspace (on the right hand side of the blue line $X = 214$, see Figure 3b) in which all the possible orientations of the medical tool are possible (i.e., the dexterous workspace). In this region of the workspace the robotic assisted medical procedure should be the least limited (due to the mechanism constraints). However, there is a point that describes the self-motion of the mechanism (of coordinates $X = 286.66$ mm, $Y = 200$ mm) which must be avoided (using the control algorithm) since in this configuration the control of the robot is lost and injury may occur.

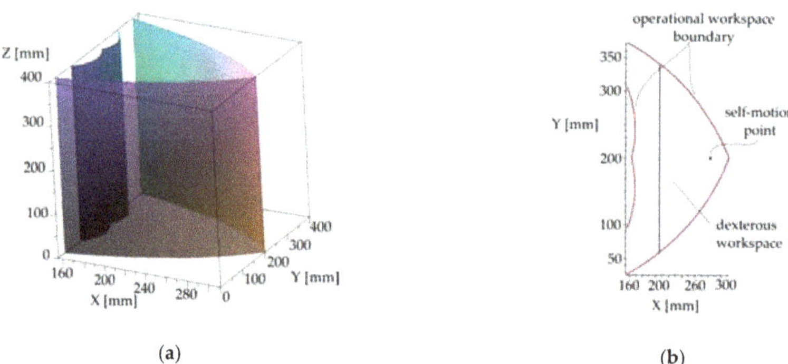

(a)   (b)

Figure 4. Pro-Hep-LCT operational workspace; (**a**) envelope; (**b**) cross-section of the dexterous work-space.

To illustrate the operational workspace while the medical tool is inserted in the patient, two entry points are chosen. First, an entry point was chosen within the dexterous workspace ($X = 260$ mm, $Y = 180$ mm, $Z = 200$ mm) and Figure 5a illustrates two spherical surfaces defining two different depths of insertion (108 mm and 70 mm, respectively). The second example was chosen outside the dexterous workspace (see Figure 5b), having the entry point at ($X = 200$ mm, $Y = 180$ mm, $Z = 200$ mm), with two insertion depths (108 mm and 70 mm, respectively).

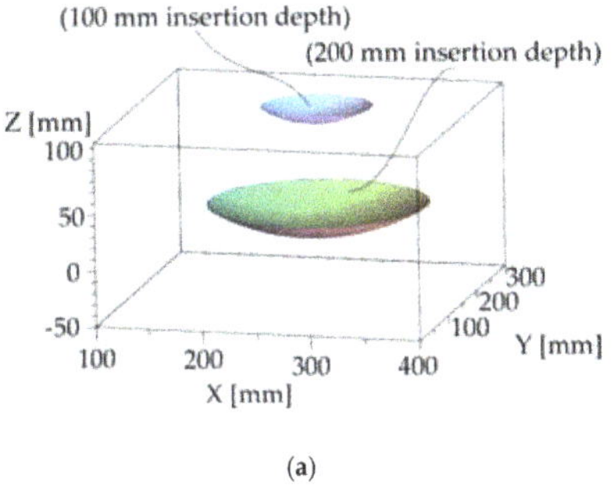
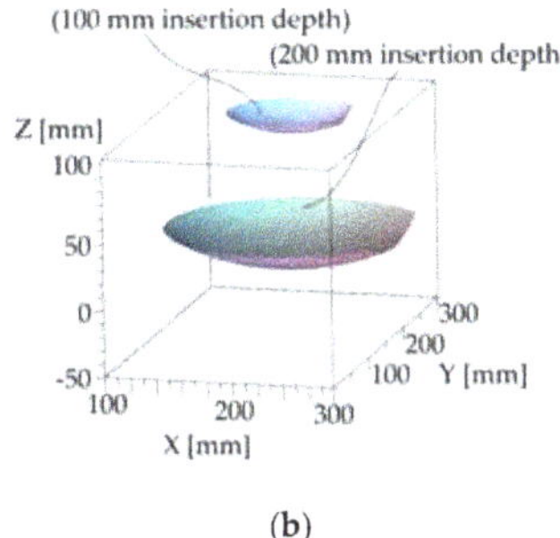

Figure 5. Pro-Hep-LCT operational workspace for the RCM manipulation; (**a**) entry point [260 mm, 180 mm, 200 mm]; (**b**) entry point [200 mm, 180 mm, 200 mm].

4. Numerical Results and Discussion

A numerical simulation is illustrated in this section to show how the results presented in Section 2 apply to the Pro-Hep-LCT robotic system functionality. To illustrate an example a Cartesian point is chosen as an entry-point E on the patient abdomen with the numerical coordinates [X_E = 300 mm, Y_E = 180 mm, Z_E = 110]. Furthermore, for this example, the ultrasound probe at the entry-point is chosen in a configuration where it points directly downwards in a vertical pose. The configuration has the following numerical values of the Study parameters [x_0 = 1, x_1 = 0, x_2 = 0, x_3 = 0, y_0 = 0, y_1 = -150, y_2 = -90, y_3 = 55] (relative to the fixed coordinate frame XYZ; to achieve this q_1 = 220 mm is a necessary condition). The inverse kinematics is computed from Equations (14), (16) and (18) with the numerical values for the Study parameters substituted in, yielding the numerical values for the active joint parameters [t_1 = -0.1343, w_1 = 0.0214, q_4 = 20.802 mm, q_5 = 180 mm] (solution describing the intended working mode for the Pro-Hep-LCT robotic system). Table 5 shows the initial configuration of the ultrasound probe (O at the entry-point) and two points A and B which are the end points of two simple tool paths (at 100 mm insertion depths). The Study parameters required for the RCM manipulation are also shown for the two tool paths with a discretization by angular values of 5° starting from 0° (the initial configuration) and ending at 30° (for point A with φ being a rotation about X' axis and for point B where θ represents a rotation about Y' axis).

Table 5. Numerical values for two paths of the ultrasound probe constrained by RCM.

Point	Study Parameters	Active Joint Numerical Values [Scalars, mm]
O	[1:0:0:0:0:-150:-90:55]	[t_1 = -0.1343, w_1 = 0.0214, q_4 = 20.802, q_5 = 180, n = 0]
	[1:φ:0:0:0:0:0:0]	
	φ = -0.0437	[t_1 = -0.1077, w_1 = 0.0492, q_4 = 39.8745, q_5 = 199.0725, n = 100.4202]
	φ = 0.0875	[t_1 = -0.0804, w_1 = 0.077, q_4 = 59.2413, q_5 = 218.4392, n = 101.6969]
A	φ = -0.1317	[t_1 = -0.0518, w_1 = 0.1052, q_4 = 79.2149, q_5 = 238.4129, n = 103.8804]
	φ = -0.1763	[t_1 = -0.0213, w_1 = 0.1344, q_4 = 100.1475, q_5 = 259.3455, n = 107.0596]
	φ = -0.2217	[t_1 = 0.012, w_1 = 0.1653, q_4 = 122.4570, q_5 = 281.6550, n = 111.3716]
	φ = -0.2679	[t_1 = 0.0496, w_1 = 0.1984, q_4 = 146.6643, q_5 = 305.8623, n = 117.0170]
	[1:0:θ:0:0:0:0:0]	
	θ = 0.0437	[t_1 = -0.1189, w_1 = 0.0070, q_4 = 23.8913, q_5 = 180, n = 100.4202]
	θ = 0.0875	[t_1 = -0.1031, w_1 = 0.0101, q_4 = 29.5461, q_5 = 180, n = 101.6969]
B	θ = 0.1317	[t_1 = -0.0835, w_1 = -0.0316, q_4 = 38.3570, q_5 = 180, n = 103.8804]
	θ = 0.1763	[t_1 = -0.0597, w_1 = -0.0596, q_4 = 51.5117, q_5 = 180, n = 107.0596]
	θ = 0.2217	[t_1 = -0.0294, w_1 = -0.1007, q_4 = 71.5689, q_5 = 180, n = 111.3716]
	θ = 0.2679	[t_1 = 0.0131, w_1 = -0.1977, q_4 = 106.0062, q_5 = 180, n = 117.0170]

The t_1 and w_1 are the tangent of the half angle of q_2 and q_3 respectively; n represents the value for the redundant DOF from the robot end-effector.

Lastly, Table 5 shows the numerical values of the active joint parameters for each ultrasound probe pose (computed by means of inverse kinematics using Equations (14), (16) and (18)).

Figure 6 illustrates the discretization of the two paths of the ultrasound probe tip starting from a point O′ (where the tool is in a vertical pose and inserted at a 100 mm depth) and ending at the points A and B respectively (with 30° angle values about X' and Y' axes respectively). These paths are simulated with the ultrasound probe constrained by the RCM. Figure 6a illustrates the Cartesian positions of each computed point from the two paths whereas Figure 6b illustrates the active joint values changing over a period of 6 s (using linear interpolation) required for the specified paths.

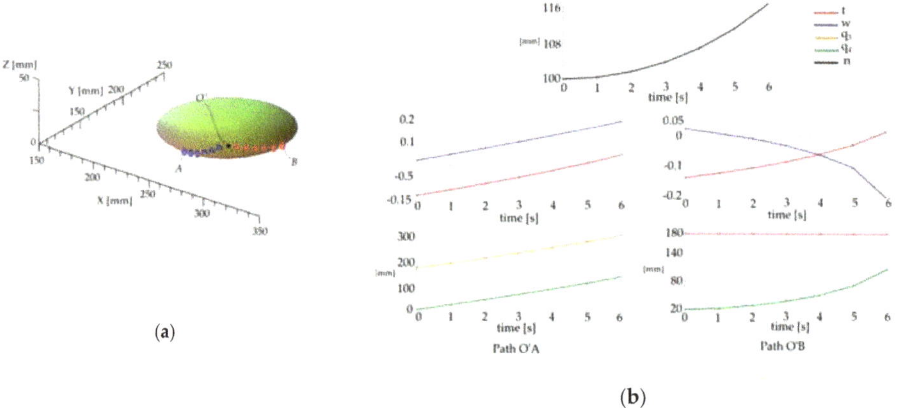

Figure 6. Ultrasound probe tip path simulation: (**a**) path discretization; (**b**) active joint space.

The computed numerical values (for the active joints) that describe the two paths were used in a kinematic simulation using the Siemens NX software. The simulation purpose was to analyze the motion of the insertion point in Cartesian coordinates during the ultrasound probe manipulation on the two paths A and B over a period of 6 s (since in between two consecutive points of a path there is no guarantee that the RCM point is unchanged). For the path O′A, Figure 7a shows the simulation from the Siemens NX environment where the variation of the entry-point E in each X, Y and Z direction relative to the predefined point $[X_E = 300$ mm, $Y_E = 180$ mm, $Z_E = 100]$ is highlighted. On the X direction, the maximum variation during the simulation was about 0.5 mm, reaching to a maximum error of 1.45 mm. (error that may be introduced by numerical errors of the interpolation). On the Y direction, the maximum variation was 0.05 mm adding to a maximum error of 0.15 mm which is less than the total error from the X displacement.

On the Z direction the maximum variation was about 0.12 mm. leading to a total error of 0.18 mm. For the path O′B, the simulation is shown in Figure 7b where the error in the displacement on X direction is about 3 mm, the error in the Y direction is about 2 mm and the error in the Z direction is about 0.2 mm. There are two types of fluctuations present in the displacement graphs illustrated in Figure 7, the first type being most likely due to the kinematic model and it was expected (see Observation 1 in Section 3.1) since the RCM is guaranteed to be kept only in the control points of the path and not in between. The second fluctuation refers to the shift in the curve (e.g., the 5 s mark for the Y displacement in Figure 7a) which is introduced by interpolation errors, most likely amplified by changing the simulation environment from Maple (where the inverse kinematics was computed) to Siemens NN (where the RCM point was studied). However, the magnitude of the errors presented so far is not significant for the medical task since a bigger error may be introduced by the tissue motion during the respiration process.

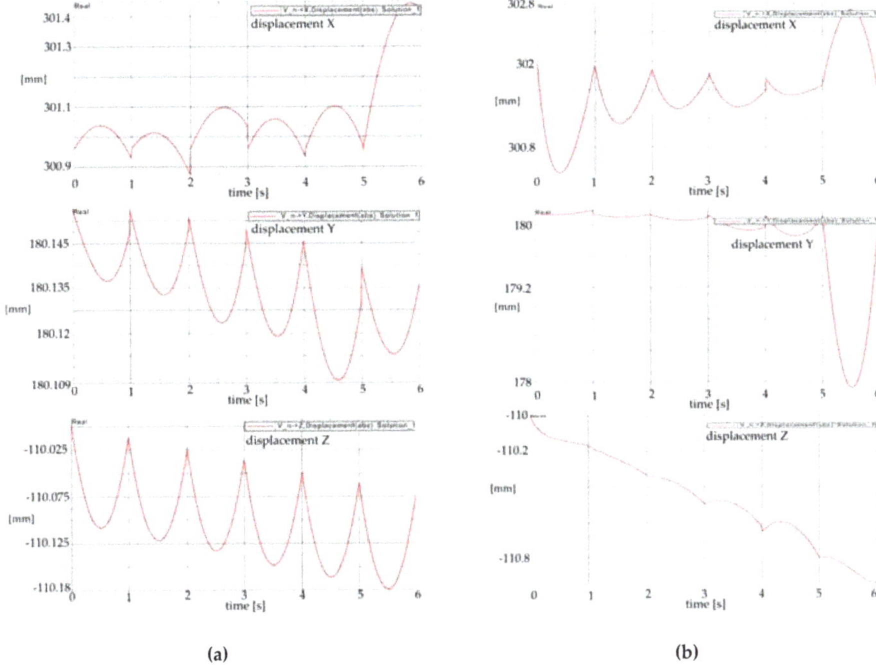

(a)   (b)

Figure 7. Simulation in the Siemens NX environment (variation of the entry-point Cartesian coordinates); (a) displacement of the entry point while the ultrasound probe was manipulated on path O′A; (b) displacement of the entry point while the ultrasound probe was manipulated on path O′B.

By utilizing the Study parameters method the authors managed to achieve a complete geometric analysis (obtaining the constraints, singularities and describing the workspace) of the Pro-Hep-LCT parallel robotic system designed for the minimally invasive task of targeted liver treatment (through brachytherapy, ablation of chemotherapy) under ultra sound imaging. Although the robotic system may appear complex, analyzing the "segments" that compose it (referred as the lower, upper and mobile platform mechanisms in Section 3) facilitated the computation of the constraints and singularities without any disadvantages (in the beginning of Section 3.2 it is explained why segmenting the robotic system is allowed). Moreover, the use of quaternion multiplication for the medical tools manipulation (Equations (8)–(12)) offered a distinct computation advantage since it complements well with the constraints computed with the Study parameters model. The dual quaternion approach for RCM manipulation is not novel (see e.g., [9]) and it has been shown that the main disadvantage (which is discussed in Observation 1 Section 3.1) may be minimized using interpolation (this method was proven on a serial manipulator in [9]). Since the solutions for the forward kinematics of the Pro-Hep-LCT parallel robotic system had the dual quaternion form (i.e., Study parameters) the manipulation of the medical instruments constrained by RCM was straightforward (see Section 2 Equations (8)–(12)). By means of linear interpolation it was also shown that the displacement of the insertion point is insignificant for the medical procedure. Of course more refined interpolation methods (or more control points) may be used if needed to increase precision but the tradeoff must be considered (computation time vs. precision since a real time control must be achieved). Besides RCM manipulation using dual quaternions, other medical applications may be considered as well, e.g., motor rehabilitation, or even determining the kinematics of the human finger in an analytical manner (in contrast to the experimental models—see e.g., [22]).

5. Conclusions

Based on a mathematical method that describes the global kinematics (using the Study parameters of the total Euclidean displacement), the Pro-Hep-LCT robotic system was completely analyzed to determine its forward and inverse kinematics, singularities and workspace. To determine the robot capabilities to perform the medical task, the operational workspace of the robotic system was described in two different ways: first, based on the medical relevant position of the robot (relative to the patient) the singularity free operational workspace was shown which also contains the dexterous workspace (which is especially relevant for the medical task); second, the manipulation of the medical instrument under the constraint of the RCM was studied creating also a description of the active joint space in a discrete manner. The analysis of the singularity free operational workspace shows that the Pro-Hep-LCT parallel robotic system is feasible for the RCM manipulation within the angular values of $\pm 30°$ (relative to the vertical direction) in all directions. The numerical values of the active joint parameters which were computed via inverse kinematics (for consecutive points on predefined tool paths) were linearly interpolated and used as inputs for a kinematic simulation in the Siemens NX environment which showed a maximum relative deviation (from the insertion point) of 1.45 mm. The preliminary numerical results show that the robot control based on the Study parameters is feasible due to the insignificant deviation of the insertion point (with respect to the imposed point).

Further work is intended for developing the robotic system, namely the development of the mechanical architecture of the robotic system, the development of a controls system which uses the constraint equations together with the instrument manipulation method based on the dual quaternions. Moreover, further work is intended to test the precision of the prototype and optimize the control to account for needed corrections during the RCM manipulation (e.g., when parasite motions are present).

Author Contributions: Investigation, I.B., M.H., C.V., N.P., A.N. and D.P.; Methodology, I.B., M.H., C.V., N.P., A.N. and D.P.; Project administration, D.P.; Validation, M.H. and N.P.; Writing—original draft, I.B.; Writing—review & editing, I.B., C.V. and D.P.

Funding: This work was supported by a grant of the Romanian Ministry of Research and Innovation, PCCCDI – UEFISCDI, project number PN-III-P1-1.2-PCCDI-2017-0221/59PCCDI/2018 (IMPROVE), within PNCDI III and by project ExNanoMat, contract no. 21 PFE-2018 within PNCDI III.

Conflicts of Interest: The authors declare no conflict of interest.

References

1. Angeles, J. Fundamentals of Robotic Mechanical Systems. In *Theory, Methods and Algorithms*; Springer: New York, NY, USA, 1997.
2. Husty, M.; Pfurner, M.; Schröcker, H.-P.; Brunnthaler, K. Algebraic Methods in Mechanism Analysis and Synthesis. *Robotica* **2007**, *25*, 661–675. [CrossRef]
3. Schadlbauer, J.; Walter, D.R.; Husty, M. The 3-RPS parallel manipulator from an algebraic viewpoint. *Mech. Mach. Theory* **2014**, *75*, 161–176. [CrossRef]
4. Pisla, D.; Birlescu, I.; Vaida, C.; Tucan, P.; Pisla, A.; Gherman, B.; Crisan, N.; Plitea, N. Algebraic modeling of kinematics and singularities for a prostate biopsy parallel robot. *Proc. Rom. Acad. Ser. A-Math. Phys. Tech. Sci. Inf. Sci.* **2018**, *19*, 489–497.
5. Husty, M. An Algorithm for Solving the Direct Kinematic of General StewartGough Platforms. *Mech. Mach. Theory* **1996**, *31*, 365–380. [CrossRef]
6. Vaida, C.; Pisla, D.; Schadlbauer, J.; Husty, M.; Plitea, N. Kinematic Analysis of an Innovative Medical Parallel Robot Using Study Parameters. *New Trends Med. Serv. Robot. Mech. Mach. Sci.* **2016**, *39*, 85–99.
7. Birlescu, I.; Pisla, D.; Gherman, B.; Pisla, A.; Vaida, C.; Carbone, G.; Plitea, N. On the Singularities of a Parallel Robotic System Used for Elbow and Wrist Rehabilitation. In *International Symposium on Advances in Robot Kinematics*; Springer: Berlin/Heidelberg, Germany, 2019; Volume 8, pp. 203–211.
8. Husty, M.; Birlescu, I.; Tucan, P.; Vaida, C.; Pisla, D. An algebraic parameterization approach for parallel robots analysis. *Mech. Mach. Theory* **2019**, *140*, 245–257. [CrossRef]

9. Simaan, N.; Taylor, R.; Flint, P. A dexterous system for laryngeal surgery. In Proceedings of the 2004 IEEE International Conference on Robotics and Automation, New Orleans, LA, USA, 26 April–1 May 2004; pp. 351–357.

10. Marinho, M.M.; Bernardes, M.C.; Bo, A.P. Using General-Purpose Serial-Link Manipulators for Laparoscopic Surgery with Moving Remote Center of Motion. *J. Med Robot. Res.* **2016**, *1*, 1650007. [CrossRef]

11. Marinho, M.M.; Bernardes, M.C.; Bó, A.P.L. A programmable remote center-of-motion controller for minimally invasive surgery using the dual quaternion framework. In Proceedings of the 5th IEEE RAS/EMBS International Conference on Biomedical Robotics and Biomechatronics, Sao Paulo, Brazil, 12–15 August 2014; pp. 339–344.

12. Sandoval, J.; Su, H.; Vieyres, P.; Poisson, G.; Ferrigno, G.; De Momi, E. Collaborative framework for robot-assisted minimally invasive surgery using a 7-DoF anthropomorphic robot. *Robot. Auton. Syst.* **2019**, *106*, 95–106. [CrossRef]

13. Hamid, S.; Fatemeh, Z.; Shahram, H.J. Constrained Kinematic Control in Minimally Invasive Robotic Surgery Subject to Remote Center of Motion Constraint. *J. Intell. Robot. Syst.* **2019**, *95*, 901–913.

14. Wu, Y.; Carricato, M. Line-symmetric motion generators. *Mech. Mach. Theory* **2018**, *127*, 112–125. [CrossRef]

15. Selig, J.M. Geometric Fundamentals of Robotics. In *Monographs in Computer Science*; Springer: New York, NY, USA, 2005.

16. Denavit, J.; Hartenberg, R.S. A kinematic notation for lower-pair mechanisms based on matrices. *ASME J. Appl. Mech.* **1955**, *2*, 215–221.

17. Hitachi, H.A. Available online: http://www.hitachi-aloka.com/products/prosound-alpha-7/surgery/laparoscopic-surgery (accessed on 20 October 2019).

18. Birlescu, I.; Craciun, F.; Vaida, C.; Gherman, B.; Pisla, D. An innovative automated instrument for robotically assisted brachytherapy used in cancer treatment. *Acta Tech. Napoc.-Ser. Appl. Math. Mech. Eng.* **2017**, *60*, 633–638.

19. Gherman, B.; Al Hajjar, N.; Burz, A.; Birlescu, I.; Tucan, P.; Graur, F.; Pisla., D. Design of an innovative medical robotic instrument for minimally invasive treatment of liver tumors. *Acta Tech. Napoc.* **2019**, *62*, 557.

20. Cox, D.; Little, J.; O'Shea, D. *Ideals, Varieties and Algorithms*, 3rd ed.; Springer: Berlin/Heidelberg, Germany, 2007.

21. Dominic, R.W.; Husty, M. On implicitization of kinematic constraint equations. *Mach. Des. Res.* **2010**, *26*, 218–226.

22. Berceanu, C.; Tarnita, D.; Filip, D. About an experimental approach used to determine the kinematics of the human finger. *J. Solid State Phenom. Robot. Autom. Syst.* **2010**, *166–167*, 45–50. [CrossRef]

Article

Research of the Operator's Advisory System Based on Fuzzy Logic for Pelletizing Equipment

Darius Andriukaitis [1,*], Andrius Laucka [1], Algimantas Valinevicius [1], Mindaugas Zilys [1], Vytautas Markevicius [1], Dangirutis Navikas [1], Roman Sotner [2], Jiri Petrzela [2], Jan Jerabek [3], Norbert Herencsar [3] and Dardan Klimenta [4]

[1] Department of Electronics Engineering, Kaunas University of Technology, Studentu St.50-438, LT-51368 Kaunas, Lithuania; andrius.laucka@ktu.edu (A.L.); algimantas.valinevicius@ktu.lt (A.V.); mindaugas.zilys@ktu.lt (M.Z.); vytautas.markevicius@ktu.lt (V.M.); dangirutis.navikas@ktu.lt (D.N.)

[2] Department of Radio Electronics, SIX Research Center, Brno University of Technology, Technicka 3082/12, 616 00 Brno, Czech Republic; sotner@feec.vutbr.cz (R.S.); petrzelj@feec.vutbr.cz (J.P.)

[3] Department of Telecommunications, SIX Research Center, Brno University of Technology, Technicka 3082/12, 616 00 Brno, Czech Republic; jerabekj@feec.vutbr.cz (J.J.); herencsn@feec.vutbr.cz (N.H.)

[4] Faculty of Technical Sciences Kosovska Mitrovica, University of Pristina in Kosovska Mitrovica, Kneza Milosa St. 7, RS-38220 Kosovska Mitrovica, Serbia; dardan.klimenta@pr.ac.rs

* Correspondence: darius.andriukaitis@ktu.lt

Received: 18 October 2019; Accepted: 11 November 2019; Published: 12 November 2019

Abstract: Fertilizer manufacturing in the chemical industry is closely related with agricultural production. More than a half of raw materials for food products are grown by fertilizing plants. The demand of fertilizers has been constantly increasing along the growth of human population. Fertilizer manufacturers face millions of losses each year due to poor quality products. One of the most common reasons is wrong decisions in control of manufacturing processes. Operator's experience has the highest influence on this. This paper analyzes the pellet measurement data, collected at the fertilizer plant by using indirect measurements. The results of these measurements are used to construct the model of equipment status control, based on the fuzzy logic. The proposed solution allows to respond to changes in production parameters in a 7–10 times faster manner. On average, a manufacturer with production volumes of up to 80 tonnes/hour could have lost about 8400 tonnes/year of high-quality production. The publication seeks symmetry between human and system decision making.

Keywords: image processing; distribution; granulation; on-line monitoring; fertilizers

---

1. Introduction

Fertilizers used in agriculture, household, and other fields usually are bulk, spherical, fine particles. The process of chemical fertilizer particle formation can be either wet or dry. Depending on the chemical composition of the fertilizer, it may be pelletizing or prilling. One of the most common methods of pelletizing [1] is wet prilling. During the prilling process, molten fertilizers are converted into liquid drops. As they fall through the prilling tower (shot tower), they are cooled and form pellets. In comparison with a mixture of chemicals, a pelleted product has the following advantages [2]:

- More convenient storage, transportation;
- Rare adhesion of fertilizer particles;
- Even distribution of chemicals during fertilization; and
- Increased distance of fertilization.

Another important parameter is the pellet size. Smaller particles are characterized by faster disintegration in solution and result in higher fluid viscosity. Furthermore, small pellets (2–5 μm) or chemical mixture more easily disperse in the environment and penetrate into the lungs. This can lead to suffocation of animals fed with supplements. In addition, the pellet size influences the effectiveness of fertilization. Geometric properties of chemical particles and their mass distribution have significant influence on spreading and transportation. At higher bulk density, the fall trajectory of a pellet is also higher [3]. However, the pellet distribution interval must also be optimal, ideally with a mean distribution in the range of 0.8 mm (width). The sphericity and size of the pellets determine the distance the pellets are thrown by a fertilizer spreader. Meanwhile, the shape and size of the pellets are influenced by the compatibility of the chemicals [4].

Vibratory granulators are common in manufacturing of ammonium nitrate and urea fertilizers [5]. This technology is characterized by high capacity and high output quality-monodisperse particle-size distribution. During manufacturing, a vibratory granulator breaks down the chemical alloy into drops. The system of rotating or stationary perforated plates, designed for forced spraying of molten fertilizer mixture is used for splitting. When falling through the prilling tower, the particles collide with the opposite air flow and crystallize [6–10]. Perforated plates have resonators, the vibration frequency of which determines the particle size [11]. The resulting pellets are spherical in shape. This shape has the largest surface area per unit volume, which is important given the mass transfer limitations. Furthermore, the larger contact area facilitates the dissolution of fertilizer [12].

During prilling, residues of the chemical melt settle on the perforated plates. It should be noted that contaminated equipment results in reduced production volumes, increased fine fraction content, and qualitative parameters below the tolerance limits. The contaminated perforated cover dampens the vibrations of the resonators. As a result, the dispersion of the pellet diameter increases. Upon identifying the changes in products, equipment should be cleaned. In most cases, these cleanings are carried out periodically without taking into account the pellet parameters, which itself is not an optimal solution. The system does not have a feedback assessment subsystem, it is based on the operator experience.

The article analyzes the relationship between the pellet measurement results and the equipment contamination parameters. The results can be used in the complex system with feedback. Accurate assessment of the system status reduces the number of low-quality products. Section 2 (Materials and Methods) analyzes the pellet measurement and the system status monitoring methods. The current methods of measurement are discussed and the choice of the measurement method is justified. The third section presents the advisory model for the operator to make objective decisions related to the pellet production. The system helps the operator to react faster to production changes. This, in turn, allows a reduction in the amount of poor-quality production. The fourth section introduces the improved combined system control model. Advisory systems that allow inexperienced operators to quickly make the right decisions have not been analyzed in pellet production. The advisory system not only reduces the quality of output but also facilitates the cleaning process. Section 6 (Discussion) provides information on the improved control of the manufacturing line. Meanwhile, conclusions are given in the final section.

2. Materials and Methods

In most cases, especially when working with older equipment, the method used to identify the contamination of vibratory granulators in the chemical industry is based on the experience of operators. This means that the operator evaluates the level of contamination based on his competence. This subjective judgment is limited as it is not expressed in any numerical terms but simply determined by personal experience. In order to make a more accurate decision, the laboratory measurements are made using the sieve method. Interpretation of additional information reduces the risk of incorrect decision making, especially for less experienced operators. However, the frequency of these measurements is too rare, i.e., measurements are made every 2–3 h, and the production amounts to

50–80 tons of fertilizer per hour. This delay of measurements and the current feedback is too high; therefore, the operator is often required to make subjective decisions.

Control of the resonant frequency influences the dispersion of the pellet diameter distribution. Adjusting the frequency with partially soiled equipment can still result in quality output, thus, extending the manufacturing period between washes.

The quality assessment of products, according to the resonant frequency, may be used for control of pellet production parameters [13]. Efficiency of the prilling process is closely related with control of the vibration frequency of the resonator. Ultrasonic sensors are useful for identification of changes in resonant frequency. After performing a Fourier transform (FT) and filtering out the noise, the frequency suppression due to contaminated equipment is analysed (Figure 1). Upon detecting these changes at an early stage, the resonant frequency may be adjusted to extend the equipment life [14]. However, this methodology does not consider other parameters, e.g., temperature and humidity. Therefore, there is a lack of data related to the assessment of raw material and environmental parameters.

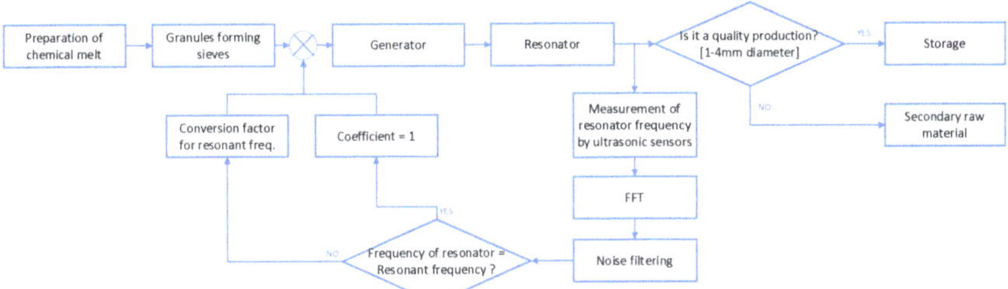

Figure 1. The algorithm of the estimation of equipment contamination.

In addition, pellet diameter measurement may be used as an alternative solution to assess the changes in production. It is possible to adjust the vibration frequency according to the change in the diameter. However, the usual laboratory measurements with the sieve technique are not frequent enough. Applied indirect pellet measurements are accelerated using computer hardware resource capabilities [15].

Dust and larger dispersions of the pellet diameter distribution lead to lower qualitative parameters. Image processing is used to analyse the size of the emerging drops adjacent to the perforated plate [16]. This method assesses the size of the pellets, according to which the operating frequency of the resonator is adjusted. The authors of the publication [16] claim to have reduced the dispersion of the particle size. At this point, however, the influence of the resonator on the formation of the final product, i.e., pellets, remains unclear. The analysis uses sophisticated and expensive equipment—a high-speed video camera. When measuring inside a high-capacity vibratory granulator, several video cameras are required. Furthermore, a special lens cleaning system is needed.

The quality of the pellets depends on their uniformity or on the elimination of fine fraction and dust from the products. The research papers tend to solve these issues by modelling the real conditions of manufacturing processes. Product viscosity, pellet velocity in the prilling tower and surface tension forces are evaluated. The latter determine the shape and brittleness of the particles [16]. These are complex processes, the analysis of which requires a detailed description of the solid structures. Research emphasizes the need of three-dimensional image processing techniques using sophisticated data assessment [17]. The surface roughness of the particle can be estimated by analysing the three-dimensional image of the granules [18].

Alternative methods of the pellet size and velocity analysis may be based on optical fibre spatial filtering. Scientific publications [19,20] report the results of the research, carried out by using Parsum Gmbh equipment. Measurements were made with particles in the range of 50–6000 μm. Real-time measurements can be helpful for development of the particle models. Meanwhile, the latter are used to control the particle size and to define the critical parameters.

Digital image processing [21] is used for the pellet size measurements. Along with indirect measurements of particles, the moisture content of the products is also measured. Measurements are made directly on the prilling tower. Video cameras and LED lights were installed on the opposite sides of the tower at different height levels. The same sample is analysed in the laboratory for evaluation of the results. The accuracy of the results, presented by the authors, depends on the speed of the manufacturing process. Particle measurement results agree with the laboratory results at a flow rate of 130 mL/min. Increasing the flow rate to 172 mL/min results in a shift of the pellet size distribution curve by ~0.03 mm to the larger side. Furthermore, the system is complex in structure. It consists of many measuring units that are integrated—embedded in production equipment. The research does not include modelling of the manufacturing process control.

Digital image processing techniques for measuring pellet products were described in the patent material long ago. With the rapid growth in computer performance, this area has been gaining popularity. It is essential to evaluate the quality of the image [22] as it affects the reliability of the metrological measurements. In summary of the literature review, several key features can be distinguished among the methods described:

- Analysis uses area scan cameras [23–27];
- Analysis uses line scan cameras [25,26,28–32];
- Continuously operating light [28,30] or strobe light [23,33] is used;
- Sample pellets are illuminated from front [33,34] or from behind (backlight) [26,31,35];
- Pellet monolayer [23,30] or the pellet structure on conveyor or other surfaces is analysed [36–41]; and
- Three-dimensional images [42,43] or complex data from hybrid (combining several measuring methods) systems are analysed [44].

Indirect measurement shortens the duration of analysis from a few hours to several minutes [45]. The duration of analysis also depends on the number of pellets in the sample. Measurements made with a computer vision system not only provide the information needed to monitor production parameters but can also be used for detailed analysis of number, shape, and volume of particles. Proper image binarization [46] is essential to evaluate the pellets with enough accuracy. Additional information on the production quality:

- Aspect ratio;
- Circularity;
- Convexity;
- Elongation;
- Sphericity;
- Surface roughness;
- Compactness;
- Symmetry;
- Length;
- Width;
- Perimeter; and
- Area.

3. System Model

The base model of the system follows the subjective opinion of the operator when assessing the level of equipment contamination. Thus, reliability of this decision is directly linked to the operator's competence. The data analysed in the paper were collected at the fertilizer plant. Measurements were made under the real production conditions. Special equipment was used for pellet measurements (more info about measurement was submitted in the previous paper [45]; the measuring system is shown in Figure 2).

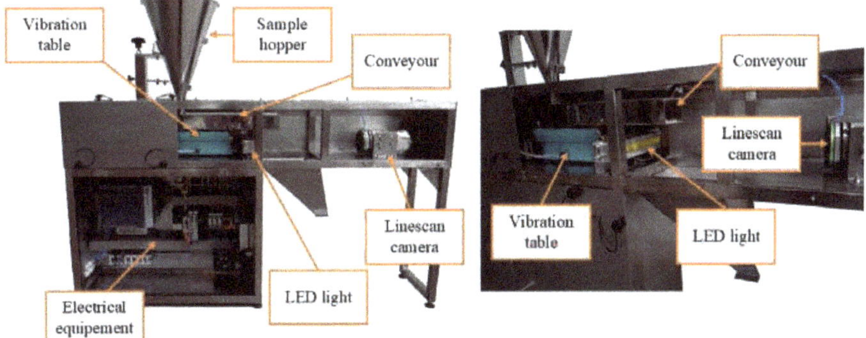

Figure 2. Pellet measuring equipment [45].

Ammonia gas and nitric acid, resulting in ammonium nitrate solution, are used for the production of ammonium nitrate. Thus, an ammonium nitrate solution of 83% is usually obtained, which is then used for evaporating to an ammonium nitrate content of 95–99.9%. In addition, various accessories are often needed. If, during the manufacturing process, the operator of the equipment observes a trendy change in the granulometric composition, it is assumed that the quantity of melt is too high in the system. This, in turn, shows that the perforated plates used for prilling are contaminated—there is no leakage of melt. The production process is then stopped for cleaning the equipment. The current control algorithm is presented in Figure 3.

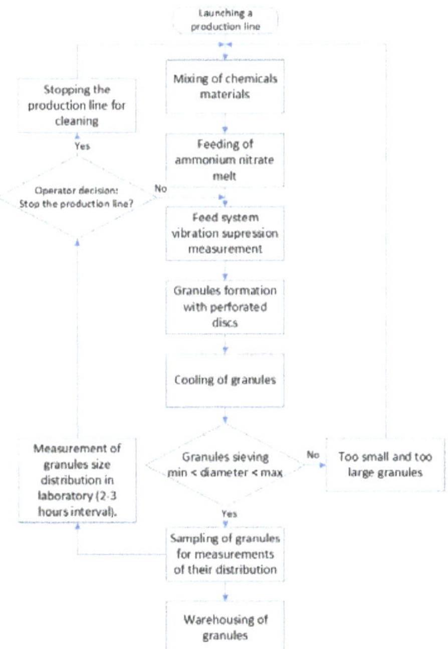

Figure 3. The existing operating algorithm.

One of the most important parameters in the prilling process is the size of the pellets. The fertilizer manufacturer defines the precise range within which the pellet size distribution is likely. The plant, where the research was carried out, used sieve sets to ensure a pellet size range of 1–4 mm.

The statistical parameters of granules size and vibrations of the melt feeding system was used to create the advisory system. It should be noted that pellet measurement calibration and mathematical models were used in the previous publication [45]. The model for indirect particle measurements was based on the evaluation of particle roundness. This paper evaluates the correlation between fertilizer granules measurement results and other indirect measurements (chemical feed system vibration). Based on the experience of the operators, a rule base for the fuzzy logic controller was created. Importantly, the final decision on contamination of the production line is made by the operator.

The paper presents an improved advisory system model. The solution involves indirect measurements of particle distribution. Moreover, the indirect digital image processing method, used for the particle analysis, increases the sampling frequency of measurements. Measurements were made by using specially designed software (one of the software screenshots is given in Figure 4 and a measurement photo is presented in Figure 5).

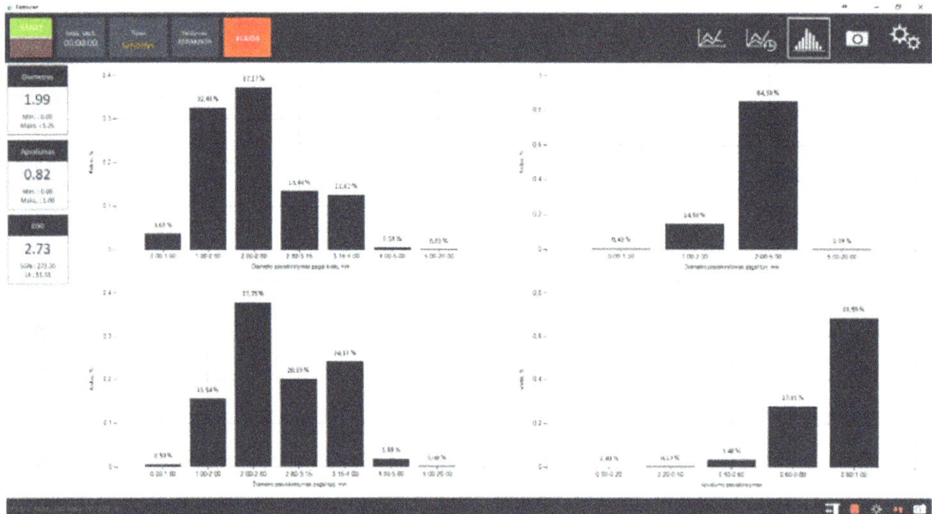

Figure 4. Screenshot of the indirect measurement software. The software was created in Elinta JSC (Kauno raj., Lithuania).

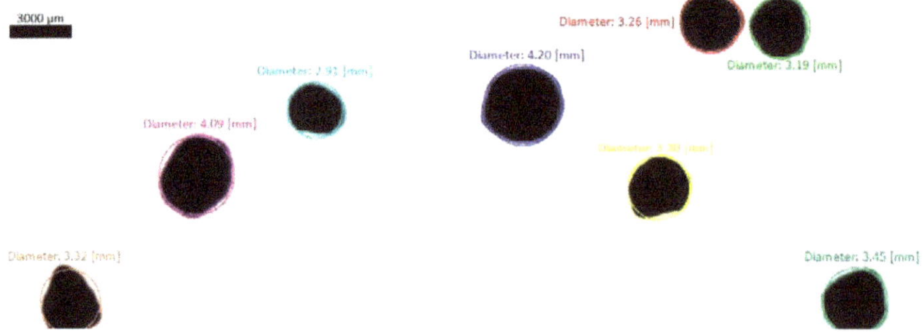

Figure 5. Shadows of granules' digital images were captured with a linear video camera [45].

The information obtained is analysed along with other input parameters and, then a decision about the system status is made. Ancillary information allows to justify the operator's actions. In addition, particle temperature and relative humidity are analysed. All measurements were made using non-interventional measurement methods: humidity was measured using NIR moisture sensor (Moistech IR-3000FP (manufacturer—Moistech, Sarasota, Florida, United States; supplier—Elintos MS, Kauno raj., Lithuania); moisture ranges: 0–0.1%, 0–100%), temperature, using PIR sensors (pyrometer integrated in the Moistech IR-3000FP moisture sensor (manufacturer—Moistech, Sarasota, Florida, United States; supplier—Elintos MS, Kauno raj., Lithuania); measuring temperature range: 0–100 °C with an accuracy of 0.1 °C). The block structure of the advanced auxiliary system assessment algorithm is presented in Figure 6.

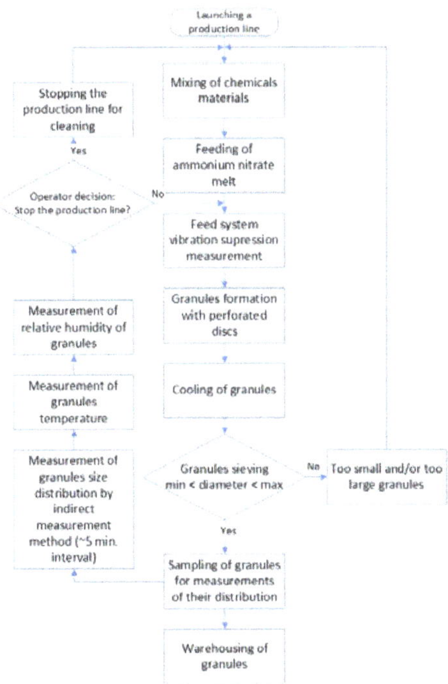

Figure 6. Algorithm of the system performance evaluation with additional measurements.

4. Evaluation of the Prilling Process

Upon analysing the data for a period of six months, the equipment washing moments were identified. The main reason for stopping the prilling process is equipment contamination without considering other technical failures. When the line is stopped, the equipment is washed. Washings can be divided into two groups: instant and long-term (preventive) ones. Instant equipment washings can be performed several times per shift. They are mainly related to the additives used in the chemicals and performed as required rather than periodically. This process cleans chemical mixing and feeding systems. Long-term washes are performed at longer intervals and last up to 4–8 h. The equipment is completely cleaned. Figure 7 presents the data of a shorter period with vibration damping (data from one of the sensors).

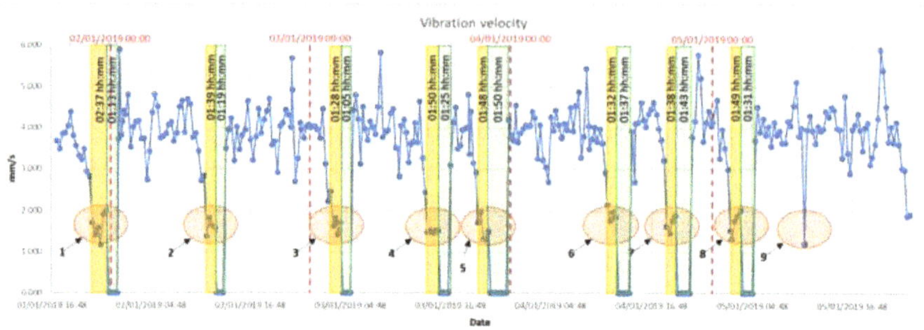

Figure 7. Vibration suppression. The orange ellipses denote the moments of equipment contamination. Yellow marking: operation of a production line with contaminated equipment (line life is marked). Green marking: line cleaning (marked line: cleaning time).

The graph shows the equipment contamination moments (numbered 1–9). This is reduced vibration frequency. In all these cases, instant washings were performed. Depending on the composition of the used raw materials, the number of washings varied from day to day. As the operator cannot accurately assess the system status and detect contamination at an early stage of the process, manufacturing lasted 1 h and 48 min longer on average. The average time of cleaning was determined of 1 h and 28 min.

During measurements, the pellet size was also analysed by indirect measurement (Figure 8). The graph shows the change in the number of pellets with a diameter in the range of 1–4 mm (see Section 3) relative to the total sample. These measurements reveal trends in the dispersion growth. This is a direct loss of profit for the manufacturer due to poor quality products.

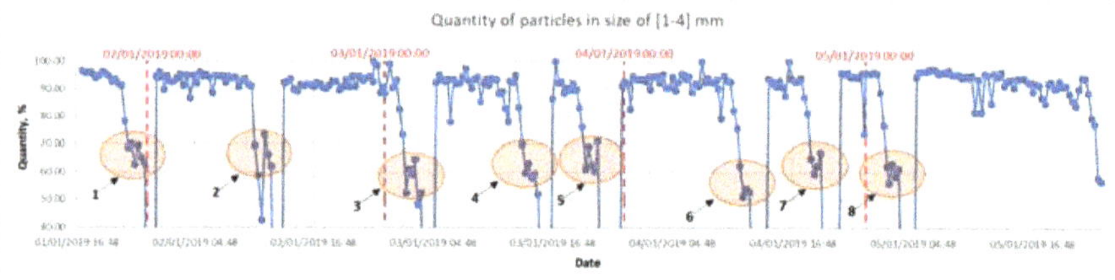

Figure 8. Particle distribution. The orange ellipses denote the moments of equipment contamination and changes in the granules' diameter.

Pellet temperature and relative humidity measurements were made at intermediate position of pellet transportation. Measurements revealed that the ambient temperature (in the range of 5–45 °C) and the relative humidity of the air (in the range of 10–90%) have no significant influence on the particle production process (Figure 9). Furthermore, the moisture content of the pellets must not exceed 0.30% of the relative humidity. The optimum value for the relative humidity of the pellets is 0.21%. However, relative humidity fluctuations within the range of 0.05% are possible at different times of the year—lower relative humidity in winter and higher in autumn. The measurements do not allow the assessment of the equipment contamination and qualitative parameters of particles. The measurement results show no relationship between the pellet size distribution and temperature and relative humidity.

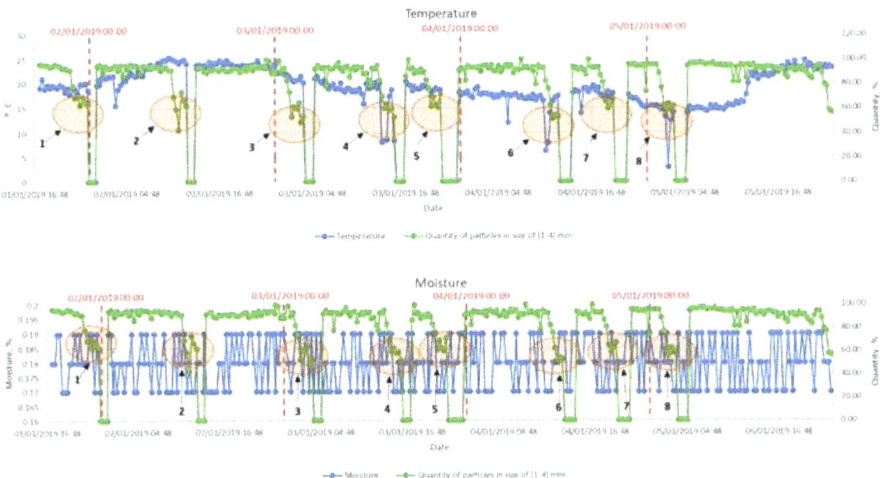

Figure 9. Temperature and moisture measurements. The orange ellipses denote the moments of equipment contamination and changes in the granules' diameter.

Every manufacturer attempts to avoid quality "pits". Thus, it is necessary to optimize the production process, i.e., to select the correct parameters for different situations. Early identification of equipment contamination and the extension of manufacturing time increases the production efficiency. The completed measurements of the pellet size and vibrations of the melt feeding system revealed a relationship between the quality of products and the equipment contamination (Figure 10). By using these measured parameters as feedback, the resonant frequency of the prilling equipment can be controlled, thus, increasing the time period between washings. In addition, it is possible to accurately determine the washing time of equipment based on the objective measurements.

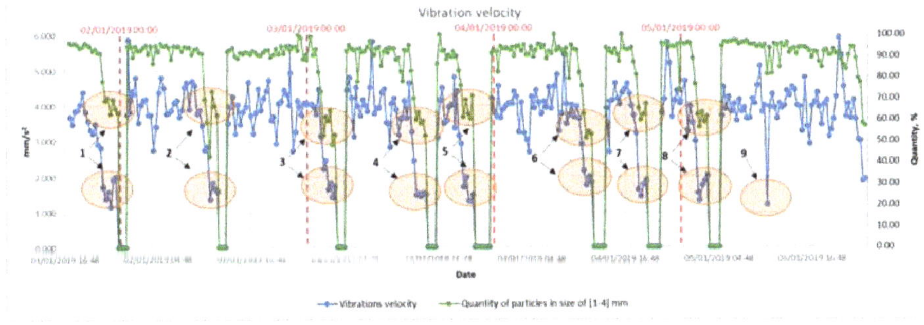

Figure 10. Vibration suppression and particle distribution. The orange ellipses denote the moments of equipment contamination and changes in the granules' diameter.

The operator can decide on the equipment contamination by taking into account not only the vibration damping, but also the changes in the pellet size. In this case, the operator's decision is based on additional information about the product. Following the results of different measurements, it is possible to avoid an incorrect assessment of the system status (ellipse 9 in Figure 10). However, the decision-making, based on the operator's competence, is still undefined. For elimination of human error, the relationship between the resonator vibration damping and the particle size distribution can be evaluated by implementing a complex feedback system. While implementing the works, the equipment control model was improved to help to identify the equipment contamination levels at an early stage of manufacturing and to inform the operator, in order to increase the time between equipment washes, i.e., to increase the system efficiency.

5. Combined Advisory System Model

Fertilizer prilling equipment is a sophisticated control system of chemical processes. Installation of auxiliary system must be considered as an advisory option to the operator. In this case, the person who oversees the process, may rely on additional comments about the status of the prilling process. The final decision is made by the operator, but with enough data on the quality of the products. The fertilizer prilling process is presented in Figure 11.

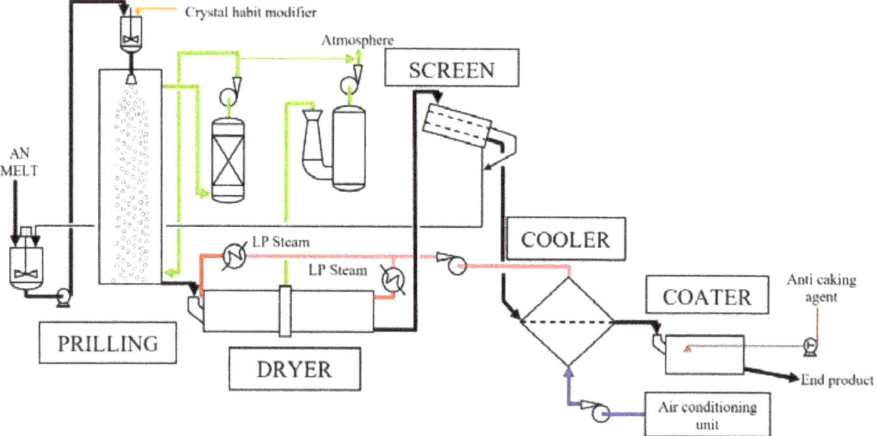

Figure 11. Technical grade ammonium nitrate process [47].

It can be argued that the results of frequent indirect particle measurements contribute to the overall assessment of the system status. Objective evaluation requires an estimate that defines the direction and magnitude of changes in system production. The measurement results for the different particle fractions are shown in Figure 12 (according to the sieve sets, used by the manufacturer).

Upon improving the system, the operator is warned about the expected degradation of the system. The final decision on equipment contamination is made by the operator. Preparation of chemical melt was not analysed because it is the same process for all manufacturing lines. Seeking to assess the process, particle measurement results were used along with vibration damping data. The results were evaluated by using the fuzzy logic module (Figure 13 shows the process control block). Fuzzy logic was chosen as one of the most popular decision-making models [48]. As a result, the estimate for perforated plate contamination was obtained. The fuzzy logic simulation results are given in Figure 13.

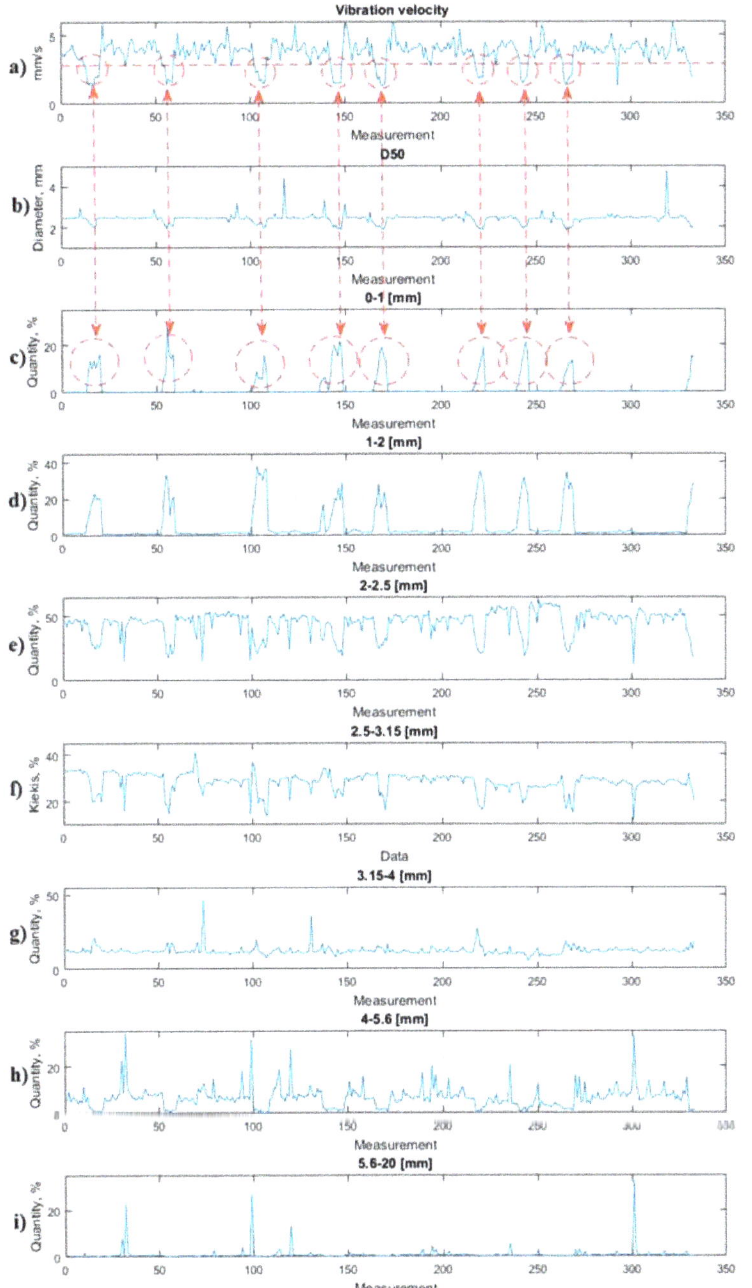

Figure 12. Measurement data of granules (by fractions). (**a**) Vibration suppression graph, (**b**) average particle size distribution graph, and (**c–i**) graphs of the distribution of granule content according to the size of the sieves used by the manufacturer. Stages of contamination of the equipment are marked in red.

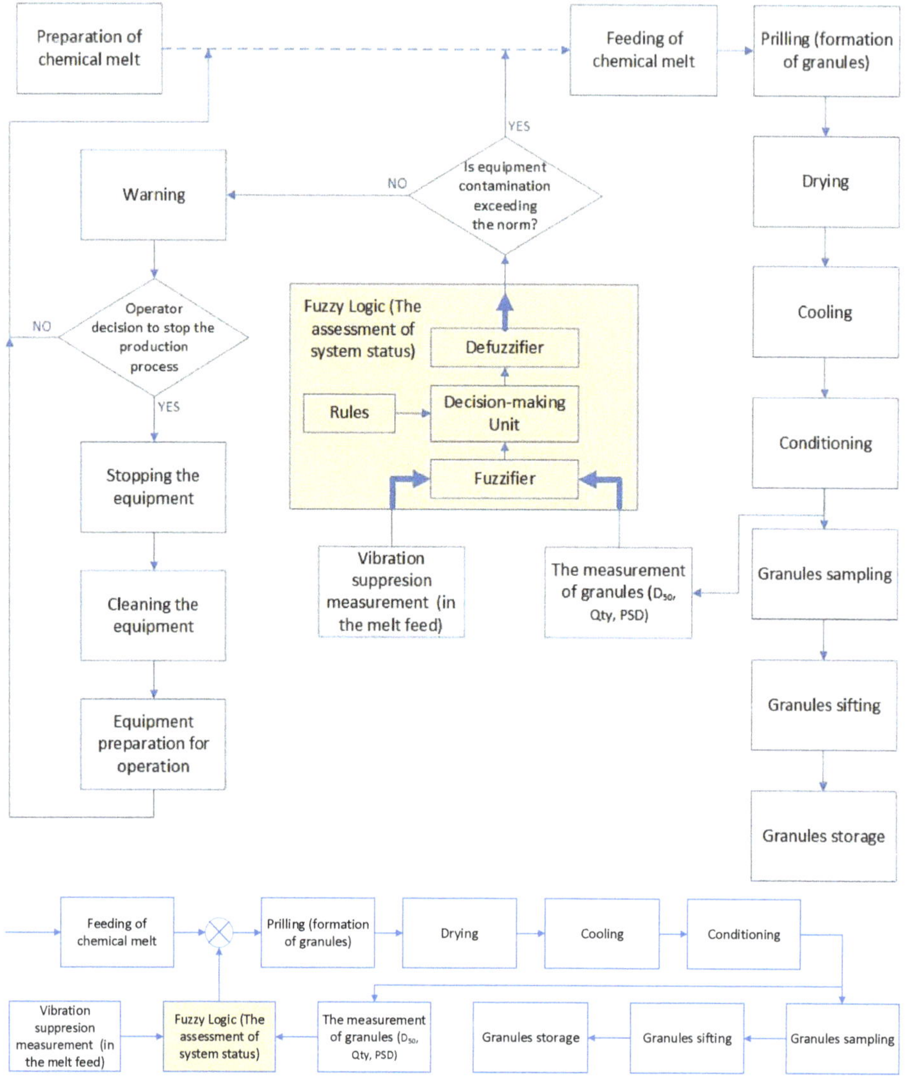

Figure 13. Block structure of advisory system with fuzzy logic.

Mean values of the pellet size, distribution of pellets according to the sieves used, damping of the sieve vibrations were presented for the fuzzy logic model:

1. [1–4] mm particles make > 97% of the total particles
2. Average particle size $D_{50} \approx 2.5$ mm
3. During the experiment, the sample analysed consists of 40,000–50,000 particles on average

The logic of the advisory model was described based on the experience and information gathered by the operators involved in the pellet manufacturing process. Fuzzy logic sets are described by S-shaped membership functions:

$$f(x;a,b) = \begin{cases} 0, x \leq a \\ 2\left(\frac{x-a}{b-a}\right)^2, a \leq x \leq \frac{a+b}{2} \\ 1 - 2\left(\frac{x-b}{b-a}\right)^2, \frac{a+b}{2} \leq x \leq b \\ 1, x \geq b \end{cases}$$

where a and b are allowable acquired values for the particle granulometric composition (Table 1).

Table 1. Acceptable limits for granulometry.

Sieve Size, mm	Minimum Value for Granules Content, %	Minimum Value for Granules Content, %
0.0–1.0	1.5	7.0
1.0–2.0	2.0	10.0
2.0–2.5	35.0	45.0
2.5–3.15	22.0	30.0
3.15–4.0	6.0	12.0
4.0–5.6	4.0	12.0
5.6–20.0	0.0	8.0

Upon applying the fuzzy logic model with the rules defined, the estimate—the coefficient, describing the system operation—is obtained. Its value remains ~1 when the equipment is operating in a stable manner. The coefficient value increases as the amount of chemical melt on the perforated plates increases. The graph of the system status estimate along with the measurement results are presented in Figure 14.

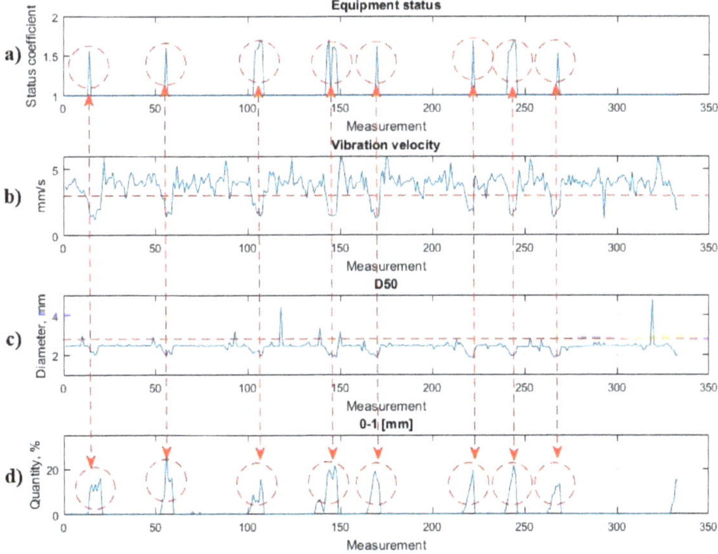

Figure 14. The evaluation of system status: (**a**) variation of the system status assessment factor, (**b**) vibration suppression graph, (**c**) average particle size distribution graph, and (**d**) graph of the distribution of granules content according to the size of the sieve: 0–1 (mm). Stages of the contamination of equipment are marked in red.

The obtained equipment contamination estimate describes the operation of the system. The graph in Figure 7 clearly shows the stops for washing the equipment (marked in yellow), which correspond to the washing times given in the fertilizer manufacturer's logs. Based on the collected information (some measurement results are given in the article), the manufacturer completed 56 stops per month to wash its equipment. On average, the equipment was stopped for cleaning after 1 h and 48 min. During this period, most of the products did not meet the quality requirements and were directed to the warehouse of second-class products. The manufacturer spent 105 h a month for making products with lower-quality parameters. With production volumes of up to 80 tonnes/h, the manufacturer could have lost about 8400 tonnes/year of high-quality production. The customized system status assessment model with fuzzy logic reduces the reaction time and equipment can be stopped for cleaning as early as after 10–15 min, after several indirect granulometric measurements. On average, losses are reduced to 745–1120 tonnes per month. By increasing the amount of measurement data, the model can be refined, thus, further reducing the losses.

6. Discussion

Errors are unavoidable in assessing the status of the manufacturing process if it is based on the human factor alone. There are no clearly defined criteria; meanwhile, laboratory granulometry is too rare. In the fertilizer industry, production amounts to approximately 50–80 tonnes per hour in one workshop. Therefore, each stop causes direct losses to the manufacturer. In addition, not all pellets that do not meet qualitative parameters are returned for recycling. These pellets are used for manufacturing of secondary products. Importantly, the results obtained reveal that the status of the prilling system can be objectively assessed. Indirect measurement, using digital image analysis, gives granulometric results at intervals of 5–8 min. Application of the fuzzy logic results in an estimate of the system status which can be used to control the operating frequency of the perforated plate resonators.

7. Conclusions

In this article, the advisory system model for the evaluation of the results was developed by analysing the pellet manufacturing information, collected at the fertilizer plant. The first model allowed the operator to make a more objective decision about the equipment contamination status. However, a person still had to rely on his competence while making this decision. The advanced advisory system model with fuzzy logic describes the status of the manufacturing process in a more precise manner. Furthermore, the model with clearly defined rules assesses the production line. Reaction time for stopping the line for cleaning decreased from ~1 h and 48 min to ~10–15 min. This results in a significant reduction in the volume of second-class production by approximately 8–10 times. On average, the manufacturer with production volumes of up to 80 tonnes/h, could have lost about 8400 tonnes/year of high-quality production. This amount of production can be matched by the manufacturer using an advisory system. At the same time, equipment is easier to clean because it is less contaminated. The advisory system was installed at the factory. Further studies will be carried out to assess the stability, accuracy and reliability of the system. In the future, the possibility to perform equipment contamination identification and production line stoppage in automatic mode will be evaluated.

Author Contributions: Conceptualization: D.A. and A.L.; methodology: D.A. and D.N.; software: A.L. and D.N.; validation: A.L., M.Z., V.M., and J.J.; formal analysis: R.S. and J.P., visualization: M.Z., N.H., and D.K.; investigation: A.V., V.M., J.J., and D.K.; resources: A.L. and A.V.; data curation: A.L.; writing—original draft preparation: D.A. and A.L.; writing—review and editing: V.M., J.P., N.H., D.K.; supervision: D.A.; funding acquisition: A.V., M.Z., and D.N.

Funding: This research received no external funding.

Conflicts of Interest: The authors declare no conflict of interest.

References

1. Rahmanian, N.; Naderi, S.; Supuk, E.; Abbas, R.K.; Hassanpour, A. Urea Finishing Process: Prilling Versus Granulation. *Procedia Eng.* **2015**, *102*, 174–181. [CrossRef]
2. Sahoo, C.K.; Rao, S.R.M.; Sudhakar, M.; Bhaskar, J. Advances in Granulation Technology. *Res. J. Pharm. Technol.* **2016**, *9*, 571–580. [CrossRef]
3. Fulton, J.; Port, K. Physical properties of granular fertilizers and impact on spreading. *Ohioline* **2016**. Available online: https://fabe.osu.edu/sites/fabe/files/imce/images/Precision_Ag/Physical%20Properties%20of%20Granular%20Fertilizers%20and%20Impact%20on%20Spreading%20_%20Ohioline.pdf (accessed on 17 August 2019).
4. Das, S.C.; Behara, S.R.B.; Morton, D.A.V.; Larson, I.; Stewart, P.J. Importance of particle size and shape on the tensile strength distribution and de-agglomeration of cohesive powders. *Powder Technol.* **2013**, *249*, 297–303. [CrossRef]
5. Wisniewski, R. Spray Drying Technology Review. In Proceedings of the 45th International Conference on Environmental Systems, Bellevue, DC, USA, 12–16 July 2015; pp. 1–46.
6. Gezerman, A.O.; Corbacioglu, B.D. A New Approach to Cooling and Prilling during Fertilizer Manufacture. *Int. J. Chem.* **2011**, *3*, 158–168. [CrossRef]
7. Mehrez, A.; Ali, A.H.H.; Zahra, W.K.; Ookawara, S.; Suzuki, M. Study on Heat and Mass Transfer During Urea Prilling Process. *Int. J. Chem. Eng. Appl.* **2012**, *3*, 347–353. [CrossRef]
8. Wu, Y.; Bao, C.; Zhou, Y. An Innovated Tower-fluidized Bed Prilling Process. *Chin. J. Chem. Eng.* **2007**, *15*, 424–428. [CrossRef]
9. Rahmanian, N.; Homayoonfard, M.; Alamdari, A. Simulation of urea prilling process: An industrial case study. *Chem. Eng. Commun.* **2013**, *200*, 1–19. [CrossRef]
10. Muhammad, A.; Pendyala, R.; Rahmanian, N. CFD Simulation of Droplet Formation Under Various Parameters in Prilling Process. *Appl. Mech. Mater.* **2014**, *625*, 394–397. [CrossRef]
11. Kjaergaard, O.G.; Vilstrup, P. Multiple-core encapsulation: Prilling. In *Microencapsulation of Food Ingredients*; Leatherhead Publishing: Surrey, UK, 2001; pp. 197–214.
12. Kozlov, V.V.; Grek, G.R.; Litvinenko, M.A.; Litvinenko, Y.A.; Kozlov, G.V. Round jet in a transverse shear flow. *Vestnik NSU Ser. Phys.* **2010**, *5*, 9–28. [CrossRef]
13. Ku, N.; Hare, C.; Ghadiri, M.; Murtagh, M.; Oram, P.; Haber, R.A. Auto-granulation of Fine Cohesive Powder by Mechanical Vibration. *Procedia Eng.* **2015**, *102*, 72–80. [CrossRef]
14. Halstensen, M.; Bakker, P.; Matveyev, I.; Esbensen, K.H. Acoustic chemometrics monitoring of chemical production processes. In Proceedings of the European Congress of Chemical Engineering (ECCE-6), Copenhagen, Denmark, 6–12 September 2007; pp. 16–20.
15. Atasoy, H.; Yildirim, E.; Yildirim, S.; Kutlu, Y. A Real-Time Parallel Image Processing Approach on Regular PCs with Multi-Core CPUs. *Res. J. Elektron. Elektrotechnika* **2017**, *23*, 64–71. [CrossRef]
16. Skydanenko, M.; Sklabinskyi, V.; Saleh, S.; Barghi, S. Reduction of Dust Emission by Monodisperse System Technology for Ammonium Nitrate Manufacturing. *Processes* **2017**, *5*, 37. [CrossRef]
17. Tsotsas, E. Multiscale Approaches to Processes That Combine Drying with Particle Formation. *Dry. Technol.* **2015**, *33*, 1859–1871. [CrossRef]
18. Fastowicz, J.; Grudziński, M.; Tecław, M.; Okarma, K. Objective 3D Printed Surface Quality Assessment Based on Entropy of Depth Maps. *Entropy* **2019**, *21*, 97. [CrossRef]
19. Dar, B.A.; Bhowmik, A.; Sharma, A.; Sharma, P.R.; Lazar, A.; Singh, A.P.; Sharma, M.; Singh, B. Ultrasound promoted efficient and green protocol for the expeditious synthesis of 1, 4 distributed 1, 2, 3-triazoles using Cu(II) doped clay as catalyst. *Appl. Clay Sci.* **2013**, *80–81*, 351–357. [CrossRef]
20. Petrak, D.; Dietrich, S.; Eckardt, G.; Köhler, M. In-line particle sizing for process control by an optical probe based on the spatial filtering technique (SFT). In Proceedings of the 6th World Congress Particle Technology, Nuremberg, Germany, 26–29 April 2011; Volume 22, pp. 203–208. [CrossRef]
21. Jaskulski, M.; Fischer, C.; Sajontz, F.; Friese, S. Particle Size Distribution and Moisture Content Inline Measurement Systems for the Spray Dryers. In Proceedings of the 22nd Polisch Conference of Chemical and Process Engineering, Spała, Poland, 5–9 September 2016; pp. 499–509.
22. Okarma, K. Current Trends and Advances in Image Quality Assesment. *Res. J. Elektron. Elektrotechnika* **2019**, *25*, 77–84. [CrossRef]

23. Sakamoto, Y.; Tamura, Y.; Kawaguchi, K.; Inada, K. Measuring Particle Size Distribution. U.S. Patent 4,288,162, 27 February 1979.

24. Tokoyama, K. Powder and Granule Inspection Apparatus. U.S. Patent 5,309,773, 10 May 1994.

25. Plate, M.; Pankratz, J. Apparatus for Determining the Particle Size Distribution of a Mixture. U.S. Patent 6,061,130, 22 January 1998.

26. Browne, I.B.; Lieber, K.J.W.; Young, L. Method and Apparatus for Sizing Particulate Material. WO Patent 1997014950A1, 16 October 1995.

27. Shanthi, C.; Porpatham, R.K.; Pappa, N. Image analysis for particle size distribution. *Eng. Technol.* **2014**, *6*, 1340–1345.

28. Jorgensen, T.; Strand, O.E.; Asbjornsen, O.A. Method and Apparatus for Performing Automatic Particle Analysis. U.S. Patent 5,011,285, 30 April 1987.

29. Schumann, M. Procedure for the Determination of Particle Size Distribution in Particle Mixtures. U.S. Patent 5,309,215, 3 May 1994.

30. Penumadu, D.; Zhao, R.; Steadman, E.F. Particle Size and Shape Distribution Analyzer. U.S. Patent 6,960,756, 1 November 2001.

31. Hodenberg, M.F.V. Device for Determining Parameters of a Bulk Material Particle Flow. WO Patent 2008046914A1, 20 October 2006.

32. Lech, P.; Okarma, K. Prediction of the Optical Character Recognition Accuracy based on the Combined Assessment of Image Binarization Results. *Res. J. Elektron. Elektrotechnika* **2015**, *21*, 62–65. [CrossRef]

33. Jorgensen, T.; Reinholt, F.; Johnsen, O.M. Automatic Particle Analyzing System. U.S. Patent 7,154,600, 7 May 2001.

34. Koh, T.K.; Miles, N.J.; Morgan, S.P.; Hayes-Gill, B.R. Improving particle size measurement using multi-flash imaging. *Miner. Eng.* **2009**, *22*, 537–543. [CrossRef]

35. Kosaka, T. Apparatus and Method for Analyzing Particle Images Including Measuring at a Plurality of Capturing Magnifications. U.S. Patent 5,721,433, 24 February 1998.

36. Uesugi, M.; Harayama, M.; Ota, K.; Kawaguchi, S.; Shibuya, H. Method of Measuring Average Particle Size of Granular Material. U.S. Patent 5,129,268, 5 April 1989.

37. Althyabat, S.; Miles, N.J. An improved estimation of size distribution from particle profile measurements. *Powder Technol.* **2006**, *166*, 152–160. [CrossRef]

38. Hamzeloo, E.; Massinaei, M.; Mehrshad, N. Estimation of particle size distribution on an industrial conveyour belt using image analysis and neural networks. *Powder Technol.* **2014**, *261*, 185–190. [CrossRef]

39. Castro, P.; Vicens, R. Grain-size measurement of fluvial gravel bars using Object-Based image analysis. *Rev. Bras. Geomorfol.* **2018**, *19*, 60–73. [CrossRef]

40. Dipova, N. Determining the grain size distribution of granular soils using image analysis. *Acta Geotech. Slov.* **2017**, *14*, 29–37.

41. Cardona, J.; Ferreira, C.; McGinty, J.; Hamilton, A.W. Image analysis framework with focus evaluation for in situ characterisation of particle size and shape attributes. *Chem. Eng. Sci.* **2018**, *191*, 208–231. [CrossRef]

42. Canty, T.M.; O'Brien, P.J.; Marks, C.P.; Owen, R.E. Granular Product Inspection Device. U.S. Patent 7,009,703, 27 March 2003.

43. Labati, R.D.; Genovese, A.; Ballester, E.M.; Piuri, V. 3D granulometry using image processing. *IEEE Trans. Ind. Inform.* **2018**, *15*, 1251–1264. [CrossRef]

44. Cepuritis, R.; Garboczi, E.J.; Ferraris, C.F.; Jacobsen, S. Measurement of particle size distribution and specific surface area for crushed concrete aggregate fines. *Adv. Powder Technol.* **2017**, *28*, 706–720. [CrossRef]

45. Laucka, A.; Adaskeviciute, V.; Andriukaitis, D. Research of the Equipment Self-Calibration Methods for Different Shape Fertilizers Particles Distribution by Size Using Image Processing Measurement Method. *Symmetry* **2019**, *11*, 838. [CrossRef]

46. Laucka, A.; Adaskeviciute, V.; Andriukaitis, D.; Valinevicius, A. Research of the Equipment Calibration Methods for Fertilizers Particles Distribution by Size Using Image Processing Measurement Method. In Proceedings of the 23rd International Conference on Methods & Models in Automation & Robotics (MMAR), Międzyzdroje, Poland, 27–30 August 2018; pp. 407–412. [CrossRef]

47. Ceamag. Technical Grade Ammonium Nitrate. Available online: https://www.ceamag.com/en/technologies/ammonium-nitrate?id=56 (accessed on 12 November 2019).

48. Zavadskas, E.K.; Bausys, R.; Antucheviciene, J. Civil Engineering and Symmetry. *Symmetry* **2019**, *11*, 501. [CrossRef]

 *symmetry*

Article

The Use of Structural Symmetries of a U12 Engine in the Vibration Analysis of a Transmission

Mircea Mihălcică [1],*, Sorin Vlase [1] and Marius Păun [2]

[1] Department of Mechanical Engineering, Transilvania University of Brașov, Brașov 500036, România; svlase@unitbv.ro

[2] Department of Mathematics and Computer Science, Transilvania University of Brașov, Brașov 500036, România; m.paun@unitbv.ro

* Correspondence: mihalcica.mircea@unitbv.ro; Tel.: +40-734-311-885

Received: 2 September 2019; Accepted: 11 October 2019; Published: 15 October 2019

Abstract: The paper focuses on the vibration analysis of a vehicle equipped with two identical engines. Such solutions are encountered in practice when less power is needed for a vehicle for a certain period of time and then greater power the rest of the time. An example of this would be a mobile drilling rig. During transport (a relatively short period of time) only one engine operates and then, in service (most of the operating time), both engines operate. A characteristic of such an aggregate is the existence, within the transmission, of two identical engines. The existence of identical parts in mechanical systems leads to properties that allow the computations to be simplified in order to obtain suggestive and rapid results, with reduced computation effort. These properties refer to the eigenvalues and eigenmodes of vibration for these types of systems and have been stated and demonstrated in the paper. It also allows for a qualitative analysis of the behavior of the system in case of vibrations. The existence of these properties allows for easier calculation and shortening of the design time. The mechanical consequences of the existence of symmetries or identical parts have begun to be studied in more detail in the last decade (see references), and the work is part of these trends. The vibration properties of a transmission of a truck with two identical engines have been stated and proven and a real example is analyzed. Two 215 hp engines were used in the application. In order to establish a useful solution in practice, two constructive variants with a different clutch position in the transmission are analyzed in parallel.

Keywords: vibrations; symmetries; two-engine vehicle; rear clutch; front clutch; vibration modes

---

1. Introduction

The structural symmetries and the properties that they give to the structures have been observed by the researchers and used especially in the static case; they are presented in the classical courses of strength of materials or structural analysis. An analysis of the different types of symmetries in applied mechanics is made in [1]. Symmetries are encountered in all aspects of human life, and engineering applications are no exception in using the properties and benefits that these symmetries bring.

Symmetries in mechanics have been studied mainly from the point of view of mathematicians [2–4] as they have effects in writing equations of motion, but with fewer applications in practice. A presentation of the application of symmetries in continuous mechanics is made in [5,6]. In January 2018, a special issue of the Symmetry magazine dedicated to applications in structural mechanics was launched (Civil Engineering and Symmetry—2018, a special issue of Symmetry—ISSN 2073-8994, see ref. [7]). A European project was also funded to study this type of problem (mechanics and symmetry in Europe: The geometry and dynamics of deformable systems. Project. HPRN-CT-2000-00113, funded under: FP5-HUMAN POTENTIAL, see ref. [8]) and courses were held at the Center for Solid Mechanics—CISM from Udine (similarity, symmetry and group theoretical methods in mechanics,

September 7, 2015–September 11, 2015. Lectures at the International Center for Mechanical Sciences, see ref. [9]).

Lately, numerous works have been published that aim to use symmetries in order to obtain properties that allow for a simplified analysis of the models used, in different fields [10–13]. Different mechanical systems with symmetries and properties induced by these symmetries are also analyzed [14,15]. In the vibrations research field, the effect of symmetries was less used but there are works that have begun to study this type of problem, sometimes with implications in other works, which focus on different other aspects [16–23].

However, there are many situations that can be studied and, therefore, the paper aims to complete some of the studied cases by offering some ideas for the application of these properties that could help a design engineer.

In the spirit of these works, some vibration properties of the considered symmetrical mechanical system (in our case the transmission of a truck) are stated in the paper, then they are proved and are applied in a real, concrete example of a truck equipped with two 215 HP identical engines. Numerical results are obtained, with real values of the parameters, which confirm the theoretical results obtained and, thus, shorten the time to calculate eigenfrequencies and eigenmodes of vibrations for these kinds of systems.

2. Materials and Methods

If we have already verified and found convenient engines, a solution used in practice in order to obtain greater power is to use two identical engines within a machine. It is possible that, for a period of time, only the power of one engine is required and, after a while, the power of both engines is needed. For example, in the case of the above mentioned mobile drilling plant, only the power of an engine is required for the movement of the vehicle, while for the actual drilling operations, the power of both engines to be required [1] (Figure 1).

Figure 1. Mobile oil drilling plant type TW125 CAA6 mounted on a chassis ROMAN 75,540 MFEG (12 × 8) [24].

A mobile drilling plant is a special commercial vehicle, designed and manufactured so as to ensure the dynamics and durability/reliability imposed by the conditions under which the drilling is performed, conditions that involve both the operation and moving in harsh conditions (unpaved roads). Another practical application is the use of two engines in certain special vehicles just to obtain higher power. Such a vehicle can be equipped with a power outlet that could be connected to a machine and can only be operated by a single engine. Such a system has been studied in our work, but the results obtained are valid and can be applied to any vehicle equipped with two identical engines from which the power is collected and summed using a combiner transmission box. The demands of the active and passive components of these types of vehicles are large and complex, including high static demands and dynamic demands through vibrations and shocks. The dynamics that such vehicles must provide and the mechanical strength (stresses and deformations) that must exist make these vehicles interesting for research, especially considering the demands and environmental conditions of exploitation [1].

The main components of the powertrain are the drive system, consisting of two diesel engines supercharged with electronic injection and the add-on box/add-on and distribution box, with the role of summing the torques transmitted by the two engines. If there is no need for both engines to produce power, only the power transmitted by an engine can be used. The gearboxes in the add-on gearbox bring the advantages of high transmission efficiency, the possibility of transmitting extremely high-value torque, the stable operation and high durability/reliability indices.

The kinematic transmission diagram of the truck and the combiner transmission box connected to the two engines arerepresented in Figures 1–3.

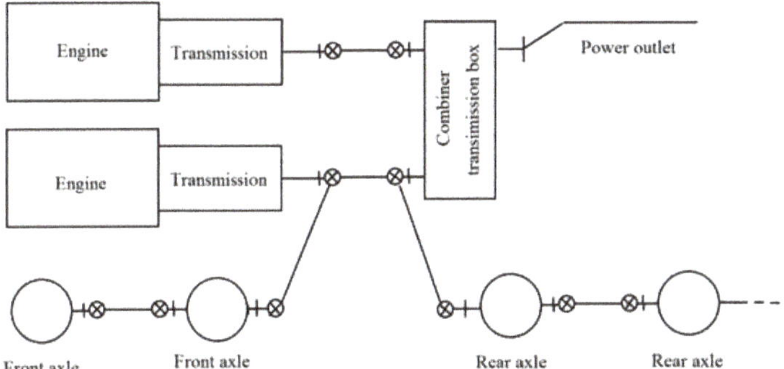

Figure 2. The kinematic transmission diagram.

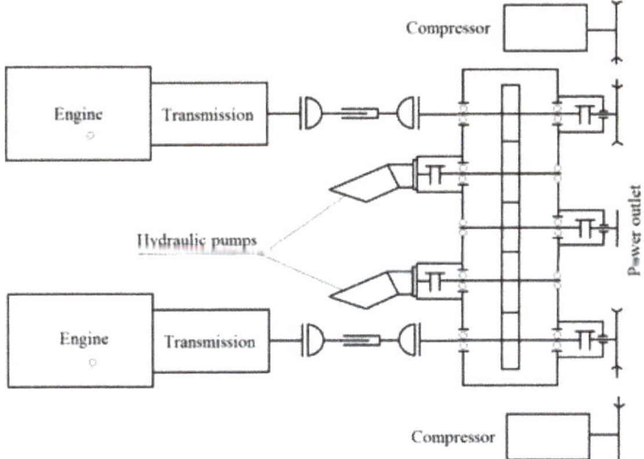

Figure 3. The combiner transmission box.

Let us consider a vehicle equipped with two identical engines. Two constructive solutions have been proposed by the designer: one solution with the clutch located behind the engines and the other solution with the clutch located in front of the engines. The constructive schemes of the two solutions are presented in Figures 4 and 5.

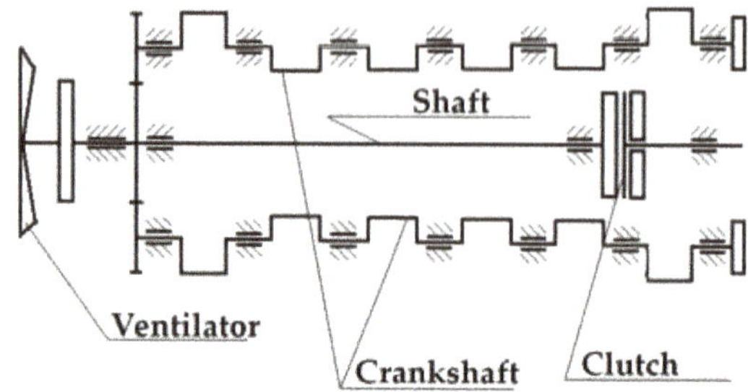

Figure 4. The variant with the clutch located behind the engines.

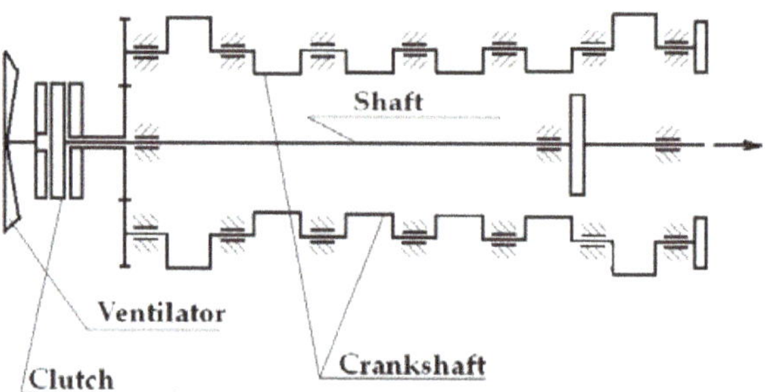

Figure 5. The variant with the clutch located in front of the engines.

The model for the study of vibrations of the system in the two situations is shown in Figure 6. (the clutch is considered coupled). The model with wheels (for one single engine) is presented in Figure 7. The fact that the two engines are identical leads to the highlighting of some properties of the equations of motion that describe the free vibration of the system that can allow for the necessary calculations to be made easier and to obtain qualitative conclusions regarding the vibrating system.

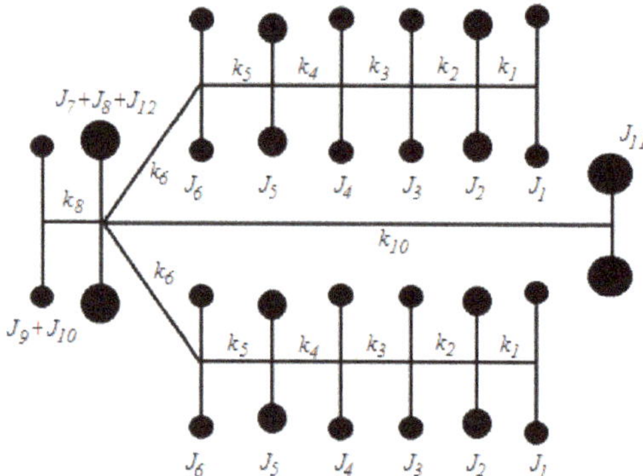

Figure 6. The two-engined vehicle, the front clutch.

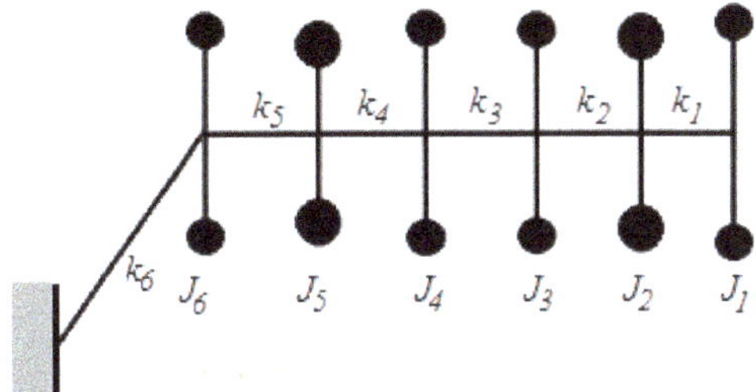

Figure 7. The mechanical model of a single engine.

The values considered in the calculus for this paper can be found in Tables 1–3.

Table 1. The moments of inertia.

No.	Moment of Inertia	Details
1	J_1	Cylinder 1
2	J_2	Cylinder 2
3	J_3	Cylinder 3
4	J_4	Cylinder 4
5	J_5	Cylinder 5
6	J_6	Cylinder 6
7	J_7, J_{12}	Gears
8	J_8	Central Gear
9	J_9	Flywheel
10	J_{10}	Ventilator
11	J_{11}	Exit steering wheel

Table 2. The moment of inertia of the wheels in the two studied cases.

No.	Rear Clutch Model		Front Clutch Model	
	Moment of Inertia	Values (kg*m^2)	Moment of Inertia	Values (kg*m^2)
1	J_1	0.1048	J_1	0.1048
2	J_2	0.0638	J_2	0.0638
3	J_3	0.1048	J_3	0.1048
4	J_4	0.1048	J_4	0.1048
5	J_5	0.0638	J_5	0.0638
6	J_6	0.1048	J_6	0.1048
7	$J_7+J_8+J_{12}$	1.81182	$J_7+J_8+J_{12}$	1.4157
8	J_9	3.41895	J_9	2.9841
9	J_{11}	3.70752	J_{11}	1.3382

Table 3. Stiffness of the shafts between two neighboring steering wheels.

Between	Rear Clutch Model		Front Clutch Model	
	Stiffness	Values (Nm/rad)	Stiffness	Values (Nm/rad)
1–2	k_1	2.56×10^6	k_1	2.56×10^6
2–3	k_2	2.56×10^6	k_2	2.56×10^6
3–4	k_3	2.53×10^6	k_3	2.53×10^6
4–5	k_4	2.56×10^6	k_4	2.56×10^6
5–6	k_5	2.56×10^6	k_5	2.56×10^6
6–7	k_6	20.87×10^6	k_6	20.87×10^6
7–8	k_7	12.67×10^6	k_7	4.683×10^6
7–9	k_8	0.045961×10^6	k_8	0.030158×10^6

3. Results

If we consider:

$$[J_e] = \begin{bmatrix} J_1 & & & & & \\ & J_2 & & & 0 & \\ & & J_3 & & & \\ & & & J_4 & & \\ & 0 & & & J_5 & \\ & & & & & J_6 \end{bmatrix}; \quad [J_r] = \begin{bmatrix} J_7 + J_8 + J_{12} & 0 & 0 \\ 0 & J_9 + J_{10} & 0 \\ 0 & 0 & J_{11} \end{bmatrix}$$

$$[K_e] = \begin{bmatrix} k_1 & -k_1 & & & & \\ -k_1 & k_1 + k_2 & -k_2 & & 0 & \\ & -k_2 & k_2 + k_3 & -k_3 & & \\ & & -k_3 & k_3 + k_4 & -k_4 & \\ & 0 & & -k_4 & k_4 + k_5 & -k_5 \\ & & & & -k_5 & k_5 + k_6 \end{bmatrix} \tag{1}$$

$$[K_c] = \begin{bmatrix} 0 & 0 & 0 \\ 0 & 0 & 0 \\ 0 & 0 & 0 \\ 0 & 0 & 0 \\ 0 & 0 & 0 \\ -k_6 & 0 & 0 \end{bmatrix}; \quad [K_r] = \begin{bmatrix} 2k_6 + k_7 + k_8 & -k_7 & -k_8 \\ -k_7 & k_7 & 0 \\ -k_8 & 0 & k_8 \end{bmatrix}$$

$$[J] = \begin{bmatrix} [J_e] & O_{6x6} & O_{6x3} \\ O_{6x6} & [J_e] & O_{6x3} \\ O_{3x6} & O_{3x6} & [J_r] \end{bmatrix}; \quad [K] = \begin{bmatrix} [K_e] & O_{6x6} & [K_c] \\ O_{6x6} & [K_e] & [K_c] \\ [K_c]^T & [K_c]^T & [K_r] \end{bmatrix}; \quad \{\theta\} = \begin{Bmatrix} \{\theta_s\} \\ \{\theta_d\} \\ \{\theta_r\} \end{Bmatrix}; \quad \{\ddot{\theta}\} = \begin{Bmatrix} \{\ddot{\theta}_s\} \\ \{\ddot{\theta}_d\} \\ \{\ddot{\theta}_r\} \end{Bmatrix}.$$

In (1) $\{\theta_s\}$ represents the vector of the rotations of the flywheels of the first engine, $\{\theta_d\}$ represents the vector of the rotations of the flywheels of the second engine and $\{\theta_r\}$ the vector of the other rotations of the system.

The motion equations for the free non-damped vibrations for the whole structure can be obtained:

$$\begin{bmatrix} [J_e] & O_{6x6} & O_{6x3} \\ O_{6x6} & [J_e] & O_{6x3} \\ O_{3x6} & O_{3x6} & [J_r] \end{bmatrix} \begin{Bmatrix} \{\ddot{\theta}_s\} \\ \{\ddot{\theta}_d\} \\ \{\ddot{\theta}_r\} \end{Bmatrix} + \begin{bmatrix} [K_e] & O_{6x6} & [K_c] \\ O_{6x6} & [K_e] & [K_c] \\ [K_c]^T & [K_c]^T & [K_r] \end{bmatrix} \begin{Bmatrix} \{\theta_s\} \\ \{\theta_d\} \\ \{\theta_r\} \end{Bmatrix} = 0. \tag{2}$$

For one single engine, the equations are:

$$[J_e]\{\ddot{\theta}_e\} + [K_e]\{\theta_e\} = 0. \tag{3}$$

Using the results presented in [9–11] the following property has been established:

3.1. Theorem T1. The Eigenvalues for the System (3) are Eigenvalues for the System (2) As Well

That means that the solutions of algebraic equations:

$$\det\left([K_e] - \omega^2 [J_e]\right) = 0 \quad \left(or \; \left|K_e - \omega^2 j_e\right| = 0\right), \tag{4}$$

are also solutions of the algebraic equation:

$$\det\left(\begin{bmatrix} K_e & 0 & K_c \\ 0 & K_e & K_c \\ K_c^T & K_c^T & K_r \end{bmatrix} - \omega^2 \begin{bmatrix} J_e & 0 & 0 \\ 0 & J_e & 0 \\ 0 & 0 & J_r \end{bmatrix}\right) = 0, \tag{5}$$

or:

$$\begin{vmatrix} K_e - \omega^2 J_e & 0 & K_c \\ 0 & K_e - \omega^2 J_e & K_c \\ K_c^T & K_c^T & K_r - \omega^2 J_r \end{vmatrix} = 0. \tag{6}$$

We could write:

$$\left| K_e - \omega^2 J_e \right| = 0 \quad \Rightarrow \quad \begin{vmatrix} K_e - \omega^2 J_e & 0 & K_c \\ 0 & K_e - \omega^2 J_e & K_c \\ K_c^T & K_c^T & K_r - \omega^2 J_r \end{vmatrix} = 0, \tag{7}$$

which implies that the polynomial in ω^2 from Equation (6) is divided by the polynomial in ω^2 from Equation (4).

We consider:

$$[P(\omega^2)] = [K] - \omega^2[J], \quad [P_e(\omega^2)] = [K_e] - \omega^2[J_e], \quad [P_r(\omega^2)] = [K_r] - \omega^2[J_r], \quad \Delta = \det[P(\omega^2)], \Delta_e = \det[P_e(\omega^2)], \Delta_r = [P$$

For the vibrating system (as a whole) the free undamped torsional vibrations will be the system of equations (2), which can be written in compact form:

$$[J]\{\ddot{\varphi}\} + [K]\{\varphi\} = 0, \tag{8}$$

where $\{\varphi\}$ represents the vector of the rotations of the flywheels, $[J]$ being the matrix of inertia and $[K]$ being the stiffness matrix. The characteristic equation for the presented system (7) can be written:

$$\Delta = P(\omega^2) = \det([K] - \omega^2[J]). \tag{9}$$

Theorem T1 will be now proven.

Proof. The characteristic polynomial for the single-engine system is:

$$\Delta_e = P_e(\omega^2) = \det([K_e] - \omega^2[J_e]).$$

If we consider $\Delta_e = 0$, we obtain the natural frequencies for a single engine, taken separately. The characteristic polynomial for the whole system is:

$$\Delta = \det([K] - \omega^2[J]) = \det\begin{bmatrix} [K_e] - \omega^2[J_e] & 0 & [K_c] \\ 0 & [K_e] - \omega^2[J_e] & [K_c] \\ [K_c]^T & [K_c]^T & [K_r] - \omega^2[J_r] \end{bmatrix} =$$
$$= \det\begin{bmatrix} [P_e(\omega^2)] & 0 & [K_c] \\ 0 & [P_e(\omega^2)] & [K_c] \\ ([K_c]^I & [K_c]^T & [P_r(\omega^2)] \end{bmatrix} \tag{10}$$

By direct calculus, applying Laplace's rule for determinants, we obtain in this case:

$$\Delta = \Delta_e P * (\omega^2). \tag{11}$$

If $\Delta_e = 0$, then we immediately have $\Delta = 0$, so the natural frequencies for the system consisting of a single engine are also natural frequencies for the whole system. $\square$

We will consider $_1\omega_i^2$, $\quad i = \overline{1, n_1}$ being the natural frequencies of a single engine and ω_i^2, $\quad i = \overline{1, 2n_1 + n_2}$ being the natural frequencies of the whole system. In our situation, $n_1 = 6$, $n_2 = 3$.

This property is valid in a more general context. This fact will be demonstrated as follows using [9].

Preliminary considerations:

Let us consider $M = \left(m_{ij}\right)_{i,j\in\{1,2,\dots,n\}}$, $U = \left(m_{ij}\right)_{i\in\{i1,i2,\dots,ik\}\subset\{1,2,\dots,n\},\ j\in\{j1,j2,\dots,jk\}\subset\{1,2,\dots,n\}}$ a sub-matrix of M and $\alpha = \det(U)$. For the chosen matrix U we consider the complementary matrix $\overline{U} = \left(m_{ij}\right)_{i\in\{1,2,\dots,n\}\setminus\{i1,\dots,ik\},\ j\in\{1,2,\dots,n\}\setminus\{j1,\dots,jk\}}$. We call an algebraic complement of the minor α (the cofactor of α) the determinant with the sign $\overline{\alpha} = (-1)^{i1+\dots+ik+j1+\dots+jk}\det(\overline{U})$.

Considering the lines $i1,\dots,ik$ being fixed we have:

$$\det(M) = \sum_{1\le j1<\dots<jk\le n} \alpha\overline{\alpha}. \tag{12}$$

This formula generalizes the Laplace expansion formula according to a line of the matrix M determinant.

$\overline{U}$ is a matrix $(n-k) \times (n-k)$ and considering V its square sub-matrix of indices $\{m1,\dots,mp\}$, $\{l1,\dots,lp\}$ of determinant β and correspondingly matrix $\overline{V}$ we may write:

$$\det(\overline{U}) = \sum_{1\le l1;\dots;lp\le n-k} \beta\cdot(-1)^{m1+\dots+mp+l1+\dots lp}\det(\overline{V}),$$

therefrom we have:

$$\det(M) = \sum_{1\le j1<\dots<jk\le n}\ \sum_{1\le l1<\dots<lp\le n-k} \alpha\beta(-1)^{i1+\dots+ik+j1+\dots+jk+m1+\dots+mp+l1+\dots+lp}\det(\overline{V}) = \sum_j\sum_l \alpha\beta\gamma \tag{13}$$

3.2. If We Consider the Square Polynomial Matrices with Complex Coefficients, of Size n, Noted A, B, C, L, Z =
O_n and matrix $M = \begin{pmatrix} A & Z & B \\ Z & A & B \\ L & L & C \end{pmatrix}$, then det(M) is Dividable by det(A)

Proof. Considering an expansion of type (13) with minors of the nth order having elements on the first n lines we have [25]:

$$\det(M) = \sum_{1\le j1;\dots;jn\le 3n}\ \sum_{1\le l1;\dots;ln\le 2n} \alpha\beta\gamma.$$

We shall prove the sentence demonstrating that for this special type of matrix if a term of the previous sum is not dividable by det(A), then there is a term $\alpha'\beta'\gamma'$ in the expansion so that $\alpha\beta\gamma + \alpha'\beta'\gamma' = 0$.

We shall analyze the possible cases, one by one. We shall highlight for the start the columns of the blocks which intervene in matrix M namely A = (A1..An), B = (B1..Bn), C = (C1..Cn), L = (L1..Ln) and Z = (Z1..Zn) = (0..0).

- For j1 = 1, jn = n we have $\alpha = \det(A)$.

- For j1 = 2n + 1, jn = 3n we have $\alpha = \det(B)$ and $\overline{\alpha} = \det\begin{pmatrix} Z & A \\ L & L \end{pmatrix} = -\det(L)\cdot\det(A)$.

- For the rest, we notice that:

  - if there is an index $jk \in \{n+1,\dots,2n\}$ then the column k from α is null thus $\alpha\beta\gamma = 0$.
  - α is non-null if $\{j1,\dots,jk\} \subset \{1,2,\dots,n\}$ and $\{jk+1,\dots,jn\} \subset \{2n+1,\dots,3n\}$ in this case $\alpha = \det(Aj1..AjkBik+1..Bin)$ where $ik+l = jk+l-2n$. For such a fixed α we have three possibilities for β namely:

    - β has a column 0 thus $\beta = 0$;
    - $\beta = \det(A)$;

- $\beta = \det(As1..AslBrl + 1..Brn)$. In this case, we can determine in a unique way the matrix $\overline{V}$ for each of the two possible versions:

  - If there is $t \in \{1,\dots,n\}\backslash\{j1,\dots jk,s1,\dots,sl\}$ then $\overline{V}$ contains twice the column Lt thus $\alpha\beta\gamma = 0$;
  - If $\{1,\dots,n\} = \{j1,\dots jk,s1,\dots,sl\}$ then we consider $\alpha' = (\det(As1..AslBrl + 1..Brn)$ $\beta' = \det(Aj1..AjkBik + 1..Bin)$ and γ' will be a determinant having the same C type columns located in the same position as in γ and the L type columns will be the same but permutated as far as the position is concerned. A direct calculation of signs will lead to $\alpha\beta\gamma + \alpha'\beta'\gamma' = 0$.

Thus the sentence has been proved. $\square$

In our case we have:

$$M = \begin{pmatrix} A & Z & B \\ Z & A & B \\ L & L & C \end{pmatrix} = \begin{bmatrix} [P_e(\omega^2)] & 0 & [K_c] \\ 0 & [P_e(\omega^2)] & [K_c] \\ ([K_c]^T & [K_c]^T & [P_r(\omega^2)] \end{bmatrix}, \tag{14}$$

that is a more particular case.

3.3. The Natural Modes of Vibration

To find the natural modes of vibration for this problem is the same with solving the linear homogenous system:

$$\begin{bmatrix} [P_e(\omega_i^2)] & 0 & [K_c] \\ 0 & [P_e(\omega_i^2)] & [K_c] \\ [K_c]^T & [K_c]^T & [P_r(\omega_i^2)] \end{bmatrix} \begin{Bmatrix} \{\Phi_s\} \\ \{\Phi_d\} \\ \{\Phi_r\} \end{Bmatrix}_i = 0, \quad i = \overline{1,n}, \tag{15}$$

where the eigenvector was partitioned according to the subsystems composed of the two engines and the rest of the flywheels.

Theorem T2. The system (15) has, for $\omega_i^2 =_1 \omega_i^2$, solutions such as (skew-symmetrical eigenmodes):

$$\begin{Bmatrix} \{\Phi_s\} \\ -\{\Phi_s\} \\ 0 \end{Bmatrix}_i, \quad i = \overline{1,n_1}. \tag{16}$$

Proof.

$$\begin{bmatrix} [P_e(_1\omega_i^2)] & [K_c] \end{bmatrix} \begin{Bmatrix} \{\Phi_s\} \\ \{\Phi_r\} \end{Bmatrix}_i = \{0\}. \tag{17}$$

$$\begin{bmatrix} [P_e(_1\omega_i^2)] & [K_c] \end{bmatrix} \begin{Bmatrix} \{\Phi_d\} \\ \{\Phi_r\} \end{Bmatrix}_i = \{0\}. \tag{18}$$

$$\begin{bmatrix} [K_c] & [K_c] & [P_r(_1\omega_i^2)] \end{bmatrix} \begin{Bmatrix} \{\Phi_s\} \\ \{\Phi_d\} \\ \{\Phi_r\} \end{Bmatrix}_i = \{0\}. \tag{19}$$

From equations (17) and (18) we have:

$$[P_e(_1\omega_i^2)]\{\Phi_s\}_i = [P_e(_1\omega_i^2)]\{\Phi_d\}_i = -[K_c]\{\Phi_r\}_i, \tag{20}$$

whereas, by the statement:

$$\Delta_e({}_1\omega_i^2) = \det[P_e({}_1\omega_i^2)] = 0, \tag{21}$$

then (20) can only take place if:

$$\{\Phi_r\}_i = 0. \tag{22}$$

From (19) we have:

$$[K_c]\{\Phi_s\}_i + [K_c]\{\Phi_d\}_i + \left[P_r({}_1\omega_i^2)\right]\{\Phi_r\}_i = [K_c]\{\Phi_s\}_i + [K_c]\{\Phi_d\}_i = \{0\}, \tag{23}$$

so that:

$$\{\Phi_s\}_i = -\{\Phi_d\}_i. \tag{24}$$

$\square$

Theorem T3. The system (15) has, for $\omega_i^2 \neq_1 \omega_i^2$ $(i = \overline{n_1 + 1, 2n_1 + n_2})$ solutions such as (symmetrical eigenmodes):

$$\left\{ \begin{array}{c} \{\Phi_s\} \\ \{\Phi_s\} \\ \{\Phi_r\} \end{array} \right\}_i. \tag{25}$$

Proof. In this case $\det[P_e(\omega_i^2)] \neq 0$, so (17) and (18) from Theorem T2 can be written as:

$$[P_e(\omega_i^2)]\{\Phi_s\}_i = -[K_c]\{\Phi_r\}_i, \tag{26}$$

$$[P_e(\omega_i^2)]\{\Phi_d\}_i = -[K_c]\{\Phi_r\}_i, \tag{27}$$

having the solution:

$$\{\Phi_s\} = \{\Phi_d\} = [P_e(\omega_i^2)]^{-1}[K_c]\{\Phi_r\} \tag{28}$$

$\square$

In Table 4, the eigenvalues for the two constructive solutions of a motor vehicle with two identical engines are presented and, for the sake of compassion, the eigenvalues of a single engine are also presented. We observed the six values of the natural frequencies of a single engine that coincide with six of the natural frequencies of the whole assembly.

Table 4. Eigenvalues for the two models.

No	Rear Clutch Model Eigenvalues (rpm)	Front Clutch Model Eigenvalues (rpm)	Single Engine Model Eigenvalues (rpm)
1	0	0	
2	1.338	1.596	
3	14.062	13.449	
4	14.564	14.062	14.062
5	30.417	22.397	
6	40.278	40.278	40.278
7	41.483	41.329	
8	67.067	67.067	67.067
9	67.561	67.632	
10	92.764	92.764	92.764
11	93.394	93.553	
12	100.959	100.959	100.959
13	101.039	101.056	
14	144.921	144.921	144.921
15	151.079	152.676	

The representation of the vibration modes for the branched system is suggestive for presenting the results (Figures 8 and 9). If the natural frequencies of the branched system coincide with the natural frequencies of the single engine, the vibration modes will be skew-symmetric, and for the other natural frequencies, the natural modes will be symmetrical.

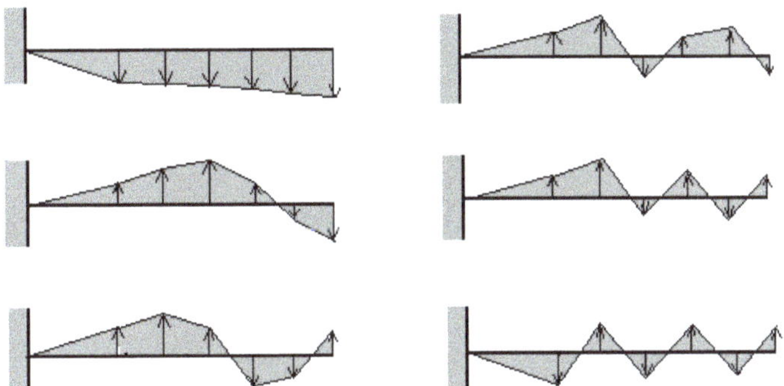

Figure 8. The natural modes of vibration for one engine.

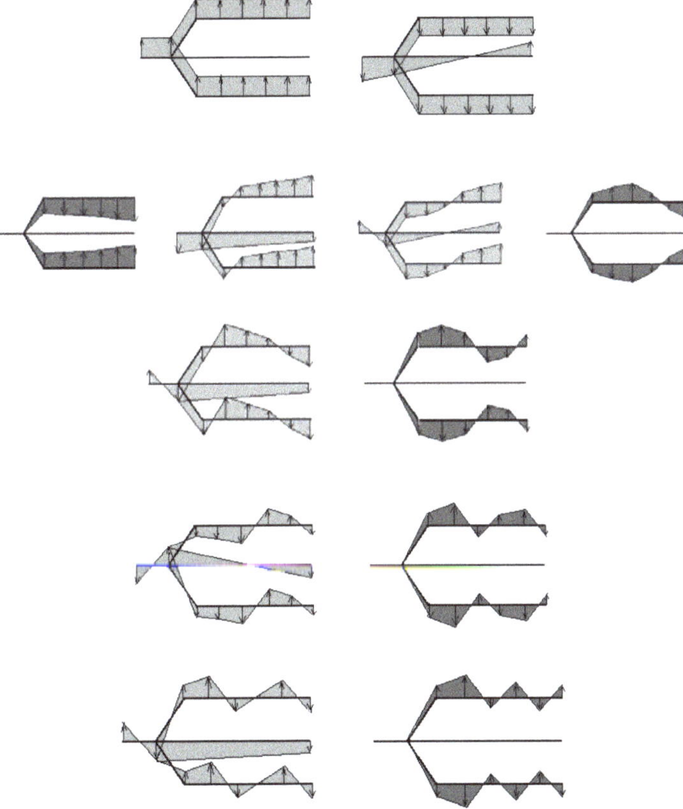

Figure 9. *Cont.*

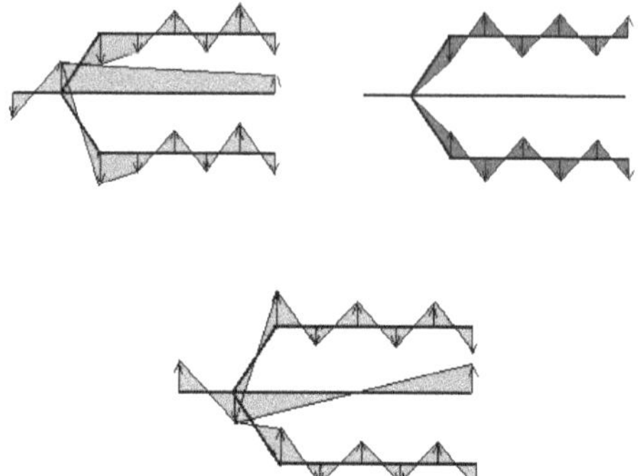

Figure 9. The natural modes of vibration for the whole system.

4. Discussion

For a mechanical transmission presenting symmetry, some properties regarding eigenvalues and eigenmodes were stated and demonstrated. These properties suggest a method that would facilitate the calculation of eigenvalues. Thus, in Section 3 we have shown that eigenvalues for a single engine were among the eigenvalues of the whole mechanical system. From this, a method to simplify the calculus was presented: First, we calculated the eigenvalues for one single engine; then, these eigenvalues (which we already determined) were eliminated from the characteristic equation of the whole system. In this way, the size of the system was reduced and the calculation became easier. We have also shown that eigenmodes could be classified into symmetric and skew-symmetric modes of vibration and the skew-symmetric modes could be immediately built if one knows the eigenmodes for a single engine.

In the paper, two variants of mechanical systems that were equipped with two identical engines were analyzed, from the point of view of vibrations. The results show that the differences, from this point of view, were insignificant in terms of performance. Considering this, the choice of the optimum solution will be made according to criteria other than the vibrations of the transmission. In the paper, it was shown that the constructive symmetries that exist could help ease the calculation. Thus it was shown that the vibrations of the symmetrical parts could also be found among the natural frequencies of the whole system. The vibration modes were of two types: some were skew-symmetric, while others respected the symmetry of the system (symmetrical vibration modes). The problem studied allowed to highlight the properties of some symmetrical systems, applicable in any type of system with symmetrical parts, which allowed the calculation to be made in an easier way.

The method could also be applied for other mechanical systems presenting some symmetries. For small systems, with a reduced number of degrees of freedom, the method had no obvious advantages. For large systems, with a large number of degrees of freedom (as we could find in the Finite Element Method), the use of the presented properties for calculating eigenvalues and eigenmodes of vibration could significantly reduce the computation time by reducing the dimensions of the studied system. Thus, instead of calculating the eigenvalues and the eigenmodes for the whole system, we first calculated them for a symmetrical part and then we eliminated the determined values from the written equations for the whole system, in this way reducing the complexity of the problem, making it easier to be solved.

Author Contributions: Conceptualization, S.V., M.M. and M.P.; methodology, S.V. and M.P.; software, M.M.; validation, S.V. and M.P.; formal analysis, M.M. and M.P.; investigation, M.M. and S.V.; resources, S.V.; data curation, S.V.; writing—original draft preparation, S.V.; writing—review and editing, M.M.; visualization, M.M. and S.V.; supervision, S.V.; project administration, S.V. and M.M.

Funding: This research received no external funding.

Conflicts of Interest: The authors declare no conflict of interest.

References

1. Ambrus, C. Analiza Dinamică a Solicitărilor din Ansamblul Motor-Transmisie al Instalațiilor Mobile de Foraj de Mare Putere (Dynamic Analysis of Stresses in the Motor-Transmission Assembly of Mobile Power Drilling Installations). Ph.D. Thesis, Transylvania University of Brasov, Brasov, Romania, 2017.
2. Holm, D.D.; Stoica, C.; Ellis, D.C.P. *Geometric Mechanics and Symmetry*; Oxford University Press: Oxford, UK, 2009.
3. Marsden, J.E.; Ratiu, T.S. *Introduction to Mechanics and Symmetry: A Basic Exposition of Classical Mechanical Systems*; Springer: Berlin/Heidelberg, Germany, 2003; ISBN 13 978-0387986432.
4. Singer, S.F. *Symmetry in Mechanics*; Springer: Berlin/Heidelberg, Germany, 2004; ISBN 978-1-4612-0189-2.
5. Celep, Z. On the axially symmetric vibration of thick circular plates. *Ingenieur-Archiv* **1978**, *47*, 411–420. [CrossRef]
6. Chen, Y.; Feng, J. Generalized Eigenvalue Analysis of Symmetric Prestressed Structures Using Group Theory. *J. Comput. Civ. Eng.* **2012**, *26*, 488–497. [CrossRef]
7. Zavadskas, E.K.; Bausys, R.; Antucheviciene, J. Civil Engineering and Symmetry. *Symmetry* **2019**, *14*, 501. [CrossRef]
8. Mechanics and symmetry in Europe: The geometry and dynamics of deformable systems. Project. HPRN-CT-2000-00113, Funded under: FP5-HUMAN POTENTIAL, University of Surrey, United Kingdom, Centre National de la Reserche Scientifique, France, Instituto Superior Tecnico, Portugal, Swiss Federal Institute of Technology of Nottingham, Switzerland, Universita degli studi di Padova, Italy, University of Nottingham, United Kingdom, Universite du Litoral, France, Utrecht University, Netherlands. Available online: https://cordis.europa.eu/project/rcn/53964/factsheet/en (accessed on 12 October 2019).
9. Ganghoffer, J.F.; Mladenov, I. *Similarity, Symmetry and Group Theoretical Methods in Mechanics*; International Centre for Mechanical Sciences: Udine, Italy, 2015.
10. Mangeron, D.; Goia, I.; Vlase, S. Symmetrical Branched Systems Vibrations. *Sci. Mem. Rom. Acad.* **1991**, *12*, 232–236.
11. Weimann, S.; Kremer, M.; Plotnik, Y.; Lumer, Y.; Nolte, S.; Makris, K.G.; Segev, M.; Rechtsman, M.C.; Szameit, A. Topologically protected bound states in photonic parity-time-symmetric crystals. *Nat. Mater.* **2017**, *16*, 433–438. [CrossRef] [PubMed]
12. Wang, B.; Yang, H.L.; Meng, F.W. Sixth-order symplectic and symmetric explicit ERKN schemes for solving multi-frequency oscillatory nonlinear Hamiltonian equations. *Calcolo* **2017**, *54*, 117–140. [CrossRef]
13. Niiranen, J.; Balobanov, V.; Kiendl, J.; Hosseini, S.B. Variational formulations, model comparisons and numerical methods for Euler-Bernoulli micro- and nano-beam models. *Math. Mech. Solids* **2019**, *24*, 312–335. [CrossRef]
14. Bourada, F.; Amara, K.; Bousahla, A.A.; Tounsi, A.; Mahmoud, S.R. A novel refined plate theory for stability analysis of hybrid and symmetric S-FGM plates. *Struct. Eng. Mech.* **2018**, *68*, 661–675.
15. Sun, X.J.; Zhang, H.; Meng, W.J.; Zhang, R.H.; Li, K.N.; Peng, T. Primary resonance analysis and vibration suppression for the harmonically excited nonlinear suspension system using a pair of symmetric viscoelastic buffers. *Nonlinear Dyn.* **2018**, *94*, 1243–1265. [CrossRef]
16. Vlase, S.; Păun, M. Vibration analysis of a mechanical system consisting of two identical parts. *Rom. J. Tech. Sci. Appl. Mech.* **2015**, *60*, 216–230.
17. Vlase, S.; Marin, M.; Scutaru, M.L.; Munteanu, R. Coupled transverse and torsional vibrations in a mechanical system with two identical beams. *AIP Adv.* **2017**, *7*, 065301. [CrossRef]
18. Othman, M.I.A.; Marin, M. Effect of thermal loading due to laser pulse on thermoelastic porous medium under G-N theory. *Results Phys.* **2017**, *7*, 3863–3872. [CrossRef]
19. Vlase, S.; Năstac, D.C.; Marin, M.; Mihălcică, M. A method for the study of the vibration of mechanical bars systems with symmetries. *Acta Tech. Napoc. Ser. Appl. Math. Mech. Eng.* **2017**, *60*, 539.
20. Hassan, M.; Marin, M.; Ellahi, R.; Alamri, S.Z. Exploration of convective heat transfer and flow characteristics synthesis by Cu–Ag/water hybrid-nanofluids. *Heat Transf. Res.* **2018**, *49*, 1837–1848. [CrossRef]
21. Zingoni, A. Symmetry recognition in group-theoretic computational schemes for complex structural systems. *Comput. Struct.* **2012**, *94–95*, 34–44. [CrossRef]
22. Zingoni, A. Group-theoretic exploitations of symmetry in computational solid and structural mechanics. *Int. J. Numer. Methods Eng.* **2009**, *79*, 253–289. [CrossRef]

23. Zingoni, A. Group-theoretic insights on the vibration of symmetric structures in engineering. *Philos. Trans. R. Soc. A Math. Phys. Eng. Sci.* **2014**, *372*. [CrossRef]

24. Catalog. Roman Autocamioane S.A Brașov. Roman Trucks & Buses—Special vehicles. 2012. Available online: http://www.roman.ro/index.php?lang=ro&showlang=&cat=PRODUSE%20/%20AUTOVEHICULE%20CIVILE&subcat=Autovehicule%20pentru%20industria%20petroliera&subid=3112&main=produse&end=fnpr (accessed on 12 October 2019).

25. Horn, R.A.; Johnson, C.R. *Matrix Analysis*; Cambridge University Press: Cambridge, UK, 1985.

Article

Research on a Real-Time Monitoring Method for the Wear State of a Tool Based on a Convolutional Bidirectional LSTM Model

Qipeng Chen, Qingsheng Xie, Qingni Yuan *, Haisong Huang and Yiting Li

Key Laboratory of Advanced Manufacturing Technology, Ministry of Education, Guizhou University, Guiyang 550025, China
* Correspondence: qnyuan@gzu.edu.cn; Tel.: +86-189-851-07557

Received: 14 August 2019; Accepted: 20 September 2019; Published: 2 October 2019

Abstract: To monitor the tool wear state of computerized numerical control (CNC) machining equipment in real time in a manufacturing workshop, this paper proposes a real-time monitoring method based on a fusion of a convolutional neural network (CNN) and a bidirectional long short-term memory (BiLSTM) network with an attention mechanism (CABLSTM). In this method, the CNN is used to extract deep features from the time-series signal as an input, and then the BiLSTM network with a symmetric structure is constructed to learn the time-series information between the feature vectors. The attention mechanism is introduced to self-adaptively perceive the network weights associated with the classification results of the wear state and distribute the weights reasonably. Finally, the signal features of different weights are sent to a Softmax classifier to classify the tool wear state. In addition, a data acquisition experiment platform is developed with a high-precision CNC milling machine and an acceleration sensor to collect the vibration signals generated during tool processing in real time. The original data are directly fed into the depth neural network of the model for analysis, which avoids the complexity and limitations caused by a manual feature extraction. The experimental results show that, compared with other deep learning neural networks and traditional machine learning network models, the model can predict the tool wear state accurately in real time from original data collected by sensors, and the recognition accuracy and generalization have been improved to a certain extent.

Keywords: tool wear state; CNN; BiLSTM; attention mechanism; signal features

---

1. Introduction

As a critical component of intelligent manufacturing, mechanical intelligent fault diagnosis has become an essential part of "Made in China 2025" [1]. In mechanical processing, cutting is the most important means of manufacturing. At present, research in this field mainly focuses on tool cutting parameter optimization [2,3] and tool wear condition monitoring [4,5]. Real-time monitoring of the tool wear state is an essential part of the computerized numerical control (CNC) machining process in a manufacturing workshop. The wear state of a tool is affected by the processing procedures, workpiece materials, cutting parameters, and other factors. The whole system exhibits strong nonlinearity and uncertainty. The tool wear will not only reduce the processing quality of the CNC machining equipment but also affect the surface roughness and machining accuracy of the workpiece and seriously affect the overall stability and processing efficiency of the CNC machining equipment. The wear state of a tool will directly affect the machining accuracy, surface quality, and production efficiency of the parts. Therefore, the technology of tool condition monitoring (TCM) is of great significance for ensuring the quality of processing and realizing continuous automatic processing [6–9].

TCM methods are divided into direct measurement methods and indirect measurement methods. Direct measurement methods include resistance measurement methods, optical measurement methods, discharge current measurement methods, ray measurement methods, and computer image processing methods. The tool wear state can be obtained directly, but due to the influence of the coolant and other disturbances in the production process, the tool wear state in the mechanical processing stage cannot be detected in real time, which is rarely used in actual industrial production [10]. Indirect measurement methods include the cutting force measurement method, acoustic emission method, mechanical power measurement method, vibration signal and multi-information fusion detection [11–15]. Indirect measurement methods can acquire signals in real time through a sensor during tool cutting. After data processing and feature extraction, hidden Markov model (HMM), fuzzy neural network (FNN), back propagation neural network (BPNN), support vector machine (SVM), and other machine learning (ML) models can be used to monitor tool wear [16–18]. For example, Zhang Xiang et al. proposed micro-milling tool wear identification as the research object and established the HMM of tool wear. Eight optimal cutting forces were extracted as the HMM training input vectors by Fisher's linear discriminant. The method can identify the micro-milling tool's wear state with an accuracy rate of 85% [16]. X. Li et al. proposed an FNN designed and developed for machinery prognostic monitoring. The FNN is basically a multi-layered fuzzy-rule-based neural network that integrates a fuzzy logic inference into a neural network structure. This method is helpful to accelerate the learning process of the complex conventional neural network structure, and the accuracy in prediction and rate of convergence are better than those of similar ML models [17]. Liao Zhirong et al. proposed a tool wear condition monitoring system based on acoustic emission technology. By analysingrepresentative acoustic signals, the energy ratios from six different frequency bands are selected from the time–frequency domain. These are used as a classification feature to determine the amount of tool wear. In this method, the SVM is used as the classification method, which can ultimately achieve an accuracy rate of 93.3% [18]. The traditional ML model adopts shallow learning. Since ML is affected by the quality instability of the manual extraction feature, a random initialization of the weights can easily enable the objective function to converge to the local minimum. When the number of layers is too large, the forward propagation of the residuals will be lost, leading to gradient diffusion. At the same time, ML is limited by the inability to capture the dependence of long-distance signals on the sequential input. Deep learning (DL) can effectively avoid these problems.

DL was first introduced into machine learning (ML) in 1986 and then used in an artificial neural network (ANN) [19] in 2000. DL uses multi-level non-linear information to process low-level features to form more abstract high-level representations for supervised or unsupervised feature learning, representation, classification, and pattern recognition [20]. The DL model is an "end-to-end learning" model, which does not require complex data pre-processing of the original data, making the construction of the model more concise (Figure 1). At present, the DL method has emerged in the industrial field. DL models represented by a CNN have been gradually applied to the study of tool wear condition monitoring and achieved specific results [21–23]. For example, Zhang Cunji et al. proposed transforming the vibration signal of a tool in the process of machining into an energy spectrum by a wavelet packet transform (WPT) and inputting the spectrum into a CNN to extract the features automatically and classify them accurately [21]. German Terrazas et al. proposed that based on the gramian angular summation fields (GASF) module, a large number of continuous force signals generated by cutting tools in a high-speed milling process can be automatically converted into two-dimensional images, which are input into a CNN to obtain the tool wear status [22]. Cao Dali et al. proposed the construction of a DenseNet using the dense connection, which adaptively extracts hidden high-dimensional features from original time series signals. The results showed that deepening the network layers is helpful for improving the accuracy of the tool wear monitoring model [23]. The above methods adopt DL to extract features adaptively, which basically solves the shortcoming of a manual extraction of the signal features. However, the convolution neural network (CNN) used relies too heavily on high-dimensional feature extraction. The excessive number of convolutional layers is prone to gradient dispersion, and the number of convolutional layers is too small to grasp the global features and does not take into account the critical feature of the correlation between the timing signal samples generated during tool processing.

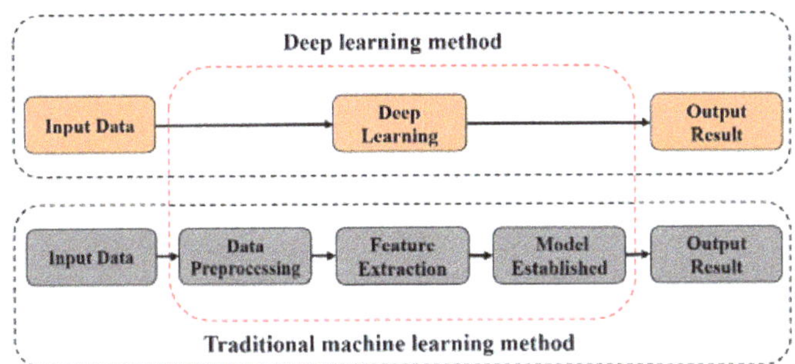

Figure 1. Comparison of deep learning and traditional machine learning methods.

Therefore, this paper proposes a method for real-time monitoring of a tool wear state based on a CNN and bidirectional long short-term memory (BiLSTM) network model with an attention mechanism (CABLSTM). The sensor acquires the signals generated during tool processing in real time, which are directly fed into the CNN for parallel local feature extraction and then into the BiLSTM network for feature extraction of the long-distance dependence information. The attention mechanism is used to calculate the network weights and distribute them reasonably. Finally, the signal feature information with different weights is sent to a Softmax classifier to classify the tool wear status, avoiding the complexity and limitation caused by a manual feature extraction. This method can meet the real-time and accuracy requirements of tool monitoring in actual industrial production.

The remainder of this paper is organized as follows. Section 2 presents the CABLSTM algorithm. Section 3 presents the monitoring process of tool wear. Section 4 presents the experimental results of the tool wear condition monitoring. Section 5 concludes the article.

2. CABLSTM Model

Inspired by the literature [24], this paper applied a CNN and recurrent neural network (RNN) fusion to the real-time monitoring task of a tool wear state, constructs two network models of convolutional long short-term memory (CLSTM) and convolutional bi-directional long short-term memory (CBLSTM), effectively solves the problem of the correlation between the ignored time-series signals in a single CNN, and avoids the problem of gradient dispersion and gradient explosion in a circular neural network. Meanwhile, the attention mechanism is introduced on the basis of the CBLSTM network. Finally, the CABLSTM network is proposed, which further improves the accuracy of model prediction.

The CABLSTM model mainly includes four parts: The first part involves the local feature extraction of the single time step timing signal, which mainly uses a one-dimensional CNN for neighborhood filtering, uses a sliding window for the convolution calculation, and finally obtains the high-dimensional features of the single time step timing signal. The second part involves the extraction of the time series of time-series signals, and the BiLSTM network is used to process the high-dimensional features generated by the continuous time step timing signals and gradually synthesize the vector feature representation of the input signals. The third part uses the attention mechanism to calculate the importance distribution of sequential signal features in continuous time steps and generate the feature model of sequential signals with an attention probability distribution. The fourth part is the classifier, which uses dropout technology to prevent overfitting and uses the Softmax classifier to predict the tool wear states. The neural network framework for real-time monitoring of the tool wear state based on CABLSTM is shown in Figure 2.

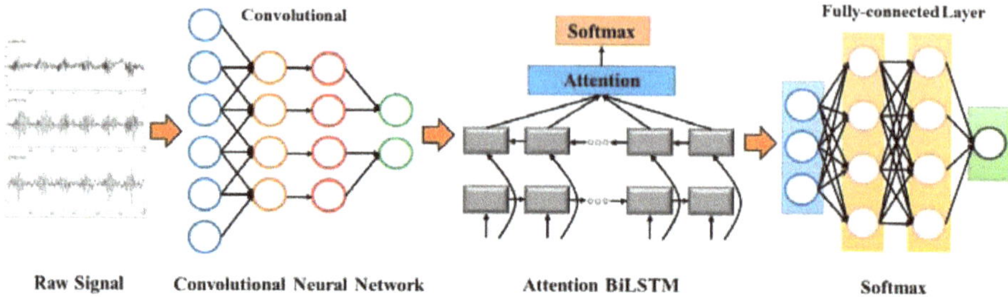

Figure 2. Neural network framework for real-time monitoring of tool wear state based on convolutional neural network (CNN) and bidirectional long short-term memory (BiLSTM) network with an attention mechanism (CABLSTM).

2.1. Local Feature Extraction of Single Time Step Timing Signals

The one-dimensional CNN can be applied to a time-series analysis of sensor data [24–26]. In the one-dimensional convolutional layer, multiple filters are used to perform neighborhood filtering of the input time-series data, and the acquired feature maps are superimposed to form an output feature map of the convolutional layer. Then, the pooling layer extracts the fixed-length feature vectors from feature maps of each candidate frame for a feature dimension reduction, thereby extracting critical features in the time-series data and simplifying the complexity of the network calculation.

In this paper, a one-dimensional CNN was used to directly process the timing signals generated during tool processing. The CNN includes two layers: A convolutional layer and a pooling layer. The convolution layer performs neighborhood filtering of the time-series signals of each dimension using a one-dimensional convolution operation to generate feature maps, and each feature map can be regarded as a convolution operation of different filters on the current time step timing signals [27]. When the input timing signal is x, the weight vector of the convolution kernel is w, the total number of samples is m, the size of the convolution kernel is n, $*$ is the convolution operation, and the output feature map of the convolutional layer y can be expressed as follows:

$$y = x * w = \sum_{m=0}^{m} x(m) \cdot w(n - m).\tag{1}$$

In the convolutional layer, each neuron of the l layer is only connected to a local window neuron in the $l - 1$ layer to form a local connection network. The calculation formula for the one-dimensional convolution layer is as follows:

$$x_j^l = f\left(\sum_{i \in M_j} x_i^{l-1} \cdot w_{ij}^l + b_j^l\right),\tag{2}$$

where x_j^l is the j feature map of the l layer, $f(\cdot)$ is the activation function, M_j is the input feature vector, x_i^{l-1} is the i feature map of the $l - 1$ layer, w_{ij}^l is a trainable convolution kernel, and b_j^l is the bias parameter. Considering the convergence speed and overfitting problems, the rectified linear unit (Relu) is chosen for the non-linear activation function, which converges faster to improve the sparsely of the network in this paper, reduces the interdependence of the parameters, and alleviates the occurrence of overfitting. The formula for the Relu activation function is as follows:

$$a_i^{(l+1)}(j) = f(y_i^{l+1}(j)) = \max\left\{0, y_i^{l+1}(j)\right\},\tag{3}$$

where $y_i^{l+1}(j)$ is the output value of the volume and operation and $a_i^{l+1}(j)$ is the activation value of $y_i^{l+1}(j)$.

The convolutional layer is connected to the pooling layer for the local maximum or local mean, namely, max pooling and mean pooling [28]. The pooling layer has the function of feature selection, which can ensure that the feature can resist a deformation; at the same time, the pooling layer can reduce the feature dimension, speed up the network training, reduce the number of parameters, and improve the robustness of the feature. In this paper, max pooling was used to obtain the maximum value of the feature points in the neighborhood. The formula is as follows:

$$P_i^{l+1}(j) = \max_{(j-1)W+1 \leq t \leq jW} \left\{ q_i^l(t) \right\}, \tag{4}$$

where $q_i^l(t)$ is the value of the t neuron in the i feature vector of the l layer and $t \in [(j-1)w+1, jw]$. w is the width of the pooled region, and $P_i^{l+1}(j)$ is the value corresponding to the $l+1$ layer neuron.

The one-dimensional CNN performs the feature extraction of the original data, and the three-dimensional features of the time-series signal are better expressed as high-dimensional features, which facilitate the subsequent time-series feature extraction of the BiLSTM network. The basic structure of the one-dimensional CNN is shown in Figure 3.

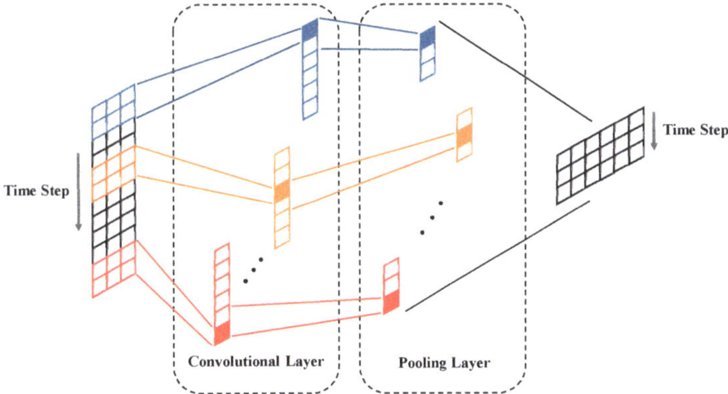

Figure 3. The basic structure of the one-dimensional convolutional neural network (CNN).

2.2. Time-Series Feature Extraction of Time-Series Signals

Long short-term memory (LSTM) is an exclusive self-connected recurrent neural network (RNN). LSTM introduces a gate function to generate the path of continuous gradient flow for a long time, which effectively avoids the problem of gradient disappearance and gradient explosion caused by the chain rule in the gradient calculation of hidden layers in RNN [29]. LSTM can mine the temporal variation law of relatively long intervals in time series, and it is particularly used to process time-series data. The original signal generated during tool processing has a timing relationship. The LSTM network can encode the time series of time-series signals and mine the timing variation in relatively long intervals in the time series [30]. To ensure that the real-time monitoring model of tool wear can better learn the dependence of time-series features between time-series signals and improve the accuracy of the model classification, this paper improves the existing LSTM network [31] and builds a BiLSTM network with a symmetric structure by constructing two directions of LSTM networks [32]. At the same time, the attention mechanism is introduced into the BiLSTM network to increase the attention layer, which enables the model to both extract temporal signal features from both the positive and negative directions and selectively learn the critical information of the signal features.

The constructed BiLSTM network contained 256 neurons in this paper. The forward and reverse LSTM networks consisted of 128 neurons. Each BiLSTM neuron included an input gate, a forget gate and an output gate, which are represented by i, f, and o, respectively. The internal structure of the BiLSTM neurons is shown in Figure 4.

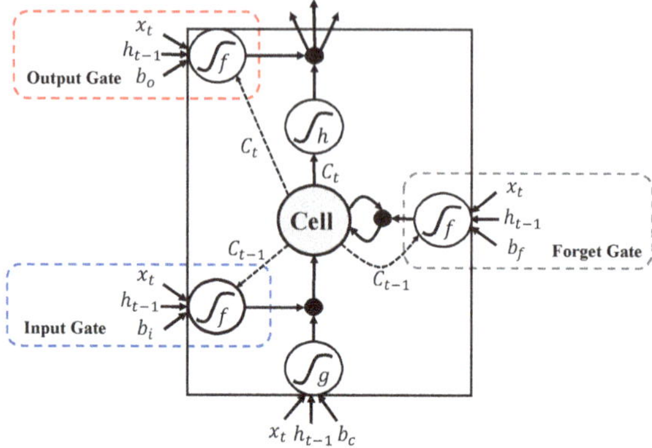

Figure 4. The internal structure of the BiLSTM neurons.

The input gate i is used to control the amount of current input information x_t of the network that can be saved to the memory unit C_t, uses the sigmoid function to determine new information to be saved, uses the tanh function to generate a new candidate vector $\widetilde{C}_t$, and sends the information to be saved to the memory. The unit completes the update. The forget gate f is used to control the self-connecting unit, filters the information in the memory unit C_{t-1} at the previous moment to determine the amount of valid information that needs to be retained in the current memory unit C_t, and forgets the useless information. The output gate o controls the influence of the memory unit C_t on the current output value h_t and determines the amount of information that the memory unit C_t outputs at time step t. The formula is as follows:

$$i_t = \sigma(W_{xi}x_t + W_{hi}h_{t-1} + b_i), \tag{5}$$

$$\widetilde{C}_t = \tanh(W_{xc}x_t + W_{hc}h_{t-1} + b_c), \tag{6}$$

$$f_t = \sigma(W_{xf}x_t + W_{hf}h_{t-1} + b_f), \tag{7}$$

$$C_t = f_t \odot c_{t-1} + i_t \odot \widetilde{C}_t, \tag{8}$$

$$o_t = \sigma(W_{xo}x_t + W_{ho}h_{t-1} + b_o), \tag{9}$$

$$h_t = o_t \odot \tanh(C_t), \tag{10}$$

where C is the memory unit, which is called the cell state, C_t is the memory cell state at time step t, $\widetilde{C}_t$ is the candidate vector of the memory cell at time step t, x_t is the input vector at time step t, h_t is the output vector at time step t, W is the weight vector of the network, b is the offset vector, $\odot$ represents a multiplication of vector elements, $\sigma(\cdot)$ is the sigmoid function, and the tanh function is the hyperbolic tangent activation function.

The high-dimensional feature of the input timing signal is outputted by the forward LSTM network vector $\overrightarrow{h}_t$, the inverse LSTM network output vector is $\overleftarrow{h}_t$, and the BiLSTM network output eigenvector is P_t at time step t. The formula is as follows:

$$\overrightarrow{h}_t = \overrightarrow{LSTM}(h_{t-1}, x_t, C_{t-1}). \tag{11}$$

$$\overleftarrow{h}_t = \overleftarrow{LSTM}(h_{t+1}, x_t, C_{t+1}). \tag{12}$$

$$P_t = [\overrightarrow{h}_t, \overleftarrow{h}_t].\tag{13}$$

In this paper, the attention mechanism was used to assign weights to each time step output vector of the BiLSTM layer by assigning different initialization probability weights. Finally, the values were calculated by the sigmoid function. The attention mechanism achieves selective filtering and focusing of some critical information from a large number of signal features. The focusing process was embodied in the calculation of the weight coefficients. Different weights were allocated to different critical pieces of information, and the proportion of critical information was enhanced by lifting the weights to reduce the loss of critical information of long sequence timing signals. The calculation formula for the attention mechanism [30] is as follows:

$$u_t = \tanh(W_s P_t + b_s),\tag{14}$$

$$\alpha_t = softmax(u_t^T, u_s),\tag{15}$$

$$v = \sum \alpha_t P_t,\tag{16}$$

where P_t is the output eigenvector of the BiLSTM layer at time step t, u_t is the hidden layer representation of P_t through the neural network layer, u_s is the randomly initialized context vector, α_t is the importance weight of u_t normalized by the Softmax function, and v is the feature vector of the final text message. u_s is generated randomly during the training process, and finally, the output value v of the attention layer is mapped via the Softmax function to obtain a real-time classification result of the tool wear state. The partial expansion of the BiLSTM network model with the attention mechanism along the time axis is shown in Figure 5.

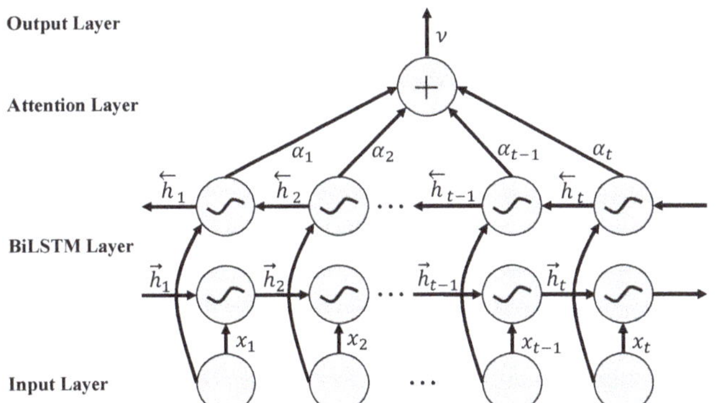

Figure 5. Partial expansion of the BiLSTM network model with the attention mechanism along the time axis.

2.3. Network Model Training

In this paper, dropout technology was introduced into the real-time monitoring model of the tool wear state to prevent the model from overfitting during training. The activation function of the network model uses Softmax, and the loss function uses Categorical_crossentropy, which was used to classify the wear features of the acquired time-series signals. The formula is as follows:

$$y = softmax(v) = \frac{e^{vi}}{\sum_{m=1}^{M} e^{vm}}.\tag{17}$$

y is a vector whose dimensions are the number of categories, each of which has a value between [0,1], and the sum of all dimensions is 1, which is the probability that the tool wear state belongs to a category. M is the number of possible categories. During the training of the model, the entire model was trained by the Categorical_crossentropy loss. The calculation formula for the cross-entropy error is as follows:

$$loss = -\sum_{i=1}^{n} \hat{y}_{i1} \log y_{i1} + \hat{y}_{i2} \log y_{i2} + \cdots + \hat{y}_{im} \log y_{im}, \tag{18}$$

$$\frac{\partial loss}{\partial y_{i1}} = -\sum_{i=1}^{n} \frac{\hat{y}_{i1}}{y_{i1}}, \tag{19}$$

$$\frac{\partial loss}{\partial y_{i2}} = -\sum_{i=1}^{n} \frac{\hat{y}_{i2}}{y_{i2}}, \tag{20}$$

$$\frac{\partial loss}{\partial y_{im}} = -\sum_{i=1}^{n} \frac{\hat{y}_{im}}{y_{im}}, \tag{21}$$

where m is the number of classifications, n is the number of samples, $\hat{y}_{im}$ is the i value in the tool wear state real category label vector, and y_{im} is the i value of the output vector y of the Softmax classifier. For the obtained cross-entropy error, the average was taken as the loss function of the model. The Adam method was used to minimize the objective function when training the model. The Adam method is essentially the RMSprop method with a momentum term. The Adam method dynamically adjusts the learning rate of each parameter by using a first-order moment estimation and a second-order moment estimation of the gradient. The main advantage of the Adam method was that after the offset correction, the learning rate of each iteration had a specific range, which makes the parameter change relatively stable.

3. Real-Time Monitoring Method of the Tool Wear State

An acceleration sensor is used to collect the vibration signal generated by a computerized numerical control (CNC) machining device in the process of machining the workpiece in real time. The input signal of the real-time monitoring model of the tool wear state is the α_x, α_y, and α_z vibration signals, and the output of the model is the predicted value of the tool wear state. In this paper, after continuous sampling of the original vibration signal generated by each milling cutter feed, the sampling points with a length of 2000 were cut to form multiple tensors (3×2000), which were taken as the input data of the model for the DL neural network. The schematics diagram of the CABLSTM network is shown in Figure 6. The CBLSTM network did not have an attention block, while the CLSTM network was similar to the CBLSTM network but with an LSTM block instead of a BiLSTM block.

The input data of the CABLSTM network included the time-series signal (data type) and the wear classification (label type). The feature extraction and expression of the time-series signal were achieved by two convolution layers, one pooling layer, one flatten layer, one BiLSTM layer, one attention layer, and two fully-connected layers. The parameters of each layer of the network are shown in Table 1.

Table 1. CABLSTM: The network parameters settings.

Layer Name	Output Feature Size	Quantity	CABLSTM Network
Input layer	3×2000	1	/
Convolution layer 1	$20 \times 98 \times 128$	1	Conv1D, 1; kernel size = 3; stride = 1
Convolution layer 2	$20 \times 96 \times 128$	1	Conv1D, 1; kernel size = 3; stride = 1
Pooling layer	$20 \times 48 \times 128$	1	MaxPooling1D, 1; stride = 2
Flatten layer	20×6144	1	/
Bidirectional layer	20×256	1	/
Attention layer	256	1	/
Fully-connected layer	128	2	Dense, 128, 3
Output layer	3	1	Softmax, Loss: Categorical_crossentropy

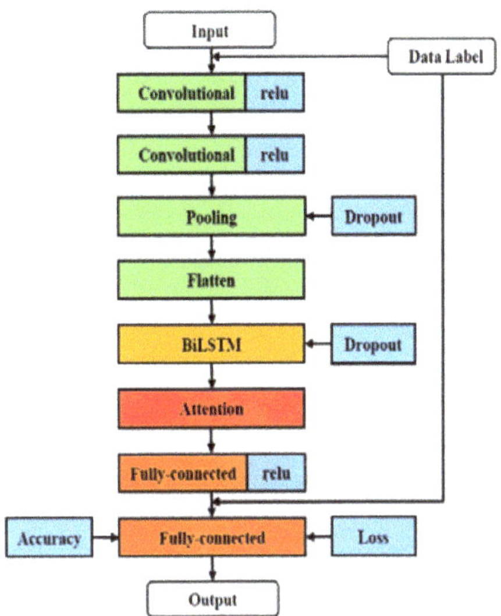

Figure 6. Schematic diagram of the CABLSTM.

4. Experimental

4.1. Experimental Design

A real-time monitoring system for the tool wear state includes a condition monitoring facility and a data analysis unit. The condition monitoring facilities include the basic equipment used to process the workpiece, the equipment to collect the vibration signals generated during the processing, and the equipment to measure the value of tool wear. The data analysis facility included high-performance computers and DL platforms for analyzing and processing the data and classifying and reporting the tool wear status in real time.

4.1.1. Condition Monitoring

The experimental platform of this paper was provided by the Engineering Training Center of Guizhou University. A high-precision CNC vertical milling machine (Model: VM600) was used for the milling workpiece. No coolant was added during milling. The workpiece was milled steel (S136). The milling tool had a cemented carbide 4-edge milling cutter, and its surface was covered with layers of a titanium aluminum nitride coating. The diameter of the tool was 6 mm, the rake angle was 4°, the clearance angle was 8°, and the helix angle was 30°. The cutting parameters of the milling experiment are shown in Table 2.

Table 2. Cutting parameters of the milling experiment.

Spindle Speed	Feed Rate	Cutting Width	Cutting Depth	Tool Overhang	Processing Mode	Cooling Condition
8000 (RPM)	1000 (mm/min)	0.5 (mm)	1 (mm)	15 (mm)	Up milling	Dry milling

In the experiment, three accelerometers (Model: INV9822; Range: ±50 g) were magnetically attracted to the machine tool fixture in the x, y, and z directions for real-time acquisition of the original vibration signals generated during tool machining. A high-precision digital acquisition instrument (model: INV3018CT) from the Beijing Oriental Institute of Vibration and Noise was used to process the real-time signals and transmit them to a computer. The sampling frequency of the signal was 20 kHz, 200 mm of milling in each direction of the tool was recorded as a milling stroke, and each tool was milled for 330 strokes. After each milling stroke, the milling cutter was removed from the milling machine and photographed. A pre-calibrated high-precision digital microscope (EVDM-101) was used for the measurement, the optical magnification was 0.7×–4.5×, the electronic magnification was 35×–235×, and the measuring accuracy was 0.1 μm. During the measurement process, the position of the wear zone of the minor flank surface of the milling cutter, which was the most easily worn, was selected as the measurement position, and the same reference line was taken as the standard to ensure that the position remains unchanged during the measurement. The wear value (VBmax) was calculated by subtracting the current cutting edge length from the initial length of the cutting edge of the milling cutter. The real-time monitoring experimental device of the tool wear state is shown in Figure 7.

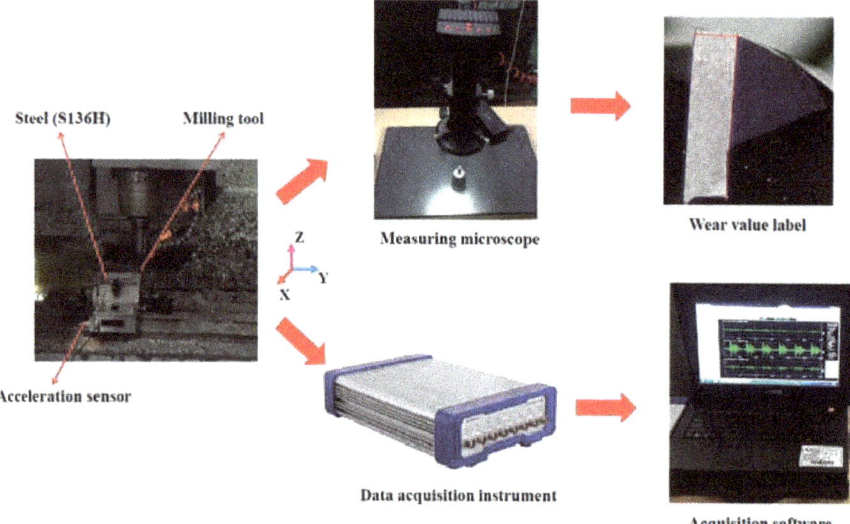

Figure 7. Real-time monitoring experimental device of the tool wear state.

4.1.2. Data Analysis

The DL hardware platform of the experiment used high-performance servers: An Intel Xeon E5-2650 processor, with a frequency of 2.3 GHz, 256 GB of memory, and an NVIDIA GeForce TITAN X graphics processing unit (GPU). The software platform used the Ubuntu 16.04.4 operating system with Keras as the front-end of the in-depth learning framework and TensorFlow as the back-end for data analysis.

The milling operation was carried out with four milling cutters (C1, C2, C3, and C4). Each milling cutter was performed 330 times, and 1320 original signal samples were obtained. The data of three milling cutters (C1, C2, and C3) were used for the training set and verification set of the model, and one milling cutter (C4) data was used for the test set of the model. The training set was used for model fitting the data samples, the verification set was used for adjusting the hyperparameters of the model, the initial ability of the model was evaluated, and the test set was used to evaluate the generalization ability of the final model. In the DL training process, a sufficient number of samples were needed to improve the learning quality of the neural network. The data samples of the original processed signals were long sequences of periodic timing signals. According to the principle of signal sampling, in this paper, 100,000 points of each sample were sampled continuously, and 50 short sequence timing signals with a length of 2000 were cut to be used for model input after data normalization to reduce the computational intensity of the network training. At the same time, data expansion could increase the experimental data based on the original magnitude data, improve the robustness of the network, and reduce the risk of overfitting.

The processing conditions in the experiment had the following characteristics: 1. Finishing milling and small back engagement were performed; 2. the workpiece was milled steel (S136) with high hardness after heat treatment; and 3. the experiment needed to produce tool data set quickly and accurately. This paper referred to references [33–35] and the measurement methods of milling tool wear in 2010 prognostics and health management (PHM) competition. The following method was used as the blunt standard for the milling cutter in this experiment: The maximum value (VBmax) of the wear zone of the minor flank surface of the milling cutter was selected as the quantified value reflecting the wear state. It was specified that failure of the milling cutter occurred when the wear value of the milling cutter was greater than 0.13 mm. The wear process of the milling cutters (C1, C2, C3, and C4) is shown in Figure 8.

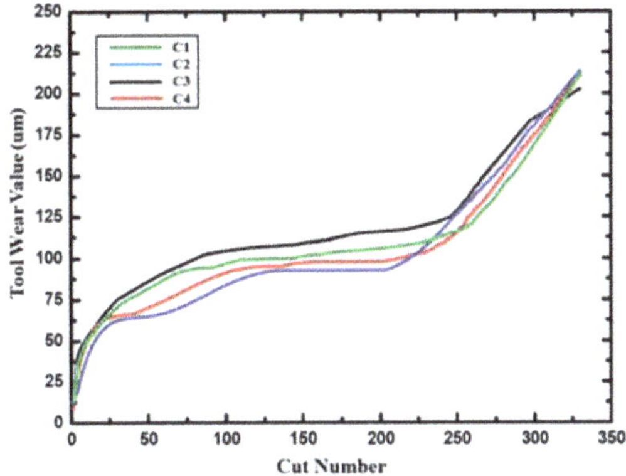

Figure 8. Wear process of the milling cutters.

Each sample contains three-dimensional vibration signals and the wear values of the four rear blades. To prevent mutual interference of the different blade wear values, the maximum wear value of the four blades was selected as the label of the milling stroke. The wear state of the tool was divided into initial wear, normal wear, and rapid wear. In this paper, the wear state of the tool was defined according to the actual wear curve of each milling cutter. The actual wear curve was used to determine the wear degree of the tool. The tool wear degree was divided into three types of label data, and the label data were converted by a one-hot coding form to facilitate the classification of the final tool wear state. The classification of the final tool wear state is shown in Table 3.

Table 3. Classifications of the final tool wear state.

Label Classification	Tool Wear Value/mm	Tool Wear State
0	0–0.06	Initial wear
1	0.06–0.13	Normal wear
2	0.13–0.22	Rapid wear

4.2. Comparison of the Experimental Results of the Deep Learning Model

The original signal generated by the milling process was sampled and then sent to the DL neural network model. The model adaptively extracted the high-dimensional features implied in the time-series signal and calculated the actual output value and reality of the model. The Adam algorithm reduced the error distance between the values, and the network weight was continuously updated so that the actual output value of the model was closer to the real value. To further verify the performance of the proposed algorithm, we implemented the bearing fault diagnosis algorithm of the CNN model in [25] and the turbofan engine life prediction algorithm of the BiLSTM model in [26]. The above model was compared with our proposed CLSTM, CBLSTM, and CABLSTM networks. The five training models used the same training parameters. The specific training parameters of the model are shown in Table 4.

Table 4. Specific training parameters of the model.

Parameter	Model
Learning rate	0.001
Dropout	0.5
Epoch	100
Batch Size	16
Optimizer	Adam

After the training and verification of the DL neural network, different loss function values and accuracies were obtained. The loss function values of the CNN [25], BiLSTM [26], CLSTM, CBLSTM, and CABLSTM models and the accuracy of the verification set are shown in Figures 9–13, where the x axis was used to represent the number of iterations of the milling data set, and the double y axis was used to represent the loss function value and the model verification accuracy.

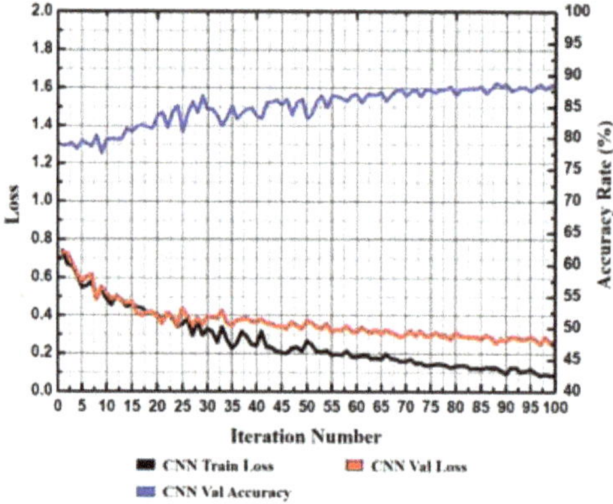

Figure 9. Loss function and accuracy of CNN model training and verification.

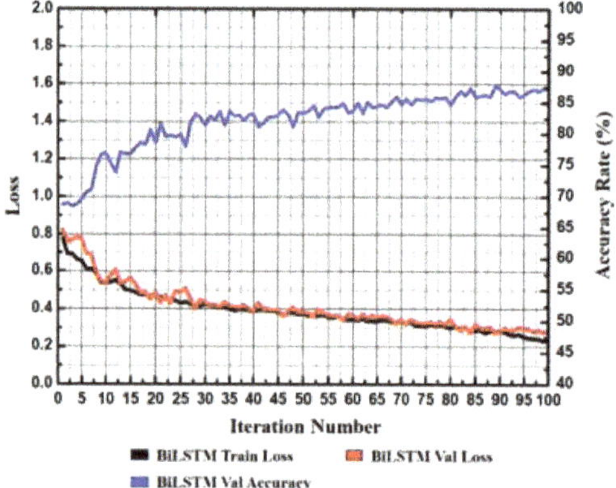

Figure 10. Loss function and accuracy of BiLSTM model training and verification.

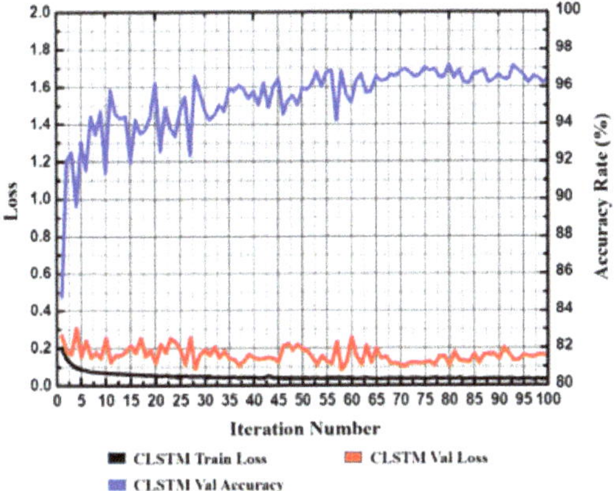

Figure 11. Loss function and accuracy of convolutional long short-term memory (CLSTM) model training and verification.

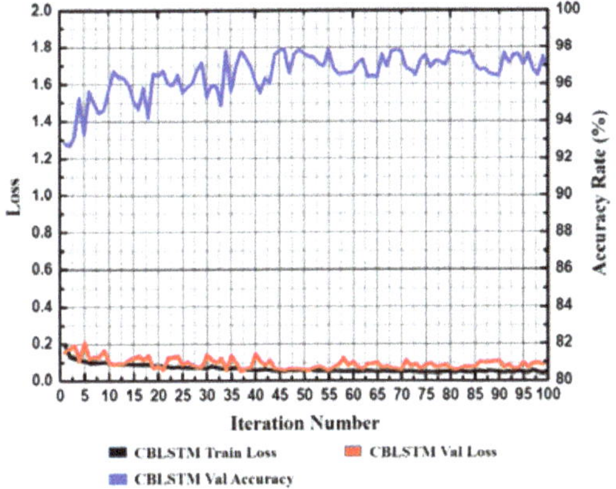

Figure 12. Loss function and accuracy of convolutional bi-directional long short-term memory (CBLSTM) model training and verification.

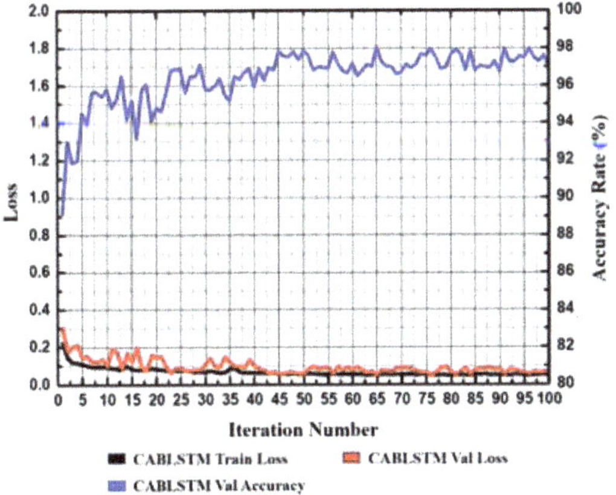

Figure 13. Loss function and accuracy of CABLSTM training and verification.

It can be concluded from the figure that the loss function value of the network model training set decreased with an increase in the number of iterations and finally stabilized. The loss function value of the verification set fluctuated periodically, and the loss function of the CLSTM model had a large amplitude. The CNN, BiLSTM, CBLSTM, and CABLSTM models were relatively stable, the overall trend of the loss function was decreasing and finally converging, there was no gradient explosion or dispersion phenomenon, and the network convergence speed was faster. The accuracy rates of the CNN and BiLSTM model validation sets were 87.57% and 86.36%, respectively, and the prediction accuracy was low. This result indicates that the individual DL network could predict the tool wear state, but deeper features could not be captured due to the limitation of the network model capability. There were deeper features hidden in the tool vibration signal. The network model proposed in this paper was superior to the CNN and BiLSTM network. This is because the network structure was relatively deep, which is conducive to mining deeper features. First, the CNN was used to extract the local features of the timing signals, which could effectively filter the noise in the original signal. At the same time, the length of the timing signal was reduced, which facilitates subsequent network learning depending on the time-series characteristics of the time-series signals and improved the ability of the model prediction.

In the network model proposed in this paper, the CABLSTM model had the best performance, which ewas superior to that of the CLSTM and CBLSTM models, and achieved high prediction accuracy. The initial prediction accuracy of the CLSTM model was relatively low. After 65 iterations, the accuracy of the verification set was basically stable and above 96%, and the accuracy was 96.42% after 100 iterations. The CBLSTM model used a two-way LSTM network to access past and future information; that is, it could extract timing signal features from both the forward and reverse directions and extract more abundant information features. After 42 iterations, the accuracy rate of the verification set was basically stable at over 96%, and the accuracy rate was 97.04% after 100 iterations. The CABLSTM model introduced the attention mechanism on the basis of CBLSTM, which selectively filtered out some key information from a large amount of information and focused on the key information, reducing the loss of key information features of long sequence texts. After 35 iterations, the accuracy of the verification set was basically stable and above 96%, the accuracy was 97.50% after 100 iterations, the loss function value reached 0.0651, and the network stability was higher. The loss function and the accuracy of the verification set and test set are shown in Table 5.

Table 5. Loss function and the accuracy of the verification set and test set.

Parameter	Loss	Verification Accuracy Rate (%)	Single Test Time/ms	Test Accuracy Rate (%)
CNN [25]	0.2688	88.34%	2	87.57%
BiLSTM [26]	0.2857	87.13%	20	86.36%
CLSTM	0.1608	96.42%	4	93.64%
CBLSTM	0.0931	97.04%	5	95.15%
CABLSTM	0.0651	97.50%	6	96.97%

The data of the milling cutter (C4) were selected as the test set of the DL network model to evaluate the generalization ability of the final model. The total number of test samples was 330, including 23 initial wear samples, 232 standard wear samples, and 75 sharp wear samples. The samples were randomly fed into the trained DL network model. The CABLSTM model had high precision and recall. The F1-score reaches the optimum value at 1 (perfect precision and recall), and the worst is 0. The F1-score in this paper was 0.9697. The evaluation indices of the CABLSTM model are shown in Table 6. The test results show that the CABLSTM model proposed in this paper hade a strong generalization ability. Although the test time was not as good as that of the partial comparison model, the algorithm found a good balance between time and precision.

Table 6. Evaluation indices of the CABLSTM model.

Label Classification	Precision	Recall	F1-Score	Support
0	0.9130	0.9130	0.9130	23
1	0.9703	0.9870	0.9786	232
2	0.9859	0.9333	0.9589	75
avg/total	0.9697	0.9697	0.9697	330

It can be concluded from the figure that the CABLSTM model proposed in this paper completed the inspection of the milling cutter (C4) with an accuracy of 96.97%. The predicted results of normal wear were more accurate. There were some deviations between the initial wear and sharp wear, but the deviations were within a reasonable range. The incorrect prediction results mainly occurred in the transition stage of the wear degree. This is because the tool was in the normal wear state for a long time during the machining process, the amount of data that could be learned by the model was relatively large, and the features were relatively distinct; in addition, the tool had a short period of initial wear and rapid wear, and the amount of data that could be obtained was insufficient. The confusion matrix of the wear test results of the tool test set is shown in Figure 14.

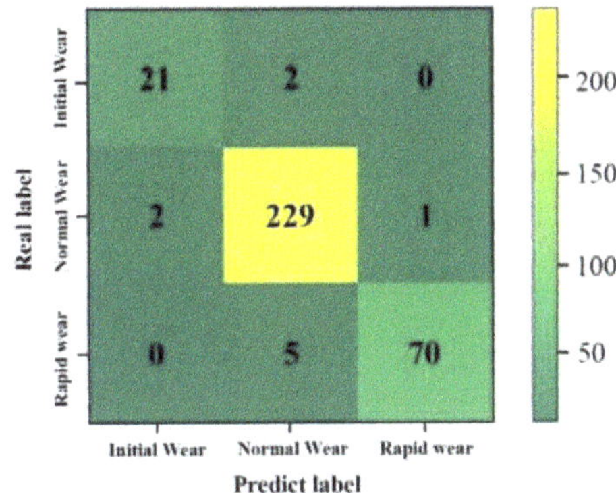

Figure 14. Confusion matrix of the wear test results of the tool test set.

When the real-time monitoring system of tool wear state was working, the acceleration sensors would bring a three-axis vibration signal of length 2000 to the monitoring model of the CABLSTM network. The model performed a forward calculation to identify the current tool wear state and achieve real-time monitoring of the tool wear state.

4.3. Comparison of Deep Learning and Machine Learning

To further validate the feasibility of the proposed model, a comparative experiment was designed with alternative ML models. The same data set used for DL was used in the experiment. More specifically, the commonly used models in traditional tool wear value detection approaches, including the BPNN, the SVM, the HMM, and the FNN, were compared with the CABLSTM model proposed in this paper. The wavelet threshold denoising method was used to perform noise reduction processing on the original signal collected by the acceleration sensor. The data features of the time domain, frequency domain, and time-frequency domain were extracted, and the specific extraction method is shown in Table 7. Pearson's correlation coefficient (PCC) was used to reflect the correlation between the feature and the wear value, and the feature with a correlation coefficient greater than 0.9 was selected as the extraction object to achieve a feature dimensionality reduction. The extracted features were used as the input of the ML model.

Table 7. Feature extraction category table of the machine learning (ML) models.

Feature Attribute	Feature Category	Extraction Method
Time domain feature	Maximum, Mean, Root mean square, Variance, Standard deviation, Skewness, Kurtosis, Peak, Peak factor	Statistical calculation
Frequency domain feature	Power spectrum maximum, Band energy value, Mean, Variance, Skewness, Kurtosis, Band peak	Fourier transform
Time-frequency domain feature	Node energy value	Wavelet packet transform

It can be concluded from Table 7 that the accuracy of traditional ML models varied greatly, which was due to the instability of the artificial extraction features, and the construction of the model would have an impact on the prediction results. The DL model proposed in this paper could achieve ideal results by adaptively extracting hidden high-dimensional features and reasonable network depth design for tool processing signals without data pre-processing. The prediction accuracy was significantly higher than that of the BPNN, SVM, and HMM. However, the prediction accuracy of the FNN reached 94.24% because the FNN used a neural network to learn the rules of the fuzzy system. According to the learning sample of the input and output, the design parameters of the fuzzy system were automatically designed and adjusted to realize the self-learning and adaptive functions of the fuzzy system. Compared with the other algorithm models, this method demonstrated a great improvement in performance. The test sample speed of the CABLSTM model could reach 6 ms, which could meet the requirements of real-time tool wear monitoring in industrial production. The accuracy of ML and DL prediction is shown in Table 8.

Table 8. Accuracy of machine learning and deep learning prediction.

Models	Parameter	Accuracy Rate (%)
Machine Learning	BPNN	84.85%
	SVM	91.21%
	HMM	85.76%
	FNN	94.24%
Deep Learning	CNN [25]	87.57%
	BiLSTM [26]	86.36%
	CLSTM	93.64%
	CBLSTM	95.15%
	CABLSTM	96.97%

5. Conclusions

In this paper, we proposed the application of a CNN and RNN fusion to real-time monitoring of a tool wear state and modified the network parameters and structure according to the characteristics of vibration signals to monitor the tool wear degree in real time. The prediction accuracy of the CBLSTM reached 96.97%. In the pre-processing stage, the wear state of the tool was defined according to the actual wear curve, which was used to determine the wear degree of the tool and improve the accuracy of the data label classification. At the same time, the experimental data were added to the original magnitude data to improve the robustness of the algorithm by employing the data expansion method. A one-dimensional CNN was used to extract the local features, and abundant high-dimensional features were extracted from the original signal, which avoided the limitation of the traditional manual feature extraction, better characterizede the hidden tool wear state information in the original signal, and shortened the network model training time. The idea of introducing the attention mechanism was innovatively applied to the improved CBLSTM network model, which effectively improved the recognition accuracy and generalization performance of the real-time monitoring. The experimental results show that the CABLSTM model had certain advantages in the real-time monitoring of tool wear, which could meet the industrial requirements in terms of recognition accuracy and recognition speed.

In the process of actual manufacturing, the processing procedures and site conditions were often complicated and variable. There were many features that could reflect the wear state of a tool. In this paper, the original signal collected by the acceleration sensor was used as the tool wear monitoring index, which was restricted by the training data volume and processing method. It might not be applicable to meet the requirements of arbitrary working conditions. In future work, multi-source data fusion technology and DL theory will be used to further study the information characterizing the wear state of the tool, improve the proposed method, and extend the method to industrial monitoring.

Author Contributions: Q.C. and Q.Y. conceived and designed the experiments; Q.C. and Y.L. performed the experiments; Q.C. and H.H. analyzed the data; Q.C. wrote the paper; Q.Y., Q.X., and Q.C. revised and polished the manuscript. All authors have read and approved the final manuscript.

Funding: This research was funded by the Guizhou Province Science and Technology Fund Project (Branch Support [2017] 2870), and Guizhou Province Education Department Science and Technology Talents Support Project (Branch Support KY [2017]062).

Acknowledgments: We gratefully acknowledge the support of NVIDIA Corporation with the donation of the Titan X GPU used for this research.

Conflicts of Interest: The authors declare no conflict of interest.

References

1. Lei, Y.G.; Jia, F.; Kong, D.T.; Lin, J.; Xing, S.B. Opportunities and Challenges of Machinery Intelligent Fault Diagnosis in Big Data Era. *J. Mech. Eng.* **2018**, *54*, 94–104. [CrossRef]
2. Mia, M.; Królczyk, G.; Maruda, R.; Wojciechowski, S. Intelligent Optimization of Hard-Turning Parameters Using Evolutionary Algorithms for Smart Manufacturing. *Materials* **2019**, *12*, 879. [CrossRef] [PubMed]
3. Andres, B.; Gorka, U.; Jose, M.P.; Octavio, M.P.; Luis, N.L. Smart optimization of a friction-drilling process based on boosting ensembles. *J. Manuf. Syst.* **2018**, *48*, 108–121.
4. Zhu, K.P.; Zhang, Y. A generic tool wear model and its application to force modeling and wear monitoring in high speed milling. *Mech. Syst. Signal Process.* **2018**, *115*, 147–161. [CrossRef]
5. Benkedjouh, T.; Zerhouni, N.; Rechak, S. Tool wear condition monitoring based on continuous wavelet transform and blind source separation. *Int. J. Adv. Manuf. Technol.* **2018**, *97*, 3311–3323. [CrossRef]
6. Zhou, Y.Q.; Xue, W. Review of Tool Condition Monitoring Methods in Milling Processes. *Int. J. Adv. Manuf.* **2018**, *96*, 2509–2523. [CrossRef]
7. Beranoagirre, A.; Urbikain, G.; Marticorena, R.; Bustillo, A.; Lacalle, L. Sensitivity Analysis of Tool Wear in Drilling of Titanium Aluminides. *Metals* **2019**, *9*, 297. [CrossRef]
8. Krahmer, D.M.; Hameed, S.; Egea, A.J.; Pérez, D.; Canales, J.; Lacalle, L.N. Wear and MnS Layer Adhesion in Uncoated Cutting Tools When Dry and Wet Turning Free-Cutting Steels. *Metals* **2019**, *9*, 556. [CrossRef]
9. Lacalle, L.N.; Fernandez-Larrinoa, J.; Rodriguez-Ezquerro, A.; Valdivielso, A.F.; Lopez-Blanco, R.; Azkona, I. On the cutting of wood for joinery applications. *Proc. Inst. Mech. Eng. Part B J. Eng. Manuf.* **2014**, *229*, 1–13.
10. Dutta, S.; Pal, S.K.; Mukhopadhyay, S.; Sen, R. Application of digital image processing in tool condition monitoring: A review. *CIRP J. Manuf. Sci. Technol.* **2013**, *6*, 212–232. [CrossRef]
11. Saglam, H.; Unuvar, A. Tool condition monitoring in milling based on cutting forces by a neural network. *Int. J. Product Res.* **2003**, *41*, 1519–1532. [CrossRef]
12. Chen, X.Z.; Li, B.Z. Acoustic emission method for tool condition monitoring based on wavelet analysis. *Int. J. Adv. Manuf. Technol.* **2007**, *33*, 968–976. [CrossRef]
13. Li, C.B.; Wan, T.; Chen, X.Z.; Lei, Y.F. On-line Monitoring Method of Tool Wear for NC Turning in Batch Processing Based on Cutting Power. *Comput. Integr. Manuf. Syst.* **2018**, *24*, 1910–1919.
14. Yesilyurt, I.; Ozturk, H. Tool condition monitoring in milling msing vibration analysis. *Int. J. Product Res.* **2007**, *45*, 1013–1028. [CrossRef]
15. Lui, Z.P. *Research on Pattern Recognition and Life Prediction of Tool Wear Based on Multi-sensor Information Fusion;* Southwest Jiaotong University: Nanjing, China, 2018.

16. Zhang, X.; Fu, H.Y.; Sun, Y.Z.; Han, Z.Y. Hidden Markov Model Based Micro-milling Tool Wear Monitoring. *Comput. Integr. Manuf. Syst.* **2012**, *18*, 141–148.
17. Li, X.H.; Lim, B.S.; Zhou, J.H.; Huang, S.; Phua, S.J.; Shaw, K.C.; Er, M.J. Fuzzy neural network modelling for tool wear estimation in dry milling operation. In Proceedings of the Annual Conference of the Prognostics and Health Management Society, PHM, Montreal, QC, Canada, 27 September-1 October 2009; pp. 1–11.
18. Liao, Z.R.; Li, S.M.; Lu, Y.; Dong, G. Tool Wear Identification in Turning Titanium Alloy Based on SVM. *Mater. Sci. Forum* **2014**, *800–801*, 446–450. [CrossRef]
19. LeCun, Y.; Bottou, L.; Bengio, Y.; Haffner, P. Gradient-based learning applied to document recognition. *Proc. IEEE* **1998**, *86*, 2278–2324. [CrossRef]
20. Minar, M.R.; Naher, J. Recent Advances in Deep Learning: An Overview. *arXiv* **2018**, arXiv:1807.08169.
21. Zhang, C.J.; Yao, X.F.; Zhang, J.M.; Liu, E.H. Tool Wear Monitoring Based on Deep Learning. *Comput. Integr. Manuf. Syst.* **2017**, *23*, 2146–2155.
22. Terrazas, G.; Martínez-Arellano, Z.; Benardos, P.; Ratchev, S. Online Tool Wear Classification during Dry Machining Using Real Time Cutting Force Measurements and a CNN Approach. *J. Manuf. Mater. Process.* **2018**, *2*, 72. [CrossRef]
23. Cao, D.L.; Sun, H.B.; Zhang, H.Z.; Mo, R. In-process Tool Condition Monitoring Based on Convolution Neural Network. *Comput. Integr. Manuf. Syst.* **2018**. Available online: https://kns.cnki.net/KCMS/detail/11.5946.TP.20180913.1536.020.html (accessed on 3 September 2019).
24. Zhao, R.; Yan, R.Q.; Wang, J.J.; Mao, K.Z. Learning to Monitor Machine Health with Convolutional Bi-Directional LSTM Networks. *Sensors* **2017**, *17*, 273. [CrossRef] [PubMed]
25. Zhang, W. *Study on Bearing Fault Diagnosis Algorithm Based on Convolutional Neural Network*; Harbin Institute of Technology: Harbin, China, 2017.
26. Zhang, A.S.; Wang, H.L.; Li, S.B.; Cui, Y.X.; Liu, Z.H.; Yang, G.C.; Hu, J.J. Transfer Learning with Deep Recurrent Neural Networks for Remaining Useful Life Estimation. *Appl. Sci.* **2018**, *8*, 2416. [CrossRef]
27. Li, X.D.; Ye, M.; Li, T. Review of Object Detection Based on Convolutional Neural Networks. *Appl. Res. Comput.* **2017**, *34*, 2881–2891.
28. Lipton, Z.C. A Critical Review of Recurrent Neural Networks for Sequence Learning. *arXiv* **2015**, arXiv:1506.00019.
29. Kolen, J.F.; Kremer, S.C. *Gradient Flow in Recurrent Nets: The Difficulty of Learning LongTerm Dependencies*; Wiley-IEEE Press: Hoboken, NJ, USA, 2007.
30. Hochreiter, S.; Schmidhuber, J. Long Short-term Memory. *Neural Comput.* **1997**, *9*, 1735–1780. [CrossRef]
31. Zhao, R.; Wang, J.J.; Yan, R.Q.; Mao, K.Z. Machine health monitoring with LSTM networks. In Proceedings of the 2016 10th International Conference on Sensing Technology (ICST), Nanjing, China, 11–13 November 2016.
32. Graves, A.; Schmidhuber, J. Framewise Phoneme Classification with Bidirectional LSTM and other Neural Network Architectures. *Neural Netw.* **2005**, *18*, 602–610. [CrossRef]
33. Andres, B.; Maritza, C.; Anibal, R. A Virtual Sensor for Online Fault Detection of Multitooth-Tools. *Sensors* **2011**, *11*, 2773–2795.
34. Mikołajczyk, T.; Nowicki, K.; Bustillo, A.; Pimenov, D.Y. Predicting tool life in turning operations using neural networks and image processing. *Mech. Syst. Signal Process.* **2011**, *104*, 503–513. [CrossRef]
35. Andres, B.; Juan, J.R. Online breakage detection of multitooth tools using classifier ensembles for imbalanced data. *Int. J. Syst. Sci.* **2013**, *45*, 2590–2602.

Article

A Study of the Effect of Medium Viscosity on Breakage Parameters for Wet Grinding

Adriana M. Osorio [1], Moisés O. Bustamante [2], Gloria M. Restrepo [1], Manuel M. M. López [3] and Juan M. Menéndez-Aguado [3,*]

1   Facultad de Ingeniería, Universidad de Antioquia, Calle 70 # 52-21, 050010 Medellin, Colombia; maradrio@gmail.com (A.M.O.); gloma@udea.edu.co (G.M.R.)
2   Instituto de Minerales CIMEX, Universidad Nacional de Colombia, Carrera 65 #63-20, 050010 Medellin, Colombia; mobustam@unal.edu.co
3   Escuela Politécnica de Mieres, Universidad de Oviedo, Calle Gonzalo Gutiérrez Quirós, s/n 33600 Mieres, Spain; mahamud@uniovi.es
*   Correspondence: maguado@uniovi.es

Received: 24 July 2019; Accepted: 19 September 2019; Published: 25 September 2019

Abstract: The rheological behavior of mineral slurries shows the level of interaction or aggregation among particles, being a process control variable in processes such as slurry transportation, dehydration, and wet grinding systems. With the aim to analyze the effect of medium viscosity in wet grinding, a series of monosize grinding ball mill tests were performed to determine breakage parameters, according to the generally accepted kinetic approach of grinding processes. A rheological modifier (polyacrylamide, PAM) was used to modify solutions viscosity. A model was proposed by means of dimensional analysis (Buckingham's Pi theorem) in order to determine the behavior of the specific breakage rate (S_j) for a ball grinding process in terms of the rheology of the system. In addition to this, a linear adjustment was established for the relationship between specific breakage rates with and without PAM addition, based on the reduced viscosity, μ_r. Furthermore, within a certain interval of viscosity, it was proved that an increment of viscosity can increase the specific breakage rate, and consequently the grinding degree.

Keywords: grinding; rheology; breakage parameters; relative viscosity

---

1. Introduction

The importance of comminution operations in the mineral processing industry has bolstered in the last decades the search for greater process knowledge. More accurate models have been proposed [1–9] with the aim to get a better phenomenological processes description and to overcome the apparent technological barriers related to energy efficiency in grinding operations. In the last decades, considerable work has been done on the optimization of energy consumption in grinding mills using phenomenological grinding kinetics models based on population balance (PB) considerations [10]. PB modeling is based on first order kinetics and uses two functions, namely the specific rate of breakage S_i and the breakage function (b_{ij}), which provide the fundamental size-mass balance equation for fully mixed batch grinding operations.

The rheological behavior of mineral slurries shows the level of interaction or aggregation among particles, being a process control variable in processes such as slurry transportation, dehydration, and wet grinding systems [11]. Ball mill grinding is one of the most used industrial comminution solutions [12–14], and it is a process that depends on different conditions such as mill dimensions, rotation speed, filling degree, ball size distribution in the charge, feed size distribution, etc. An exhaustive and broad review of the importance of rheology in mineral processing has been recently published by Cruz et al. [15].

The importance of solids concentration and slurry viscosity to determine the operating conditions has been remarked by Yin et al. [16]. It seems proven that the physic–chemical properties in the slurry can be modified in wet grinding due to changes in physical and chemical conditions (such as size distribution, concentration of solids, temperature, shear rate, pH value, the use of grinding aids, etc.). The use of polyacrylamide (PAM) in water as a viscosity modifier has been broadly studied previously [17,18].

The main goal of this study was to set a model by means of dimensional analysis in order to analyze the behavior of the specific breakage rate (S_j) for a ball grinding process in terms of the rheology of the system. In addition to this, a linear adjustment could be established for the relationship between specific breakage rates with and without PAM addition, based on the reduced viscosity, μ_r. Furthermore, within the validity interval of viscosity, it will be studied whether an increment of viscosity can increase the specific breakage rate, and consequently the grinding degree.

2. Methodology

2.1. Materials

To carry out the tests, representative samples in a quartz ore quarry were selected and characterized with X-ray fluorescence in a Bruker XRF, model S-4 Pioneer. Trace element detection was performed through ICP-OES, after digestion in aqua regia, in a Varian, model Vista-PRO. To perform the mineralogical characterization through X-ray diffraction, a Bruker XRD device, model D8 Advance was used. Results of XRF characterization are shown in Table 1, and ICP-OES results are shown in Table 2.

Mineralogical analysis results are shown in Figure 1, in which it is clear the abundance of quartz, with the presence of clay components as illite and nacrite. This sample has been used in previous research work and further information about it can be found in Menendez-Aguado et al. [19].

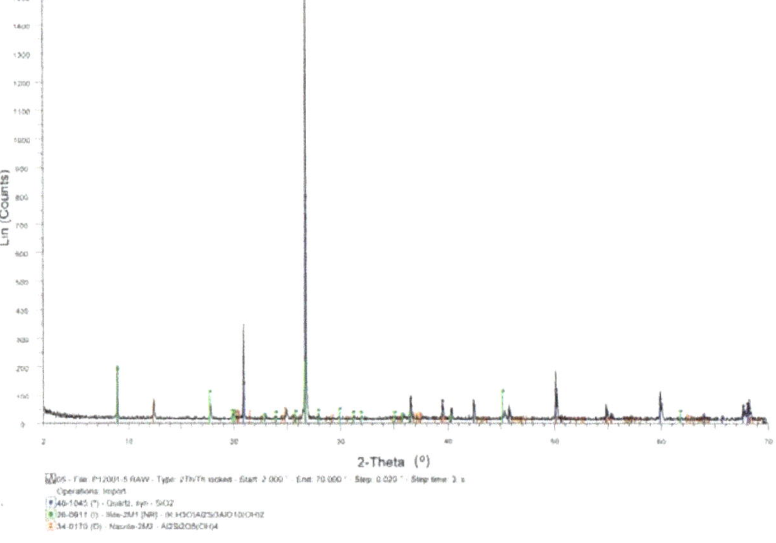

Figure 1. Results from XRD characterization.

Table 1. Mineralogical composition of the quartz sample by XRF (%).

Al$_2$O$_3$	SiO$_2$	Fe$_2$O$_3$	TiO$_2$	CaO	MgO	Na$_2$O	K$_2$O	P$_2$O$_5$	LOI
6.49	89.42	0.73	0.67	<0.1	<0.1	0.03	1.10	<0.1	1.54

Table 2. Trace elements in the quartz sample by ICP-OES (ppb).

As	Ba	Sr	Sb	Co	Cr	Cu	Cd	Hg	Pb	Zn	Zr	Ni	Mn	Sn
18	142	15	<10	14	31	22	<10	<10	<10	<10	85	<10	17	<10

2.2. Methods

To carry out the grinding tests, quartz monosize slurries were carefully prepared through sieving, and ASTM standard sieves were used to prepare the samples and determine the particle sizes. Quartz monosizes were prepared through careful wet sieving within the intervals 53/45, 45/38, and 38/30 microns, naming each monosize with the greater size in the interval. Slurries with different viscosity fluids were prepared (1, 4, 6 y 8 cP); suspension liquid viscosity was modified with different levels of PAM addition to the suspending liquid, distilled water. The PAM used was commercial grade (Sigma-Aldrich, CAS number 9003-05-8), with average molecular weight 40,000 g/mol. It is important to remark here that the low value of the molecular weight dismiss the possibility of aggregation effects on particles [20,21]. Viscosity measurements were carried out in a Brookfield viscometer with accessories, and specific gravity of suspensions was determined using a Marcy scale.

In order to determine breakage parameters, more specifically the specific breakage rate, a series of grinding tests were carried out at each fluid viscosity (including dry grinding test) and at each grinding time (0.5, 1, 3, 5, and 10 min) in laboratory a jar mill (Figure 2). Each test was repeated for each ball size diameter. A block diagram depicting the grinding tests performed for each monosize is shown in Figure 3. The grinding parameters that were fixed during the tests are shown in Table 3. Grinding tests were performed in a laboratory ball mill (0.16 m in diameter and 0.18 m long), with grinding charge made of manganese steel alloy balls, with diameters 2, 3, and 4 cm. To carry out the viscosity measurements, a Brookfield RVDV-2 +Pro device was used (Figure 4), with a shear rate of 66.93 s^{-1}. This value was experimentally determined considering a fixed rotational speed of 100 rpm, and according to the Brookfield recommendations for the selected configuration.

Figure 2. Jar mill.

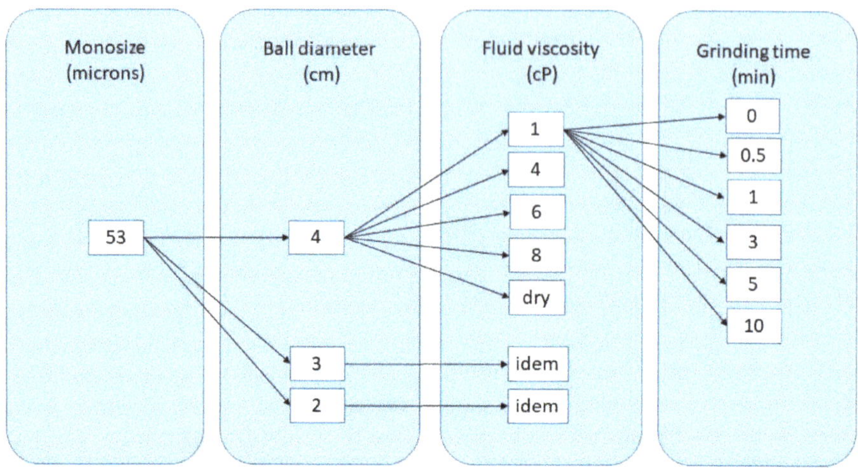

Figure 3. Block diagram, monosize 53 microns.

Table 3. Mill operating conditions.

Sample	Quartz
Solids concentration, ϕ (%*w/v*)	60
Mill lentgh, L (m)	0.18
Mill diameter, D (m)	0.16
L/D ratio	1.16
Fraction of critical speed, ϕ_C	0.75
Ball filling fraction, J	0.3
Hole fraction, U	1.0
Bed normal porosity	0.4

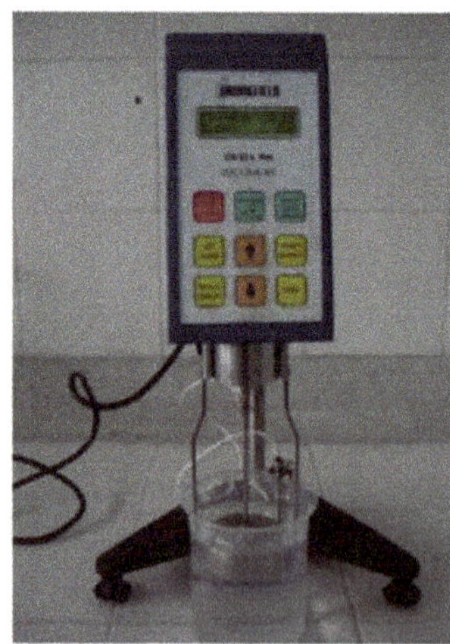

Figure 4. Viscometer.

Buckingham's Pi theorem was used to determine the final model. It states that if there are n variables in a problem and these variables contain m primary dimensions the equation relating all the variables will have $(n-m)$ dimensionless groups, which are referred to as Π groups. The Π groups must be independent of each other and no one group should be formed by multiplying together powers of other groups. [22–24].

The objective was to analyze the specific breakage rate, S_j, in terms of the variables that it is expected to have more influence on its behavior:

- viscosity of the suspending fluid, μ_l;
- viscosity of the suspension, μ_s;
- particle diameter, d_p;
- density of the grinding media, ρ_b.

3. Results and Discussion

Table 4 shows the specific breakage rates found for the different grinding tests, in the case of 2, 3, and 4 cm ball diameter. The correlation coefficients show that the results of the tests follow a first order law. These values are similar to those mentioned by Tangsathitkulchai [25,26], who found values of specific breakage rate for quartz grinding between 0.21 and 0.24 min^{-1}.

Table 4. Specific breakage rates for the different tests.

		Ball Diameter (cm)					
		2		3		4	
Monosize (μm)	Suspension Fluid Viscosity (cP)	S_j (min^{-1})	R^2	S_j (min^{-1})	R^2	S_j (min^{-1})	R^2
53	1	0.112	0.9793	0.152	0.9771	0.205	0.9826
53	4	0.215	0.9843	0.171	0.9455	0.155	0.9612
53	6	0.220	0.9739	0.175	0.9892	0.252	0.9890
53	8	0.198	0.9860	0.213	0.9796	0.271	0.9809
45	1	0.108	0.9532	0.153	0.9640	0.044	0.9820
45	4	0.141	0.9814	0.151	0.9834	0.092	0.9869
45	6	0.178	0.9619	0.158	0.9620	0.177	0.9871
45	8	0.087	0.9567	0.165	0.9829	0.199	0.9896
38	1	0.057	0.9638	0.107	0.9797	0.030	0.921
38	4	0.113	0.9824	0.130	0.9891	0.077	0.9706
38	6	0.161	0.9885	0.155	0.9870	0.100	0.9780
38	8	0.044	0.9820	0.163	0.9723	0.053	0.9891

Figures 5–7 show the effect of suspension viscosity on the specific breakage rate, S_j, in the case of each monosize. Greater values of S_j pose higher probability of particle breakage, so in general terms it seems that, within certain intervals, an increase in the viscosity of suspension can entail an increase in the breakage probability. This effect is more evident in the case of monosize 53 microns (Figure 5), while in the case of monosizes 45 and 38 microns over some value of suspension viscosity, an increase in suspension viscosity reduces the probability of particle breakage (Figures 6 and 7). This behavior is different in the case of different ball sizes in the grinding charge, although in the case of 3 cm the variations in S_j due to suspension viscosity changes is less noticeable. This probably can be caused by hydrodynamic reasons due to the combination of particle and ball dimensions, and rheological conditions, but this way of analysis evidences that the relationships among variables and their influence cannot be easily analyzed and understood, for further research should be performed to define a clear picture of the phenomenological model. This is the reason why an alternative way of analyzing these data was tried and introduced in this paper. To do this, the following notation is going to be used: We will refer as S_{jw} in the case of specific breakage rate when the fluid is just water, and S_{jd} in the rest of the cases in which the water properties have been modified with the dissolution of the abovementioned reagent.

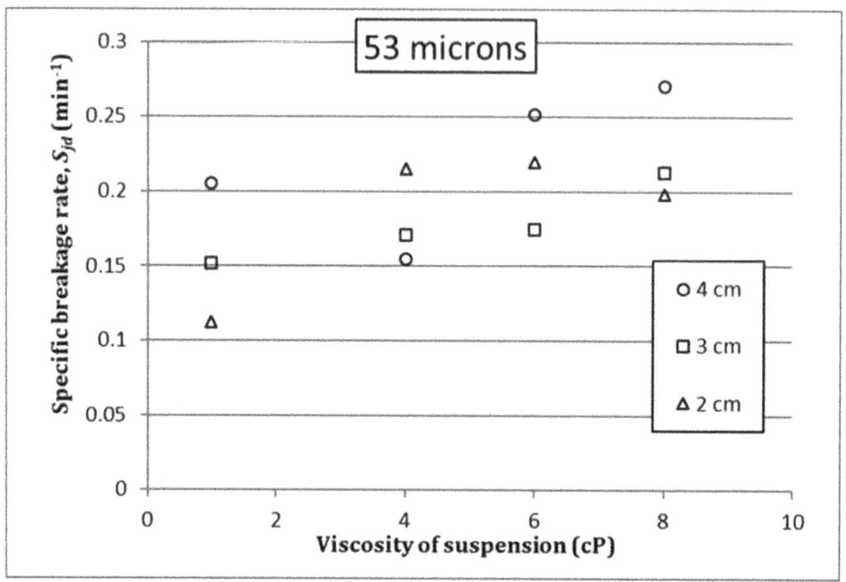

Figure 5. Effect of suspension viscosity on S_{jd}, monosize 53 microns.

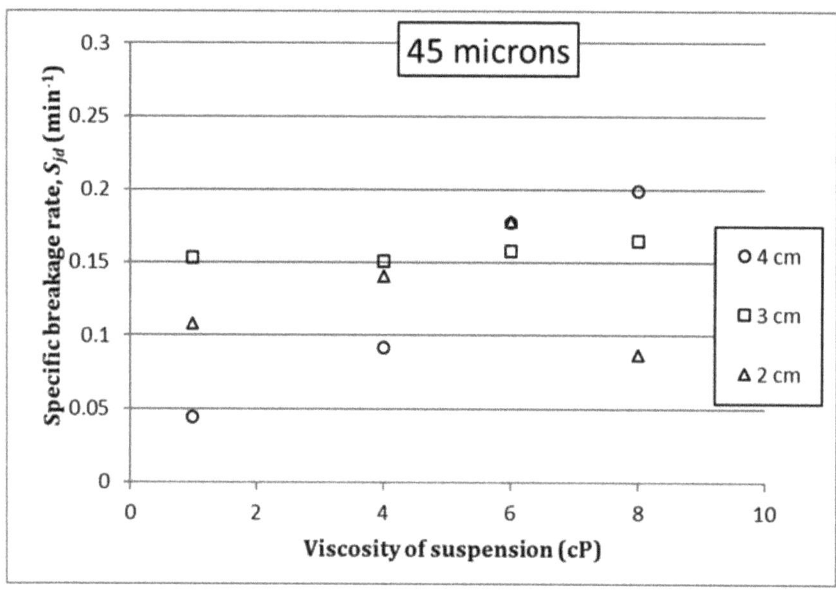

Figure 6. Effect of suspension viscosity on S_{jd}, monosize 45 microns.

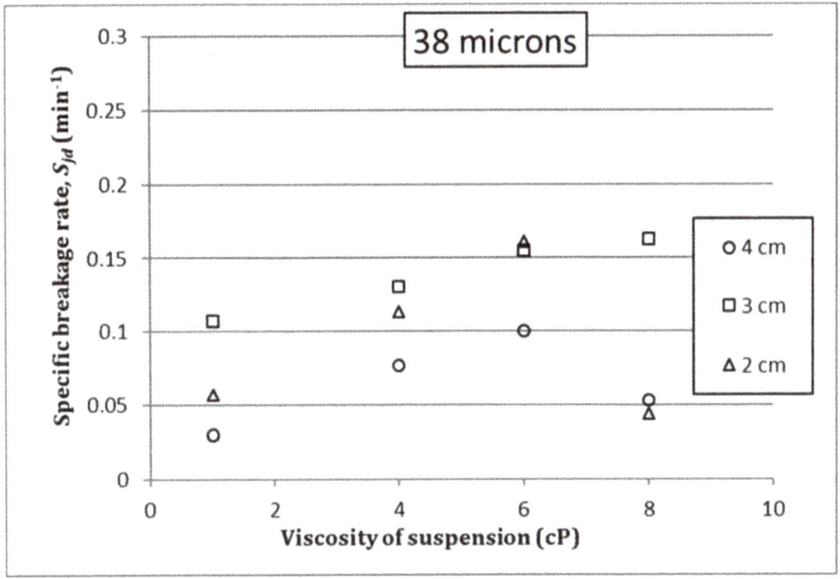

Figure 7. Effect of suspension viscosity on S_{jd}, monosize 38 microns.

With the aim of performing an analysis of the data obtained in the laboratory tests, the effect of different variables was considered when defining the Π groups. According to the Buckingham's Pi theorem, the following non-dimensional Π groups were defined:

$$\Pi_1 = \frac{S_{jw}}{S_{jd}} \tag{1}$$

$$\Pi_2 = \frac{\mu_s}{\rho_b \cdot S_{jd} \cdot d_p^2} \tag{2}$$

$$\Pi_3 = \frac{\mu_l}{\rho_b \cdot S_{jd} \cdot d_p^2} \tag{3}$$

Being dimensionless parameters, the following relationship can be considered:

$$\frac{\Pi_2}{\Pi_3} = \frac{\mu_s}{\mu_l} = \mu_r \tag{4}$$

where μ_r is the reduced viscosity.

According to the expressions above, when plotting Π_1 versus Π_2/Π_3 (see Figure 8) we can analyze the behavior the specific grinding rate at different specific gravities as a function of μ_r. Figures 8–10 show the behavior of the abovementioned relationship for the systems studied. In all cases, an increase of μ_r poses a decrease in Π_1, that is, an increase in S_{jd}. In most of the cases the linearity obtained can be satisfactory, although in the case of 2 cm diameter balls, the R^2 values are lower in the case of the monosizes 45 and 38 microns.

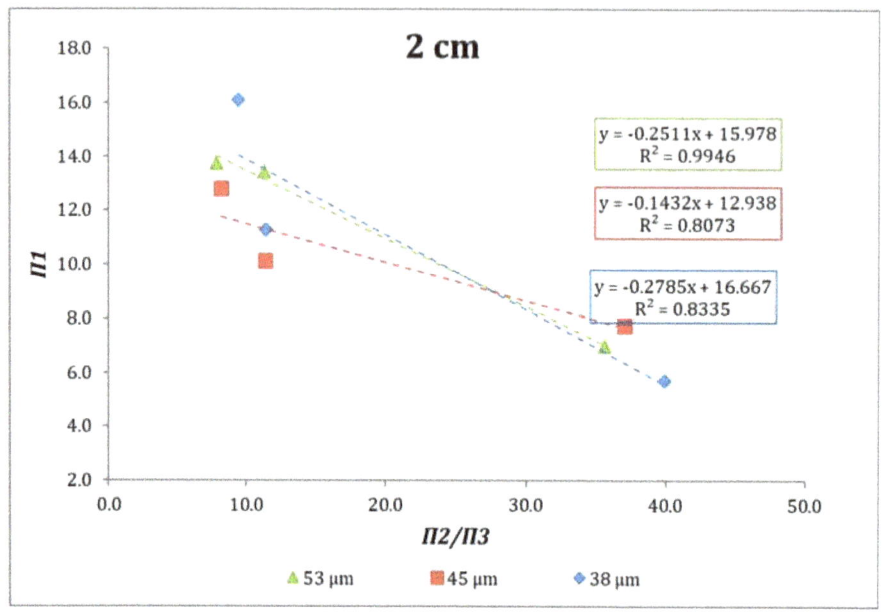

Figure 8. Fitting to the developed model, ball size 2 cm.

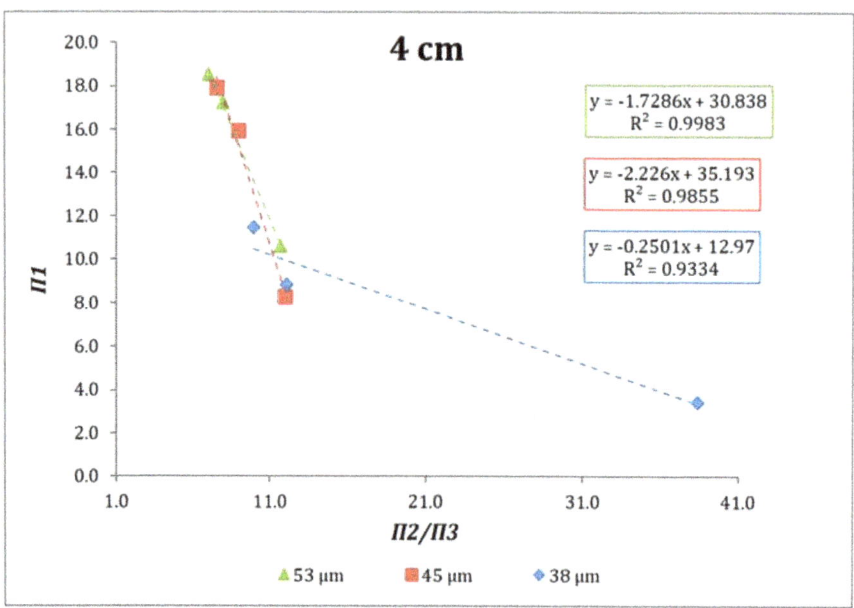

Figure 9. Fitting to the developed model, ball size 3 cm.

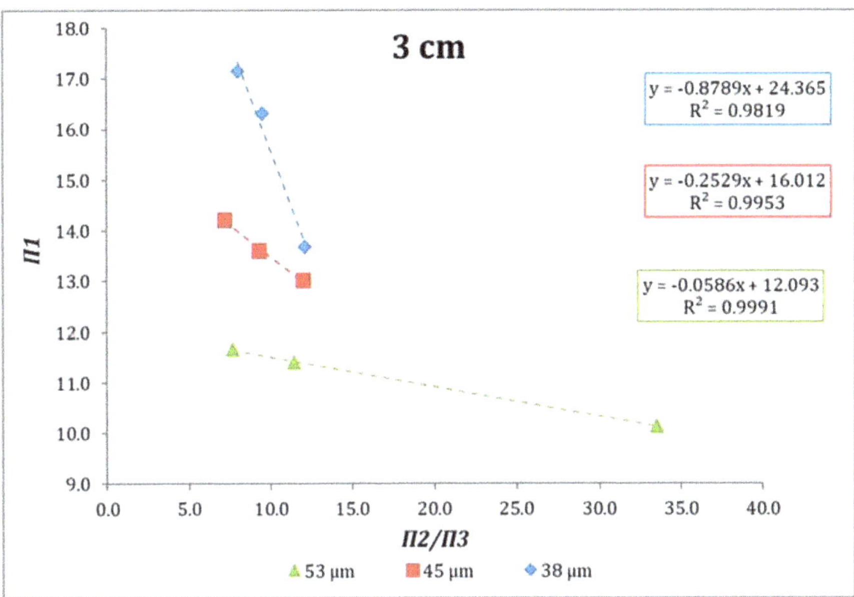

Figure 10. Fitting to the developed model, ball size 4 cm.

It can be observed that there is a reduced viscosity range for which the relationship between the specific fracture rates can be considered linear. The coefficients of the models that represent each system and the suspending fluid viscosity range for which it is valid are shown in Table 5 below.

Table 5. Linear adjustment parameters for the relationship S_{jw}/S_{jd} as a function of μ_r.

Ball Size (cm)	Monosize (µm)	a	b	R^2	Range of Linearity
	53	−0.2511	15.9780	0.9946	1–6 cp
2	45	−0.1432	12.9380	0.8073	1–6 cp
	38	−0.2785	16.6670	0.8335	1–6 cp
	53	−0.0586	12.0930	0.9991	1–6 cp
3	45	−0.2529	16.0120	0.9953	4–8 cp
	38	−0.8789	24.3650	0.9819	4–8 cp
	53	−1.7286	30.8380	0.9983	4–8 cp
4	45	−2.2260	35.1930	0.9855	4–8 cp
	38	−0.2501	12.9700	0.9334	1–6 cp

Results suggest that the viscosity of the suspension fluid influences the specific breakage rate, and that this influence is different depending on some other conditions, as is the case of ball diameter. The explanation to this phenomenon is that within a given range, an increase in the viscosity increases the chance of particle breakage as the mobility decreases within the slurry. Besides, when the viscosity is increased over a given range, the positive effect of decreasing particle mobility on grinding kinetics is reduced due to energy dissipation in ball dynamics, resulting in an inefficient fragmentation process. Thus, it could be stated that under any grinding conditions there can be defined a fluid viscosity that maximizes the breakage kinetics, and this can be an interesting strategy to optimize the grinding process from the point of view of power efficiency. Nevertheless, further research is needed to get a better understanding of the process and to set a reliable conceptual background in the development of that optimization strategy.

4. Conclusions

The influence of fluid viscosity on specific breakage rate has been studied under different test conditions. In general terms, specific breakage rate can be modified when increasing the suspension viscosity, within certain ranges. For the test conditions, that relationship could be established using the dimensional analysis by means of Pi-Buckingham's theorem for different grinding conditions. Results suggest that there is a reduced viscosity range for which the relationship between the specific fracture rates can be considered linear. Furthermore, it can be stated that under any grinding conditions a fluid viscosity that maximizes the breakage kinetics can be defined, and this can be an interesting strategy to optimize the grinding process from the point of view of power efficiency. Nevertheless, further research is needed to get a better understanding of the process and to set a reliable conceptual background in the development of that optimization strategy.

Author Contributions: Conceptualization, A.M.O. and M.O.B.; Methodology, J.M.M.-A. and A.M.O.; Software, A.M.O. and G.M.R.; Formal analysis, A.M.O.; Investigation, A.M.O. and J.M.M.-A.; Writing—original draft preparation, A.M.O., M.O.B. and G.M.R.; Writing—review and editing, J.M.M.-A. and M.M.M.L.; Supervision, J.M.M.-A.; Project administration, A.M.O. and G.M.R.; Funding acquisition, A.M.O. and J.M.M.-A.

Funding: This research received no external funding.

Acknowledgments: Authors thank to COLCIENCIAS, the University of Antioquia (Colombia) and OPTIMORE H2020 project (Grant n. 642201) for research support.

Conflicts of Interest: The authors declare no conflicts of interest.

References

1. Coello Velázquez, A.L.; Menéndez-Aguado, J.M.; Brown, R.L. Grindability of lateritic nickel ores in Cuba. *Powder Technol.* **2008**, *182*, 113–115. [CrossRef]
2. Aguado, J.M.M.; Velázquez, A.L.C.; Tijonov, O.N.; Díaz, M.A.R. Implementation of energy sustainability concepts during the comminution process of the Punta Gorda nickel ore plant (Cuba). *Powder Technol.* **2006**, *170*, 153–157. [CrossRef]
3. Umucu, Y.; Altınigne, M.Y.; Deniz, V. The effects of ball types on breakage parameters of barite. *J. Pol. Min. Eng. Soc.* **2014**, *15*, 113–117.
4. Zhang, W. Optimizing Performance of SABC Comminution Circuit of the Wushan Porphyry Copper Mine—A Practical Approach. *Minerals* **2016**, *6*, 127. [CrossRef]
5. Liang, G.; Wei, D.; Xu, X.; Xia, X.; Li, Y. Study on the Selection of Comminution Circuits for a Magnetite Ore in Eastern Hebei, China. *Minerals* **2016**, *6*, 39. [CrossRef]
6. Petrakis, E.; Komnitsas, K. Improved Modeling of the Grinding Process through the Combined Use of Matrix and Population Balance Models. *Minerals* **2017**, *7*, 67. [CrossRef]
7. Mariño-Salguero, J.; Jorge, J.; Menéndez-Aguado, J.M.; Álvarez-Rodriguez, B.; De Felipe, J.J. Heat generation model in the ball-milling process of a tantalum ore. *Miner. Metall. Process.* **2017**, *34*, 10–19. [CrossRef]
8. Pedrayes, F.; Norniella, J.G.; Melero, M.G.; Menéndez-Aguado, J.M.; del Coz-Díaz, J.J. Frequency domain characterization of torque in tumbling ball mills using DEM modelling: Application to filling level monitoring. *Powder Technol.* **2018**, *323*, 433–444. [CrossRef]
9. Lee, H.; Kim, K.; Lee, H. Analysis of grinding kinetics in a laboratory ball mill using population-balance-model and discrete-element-method. *Adv. Powder Technol.* **2019**. [CrossRef]
10. Petrakis, E.; Stamboliadis, E.; Komnitsas, K. Identification of Optimal Mill Operating Parameters during Grinding of Quartz with the Use of Population Balance Modeling. *KONA Powder Part. J.* **2017**, *34*, 213–223. [CrossRef]
11. Muster, T.H.; Prestidge, C.A. Rheological investigations of sulphide mineral slurries. *Miner. Eng.* **1995**, *8*, 1541–1555. [CrossRef]
12. Klimpel, R.R. Slurry rheology influence on the performance of mineral/coal grinding circuits. Part I. *Miner. Eng.* **1982**, *34*, 1665–1688.
13. Klimpel, R.R. Slurry rheology influence on the performance of mineral/coal grinding circuits. Part II. *Miner. Eng.* **1982**, *35*, 21–26.

14. Ding, Z.; Yin, Z.; Liu, L.; Chen, Q. Effect of grinding parameters on the rheology of pyrite-heptane slurry in a laboratory stirred media mill. *Miner. Eng.* **2007**, *20*, 701–709. [CrossRef]

15. Cruz, N.; Forster, J.; Bobicki, E.R. Slurry rheology in mineral processing unit operations: A critical review. *Can. J. Chem. Eng.* **2019**, *97*, 2102–2120. [CrossRef]

16. Yin, Z.; Peng, Y.; Zhu, Z.; Ma, C.; Yu, Z.; Wu, G. Effect of mill speed and slurry filling on the charge dynamics by an instrumented ball. *Adv. Powder Technol.* **2019**, *30*, 1611–1616. [CrossRef]

17. Barvenik, F.W. Polyacrylamide characteristics related to soil applications. *Soil Sci.* **1994**, *158*, 235–243. [CrossRef]

18. Jung, J.; Jang, J.; Ahn, J. Characterization of a Polyacrylamide Solution Used for Remediation of Petroleum Contaminated Soils. *Materials* **2016**, *9*, 16. [CrossRef] [PubMed]

19. Menéndez-Aguado, L.D.; Marina Sánchez, M.; Rodríguez, M.A.; Coello Velázquez, A.L.; Menéndez-Aguado, J.M. Recycled Mineral Raw Materials from Quarry Waste Using Hydrocyclones. *Materials* **2019**, *12*, 2047.

20. Schild, H.G. Poly(*N*-isopropylacrylamide): Experiment. theory and application. *Prog. Polym. Sci.* **1992**, *17*, 163–249. [CrossRef]

21. Nasser, M.S.; James, A.E. The effect of polyacrylamide charge density and molecular weight on the flocculation and sedimentation behaviour of kaolinite suspensions. *Sep. Purif. Technol.* **2006**, *52*, 241–252. [CrossRef]

22. Langhaar, H.L. *Dimensional Analysis and Theory of Models*; Wiley: New York, NY, USA, 1951.

23. Curtis, W.D.; David Logan, J.; Parker, W.A. Dimensional Analysis and the Pi Theorem. *Linear Algebra Appl.* **1982**, *47*, 117–126. [CrossRef]

24. Pobedrya, B.E.; Georgievskii, D.V. On the proof of the Pi-Theorem in Dimension Theory. *Russ. J. Math. Phys.* **2006**, *13*, 431–437. [CrossRef]

25. Tangsathikulchai, C. The effect of slurry rheology on fine grinding in a laboratory ball mill. *Int. J. Miner. Process.* **2003**, *69*, 29–47. [CrossRef]

26. Tangsathikulchai, C. Effect of Medium Viscosity on Breakage Parameters of Quartz in a Laboratory Ball-Mill. *Ind. Eng. Chem. Res.* **2004**, *43*, 2104–2112. [CrossRef]

Article

An Improved Butterfly Optimization Algorithm for Engineering Design Problems Using the Cross-Entropy Method

Guocheng Li [1,*], **Fei Shuang** [2], **Pan Zhao** [1] **and Chengyi Le** [3,*]

[1] School of Finance and Mathematics, West Anhui University, Lu'an 237012, China
[2] Department of Mechanical and Aerospace Engineering, University of Florida, Gainesville, FL 32611, USA
[3] School of Economics and Management, East China Jiaotong University, Nanchang 330013, China
* Correspondence: liguocheng@wxc.edu.cn (G.L.); ncycy@126.com (C.L.);
Tel.: +86-564-330-5031 (G.L.); +86-791-8704-5037 (C.L.)

Received: 4 July 2019; Accepted: 9 August 2019; Published: 14 August 2019

Abstract: Engineering design optimization in real life is a challenging global optimization problem, and many meta-heuristic algorithms have been proposed to obtain the global best solutions. An excellent meta-heuristic algorithm has two symmetric search capabilities: local search and global search. In this paper, an improved Butterfly Optimization Algorithm (BOA) is developed by embedding the cross-entropy (CE) method into the original BOA. Based on a co-evolution technique, this new method achieves a proper balance between exploration and exploitation to enhance its global search capability, and effectively avoid it falling into a local optimum. The performance of the proposed approach was evaluated on 19 well-known benchmark test functions and three classical engineering design problems. The results of the test functions show that the proposed algorithm can provide very competitive results in terms of improved exploration, local optima avoidance, exploitation, and convergence rate. The results of the engineering problems prove that the new approach is applicable to challenging problems with constrained and unknown search spaces.

Keywords: global optimization; meta-heuristic; butterfly optimization algorithm; cross-entropy method; engineering design problems

---

1. Introduction

Real-world engineering design optimization problems are very challenging to find the global optimum of a highly complex and multiextremal objective function, involving many different decision variables under complex constraints [1,2]. A general engineering design optimization problem can be stated as follows:

$$\min f(x), x = (x_1, x_2, \cdots, x_n)^T \in R^n, \tag{1}$$

subject to

$$g_j(x) \leq 0, j = 1, 2, \cdots, p, \tag{2}$$

$$h_k(x) = 0, k = 1, 2, \cdots, q, \tag{3}$$

$$lb_i \leq x_i \leq ub_i, i = 1, 2, \cdots, n, \tag{4}$$

where n is the number of search space dimensions, $g_j(x)$ and $h_k(x)$ are the jth inequality constraint and kth equality constraint, respectively. lb_i and ub_i represent the lower and upper bound of the value of x_i.

Most of the constraints of the global optimization problem are nonlinear. Such nonlinearity often results in a multimodal response landscape [3]. Due to their high complexity, nonlinearity, and multimodality, traditional optimization methods, such as gradient-based optimization algorithms, are no longer suitable for this problem [4]. Many researchers have developed various derivative-free global optimization methods for it. In general, these methods can be divided into two classes: deterministic methods and stochastic meta-heuristic algorithms [4,5]. The former, such as the Hill-Climbing [6], Newton–Raphson [7], and DIRECT algorithm [8], can obtain the same optimal results based on the same set of initial values. However, this behavior results in local optima entrapment and a loss of reliability in finding the global optimum, since most real engineering design problems have extremely high numbers of local solutions [9]. The latter, such as the Genetic Algorithm (GA) [10,11], Particle Swarm Optimization (PSO) [12], Cuckoo Search (CS) [13], Bat Algorithm (BA) [14], Grey Wolf Optimizer (GWO) [15], Forest Optimization Algorithm (FOA) [16,17], Ant Lion Optimizer (ALO) [9], Whale Optimization Algorithm (WOA) [18], Crow Search Algorithm (CSA) [19], Salp Swarm Algorithm (SSA) [20], and Butterfly Optimization Algorithm (BOA) [21,22], mostly benefit from their simplicity, flexibility, and stochastic operators, which makes them different from deterministic methods [4], and they have become very popular optimization techniques for such optimization problems in recent years.

Despite the merits of nature-inspired meta-heuristic algorithms, it has been pointed out that most of them are unable to guarantee global convergence [23]. At the same time, the No Free Lunch (NFL) theorem has proved that none of these methods are able to solve all global optimization problems [24]. They perform very well when dealing with certain optimization problems, but fail in most cases. An excellent meta-heuristic algorithm has two symmetric search capabilities: local search and global search, and achieves a proper balance between exploration and exploitation to effectively avoid it falling into a local, and find the global optimum. In order to enhance global search capabilities, many hybrid meta-heuristic algorithms have been developed by combining meta-heuristics with exact algorithms or other meta-heuristics for solving more complicated optimization problems [25–29].

Inspired by the idea of global random search algorithms [5,30], and based on co-evolution, this paper explores an improved butterfly optimization algorithm (BOA) using the cross-entropy method, which is a global stochastic optimization method based on a statistical model. The original BOA was proposed by Arora and Singh [21] in 2015, and was inspired by the food foraging behavior of butterflies, while the cross-entropy (CE) method was developed by Rubinstein [31] in 1997 to estimate the probability of rare events in complex random networks. Since the BOA has a tendency to prematurely converge to local optima [32], this paper embeds the CE method into the BOA to obtain a good balance between exploration and exploitation and improve the BOA's global search capability.

The rest of this paper is structured as follows. In Section 2, the BOA and CE method are briefly introduced, and a study of the improved BOA is presented in Section 3. The results and a discussion of benchmark functions are provided in Section 4. Section 5 solves three classical engineering design problems. In Section 6, conclusions and future research directions are presented.

2. Methods

2.1. Butterfly Optimization Algorithm

The BOA is a new nature-inspired meta-heuristic algorithm which was inspired by the food foraging behavior of butterflies [21,22]. In this method, the characteristics of butterflies are ideally summarized as follows:

1. All butterflies are assumed to emit some fragrance that makes the butterflies attractive to each other;

2. Each butterfly moves randomly or toward the best butterfly—i.e., the one which emits the most fragrance;

3. The stimulus intensity of a butterfly is determined by the landscape of the objective function.

In the BOA, the perceived magnitude of the fragrance (f) is defined as a function of the stimulus physical intensity as follows:

$$f = cI^a, \tag{5}$$

where $c \in [0, \infty]$ is the sensory modality; I is the stimulus intensity, which is associated with the encoded objective function; and $a \in [0, 1]$ is the power exponent dependent on modality, which represents the varying degree of fragrance absorption.

There are two key steps in the BOA: global search and local search. The former can make the butterflies move toward the best butterfly, which can be represented as

$$x_i^{t+1} = x_i^t + (r^2 \times g^* - x_i^t) \times f_i, \tag{6}$$

where x_i^t is the position of the ith butterfly at time t. Here, g^* represents the current best position. f_i represents the fragrance of the ith butterfly, and r is a random number in $[0, 1]$.

The latter is implemented through a local random walk, which can be represented as

$$x_i^{t+1} = x_i^t + (r^2 \times x_j^t - x_k^t) \times f_i, \tag{7}$$

where x_j^t and x_k^t are the positions of the jth and kth butterflies in the solution space, respectively, and r is a random number in $[0, 1]$.

Additionally, a switch probability p is used in the BOA to switch between a common global search and an intensive local search. Based on the above, the original BOA can be implemented with pseudo-code as shown in Algorithm 1:

Algorithm 1: Butterfly Optimization Algorithm

Begin

    Objective function $f(x)$, $x = (x_1, x_2, \ldots, x_d)^T$, here d represents the number of dimensions.

    Generate initial population P containing n butterflies $pop_i (i = 1, 2, \ldots, n)$.

    Stimulus I_i intensity at is pop_i determined by the fitness value $f(pop_i)$.

    Define sensor modality c, power exponent a, and switch probability p.

    while stop criteria not met **do**

        for each butterfly in the population P **do**

            Calculate fragrance f using Equation (5).

        end for

        Evaluate and rank the population P, and find the best butterfly.

        for each butterfly in the population P **do**

            Generate a random number $r \sim U[0, 1]$.

            if $(r < p)$

                Implement global search using Equation (6).

            else

                Implement local search using Equation (7).

            end if

        end for

        Update the value of the power exponent a.

    end while

    Output the best solution and optimal value.

End

BOA shows better performance compared to some other optimization algorithms [21,22] and has attracted the attention of many researchers due to its simplicity and interesting nature-inspired interpretations as an optimization approach for global optimization problems and for various real-life applications [32,33].

2.2. The Cross-Entropy (CE) Method

Based on Monte Carlo technology, Rubinstein [31] developed the CE method in 1997 and uses the cross-entropy or Kullback–Leibler divergence to measure the distance between two sampling distributions, solve an optimization problem by minimizing this distance, and obtain the optimal parameters of probability distribution. The CE method has good global search capability, excellent adaptability, and strong robustness. It is frequently used for many complex optimization problems such as continuous multi-extremal optimization [34], multi-objective optimization [35], combination optimization [36,37], vehicle routing problems [38], energy-efficient radio resource management [39], and complex optimization problems from other fields [40–43].

A general optimization problem can be stated as follows:

$$\min_{x \in \mathcal{X}} S(x), \tag{8}$$

where $\mathcal{X}$ is a finite set of states, and S is a real-valued performance function on $\mathcal{X}$.

Next, we can construct a probability distribution estimation problem associated with the above problem, and the auxiliary problem can be defined as follows:

$$l(\gamma) = P_u(S(X) \leq \gamma) = E_u[I_{\{S(X) \leq \gamma\}}], \tag{9}$$

where P_u is the probability measure under which the random state X has some probability density function $f(x; u)$ on $\mathcal{X}$; E_u is the corresponding expectation operator; γ represents a threshold parameter; and I represents the indicator function, whose value is 1, if $S(X) \leq \gamma$, and 0, otherwise. The importance sampling approach is used to reduce the sample size in the CE method. Consequently, Equation (9) can be rewritten as follows:

$$l(\gamma) = \frac{1}{N} \sum_{i=1}^{N} I_{S(X) \leq \gamma} \frac{f(x^i; v)}{g(x^i)}, \tag{10}$$

where x^i represents a random sample from $f(x; v)$ with importance sampling density $g(x)$. the Kullback–Leibler divergence, i.e., the cross-entropy is introduced to measure the distance between two sampling distributions for obtaining the optimal importance sampling density. Thus, the optimal density $g^*(x)$ can be found by minimizing the Kullback–Leibler divergence, which is equivalent to solving the following optimization problem [34]:

$$\min_{v} \frac{1}{N} \sum_{i=1}^{N} I_{S(X) \leq \gamma} \ln f(x^i; v). \tag{11}$$

In order to improve the convergence speed of the CE method, adaptive smoothing $\hat{v}_k$ is demoted by $\tilde{v}$, as follows:

$$\hat{v}_{k+1} = \alpha \tilde{v} + (1 - \alpha)\hat{v}_k, \tag{12}$$

where $0 \leq \alpha \leq 1$ is a smoothing parameter.

The CE method for optimization problems is described in Algorithm 2.

Algorithm 2: Cross-entropy (CE) for optimization problems

Begin
 Set $t = 0$. Initialize the probability parameter $\hat{v}_k$.
 while stop criteria for CE not met **do**
  Generate $S_1, S_2, \ldots, S_N \sim_{iid} f(x; \hat{v}_k)$. Evaluate and rank the sample S.
  Solve the problem given in Equation (11) based on the sample S. Denote the solution
 by $\tilde{v}$.
  Update the parameter $\hat{v}_k$ using $\tilde{v}$.
  Set $t = t + 1$.
 end while
 Output the best solution and optimal value.
End

---

3. Hybrid BOA-CE Method

This section presents the details of the improved algorithm, named BOA-CE. A meta-heuristic method should have two main functions—exploration and exploitation—and should try to find a proper balance between them for better performance [44]. The nature-inspired BOA has shown the advantages of excellent local search capability and fast convergence; however, it tends to fall into a local optimum rather than finding the global optimum [21,22,32,33]. In order to improve the global search capability of the BOA, and, based on a co-evolutionary technique, this paper proposes the BOA-CE algorithm by embedding the CE method into it. The new method contains BOA and CE optimization operators, and co-updates the BOA population P and the CE sample S using co-evolutionary technology in each iteration. The CE operator obtains the initial probability parameters using the BOA population P in order to improve its convergence rate while the BOA operator updates its population P and the best butterfly using the elite sample S_e of the CE operator to enhance the population diversity. The pseudo-code of the BOA-CE hybrid algorithm is described in Algorithm 3. Figure 1 presents the flow chart of the BOA-CE algorithm, and the co-evolutionary process between BOA operator and CE operator can be clearly seen from Figure 1.

In addition, as an example, we use the BOA-CE algorithm to find the global optimum of the 2D Sphere function in order to more clearly describe the co-evolutionary process of the BOA operator and the CE operator:

$$f(x) = x_1{}^2 + x_2{}^2, \quad -100 \le x_1, x_2 \le 100, \tag{13}$$

which has a global minimum $f^* = 0$ at $x^* = (0,0)^T$. Now, let us use the BOA-CE algorithm to find the optima and this process is summarized as follows: (1) the values of the parameters of the BOA operator are $c = 0.01$, $a = 0.1$, $p = 0.8$, the population size $n = 40$, and the maximum number of iteration $t_1 = 50$, while the values of the parameters of the CE operator are $\alpha = 0.8$, the sample size $N = 80$, the elite sample size $Ne = 30$, and the maximum number of iteration $l_2 = 2$; (2) initializing the population of the BOA operator P, calculating $f_i(x) = x_{i1}^2 + x_{i2}^2, i = 1, 2, \cdots, 40$, and finding the current best $f^* = 13.5036$ and $x^* = (-3.5586, -0.9164)^T$; (3) generating a rand number r, and updating x_i with $x_i^{new} = x_i^{old} + (r^2 \times f^* - x_i^{old}) \times f_i$ if $r \le p$ and $x_i^{new} = x_i^{old} + (r^2 \times x_j^{old} - x_k^{old}) \times f_i$ otherwise for each butterfly in the P; (4) calculating the mean $\mu = (-4.9571, -11.4528)$ and standard $\sigma = (60.2592, 53.9977)$ deviation of P, and using them to initialize the probability parameter $v = (-4.9571, -11.4528, 60.2592, 53.9977)$; (5) generating a sample S by the probability parameter v, and calculating $f_i(x) = x_{i1}^2 + x_{i2}^2, i = 1, 2, \cdots, 80$, and finding the new current best $f^* = 4.4290$ and $x^* = (1.8693, 0.9668)^T$; (6) updating the probability parameter v using the elite sample, and regenerating a sample S using the new parameter v, and finding the new current best $f^* = 2.4542$ and $x^* = (0.8951, 1.2857)^T$; (7) co-updating the population P using the sample S and the population P; (8) repeating the above operations until the termination condition ($t_1 > 50$) is met, and outputting the approximate optimal solution $x^* = (1.1154 \times 10^{-31}, -2.5727 \times 10^{-32})^T$ and function value $f^* = 1.3104 \times 10^{-62}$. Figure 2 clearly describes the of co-updating the current best f^* by the BOA operator and CE operator.

Algorithm 3: BOA-CE algorithm

Begin

    Objective function $f(x)$, $x = (x_1, x_2, \ldots, x_d)^T$; here, d represents the number of dimensions.

    Generate initial population P containing n butterflies $pop_i (i = 1, 2, \ldots, n)$.

    Stimulus I_i intensity at is pop_i determined by the fitness value $f(pop_i)$.

    Define sensor modality c, power exponent a, and switch probability p.

    while stop criteria not met **do**

        for each butterfly in the population P **do**

            Calculate fragrance f using Equation (5).

        end for

        Evaluate,rank P, and find the best butterfly.

        for each butterfly in the population P **do**

            Generate a random number $r \sim U[0, 1]$.

            if $(r < p)$

                Implement global search using Equation (6).

            else

                Implement local search using Equation (7).

            end if

        end for

        Update the value of the power exponent a.

        Evaluate and rank the population P.

        Initialize the probability parameter $\hat{v}_k$ using the population P.

        while stop criteria for CE not met **do**

            Generate $S_1, S_2, \ldots, S_N \sim_{iid} f(x; \hat{v}_k)$, evaluate the sample S.

            Rank the population P and the sample S together, co-update P and S, update the best butterfly.

            Calculate the probability parameter $\tilde{v}$ by the elite sample S_e.

            Update the probability parameters $\hat{v}_k$ via Equation (12).

        end while

    end while

    Output the best solution and optimal value.

End

The BOA-CE hybrid algorithm has two main operators: BOA and CE. The BOA operator has two inner loops for the butterflies population size n_1, and one outer loop for iteration t_1. While the CE operator has one inner loops for the sample size n_2, and two outer loops for iterations t_1 and t_2, respectively. Therefore, for our proposed hybrid algorithm (BOA-CE), the time complexity is $O(2n_1 t_1 + n_2 t_1 t_2)$. It is linear in terms of $t_1 t_2$, which represents the total number of iterations in the BOA-CE.

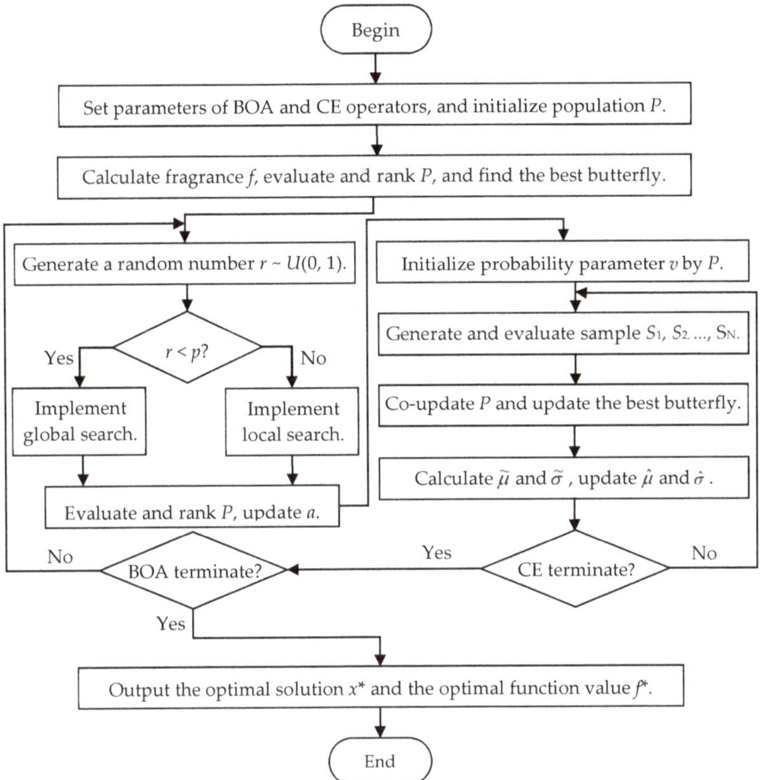

Figure 1. The flow chart of the BOA-CE algorithm.

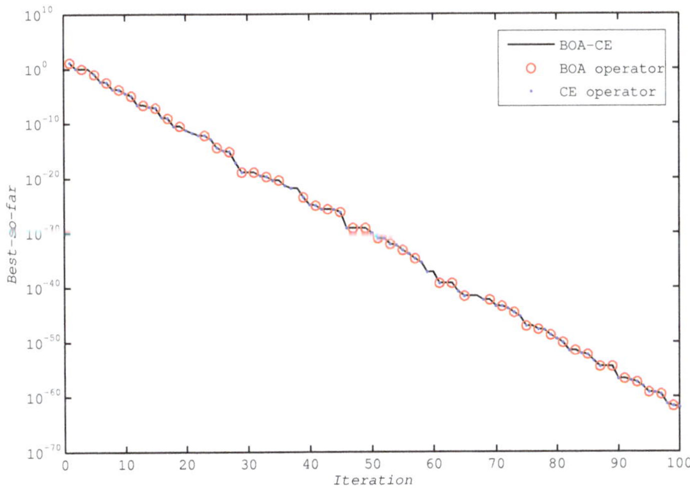

Figure 2. The BOA operator and CE operator co-update the current best f^* of the 2D Sphere function.

4. Experiment and Results

In this section, a total of 19 test functions [45,46] with different characteristics were employed to benchmark the performance of the proposed BOA-CE algorithm from different perspectives [9,15,16,20,28]. The test functions can be divided into three classes: unimodal (Table 1), multimodal (Table 2), and composite functions (Table 3). Generally, unimodal functions are employed to benchmark the exploitation and convergence of a method, while multimodal functions are selected to evaluate the performance of exploration and local optima avoidance [9,20]. In contrast, composite functions can be used to evaluate the combined performance of exploration and exploitation.

Table 1. Unimodal benchmark functions.

Function	Dim	Range	F_{min}
$F_1(x) = \sum_{i=1}^{n} x_i^2$	30	$[-100, 100]$	0
$F_2(x) = \sum_{i=1}^{n} |x_i| + \prod_{i=1}^{n} |x_i|$	30	$[-10, 10]$	0
$F_3(x) = \sum_{i=1}^{n} (\sum_{j=1}^{i} x_j)^2$	30	$[-100, 100]$	0
$F_4(x) = \max_i \{|x_i|, 1 \leq i \leq n\}$	30	$[-100, 100]$	0
$F_5(x) = \sum_{i=1}^{n-1} [100(x_{i+1} - x_i^2)^2 + (x_i - 1)^2]$	30	$[-30, 30]$	0
$F_6(x) = \sum_{i=1}^{n} ([x_i + 0.5])^2$	30	$[-100, 100]$	0
$F_7(x) = \sum_{i=1}^{n} i x_i^4 + random[0, 1)$	30	$[-1.28, 1.28]$	0

Table 2. Multi-modal benchmark functions.

Function	Dim	Range	F_{min}
$F_8(x) = \sum_{i=1}^{n} -x_i \sin(\sqrt{|x_i|})$	30	$[-500, 500]$	$-12{,}569.487$
$F_9(x) = \sum_{i=1}^{n-1} [x_i^2 - 10\cos(2\pi x_i) + 10]$	30	$[-5.12, 5.12]$	0
$F_{10}(x) = -20\exp\left(-0.2\sqrt{\frac{1}{n}\sum_{i=1}^{n} x_i^2}\right) - \exp\left(\frac{1}{n}\sum_{i=1}^{n}\cos(2\pi x_i)\right) + 20 + e$	30	$[-32, 32]$	0
$F_{11}(x) = \frac{1}{4000}\sum_{i=1}^{n} x_i^2 - \prod_{i=1}^{n}\cos\left(\frac{x_i}{\sqrt{i}}\right) + 1$	30	$[-600, 600]$	0
$F_{12}(x) = \frac{\pi}{n}\{10\sin^2(\pi y_1) + \sum_{i=1}^{n-1}(y_i - 1)^2[1 + 10\sin^2(\pi y_{i+1}) + (y_n + 1)^2]\} + \sum_{i=1}^{n} u(x_i, 10, 100, 4)$ $y_i = 1 + \frac{x_i + 1}{4}$ $u(x_i, a, k, m) = \begin{cases} k(x_i - a)^m, & x_i > a \\ 0, & -a \leq x_i \leq a \\ k(-x_i - a)^m, & x_i < -a \end{cases}$	30	$[-50, 50]$	0
$F_{13}(x) = 0.1\{\sin^2(3\pi x_1) + \sum_{i=1}^{n}(x_i - 1)^2[1 + \sin^2(3\pi x_i + 1)]\} + \sum_{i=1}^{n} u(x_i, 5, 100, 4)$	30	$[-50, 50]$	0

To verify the results, the proposed BOA-CE algorithm was compared against a number of well-known and recent algorithms: Genetic Algorithm (GA) [10], PSO [12], BA [14], GWO [15], CSA [19], SSA [20], and BOA [31]. The variants were coded in the MATLAB R2018b, running on a PC with an Intel Core i7-8700 processor (Gainesville, FL, USA), a 3.19 GHz CPU, and 16 GB of RAM. To provide a fair comparison, the following test experimental conditions and settings were used: (1) the population size of the BOA operator in the BOA-CE algorithm was set to 40, while the CE operator's sample size was 80. The maximum number of iterations of the BOA operator and CE operator were 500 and 2, respectively; (2) the other method's population sizes were 100 for fair comparison, and their maximum number of iterations was 1000; (3) all of the other parameters in each algorithm for comparison were as the same as those used in the original references [10,12,14,15,19,20,31]. Each of the algorithms was run 30 times on each of the test functions, and the results are shown in Tables 4–6. The average value and standard deviation of the best solution obtained by each method is given to compare their overall performance. The winner (best value) is highlighted in bold.

Table 3. Composite benchmark functions.

Function [1]	Dim	Range	F_{min}
F_{14}(CF1) $f_1, f_2, \ldots, f_{10} =$ Sphere Function, $[\lambda_1, \lambda_2, \ldots, \lambda_{10}] = [1, 1, \ldots, 1], [\sigma_1, \sigma_2, \ldots, \sigma_{10}] = [1/100, 5/100, \ldots, 5/100]$	10	$[-5, 5]$	0
F_{15}(CF2) $f_1, f_2, \ldots, f_{10} =$ Griewank's Function, $[\lambda_1, \lambda_2, \ldots, \lambda_{10}] = [1, 1, \ldots, 1], [\sigma_1, \sigma_2, \ldots, \sigma_{10}] = [1/100, 5/100, \ldots, 5/100]$	10	$[-5, 5]$	0
F_{16}(CF3) $f_1, f_2, \ldots, f_{10} =$ Griewank's Function, $[\lambda_1, \lambda_2, \ldots, \lambda_{10}] = [1, 1, \ldots, 1], [\sigma_1, \sigma_2, \ldots, \sigma_{10}] = [1, 1, \ldots, 1]$	10	$[-5, 5]$	0
F_{17}(CF4) $f_1, f_2 =$ Ackley's Function, $f_3, f_4 =$ Rastrigin's Function, $f_5, f_6 =$ Weierstrass Function, $f_7, f_8 =$ Griewank's Function, $f_9, f_{10} =$ Sphere Function, $[\lambda_1, \lambda_2, \ldots, \lambda_{10}] = [1, 1, \ldots, 1],$ $[\sigma_1, \sigma_2, \ldots, \sigma_{10}] = [\frac{5}{32}, \frac{5}{32}, 1, 1, \frac{5}{0.5}, \frac{5}{0.5}, \frac{5}{100}, \frac{5}{100}, \frac{5}{100}, \frac{5}{100}]$	10	$[-5, 5]$	0
F_{18}(CF5) $f_1, f_2 =$ Rastrigin's Function, $f_3, f_4 =$ Weierstrass Function, $f_5, f_6 =$ Griewank's Function, $f_7, f_8 =$ Ackley'sGriewank's Function, $f_9, f_{10} =$ Sphere Function, $[\lambda_1, \lambda_2, \ldots, \lambda_{10}] = [1, 1, \ldots, 1],$ $[\sigma_1, \sigma_2, \ldots, \sigma_{10}] = [\frac{1}{5}, \frac{1}{5}, \frac{5}{0.5}, \frac{5}{0.5}, \frac{5}{100}, \frac{5}{100}, \frac{5}{32}, \frac{5}{32}, \frac{5}{100}, \frac{5}{100}]$	10	$[-5, 5]$	0
F_{19}(CF6) $f_1, f_2 =$ Rastrigin's Function, $f_3, f_4 =$ Weierstrass Function, $f_5, f_6 =$ Griewank's Function, $f_7, f_8 =$ Ackley'sGriewank's Function, $f_9, f_{10} =$ Sphere Function, $[\lambda_1, \lambda_2, \ldots, \lambda_{10}] = [1, 1, \ldots, 1],$ $[\lambda_1, \lambda_2, \ldots, \lambda_{10}] = [0.1, 0.2, 0.3, 0.4, 0.5, 0.6, 0.7, 0.8, 0.9, 1],$ $[\sigma_1, \sigma_2, \ldots, \sigma_{10}] = [0.1 \times \frac{1}{5}, 0.2 \times \frac{1}{5}, 0.3 \times 0.4 \times \frac{5}{0.5}, 0.5 \times \frac{5}{0.5},$ $0.6 \times \frac{5}{100}, 0.7 \times \frac{5}{100}, 0.8 \times \frac{5}{32}, 0.9 \times \frac{5}{32}, 1 \times \frac{5}{100}, \frac{5}{100}]$	10	$[-5, 5]$	0

[1] Mathematical formulation of Sphere, Ackley, Rastrigin, Weierstrass, and Griewank can be found in Appendix A A1.

4.1. Results of the Algorithms Using Unimodal Test Functions

The results of the application of the algorithms to unimodal test functions are shown in Table 4. The best results are highlighted in bold. From Table 4, it can be found that: (1) the proposed method outperforms GA, PSO, BA, CSA, SSA, and BOA for the majority of the test functions; and (2) BOA-CE and GWO each provide the best results for three of these problems. This is due to the co-evolutionary technology that are adopted between the BOA and CE operators to enhance exploitation.

The progress of the average best value over 30 runs for some of the benchmark functions, such as F1, F2, F6, and F7, is illustrated in Figure 3. The figure clearly shows that the proposed BOA-CE algorithm rapidly converges towards the optimum and exploits it accurately. This benefit is mainly due to the algorithm's excellent exploitation.

4.2. Results of the Algorithms Using Multimodal Test Functions

The results of the application of the algorithms to multimodal test functions are shown in Table 5. Table 5 shows that the BOA-CE method outperforms the other approaches in the majority of the test cases. Multimodal test functions are employed to benchmark the performance in terms of exploration and local optima avoidance. Therefore, these results prove that the BOA-CE algorithm has an excellent exploration which helps it to explore the promising regions of the search space. Additionally, the local optima avoidance of the improved method is also excellent, since it is able to avoid all of the local optima and find the global optimum for the majority of the multimodal test functions.

Table 4. Results of unimodal benchmark functions.

Fun.	Metr.	GA	PSO	BA	GWO	CSA	SSA	BOA	BOA-CE
F_1	Mean	1.44×10^{-07}	1.43×10^{-15}	1.46×10^{-04}	1.97×10^{-90}	1.20×10^{-03}	6.70×10^{-09}	1.57×10^{-14}	$\mathbf{1.26 \times 10^{-95}}$
	SD	4.32×10^{-07}	3.42×10^{-15}	7.98×10^{-04}	5.36×10^{-90}	5.64×10^{-04}	1.23×10^{-09}	7.69×10^{-16}	$\mathbf{9.78 \times 10^{-96}}$
F_2	Mean	1.48×10^{-02}	7.50×10^{-07}	$2.57 \times 10^{+01}$	$\mathbf{2.92 \times 10^{-52}}$	7.16×10^{-01}	6.00×10^{-06}	9.14×10^{-12}	3.60×10^{-47}
	SD	6.05×10^{-02}	3.30×10^{-06}	$3.93 \times 10^{+01}$	$\mathbf{3.35 \times 10^{-52}}$	4.51×10^{-01}	1.33×10^{-06}	1.37×10^{-12}	2.50×10^{-47}
F_3	Mean	5.20×10^{-01}	$2.30 \times 10^{+00}$	$2.42 \times 10^{+03}$	$\mathbf{1.74 \times 10^{-26}}$	$8.30 \times 10^{+00}$	8.33×10^{-10}	1.60×10^{-14}	2.44×10^{-09}
	SD	1.95×10^{-01}	$1.18 \times 10^{+00}$	$2.18 \times 10^{+03}$	$\mathbf{5.09 \times 10^{-26}}$	$4.74 \times 10^{+00}$	3.05×10^{-10}	9.73×10^{-16}	1.83×10^{-10}
F_4	Mean	1.88×10^{-01}	2.01×10^{-01}	$3.56 \times 10^{+01}$	$\mathbf{1.74 \times 10^{-21}}$	$1.35 \times 10^{+00}$	1.23×10^{-05}	1.09×10^{-11}	8.82×10^{-07}
	SD	4.63×10^{-01}	6.33×10^{-02}	6.27×10^{-13}	$\mathbf{2.81 \times 10^{-21}}$	6.84×10^{-01}	1.99×10^{-06}	6.72×10^{-13}	1.51×10^{-07}
F_5	Mean	$1.87 \times 10^{+01}$	$4.28 \times 10^{+01}$	$1.47 \times 10^{+02}$	$2.62 \times 10^{+01}$	$4.19 \times 10^{+01}$	$\mathbf{1.26 \times 10^{+01}}$	$2.89 \times 10^{+01}$	$2.79 \times 10^{+01}$
	SD	$2.82 \times 10^{+01}$	$3.07 \times 10^{+01}$	$2.71 \times 10^{+02}$	6.30×10^{-01}	$2.82 \times 10^{+01}$	$2.86 \times 10^{+01}$	$\mathbf{3.95 \times 10^{-02}}$	2.56×10^{-01}
F_6	Mean	1.56×10^{-07}	1.47×10^{-15}	1.06×10^{-14}	1.10×10^{-01}	9.65×10^{-04}	4.74×10^{-10}	$5.05 \times 10^{+00}$	**0**
	SD	2.37×10^{-07}	2.20×10^{-15}	5.40×10^{-14}	1.58×10^{-01}	4.19×10^{-04}	1.44×10^{-10}	5.34×10^{-01}	**0**
F_7	Mean	3.79×10^{-01}	3.22×10^{-02}	1.25×10^{-01}	3.99×10^{-04}	9.87×10^{-03}	2.06×10^{-03}	6.68×10^{-04}	$\mathbf{3.29 \times 10^{-04}}$
	SD	9.79×10^{-02}	1.12×10^{-02}	4.30×10^{-02}	1.65×10^{-04}	4.50×10^{-03}	1.22×10^{-03}	3.25×10^{-04}	$\mathbf{1.52 \times 10^{-04}}$

Table 5. Results of multi-modal benchmark functions.

Fun.	Metr.	GA	PSO	BA	GWO	CSA	SSA	BOA	BOA-CE
F_8	Mean	$\mathbf{-10,486.64}$	-6500.75	-3525.02	-6941.03	-7294.61	-2824.35	-4246.74	-4017.15
	SD	626.18	819.29	$\mathbf{192.65}$	641.14	801.93	211.32	350.36	323.25
F_9	Mean	$1.33 \times 10^{+00}$	3.24×10^{-01}	$1.35 \times 10^{+02}$	$3.45 \times 10^{+01}$	$1.87 \times 10^{+01}$	$1.21 \times 10^{+01}$	$\mathbf{3.03 \times 10^{-14}}$	$1.01 \times 10^{+01}$
	SD	$1.23 \times 10^{+00}$	$1.78 \times 10^{+00}$	$3.38 \times 10^{+01}$	$1.14 \times 10^{+01}$	$7.82 \times 10^{+00}$	$5.86 \times 10^{+00}$	$\mathbf{1.46 \times 10^{-13}}$	$3.46 \times 10^{+01}$
F_{10}	Mean	1.76×10^{-04}	1.10×10^{-14}	$1.64 \times 10^{+01}$	1.89×10^{-08}	$2.56 \times 10^{+00}$	4.24×10^{-01}	1.01×10^{-11}	$\mathbf{4.44 \times 10^{-15}}$
	SD	7.59×10^{-05}	3.11×10^{-15}	$2.30 \times 10^{+00}$	1.72×10^{-08}	6.99×10^{-01}	6.81×10^{-01}	1.10×10^{-12}	**0**
F_{11}	Mean	1.23×10^{-03}	8.70×10^{-04}	$1.22 \times 10^{+02}$	1.29×10^{-02}	2.89×10^{-02}	2.47×10^{-01}	5.59×10^{-16}	**0**
	SD	3.41×10^{-03}	2.68×10^{-03}	$5.54 \times 10^{+01}$	1.45×10^{-02}	1.34×10^{-02}	1.46×10^{-01}	5.64×10^{-16}	**0**
F_{12}	Mean	5.19×10^{-02}	1.35×10^{-02}	$1.82 \times 10^{+01}$	$\mathbf{9.96 \times 10^{-18}}$	6.25×10^{-01}	8.53×10^{-02}	3.84×10^{-01}	2.02×10^{-05}
	SD	1.12×10^{-01}	9.37×10^{-03}	$5.84 \times 10^{+00}$	$\mathbf{1.13 \times 10^{-17}}$	5.39×10^{-01}	2.11×10^{-01}	1.13×10^{-01}	8.76×10^{-05}
F_{13}	Mean	5.09×10^{-03}	1.51×10^{-01}	$3.85 \times 10^{+01}$	1.10×10^{-03}	1.20×10^{-02}	7.32×10^{-04}	$2.24 \times 10^{+00}$	$\mathbf{1.35 \times 10^{-32}}$
	SD	1.23×10^{-02}	1.43×10^{-01}	$1.62 \times 10^{+01}$	3.35×10^{-03}	1.51×10^{-02}	2.79×10^{-030}	4.04×10^{-01}	$\mathbf{5.57 \times 10^{-48}}$

Table 6. Results of composite benchmark functions.

Fun.	Metr.	GA	PSO	BA	GWO	CSA	SSA	BOA	BOA-CE
F_{14}	Mean	46.67	70.00	70.00	45.90	26.67	43.33	270.28	**23.33**
	SD	50.74	95.23	83.67	71.84	44.98	67.89	58.82	**43.02**
F_{15}	Mean	57.91	128.09	139.61	115.17	116.59	**27.77**	262.27	104.76
	SD	62.53	76.97	95.29	82.14	61.98	**17.39**	110.66	72.26
F_{16}	Mean	154.18	172.96	348.41	160.34	240.86	195.18	374.19	**105.92**
	SD	42.63	75.64	111.53	48.84	66.57	**38.64**	57.17	59.17
F_{17}	Mean	305.43	378.64	452.73	360.09	424.94	319.45	602.76	**264.51**
	SD	35.95	128.17	114.34	105.96	92.62	**27.58**	85.08	96.10
F_{18}	Mean	53.09	92.41	85.22	56.75	**9.49**	14.10	130.84	71.00
	SD	79.02	136.96	150.90	62.29	**17.69**	29.53	59.39	47.42
F_{19}	Mean	627.64	765.17	745.07	698.91	**497.13**	55.26	819.40	790.79
	SD	192.16	191.11	198.03	201.69	**17.93**	152.72	112.13	153.62

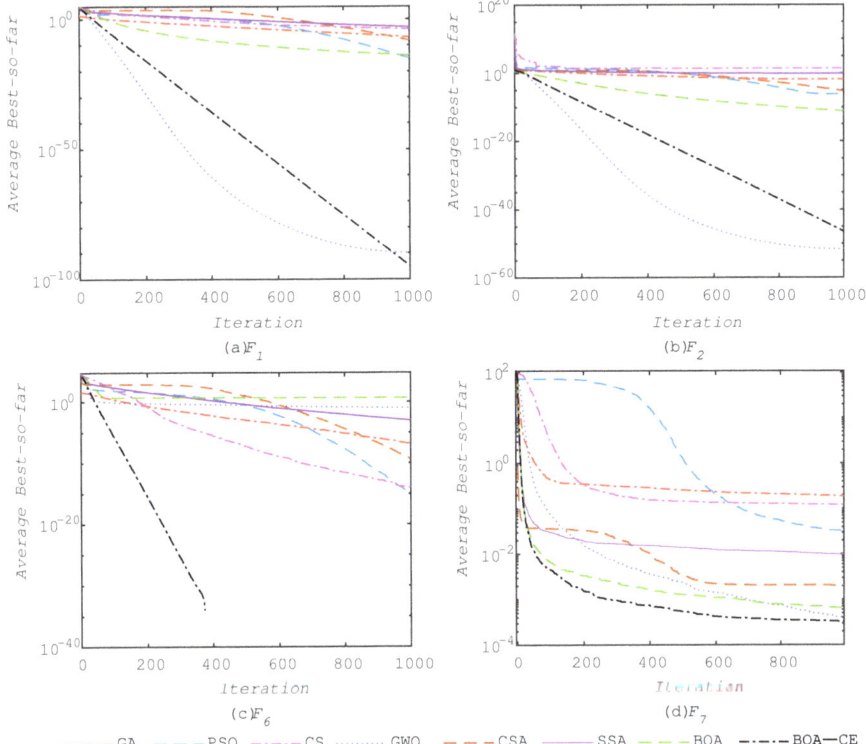

Figure 3. Convergence of algorithms when applied to some unimodal benchmark functions.

Figure 4 shows the convergence curves of the algorithms when applied to some of the multimodal test functions, such as F10, F11, F12, and F13. The figure shows that the BOA-CE algorithm has the fastest convergence for the majority of the multimodal test functions.

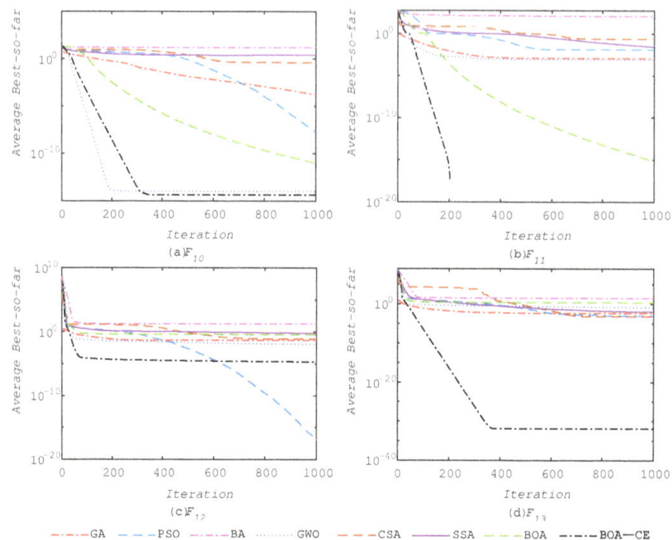

Figure 4. Convergence of algorithms on some unimodal benchmark functions.

4.3. Results of the Algorithms on Composite Test Functions

The results of composite test functions are reported in Table 6. From Table 6, we can find that the BOA-CE algorithm outperforms others in most of the composite test functions. Considering the characteristics of composite test functions and these results, it may be stated that the BOA-CE algorithm appropriately balances exploration and exploitation to focus on the high-performance areas of the search space.

The convergence curves of the methods for some of the composite test functions are shown in Figure 5. Figure 5 clearly shows that the BOA-CE algorithm has the fastest convergence on the composite test functions.

In addition, Table 7 presents the time consumption of all the algorithms to solve the 19 benchmark functions. From Table 7, it can be seen that the least total time cost is PSO to solve all of the benchmark functions, followed by CSA, and the BOA-CE is ranked third, which is better than BOA.

The results of three test functions show that the local optima avoidance of the improved method was enhanced by embedding the cross-entropy method. This global stochastic optimization method enables the BOA-CE algorithm to achieve a proper balance between exploration and exploitation, and enhance its global search capability. Furthermore, the new method is particularly outstanding in solving multimodal function optimization problems.This provides a new method for solving complex engineering design optimization problems.

4.4. Analysis of the Hybrid BOA-CE Algorithm

The main reasons for the excellent performance of the proposed BOA-CE algorithm in solving global optimization problems may be stated as follows:

- The BOA, which mimics the food foraging behavior of butterflies in nature, has the advantage of a fast convergence rate. Based on co-updating, the BOA-CE uses the excellent individuals obtained by the BOA operator to update the CE operator's probability parameters during the iterative process, which speeds up the convergence rate of the CE operator.
- The CE method is a global stochastic optimization method based on statistical model, and has the advantages of randomness, adaptability, and robustness, which bring a good population

diversity to the BOA operator so that it can effectively avoid falling into a local optimum and improve its global search capability.

- The BOA-CE algorithm employs a co-evolutionary technique to co-update the BOA operator's population and the CE operator's probability parameters. This enables the improved method to obtain an appropriate balance between exploration and exploitation and have more superior performance in terms of exploitation, exploration, and local optima avoidance when solving complex optimization problems.

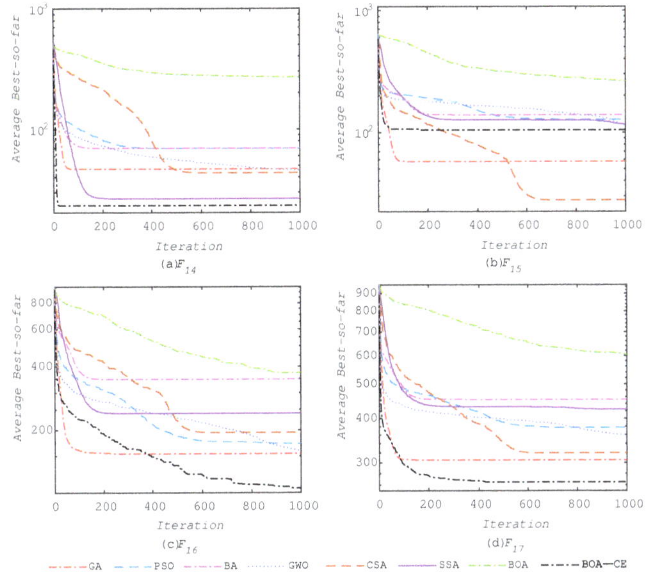

Figure 5. Convergence of algorithms when applied to some composite benchmark functions.

Table 7. Time consumption of BOA-CE solving 19 benchmark functions.

Fun.	GA	PSO	BA	GWO	CSA	SSA	BOA	BOA-CE
F_1	8.96	0.20	0.33	0.55	0.10	0.66	0.46	0.41
F_2	8.80	0.21	0.35	0.57	0.11	0.43	0.49	0.42
F_3	10.75	0.76	2.52	1.13	2.02	0.56	1.61	1.15
F_4	8.44	0.20	0.41	0.55	0.14	0.38	0.46	0.40
F_5	9.11	0.27	0.59	0.62	0.29	0.45	0.59	0.51
F_6	8.57	0.20	0.34	0.55	0.10	0.38	0.45	0.40
F_7	9.29	0.54	0.91	1.14	0.58	0.52	0.84	0.90
F_8	8.72	0.31	0.87	0.65	0.19	0.44	0.68	0.62
F_9	8.81	0.26	0.41	0.57	0.15	0.40	0.58	0.48
F_{10}	8.66	0.26	0.43	0.59	0.17	0.41	0.55	0.45
F_{11}	8.97	0.31	0.69	0.64	0.34	0.47	0.66	0.54
F_{12}	10.14	1.17	1.62	1.52	1.16	0.91	2.45	1.67
F_{13}	10.10	1.17	1.62	1.52	1.17	0.91	2.46	1.66
F_{14}	264.89	249.55	256.03	304.78	260.24	266.09	263.94	265.80
F_{15}	261.13	232.07	261.98	300.36	258.01	255.73	263.67	259.32
F_{16}	259.28	210.24	269.61	250.40	263.24	258.70	259.05	252.60
F_{17}	313.15	212.06	287.05	273.28	282.19	283.91	286.06	282.38
F_{18}	293.31	262.41	289.38	272.57	286.17	286.81	288.63	286.88
F_{19}	287.42	172.65	278.74	288.92	282.58	297.25	291.06	284.31
Total	1798.50	1344.82	1653.88	1700.91	1638.95	1655.41	1664.70	1640.89
Ranking	8	1	5	7	2	6	4	3

5. Using the BOA-CE Algorithm for Classical Engineering Design Problems

In this section, the BOA-CE algorithm was tested on three classical constrained engineering design problems: a tension/compression spring, a welded beam, and a pressure vessel [9,15,19,20,28,31,47–52]. Constraint handling is a challenging task for a method when solving these problems. Penalty functions are divided into different types: static, dynamic, annealing, adaptive, co-evolutionary, and death penalty [47]. For the sake of simplicity, we equipped the BOA-CE algorithm with a death penalty function to handle constraints. Table 8 shows the BOA-CE parameters chosen to solve these design problems, where N is the population sizes of the BOA and CE operator, and $Iter_{max}$ represents the maximum number of iterations.

Table 8. Parameters of the BOA-CE algorithm for solving design problems.

Design Problem	BOA Operator		CE Operator	
	N	$Iter_{max}$	N	$Iter_{max}$
Tension/compression spring	100	50	100	50
Welded beam	100	100	100	50
Pressure vessel	100	100	100	50

5.1. Tension/Compression Spring Design

The objective of this problem is to minimize the weight of the tension/compression spring shown in Figure 6. It contains three design variables: the wire diameter (d), mean coil diameter (D), and number of active coils (N), which are subject to one linear and three nonlinear inequality constraints on shear stress, surge frequency, and deflection. The optimization problem is formulated as follows:

$$\min f(x) = (x_3 + 2)x_2 x_1^2, \tag{14}$$

subject to

$$g_1(x) = 1 - \frac{x_2^3 x_3}{71785 x_1^4} \leq 0, \tag{15}$$

$$g_2(x) = \frac{4x_2^2 - x_1 x_2}{12566(x_2 x_1^3 - x_1^4)} + \frac{1}{5108 x_1^2} - 1 \leq 0, \tag{16}$$

$$g_3(x) = 1 - \frac{140.45 x_1}{x_2^2 x_3} \leq 0, \tag{17}$$

$$g_4(x) = \frac{x_1 + x_2}{1.5} - 1 \leq 0, \tag{18}$$

where $0.05 \leq x_1 \leq 2.00$, $0.25 \leq x_2 \leq 1.30$, and $2.00 \leq x_3 \leq 15.00$.

Figure 6. The tension/compression spring design problem.

This problem has been popular among researchers and has been optimized using various meta-heuristic algorithms. To solve this problem, Coello employed GA [48], He and Wang employed PSO [49], Gandomi et al. used BA [50], Mirjalili employed GWO [15] and SSA [20], Lee and Geem [51] used HS, and Askarzadeh [19] used CSA. The best weight values are highlighted in bold. Table 9 shows a comparison of the results obtained using the BOA-CE algorithm and those obtained using the other algorithms. It can be seen that the BOA-CE and CSA algorithms find a design with the minimum weight for this problem and outperform all other algorithms.

Table 9. A comparison of results for the tension/compression spring design problem.

Algorithm	Optimum Variables			Optimum Weight
	d	*D*	*N*	
GA [48]	0.051480	0.351661	11.632201	0.0127048
PSO [49]	0.051728	0.357644	11.244543	0.012675
BA [50]	0.051690	0.356730	11.288500	0.012670
GWO [15]	0.051690	0.356737	11.288850	0.012666
HS [51]	0.051154	0.349871	12.076432	0.012671
CSA [19]	0.051689	0.356717	11.289012	**0.012665**
SSA [20]	0.051207	0.345215	12.004032	0.012676
BOA-CE	0.051618	0.355004	11.390144	**0.012665**

5.2. Welded Beam Design

The objective of this problem is to minimize the fabrication cost of the welded beam shown in Figure 7. It contains four design variables: the thickness of the weld (h), length of the clamped bar (l), height of the bar (t), and thickness of the bar (b), which are subject to two linear and five nonlinear inequality constraints on shear stress, bending stress in the beam, buckling load, and end deflection of the beam. The optimization problem can be stated as follows:

$$\min f(x) = 1.1047x_1^2 x_2 + 0.04811x_3 x_4(14.0 + x_2), \tag{19}$$

subject to

$$g_1(x) = \tau(x) - \tau_{max} \leq 0, \tag{20}$$

$$g_2(x) = \sigma(x) - \sigma_{max} \leq 0, \tag{21}$$

$$g_3(x) = \delta(x) - \delta_{max} \leq 0, \tag{22}$$

$$g_4(x) = x_1 - x_4 \leq 0, \tag{23}$$

$$g_5(x) = P - P_c(x) \leq 0, \tag{24}$$

$$g_6(x) = 0.125 - x_1 \leq 0, \tag{25}$$

$$g_7(x) = 1.10471x_1^2 + 0.04811x_3 x_4(14.0 + x_2) - 5.0 \leq 0, \tag{26}$$

where $\tau(x) = \sqrt{(\tau')^2 + 2\tau'\tau''\frac{x_2}{2R} + (\tau'')^2}$, $\tau' = \frac{P}{\sqrt{2}x_1 x_2}$, $\tau'' = \frac{MR}{J}$, $M = P(L + \frac{x_2}{2})$, $R = \sqrt{\frac{x_2^2}{4} + (\frac{x_1 + x_3}{2})^2}$, $J = 2\{\sqrt{2}x_1 x_2[\frac{x_2^2}{4} + (\frac{x_1 + x_3}{2})^2]\}$, $\sigma(x) = \frac{6PL}{x_4 x_3^2}$, $\delta(x) = \frac{6PL^3}{Ex_3^3 x_4}$, $P_c(x) = \frac{4.013E\sqrt{\frac{x_3^2 x_4^6}{36}}}{L^2}(1 - \frac{x_3}{2L}\sqrt{\frac{E}{4G}})$, $P = 6000$ lb, $L = 14$ in, $\sigma_{max} = 0.25$ in, $E = 30 \times 10^6$ psi, $G = 12 \times 10^6$ psi, $\tau_{max} = 30{,}000$ psi, $0.1 \leq x_1, x_4 \leq 2$, and $0.1 \leq x_2, x_3 \leq 10$.

The welded beam design problem has been tackled using many meta-heuristic algorithms, such as GA [48], PSO [49], BA [50], GWO [15], HS [51], WOA [18], and CSA [19]. The optimization results using the BOA-CE algorithm are compared with those from the literature in Table 10. The minimum cost is highlighted in bold. The table shows that BOA-CE outperforms all other algorithms except CSA, with only a slight difference in its result.

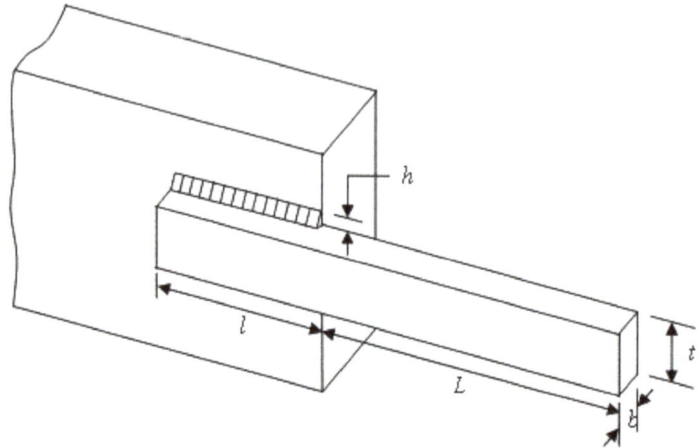

Figure 7. The welded beam design problem.

Table 10. A comparison of results for the tension/compression spring design problem.

Algorithm	Optimum Variables				Optimum Cost
	h	*l*	*t*	*b*	
GA [48]	0.205986	3.471328	9.020224	0.206480	1.728226
PSO [49]	0.202369	3.544214	9.048210	0.205723	1.731485
BA [50]	0.2015	3.562	9.0414	0.2057	1.7312
GWO [15]	0.205676	3.478377	9.03681	0.205778	1.72624
HS [51]	0.2442	6.2231	8.2915	0.2443	2.3807
WOA [18]	0.205396	3.484293	9.037426	0.206276	1.730499
CSA [19]	0.205730	3.470489	9.036624	0.205730	**1.724852**
BOA-CE	0.205730	3.470481	9.036611	0.205730	1.724854

5.3. Pressure Vessel Design

The objective of this problem is to minimize the total cost of a pressure vessel considering the cost of material, forming and welding shown in Figure 8. There are two discrete and two continuous design variables: thickness of the shell (T_s), thickness of the head (T_h), inner radius (R) and length of the cylindrical section of the vessel (L), not including the head, which are subjected to three linear and one nonlinear inequality constraints. The thicknesses of the variables are integer multiples of 0.0625 inches. This optimization problem can be mathematically formulated as follows:

$$\min f(x) = 0.6224x_1 x_3 x_4 + 1.7781 x_2 x_3^2 + 3.1661 x_1^2 x_4 + 19.84 x_1^2 x_3, \tag{27}$$

subject to

$$g_1(x) = -x_1 + 0.0193 x_3 \leq 0, \tag{28}$$

$$g_2(x) = -x_3 + 0.00954 x_4 \leq 0, \tag{29}$$

$$g_3(x) = -\pi x_3^2 x_4 - \frac{4}{3}\pi x_3^3 + 1296000 \leq 0, \tag{30}$$

$$g_4(x) = x_4 - 240 \leq 0, \tag{31}$$

where $0 < x_1, x_2 \leq 99, 0 < x_3, x_4 \leq 200$.

In the past, this problem is solved by GA [48], PSO [49], BA [50], GWO [15], HS [51], WOA [18], and DE [52]. Table 11 shows optimization results of BOA-CE are compared with those found by other methods. It can be seen that BOA-CE and BA outperform all other algorithms except GWO. However the value of T_h obtained by GWO is not an integer multiple of 0.0625 inches and thus does not satisfy the constraint.

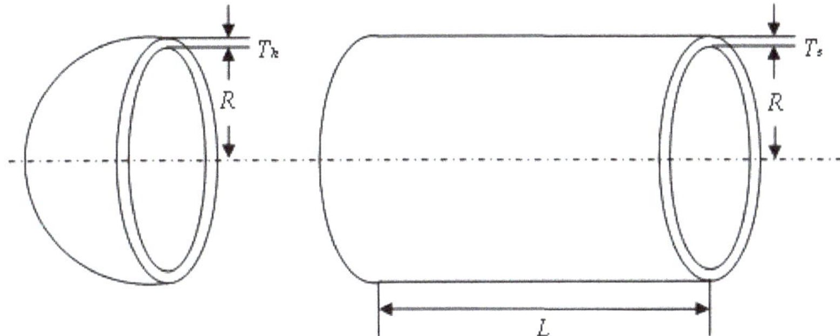

Figure 8. The tension/compression spring design problem.

Table 11. A comparison of results for the tension/compression spring design problem.

Algorithm	Optimum Variables				Optimum Cost
	T_s	T_h	R	L	
GA [48]	0.8125	0.4375	42.097398	176.654050	6059.9463
PSO [49]	0.8125	0.4375	42.091266	176.746500	6061.0777
BA [50]	0.8125	0.4375	42.0984456	176.636596	**6059.7143**
GWO [15]	0.8125	0.4345	42.089181	176.758731	6051.5639
HS [[51]	1.1250	0.6250	58.2789	43.7549	7198.433
WOA [18]	0.8125	0.4375	42.0982699	176.638998	6059.7410
DE [52]	0 .8125	0.4375	42.098411	176.63769	6059 .7341
BOA-CE	0.8125	0.4375	42.0984456	176.6365958	**6059.7143**

6. Conclusions

In order to improve the global search ability of the BOA, an improved BOA algorithm, named BOA-CE, was constructed by embedding the CE method into the BOA using a co-evolutionary technique. A total of 19 test functions were used to evaluate the performance of the new method in terms of its exploration, exploitation, local optima avoidance, and convergence rate. The results of the test functions demonstrated that the proposed algorithm can effectively avoid falling into a local optima and has excellent local and global search capacity. This is mainly due to the co-evolutionary technique, which enables the new method to obtain an appropriate balance between exploration and exploitation and has more superior performance when solving complex function optimization problems. Furthermore, the paper also considered the solving of three classical engineering problems using the hybrid algorithm. The comparative results show that the BOA-CE algorithm provides very competitive results when solving real problems with unknown search spaces. In future research, a discrete version of the BOA-CE algorithm will be constructed to solve combinatorial optimization problems.

Author Contributions: G.L. designed the algorithm, implemented all experiments, and wrote the manuscript. F.S. revised the manuscript. P.Z. conducted the literature review and analyzed the data. C.L. revised the manuscript.

Funding: This research was funded by the National Natural Science Foundation of China (Grant No. 71761012), the Natural Science Foundation of Anhui Province (Grant No. 1808085MG224), and the Natural Science Foundation of West Anhui University (Grant No. WXZR201818).

Acknowledgments: The authors are grateful for the support provided by the Risk Management and Financial Engineering Lab at the University of Florida, Gainesville, FL 32611, USA.

Conflicts of Interest: The authors declare no conflict of interest.

Appendix A

See Table A1.

Table A1. Mathematical formulation of the primitive functions in Table 3.

Name	Formulation
Sphere	$f(x) = \sum_{i=1}^{D} x_i^2$
Ackley	$f(x) = -20 \exp\left(-0.2\sqrt{\frac{1}{D}\sum_{i=1}^{D} x_i^2}\right) - \exp\left(\frac{1}{D}\sum_{i=1}^{D}\cos\left(2\pi x_i\right)\right) + 20 + e$
Griewank	$f(x) = \frac{1}{4000}\sum_{i=1}^{D} x_i^2 - \prod_{i=1}^{D}\cos\left(\frac{x_i}{\sqrt{i}}\right) + 1$
Weierstrass	$f(x) = \sum_{i=1}^{D}\left(\sum_{k=0}^{k_{max}}\left[a^k\cos(2\pi b^k(x_i + 0.5))\right]\right) - D\sum_{k=0}^{k_{max}}\left[a^k\cos(2\pi b^k \cdot 0.5)\right], a = 0.5, b = 3, k_{max} = 20$
Rastrigin	$f(x) = \sum_{i=1}^{D-1}\left[x_i^2 - 10\cos\left(2\pi x_i\right) + 10\right]$

D: dimensions.

References

1. Hu, X.; Eberhart, R.C.; Shi, Y. Engineering optimization with particle swarm. In Proceedings of the 2003 IEEE Swarm Intelligence Symposium, Indianapolis, IN, USA, 26–26 April 2003; pp. 53–57.
2. Sergeyev, Y.D.; Kvasov, D.E. A deterministic global optimization using smooth diagonal auxiliary functions. *Commun. Nonlinear Sci. Numer. Simul.* **2015**, *21*, 99–111. [CrossRef]
3. Lera, D.; Sergeyev, Y.D. GOSH: Derivative-free global optimization using multi-dimensional space-filling curves. *J. Glob. Optim.* **2018**, *71*, 193–211. [CrossRef]
4. Sergeyev, Y.D.; Kvasov, D.E.; Mukhametzhanov, M.S. On the efficiency of nature-inspired metaheuristics in expensive global optimization with limited budget. *Sci. Rep.* **2018**, *8*, 453. [CrossRef] [PubMed]
5. Zilinskas, A.; Zhigljavsky, A. Stochastic Global Optimization: A Review on the Occasion of 25 Years of Informatica. *Informatica* **2016**, *27*, 229–256. [CrossRef]
6. Goldfeld, S.M.; Quandt, R.E.; Trotter, H.F. Maximization by quadratic hill-climbing. *Econometrica* **1966**, *34*, 541–551. [CrossRef]
7. Abbasbandy, S. Improving Newton–Raphson method for nonlinear equations by modified Adomian decomposition method. *Appl. Math. Comput.* **2003**, *145*, 887–893. [CrossRef]
8. Jones, D.R.; Perttunen, C.D.; Stuckman, B.E. Lipschitzian optimization without the Lipschitz constant. *J. Optim. Theory Appl.* **1993**, *79*, 157–181. [CrossRef]
9. Mirjalili, S. The Ant Lion Optimizer. *Adv. Eng. Softw.* **2015**, *83*, 80–98. [CrossRef]
10. Whitley, D. A genetic algorithm tutorial. *Stat. Comput.* **1994**, *4*, 65–85. [CrossRef]
11. Wieczorek, L.; Ignaciuk, P. Continuous Genetic Algorithms as Intelligent Assistance for Resource Distribution in Logistic Systems. *Data* **2018**, *3*, 68. [CrossRef]
12. Kennedy, J.; Eberhart, R.C. Particle swarm optimization. In Proceedings of the 1995 IEEE International Conference on Neural Networks, Perth, Australia, 27 November–1 December 1995; pp. 1942–1948.
13. Yang, X.S.; Deb, S. Engineering Optimisation by Cuckoo Search. *Int. J. Math. Model. Numer. Optim.* **2010**, *1*, 330–343. [CrossRef]
14. Yang, X.S. A new metaheuristic bat-inspired algorithm. In *Nature Inspired Cooperative Strategies for Optimization (NICSO 2010)*; Springer: Berlin/Heidelberg, Germany, 2010; pp. 65–74.
15. Mirjalili, S.; Mirjalili, S.M.; Lewis, A. Grey Wolf Optimizer. *Adv. Eng. Softw.* **2014**, *69*, 46–61. [CrossRef]
16. Ghaemi, M.; Feizi-Derakhshi, M.R. Forest Optimization Algorithm. *Expert Syst. Appl.* **2014**, *41*, 6676–6687. [CrossRef]
17. Naz, M.; Zafar, K.; Khan, A. Ensemble Based Classification of Sentiments Using Forest Optimization Algorithm. *Data* **2019**, *4*, 76. [CrossRef]
18. Mirjalili, S.; Lewis, A. The Whale Optimization Algorithm. *Adv. Eng. Softw.* **2016**, *95*, 51–67. [CrossRef]
19. Askarzadeh, A. A novel metaheuristic method for solving constrained engineering optimization problems: Crow search algorithm. *Comput. Struct.* **2016**, *169*, 1–12. [CrossRef]

20. Mirjalili, S.; Gandomi, A.H.; Mirjalili, S.Z.; Saremi, S.; Faris, H.; Mirjalili, S.M. Salp Swarm Algorithm: A bio-inspired optimizer for engineering design problems. *Adv. Eng. Softw.* **2017**, *114*, 163–191. [CrossRef]
21. Arora, S.; Singh, S. Butterfly algorithm with Lèvy Flights for global optimization. In Proceedings of the 2015 International Conference on Signal Processing, Computing and Control (ISPCC), Waknaghat, India, 24–26 September 2015; pp. 220–224.
22. Arora, S.; Singh, S. Butterfly optimization algorithm: A novel approach for global optimization. *Soft Comput.* **2019**, *23*, 715–734. [CrossRef]
23. Yang, X.S.; Deb, S. Cuckoo search: Recent advances and applications. *Neural Comput. Appl.* **2014**, *24*, 169–174. [CrossRef]
24. Wolpert, D.H.; Macready, W.G. No Free Lunch Theorems for Optimization. *IEEE Trans. Evol. Comput.* **1997**, *1*, 67–82. [CrossRef]
25. Lai, X.S.; Zhang, M.Y. An Efficient Ensemble of GA and PSO for Real Function Optimization. In Proceedings of the 2009 2nd IEEE International Conference on Computer Science and Information Technology, Beijing, China, 8–11 August 2009; pp. 651–655.
26. Mirjalili, S.; Hashim, S.Z.M. A New Hybrid PSOGSA Algorithm for Function Optimization. In Proceedings of the 2010 International Conference on Computer and Information Application (2010 ICCIA), Tianjin, China, 3–5 December 2010; pp. 374–377.
27. Abdullah, A.; Deris, S.; Mohamad, M.S.; Hashim, S.Z.M. A New Hybrid Firefly Algorithm for Complex and Nonlinear Problem. In *Distributed Computing and Artificial Intelligence*; Springer: Berlin/Heidelberg, Germany, 2012; pp. 673–680.
28. He, X.S.; Ding, W.J.; Yang, X.S. Bat algorithm based on simulated annealing and Gaussian perturbations. *Neural Comput. Appl.* **2014**, *25*, 459–468. [CrossRef]
29. Mafarja, M.M.; Mirjalili, S. Hybrid Whale Optimization Algorithm with simulated annealing for feature selection. *Neurocomputing* **2017**, *260*, 302–312. [CrossRef]
30. Pepelyshev, P.; Zhigljavsky, A.; Zilinskas, A. Performance of global random search algorithms for large dimensions. *J. Glob. Optim.* **2018**, *71*, 57–71. [CrossRef]
31. Rubinstein, R.Y. Optimization of Computer Simulation Models with Rare Events. *Eur. J. Oper. Res.* **1997**, *99*, 89–112. [CrossRef]
32. Arora, S.; Singh, S. An Improved Butterfly Optimization Algorithm for Global Optimization. *Adv. Sci.* **2017**, *8*, 711–717. [CrossRef]
33. Arora, S.; Singh, S. Node Localization in Wireless Sensor Networks Using Butterfly Optimization Algorithm. *Arab. J. Sci. Eng.* **2017**, *42*, 3325–3335. [CrossRef]
34. Kroese, D.P.; Portsky, S.; Rubinstein, R.Y. The Cross-Entropy Method for Continuous Multi-extremal Optimization. *Methodol. Comput. Appl. Probab.* **2006**, *8*, 383–407. [CrossRef]
35. Bekker, J.; Aldrich, C. The cross-entropy method in multi-objective optimisation: An assessment. *Eur. J. Oper. Res.* **2011**, *211*, 112–121. [CrossRef]
36. Rubinstein, R.Y. The Cross-Entropy Method for Combinatorial and Continuous Optimization. *Methodol. Comput. Appl. Probab.* **1999**, *1*, 127–190. [CrossRef]
37. Rubinstein, R.Y.; Kroese, D.P. *The Cross-Entropy Method: A Unified Approach to Combinatorial Optimization, Monte Carlo Simulation and Machine Learning*; Springer: New York, NY, USA, 2004.
38. Chepuri, K.; Homem-De-Mello, T. Solving the vehicle routing problem with stochastic demands using the cross-entropy method. *Ann. Oper. Res.* **2005**, *134*, 153–181. [CrossRef]
39. Yu, J; Konaka, S.; Akutagawa, M.; Zhang, Q. Cross-Entropy-Based Energy-Efficient Radio Resource Management in HetNets with Coordinated Multiple Points. *Information* **2016**, *7*, 3. [CrossRef]
40. Joseph, A.G.; Bhatnagar, S. An online prediction algorithm for reinforcement learning with linear function approximation using cross entropy method. *Mach. Learn.* **2018**, *107*, 1385–1429. [CrossRef]
41. Peherstorfer, B.; Kramer, B.; Willcox, K. Multifidelity preconditioning of the cross-entropy method for rare event simulation and failure probability estimation. *SIAM/ASA J. Uncertain. Quantif.* **2018**, *6*, 737–761. [CrossRef]
42. Wang, Y.; Yang, H.; Qin, K. The Consistency between Cross-Entropy and Distance Measures in Fuzzy Sets. *Symmetry* **2019**, *11*, 386. [CrossRef]
43. Pramanik, S.; Dalapati, S.; Alam, S.; Smarandache, F.; Roy, T.K. NS-Cross Entropy-Based MAGDM under Single-Valued Neutrosophic Set Environment. *Information* **2018**, *9*, 37. [CrossRef]
44. Eiben, A.E.; Schipper, C.A. On evolutionary exploration and exploitation. *Fund. Inform.* **1998**, *35*, 35–50.

45. Yao, X.; Liu, Y.; Lin, G.M. Evolutionary Programming Made Faster. *IEEE Trans. Evol. Comput.* **1999**, *3*, 82–102.
46. Liang, J.; Suganthan, P.; Deb, K. Novel composition test functions for numerical global optimization. In Proceedings of the 2005 IEEE Swarm Intelligence Symposium, Pasadena, CA, USA, 8–10 June 2005; pp. 68–75.
47. Coello, C.C.A.; Mezura, M.E. Constraint-handling in genetic algorithms through the use of dominance-based tournament selection. *Adv. Eng. Inform.* **2002**, *16*, 193–203. [CrossRef]
48. Coello, C.C.A. Use of a Self-Adaptive Penalty Approach for Engineering Optimization Problems. *Comput. Ind.* **2000**, *41*, 113–127. [CrossRef]
49. He, Q.; Wang, L. An effective co-evolutionary particle swarm optimization for constrained engineering design problems. *Eng. Appl. Artif. Intell.* **2007**, *20*, 89–99. [CrossRef]
50. Gandomi, A.H.; Yang, X.S.; Alavi, A.H.; Talatahari, S. Bat algorithm for constrained optimization tasks. *Neural Comput. Appl.* **2013**, *22*, 1239–1255. [CrossRef]
51. Lee, K.S.; Geem, Z.W. A new meta-heuristic algorithm for continuous engineering optimization: Harmony search theory and practice. *Comput. Methods Appl. Mech. Eng.* **2005**, *194*, 3902–3933. [CrossRef]
52. Huang, F.; Wang, L.; He, Q. An effective co-evolutionary differential evolution for constrained optimization. *Appl. Math. Comput.* **2007**, *186*, 340–356. [CrossRef]

Article

A New Second-Order Tristable Stochastic Resonance Method for Fault Diagnosis

Lu Lu [1], Yu Yuan [1,2,*], Heng Wang [1], Xing Zhao [1] and Jianjie Zheng [1]

[1] College of Locomotive and Rolling Stock Engineering, Dalian Jiaotong University, Dalian 116028, China
[2] Traction Power State Key Laboratory, Southwest Jiaotong University, Chengdu 610031, China
* Correspondence: yuany@djtu.edu.cn; Tel.: +86-1550-985-0752

Received: 8 July 2019; Accepted: 24 July 2019; Published: 1 August 2019

Abstract: Vibration signals are used to diagnosis faults of the rolling bearing which is symmetric structure. Stochastic resonance (SR) has been widely applied in weak signal feature extraction in recent years. It can utilize noise and enhance weak signals. However, the traditional SR method has poor performance, and it is difficult to determine parameters of SR. Therefore, a new second-order tristable SR method (STSR) based on a new potential combining the classical bistable potential with Woods-Saxon potential is proposed in this paper. Firstly, the envelope signal of rolling bearings is the input signal of STSR. Then, the output of signal-to-noise ratio (SNR) is used as the fitness function of the Seeker Optimization Algorithm (SOA) in order to optimize the parameters of SR. Finally, the optimal parameters are used to set the STSR system in order to enhance and extract weak signals of rolling bearings. Simulated and experimental signals are used to demonstrate the effectiveness of STSR. The diagnosis results show that the proposed STSR method can obtain higher output SNR and better filtering performance than the traditional SR methods. It provides a new idea for fault diagnosis of rotating machinery.

Keywords: rolling bearings; Fault diagnosis; second-order tristable stochastic resonance; seeker optimization algorithm; output signal-to-noise ratio

---

1. Introduction

The rolling bearings are key components of rotary machines, but harsh working conditions often make them suffer from different failure, which may lead to the breakdown of the whole machinery and huge economic loss [1,2]. Therefore, it is of great significance to monitor the condition of the bearing. Vibration analysis has been widely applied to diagnose bearing faults. However, the faulty signal acquired from the bearing is usually weak or submerged in strong noise [3,4]. Traditional weak signal detection methods, such as empirical mode decomposition (EMD) [5], wavelets transform (WT) [6], singular value decomposition (SVD) [7], and variational mode decomposition (VMD) [8], mainly reduced noise to improve signal-to-noise ratio (SNR) and extract fault characteristics, which inevitably weakened useful fault signal characteristic information. In contrast, SR can utilize noise to enhance weak signal energy.

Stochastic resonance (SR) was proposed by Benzi in 1981 [9], which have been applied in various fields. At present, SR has been used in bearing fault diagnosis. The adiabatic approximation theory of SR requires that the input signal frequency of the SR system is less than 1 Hz, while the fault characteristic frequency of the bearing is much larger than 1 Hz in practical engineering applications. To deal with this limitation, Leng [10] proposed the re-scaling frequency SR and Tan [11] presented the frequency-shifted and re-scaling SR. However, beyond that, He [12] proposed a multi-scale noise tuning technique to overcome the limitation of small parameter requirement of the classical SR, and took advantage of the multiscale noise for an improved SR performance. These methods promoted the application of SR in practical engineering.

Scholars have proposed different SR models and parameter optimization algorithms in the field of the classical first-order SR system. Several bistable potentials have been proposed and applied in the detection of weak signals. Reference [13] proposed the use of particle swarm optimization to achieve the selection of bistable stochastic resonance system parameters (CSR). Qiao [14] established a new piecewise bistable potential model to overcome the disadvantage of output saturation of the CSR. Zhang [15,16] introduced a joint with Woods–Saxon and Gaussian potential, and step-varying asymmetric stochastic for bearing fault diagnosis. In multistable potentials, Li [17] presented overdamped multistable SR with a tristable potential to extract the fault characteristics of the rolling mill gearbox and the detection results show that the multistable SR method has better processing effects than the classical bistable SR method for weak signals. Han [18] proposed a multi-stable SR system for multi-frequency weak signal detection and Liu [19] improved an adaptive SR with the periodic potential system using a trigonometric function. In short, SR with multistable potential has a better filtering performance than bistable SR. The other intelligent methods are proposed to apply in the field of fault diagnosis [20–22].

Most of the research focused on the first-order SR system, however the filtering performance of the underdamped (second-order) SR system is much better than that of the first-order SR system, and the output SNR is higher. Lu [23] proposed an underdamped step-varying second-order SR method (USSSR), verified the practicability by analyzing a set of defective bearing signals, and confirmed the effectiveness in comparison with a traditional SR method. López [24] explored the second order SR with a FitzHug–Nagumo potential for weak signal detection. Lei [25] presented an underdamped SR method with stable-state matching for incipient bearing fault diagnosis. Xia [26] proposed an improved SR method with an arbitrary stable-state potential for incipient bearing fault diagnosis. Liu [27] proposed a step-varying vibrational resonance (SVVR) method based on the duffing oscillator nonlinear system and concluded that the SVVR is more effective in extracting weak characteristic information than other methods. All in all, the second-order SR method is able to obtain higher output SNR and better performance than the first-order SR method.

Based on the abovementioned, a new second-order tristable SR method (STSR) method is proposed in this study based on the classical bistable potential and Woods–Saxon potential. This paper uses the re-scaling frequency method to detect large frequency weak signals, in which an R parameter is introduced. Since the parameters of the SR system determines its performance, we can use Seeker Optimization Algorithm (SOA) to solve multiple parameters synchronously. To evaluate the performance of the proposed STSR method, compare this method with traditional SR methods. It is proved by a simulation signal that the STSR method is more conducive to the detection of weak signals than traditional SR methods and obtains larger output SNR. Finally, the proposed method and traditional SR method are applied to fault diagnosis of bearings, respectively.

The remainder of this paper is organized as follows. The STSR model is introduced in Section 2, the influence of the noise intensity D and damping ratio k on the STSR output signal are analyzed, and the process weak fault characteristic extraction based on SOA is presented. Sections 3 and 4 use some data to verify the effectiveness of the proposed method. Section 5 provides a summary of this paper.

2. STSR System

The bistable potential system is one of the classical SR models, where the function expression can be expressed as:

$$U_v(x) = -\frac{a}{2}x^2 + \frac{b}{4}x^4 \tag{1}$$

where a and b are real positive values. Figure 1 shows the fixed parameters a or b, respectively, in order to analyze their influence on the potential barrier and the potential well.

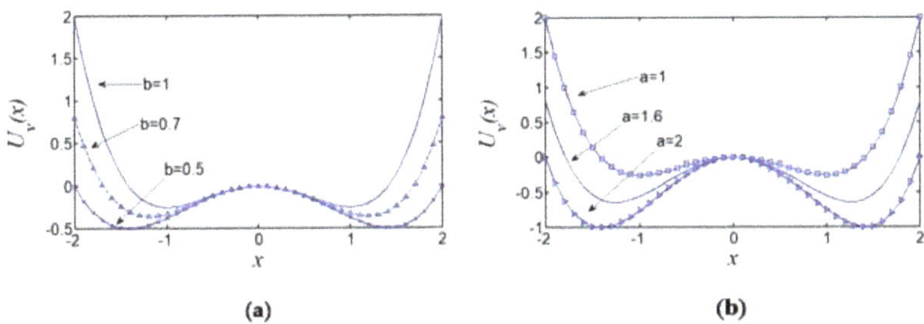

Figure 1. The bistable potential function: (**a**) When $a = 1$, the potential model for b changes; (**b**) when $b = 1$, the potential model for a changes.

It can be found that, when the fixed parameter $a = 1$, the barrier height decreases with the increase of b. Additionally, when the fixed parameter $b = 1$, the barrier height increases with the increase of a from Figure 1. Thus, the parameters a and b affect the barrier height together. The traditional method of selecting parameters a and b manually, and fixing one parameter adjustment of another parameter neglects the coupling effect between the parameters, which may directly lead to the performance of SR detection in weak features detection.

The Woods–Saxon potential (U_{ws}) [28] is a nonlinear symmetric potential, proposed by Woods and Saxon. U_{ws} can be expressed as follows:

$$U_{ws}(x) = -\frac{V}{1 + \exp(\frac{|x|-r}{c})} \tag{2}$$

where parameter V affects the depth of the potential and parameter r works on the width of the potential, while parameter c determines the wall steepness of the potential. These three parameters together determine the shape and performance of the potential. Figure 2 shows the effect of different parameters on the shape of the potential function. When the parameters V and r are fixed, the potential well becomes slower as the c increases; parameters V and c are fixed, the potential well function becomes wider as the r increases; and parameters r and c are fixed, the depth of the well becomes deeper as the V increases. Thus, the shape of the potential well can be changed by adjusting the parameters.

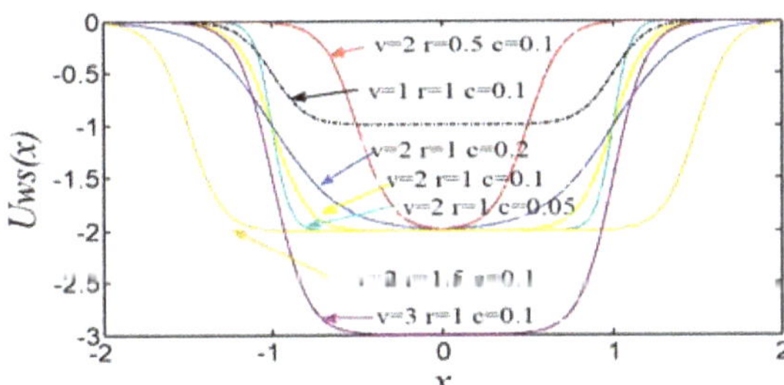

Figure 2. The Woods–Saxon potential function with different parameters.

2.1. Tristable Potential

We established a new tristable potential model by the combination of Woods–Saxon potential and bistable potential based on the above introduction, which can be expressed as follows:

$$\begin{aligned} U(x) &= U_v(x) + U_{ws}(x) \\ &= -\tfrac{1}{2}ax^2 + \tfrac{1}{4}bx^4 - \frac{V}{1+\exp(\frac{|x|-r}{c})} \end{aligned} \tag{3}$$

Set the potential parameters $a = 2$, $b = 1$, $V = 1$, $r = 0.25$, and $c = 0.05$ to construct a new tristable potential function that is still a symmetric potential shown in Figure 3. The parameters a, b, c, V and r together determine the shape of the potential function. We can adjust the parameters in order to realize the potential function to become a monostable potential or a bistable potential and a tristable potential. Therefore, the proposed potential has the advantages of monostable potential, bistable potential, and tristable potential.

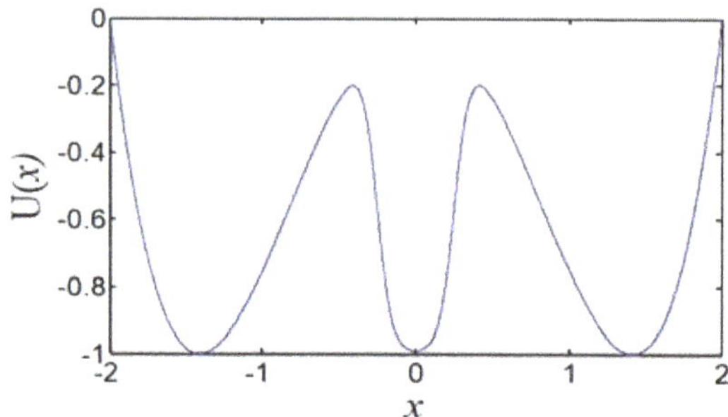

Figure 3. The tristable potential function.

2.2. The STSR

The second-order SR system can be described as:

$$\frac{d^2x}{dt^2} = -\frac{dU(x)}{dx} - k\frac{dx}{dt} + s(t) + n(t) \tag{4}$$

$s(t) + n(t)$ is the input signal of the SR system, $s(t)$ is a periodic signal, $n(t) = \sqrt{2D}w(t)$ is the background noise, $w(t)$ is the Gaussian noise, k is damping ratio, x represents the output signal of SR system, and $U(x)$ is the potential proposed in this paper.

Substituting Equation (3) in Equation (4), the Equation of second-order SR based on the $U(x)$ potential can be expressed as follows:

$$\frac{d^2x}{dt^2} = ax - bx^3 - \frac{V}{c}\mathrm{sgn}(x)\exp(\frac{|x|-r}{c})(1+\exp(\frac{|x|-r}{c}))^{-2} - k\frac{dx}{dt} + s(t) + n(t) \tag{5}$$

Equation (5) is the second-order tristable SR (STSR).

Figure 4 is a sketch that describes the STSR model. One integration can be regarded as a filtering process, so the output signal $x(t)$ is obtained by second filtering, which provides better denoising performance than the first-order SR. The STSR system can be regarded as the output signal $x(t)$ obtained by secondary filtering of the input signal $s(t) + n(t)$, and the system parameters determine the filtering effect.

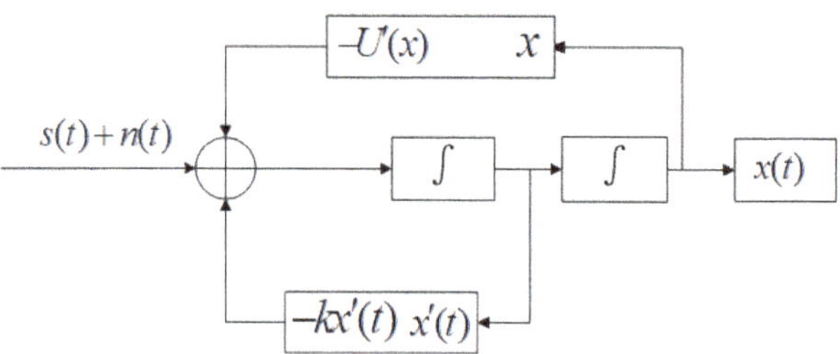

Figure 4. The second-order tristable SR (STSR) model.

This paper splits Equation (5) into two first-order differential equations, and introduces the parameter y, $\frac{dx}{dt} = y$, so Equation (4) can be rewritten as follows:

$$\begin{cases} \frac{dx}{dt} = y \\ \frac{dy}{dt} = ax - bx^3 - \frac{V}{c}\mathrm{sgn}(x)\exp(\frac{|x|-r}{c})(1 + \exp(\frac{|x|-r}{c}))^{-2} - ky + s(t) + n(t) \end{cases} \tag{6}$$

The solution process of Equation (6) can be expressed as follows:

$$\begin{cases} k_1 = y_n \\ l_1 = S_n - kk_1 - U'(x_n) \\ k_2 = y_n + 0.5hl_1 \\ l_2 = S_n - kk_2 - U'(x_n + 0.5hk_1) \\ k_3 = y_n + 0.5hl_2 \\ l_3 = S_{n+1} - kk_3 - U'(x_n + 0.5hk_2) \\ k_4 = y_n + hl_3 \\ l_4 = S_{n+1} - kk_4 - U'(x_n + hk_3) \\ x_{n+1} = x_n + (1/6)(k_1 + 2k_2 + 2k_3 + k_4) \\ y_{n+1} = y_n + (1/6)(l_1 + 2l_2 + 2l_3 + l_4) \end{cases} \tag{7}$$

where x is STSR output signal and h is the sampling interval.

2.3. Influence of Parameters on STSR and Procedure of STSR

Firstly, the effect of noise intensity D on STSR output signal is analyzed. Due to the adiabatic approximation theory of SR, the input signal amplitude A, frequency f_0 meets the requirements of small parameters, $A < 1, f_0 < 1$. Thus, we can simulate a small parameter signal to explore the effect of noise intensity D on STSR. The simulation signal is defined as:

$$x(t) = A\sin(2\pi f_0) + n(t) \tag{8}$$

where $A = 0.1, f_0 = 0.01$ Hz and $n(t)$ is the zero mean Gaussian white noise.

Set the other parameters of the signal as follows: The sampling frequency is $f_s = 5$ Hz and the length of the signal $N = 5000$. The time domain and amplitude spectrum of the signal without noise are shown in Figure 5a. Here, given a set of parameters that can produce SR, $a = 0.4$, $b = 2.7$, $V = 11$, $r = 39$, $c = 20$, and $k = 0.47$. The time domain waveform and amplitude spectrum of the STSR output signal are shown in Figure 5b when $D = 0.3$, and the frequency 0.01 Hz cannot be seen. Further, adding noise to the input signal, as the noise increases, STSR obtains the output signal, as shown in Figure 5c when $D = 0.5$. The time domain waveforms can see obviously periodic components, and spectral peaks appear at 0.01 Hz in the amplitude spectrum. When the noise $D = 0.8$, and the frequency 0.01 Hz cannot be recognized from Figure 5d. It is found that the signal amplitude at the noise intensity $D = 0.5$ is higher than the amplitude of the signal without noise comparing Figures 5c and 5a. It shows that, as the noise increases, the characteristic components of effective signal increases greatly. On the other hand, the energy of the noise transfers to the signal. However, the characteristic components of signal attenuate, while the noise increase further, which is a typical feature of the SR system.

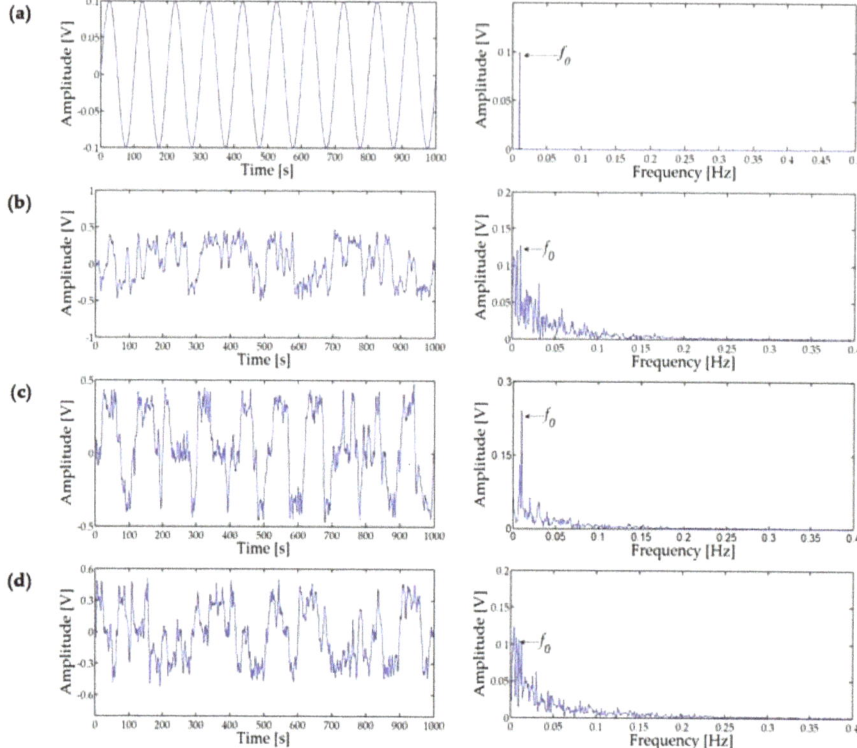

Figure 5. Effect of noise intensity D on STSR: (**a**) Input signal without noise and its amplitude spectrum; (**b**) STSR output signal and its amplitude spectrum when $D = 0.3$; (**c**) STSR output signal and its amplitude spectrum when $D = 0.5$; and (**d**) STSR output signal and its amplitude spectrum when $D = 0.8$.

Then, we explore the effect of damping ratio k on the STSR performance. In the same way, we use the signal of Equation (8) as the input of STSR to analyze the influence of k on the output signal. We cannot see the 0.01 Hz from Figure 5b, when $D = 0.3$, $a = 0.4$, $b = 2.7$, $V = 11$, $r = 39$, $c = 20$, and $k = 0.47$. We can fix parameters a, b, V, r, c, and make $k = 0.5334$, seeing 0.01Hz clearly from Figure 6a that shows the result of STSR. When the damping ratio k is increased to 0.885, the time-domain waveform and amplitude spectrum of the output signal are as shown in Figure 6b. The 0.01 Hz cannot be discerned from the amplitude spectrum. The signal with noise intensity $D = 0.8$ is analyzed to fully explain the effect of damping ratio k on STSR. When the damping ratio $k = 0.47$, the characteristic frequency of 0.01 Hz cannot be recognized from the amplitude spectrum of Figure 5d. Figure 7a displays the time-domain waveform and amplitude spectrum of STSR output signal when k is decreased to 0.375. It can be observed that 0.01 Hz is very obvious in the whole amplitude spectrum, although there are many frequency components in the whole spectrum. Continue to reduce the damping ratio k to 0.235, and the time-domain waveform and amplitude spectrum are as shown in Figure 7b, from which the frequency of 0.01 Hz cannot be observed. Thus, the damping ratio k affects the characteristics of the STSR system.

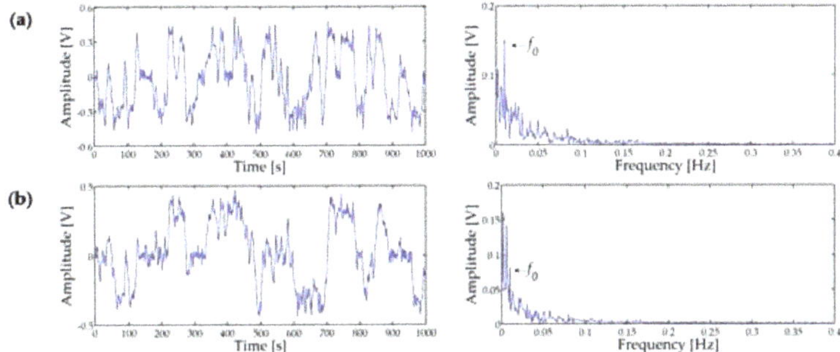

Figure 6. Effect of k on STSR when $D = 0.3$: (**a**) STSR output signal and its amplitude spectrum when $k = 0.534$; (**b**) STSR output signal and its amplitude spectrum when $k = 0.885$.

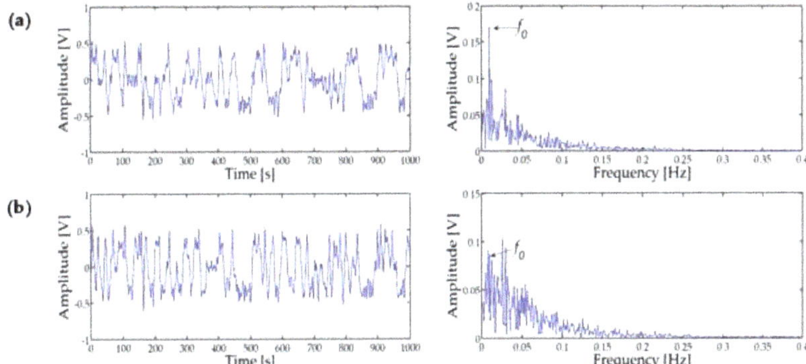

Figure 7. Effect of k on STSR when $D = 0.8$: (**a**) STSR output signal and its amplitude spectrum when $k = 0.375$; (**b**) STSR output signal and its amplitude spectrum when $k = 0.235$.

2.4. Procedure of Proposed Method

We can see the effect of noise intensity D and damping ratio k on STSR based on the above analysis. Since system parameters a, b, V, r, and c are artificially selected, they are no longer analyzed. While the noise intensity D has a certain influence on the output signal of STSR, it cannot be controlled. Thus, it can be understood that the system parameters a, b, V, r, c, and the damping ratio k affects the performance of the STSR. Due to the fault characteristic frequency of the bearing is far larger than 1 Hz, the frequency compression scale (R) is introduced by the re-scaling frequency technique to satisfy the small parameter limitation in this paper. This paper uses the Seeker Optimization Algorithm (SOA) to optimize seven parameters synchronously [29,30]. Output SNR is often used to evaluate the performance of SR, which specifies the calculation as follows:

$$\begin{cases} X(k) = \sum_{n=1}^{N} x(n)e^{-j\pi(k-1)(n-1)/N}, 1 \leq k \leq N \\ Y(k) = 2\dfrac{|X(k)|}{N}, 1 \leq \dfrac{N}{2} \end{cases} \tag{9}$$

where $x = \{x_1, x_2, \cdots, x_N\}$ is a discrete signal.

$$K_0 = \frac{f_0 \times N}{f_s} + 1 \tag{10}$$

where f_0 represents the frequency at the highest spectral peak and f_s is the sampling frequency of the signal. SNR can be defined as:

$$\text{SNR} = 10 \log_{10} \frac{Y(K_0)}{\sum_{K=1}^{N/2} Y(K) - Y(K_0)} \tag{11}$$

The specific procedure of the STSR method based on SOA can be expressed as follows:

(1) Signal preprocessing. The vibration signal is filtered by band-pass filter to eliminate some noise components. Hilbert transform is used to demodulate the filtered signal, in which the obtained signal is marked as S_1. $S_2 = S_1 - mean(S_1)$, $S = \frac{S_2}{2 \times \max(abs(S_2))}$. S is the input signal of STSR system.

(2) Parameter initialization. Setting population size, number of iterations, and range of parameters including a, b, V, R, c, k, and R.

(3) Calculate the objective function value of each location according to Equation (11).

(4) If the number of iterations reaches the set value, the process goes to step (5), otherwise it returns to step (3).

(5) The optimal record of output is the value of a, b, V, r, c, R, and k at the maximum SNR.

(6) Weak signal detection. The optimal parameters are substituted for the STSR system. Additionally, the pre-processed signal is used as the input signal of the STSR to obtain the output signal. Restore the signal amplitude and frequency scale to plot time-domain waveform and amplitude spectrum.

3. Simulation Analysis

The simulation analysis is used to prove the effectiveness of the proposed STSR method. The simulation signal is defined as:

$$s(t) = A_0 \sin(2\pi f_0 t) + n(t) \tag{12}$$

where $A_0 = 0.2$, $f_0 = 50$ Hz, $n(t)$ is Gaussian white noise, and the intensity is $D = 2.4$.

The sampling frequency $f_s = 5$ kHz, and sampling number is 5000. Experimental environments are Intel (R) core (TM) i5-7400 (Beijing, China) CPU 3.00 GHz, 8 G RAM, Windows 10, and MATLAB R2018a. The time-domain waveform and amplitude spectrum of signal are shown in Figure 8a. The periodicity of the signal cannot be seen from the time domain waveform and 50 Hz cannot be discovered from the amplitude spectrum. According to Equation (11), the SNR of the signal with noise is −29.82 dB. The optimal output result of CSR [13] is shown in Figure 8b where the parameters are $a = 19.72$, $b = 7217.256$, $R = 464.03$, and SNR = −20.53 dB. While the SNR is improved, the noise energy is also enhanced. Additionally, the low-frequency components are amplified. Figure 8c shows the optimal output of the frequency-shifted and re-scaling SR [11]. The cutoff frequency of the high-pass filter is set to 35 Hz, the modulation frequency is 40 Hz, and the target frequency is 10 Hz in this method with optimization results are $a = 1.86$, $b = 4794.73$, $R = 140.97$, and SNR = −13.38 dB. It can be seen that the SNR of frequency-shifted and re-scaling SR is further improved than CSR, but the interference frequency f_1 of the filter can be seen from Figure 8c. Additionally, the artificially set cutoff frequency of the filter may also affect the stochastic resonance system performance.

The USSSR [23] is used to deal with simulation signal and the result shows in Figure 8d where optimization parameters are $a = 2.343$, $b = 10{,}431.066$, $k = 0.13$, $R = 131.44$, and SNR = −11.49 dB. It can be seen that the characteristic frequency is significantly enhanced. The simulation signal further demonstrates that the filtering performance of the second-order SR system is better than that of the first-order SR system. The same signal is analyzed by the proposed STSR method, and the result is displayed in Figure 8e. Obvious period components can be found from the time domain waveform, and the characteristic frequency f_0 can be clearly seen from the amplitude spectrum, while other frequencies are almost nonexistent. The optimization parameters are $a = 10.23$, $b = 17186.47$, $k = 0.179$, $V = 31.56$, $r = 7.254$, $c = 6.5$, $R = 60.508$, and SNR = −6.53 dB at this time. Therefore, the proposed method can obtain larger output SNR and better filtering performance than traditional SR methods for detecting weak signals.

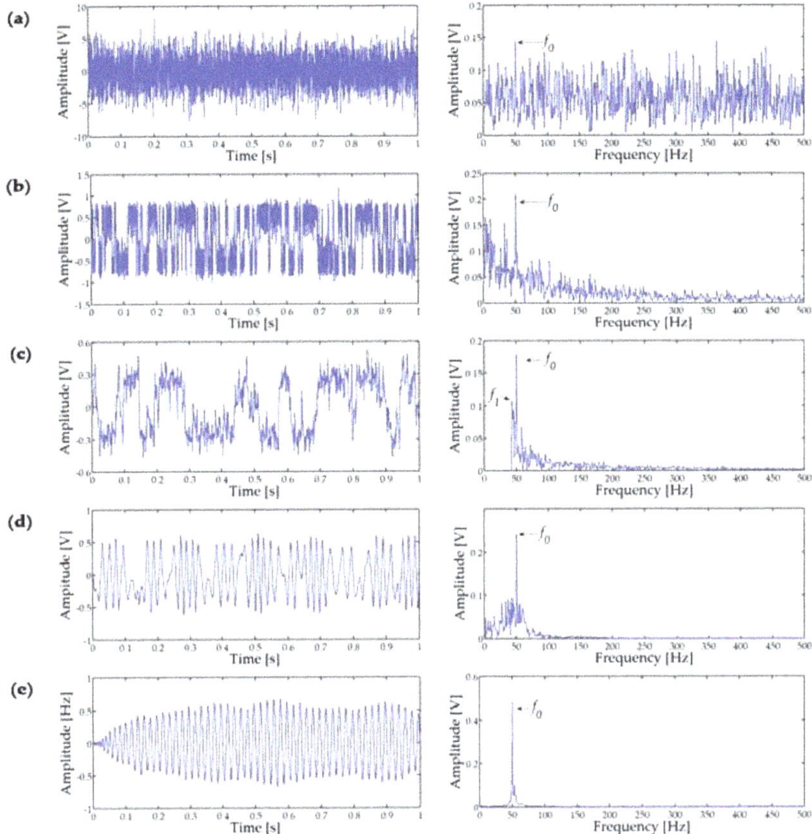

Figure 8. Analysis results of simulation signal: (**a**) Original signal and its amplitude spectrum; (**b**) stochastic resonance system parameters (CSR) output signal and its amplitude spectrum; (**c**) frequency-shifted and re-scaling SR output signal and its amplitude spectrum; (**d**) USSSR output signal and its amplitude spectrum; (**e**) STSR output signal and its amplitude spectrum.

4. Experimental Verification and Analysis

To confirm the effectiveness of proposed method in practical engineering applications, a set of vibration signals are analyzed. Meanwhile, to indicate the better STSR method, the same signals are processed using the CSR method and USSSR method. The experimental data comes from the Case Western Reserve University (CWRU) Bearing Center website [31–34]. The experimental test platform is shown in Figure 9. The bearing is the deep groove ball bearing with the type of 6205-2RS JEM SKF, where a physical dimension is shown in Table 1. The sampling frequency f_s is 12 kHz and sampling points are 4096. The fault characteristic frequencies for outer ring and inner ring are expressed by f_{BPFI} and f_{BFO}, respectively, that can be theoretically calculated as follows:

$$f_{BFO} = \frac{n}{2}\left(1 - \frac{d}{D}\cos\alpha\right)f_r \tag{13}$$

$$f_{BPFI} = \frac{n}{2}\left(1 + \frac{d}{D}\cos\alpha\right)f_r \tag{14}$$

where n is the number of rolling elements, f_r is rotating frequency of shaft, d is ball diameter, D is pitch diameter, and α is the contact angle, respectively.

Figure 9. Experimental test platform.

Table 1. Size parameters of 6205-2RS.

Inner Diameter (Inch)	Outsider Diameter (Inch)	Pitch Diameter (Inch)	Ball Diameter (Inch)	Number of the Rollers
0.9843	2.0472	1.537	0.3126	9

4.1. Outer Race Fault Signal Detection

In the first case, the bearing signal with an outer fault is tested to verify the performance of STSR. According to Equation (13), the fault characteristic frequencies of outer ring is theoretically $f_{BFO} = 105.9$ Hz with an approximate rotating speed of 1772 r/min. Figure 10a shows the original vibration signal wave and its power spectrum. It is found that the periodic component cannot be seen from time domain and the low frequency signal is modulated to the high frequency band. Firstly, the fault signal is filtered using the band-pass filter at 2000–4000 Hz. Then, Hilbert transform demodulated the filtered signal. Additionally, the enveloped signal and its amplitude spectrum are displayed in Figure 10b. While the characteristic frequency f_{BFO} can be seen from the amplitude spectrum, and the noise is still obvious in spectrum with output SNR = −15.02 dB.

The output result of CSR is shown in Figure 10c, where the parameters are $a = 0.002$, $b = 5656$, $R = 579$, and output SNR = −13.21 dB. We can see that the energy of the high frequency is significantly reduced, while the energy of f_{BFO} is amplified and the energy of the rotating frequency is greatly amplified. Then, the enveloped signal is analyzed by the USSSR method. Additionally, the results are exhibited in Figure 10d where $a = 0.46$, $b = 11471$, $R = 221.5$, $k = 0.434$, and output SNR = −8.357 dB. Finally, the proposed STSR method deals with the same signal and Figure 10e display the time domain waveform and amplitude spectrum of the output signal. The fault characteristic frequency f_{BFO} can be clearly seen in the whole spectrum diagram where $a = 8.282$, $b = 14244.13$, $k = 0.434$, $V = 35.725$, $r = 15.451$, $c = 14.328$, $R = 143.929$, and the output SNR = −5.09 dB.

The analysis results of the bearing outer ring fault signal shows that the filtering performance of STSR is significantly better than CSR and USSSR. In other words, it indicates the validity and superiority of the proposed STSR method.

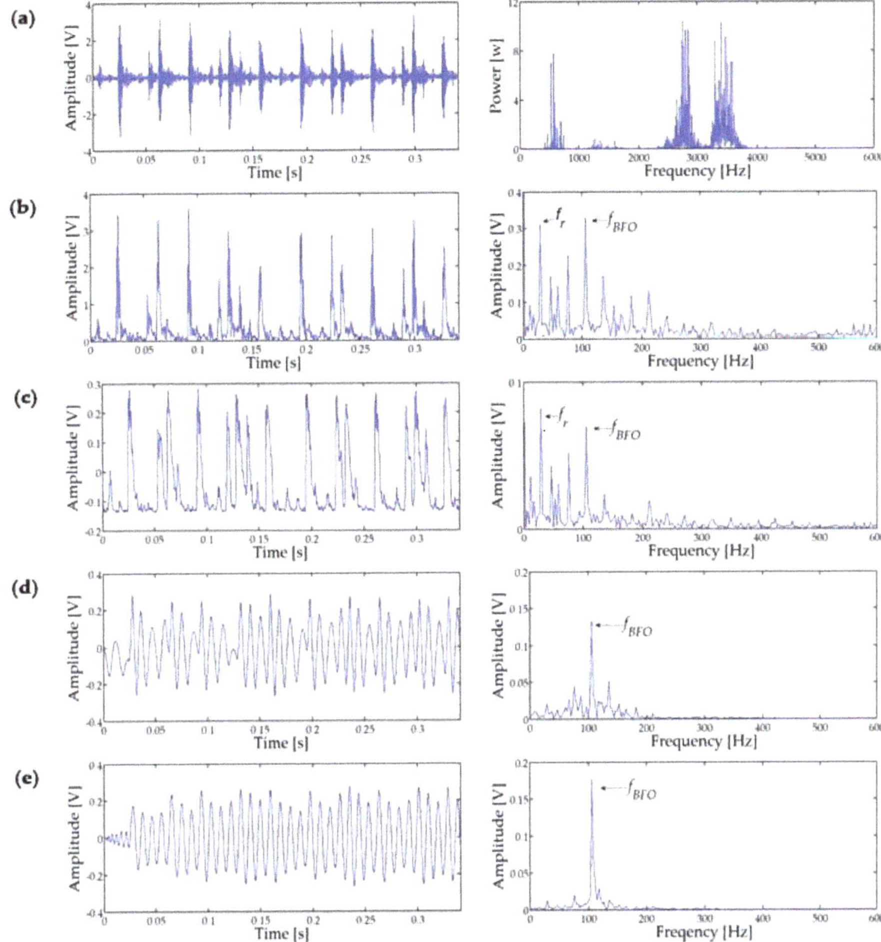

Figure 10. Analysis results with outer-race fault signal: (**a**) Original signal and its power spectrum; (**b**) envelope signal and its amplitude spectrum; (**c**) CSR output signal and its amplitude spectrum; (**d**) USSSR output signal and its amplitude spectrum; and (**e**) STSR output signal and its amplitude spectrum.

4.2. Inner Race Fault Signal Detection

To further verify the reliability of proposed STSR method, the inner race fault signal is analyzed. According to Equation (14), the characteristic frequency of inner race is theoretically $f_{BPFI} = 156.1$ Hz with approximate rotating speed of 1730 r/min. Figure 11a shows the time waveform of inner race fault and its power spectrum. It is hard to see the fault characteristic frequency from Figure 11a, due to the strong noise. The origin vibration signal was filtered by a band-pass filter in a bandwidth of 2000–4000 Hz and demodulated by Hilbert transform, then we got the envelop signal which is displayed in Figure 11b with output SNR = −17.53 dB. The CSR method was employed to test the envelop signal and the results are shown in Figure 11c. It can be observed that most of the high frequency noises are suppressed, but low frequency noises are enhanced where $a = 0.0015$, $b = 5235.68$, $R = 672.536$, and output SNR = −15.511 dB. Output results of the USSSR system are shown in Figure 11d with parameters of $a = 0.908$, $b = 12476.8$, $k = 0.222$, and $R = 304.395$. It can be seen the f_{BPFI} clearly in the amplitude spectrum and the output SNR is improved to −10.42 dB. The proposed STSR method is utilized to analyze the same signal. The optimal diagnosis result is displayed in Figure 11e where $a = 7.534$, $b = 9801.33$, $c = 10.97$, $V = 34.07$, $R = 206.5$, $r = 12.71$, $k = 0.222$, and the output SNR = −6.23 dB. It can be seen that fault characteristic frequency f_{BPFI} is prominent in the whole spectrum, which also demonstrate the effectiveness of the proposed STSR method than the traditional SR methods in bearing fault diagnosis.

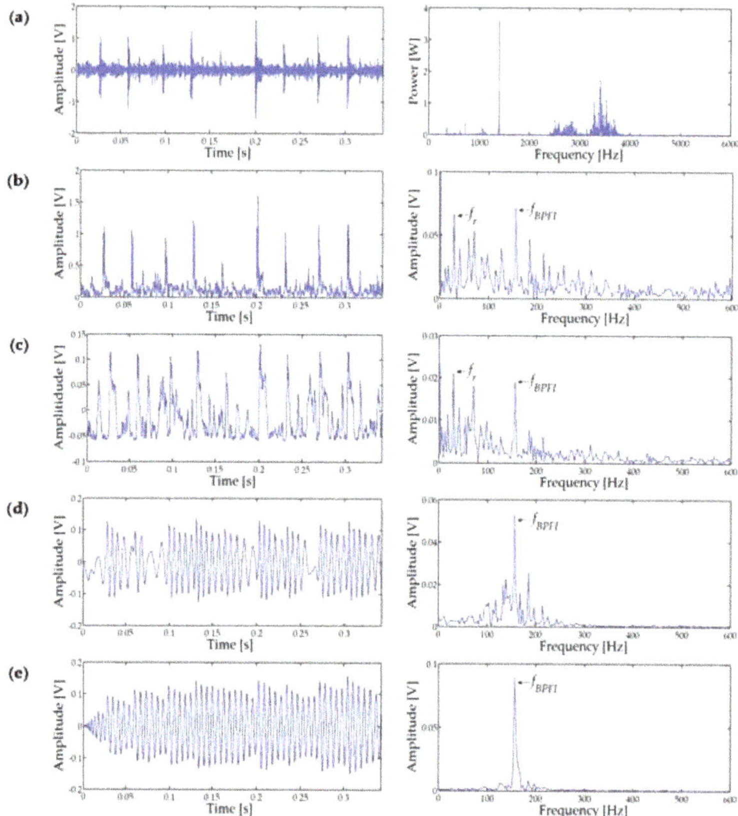

Figure 11. Analysis results with inner-race fault signal: (**a**) Original signal and its power spectrum; (**b**) envelope signal and its amplitude spectrum; (**c**) CSR output signal and its amplitude spectrum; (**d**) USSSR output signal and its amplitude spectrum; and (**e**) STSR output signal and its amplitude spectrum.

4.3. Discussion and Anlysis

Table 2 lists SNRs using different SR methods based on the above analysis. The comparison results show that the SNR of fault signals processed by STSR are higher than those of other methods. It also proves that STSR has better performance in the extraction of weak signals.

Table 2. The output SNR of different SR methods in process bearing fault signals.

Fault Type	Envelop Signal SNR (dB)	Output SNR of CSR (dB)	Output SNR of USSSR (dB)	Output SNR of STSR (dB)
Outer-race fault	−15.02	−13.21	−8.537	−5.09
Inner-race fault	−17.53	−15.511	−10.42	−6.23

The SR acts like a specific filter. The second-order SR system [23] provides a secondary filtering effect, while the first-order SR system corresponds to a filtering effect and it can be observed that USSSR output SNRs are higher than the CSR from Table 2. Thus, this paper explores second-order SR systems. Additionally, the tristable SR method [17] has a better effect than the classical bistable SR method in weak signal detection using first-order SR system. Therefore, the STSR method is proposed in this paper. The experimental results show that the proposed STSR method obtains higher than the CSR method and USSSR method and has a better filtering effect for weak signals.

The SR theory is developed with the regime of small parameter limitation (the input signal of SR model amplitude and frequency are less than 1); therefore, the rescaling frequency is utilized to detect bearing fault signals. The system parameters are crucial in determining the SR output. Hence, this work adjusts parameters synchronously using SOA. While STSR improve output signals with less interference components, SNR is the criterion to select system parameters. SNR criterion requires the precise frequency of detecting signal, the criterion establishment of SR is a key research direction in the future.

5. Conclusions

The STSR method is proposed to extract weak fault characteristic information in this paper. The proposed method has the following advantages: (1) The output signal exhibits significant periodicity in time domain and clear spectral peak line in frequency domain. (2) The STSR output SNR is higher than traditional SR methods. (3) The parameters of STSR system are optimized synchronously using SOA. Simulation and real vibration signals are used to verify the effectiveness of the STSR method. The results show that the proposed STSR has better filtering ability than traditional SR methods.

Author Contributions: Methodology, Y.Y. and L.L.; validation, L.L.; data curation, H.W. and J.Z.; writing—original draft preparation, L.L. and Y.Y.; writing—review and editing, L.L. and X.Z.; visualization, H.W. and J.Z.; funding acquisition, Y.Y.

Funding: This research was funded by the National Natural Science Foundation of China grant number 51605068, grant number 61771087, grant number 51879027, grant number 51579024, the Open Project Program of State Key Laboratory of Mechanical Transmissions of Chongqing University grant number SKLMT-KFKT-201803. Open Project Program of the Traction Power State Key Laboratory of Southwest Jiaotong University under Grant TPL1803, Liaoning BaiQianWan Talents Program and Liaoning Natural Science Foundation grant number 1553736724672.

Conflicts of Interest: The authors declare no conflict of interest.

References

1. Zhao, H.M.; Yao, R.; Xu, L.; Yuan, Y.; Li, G.Y.; Deng, W. Study on a novel fault damage degree identification method using high-order differential mathematical morphology gradient spectrum entropy. *Entropy* **2018**, *20*, 682. [CrossRef]
2. Zhao, H.M.; Sun, M.; Deng, W.; Yang, X.H. A new feature extraction method based on EEMD and multi-scale fuzzy entropy for motor bearing. *Entropy* **2017**, *19*, 14. [CrossRef]
3. Yuan, Y.; Zhao, X.; Fel, J.Y.; Zhao, Y.L.; Wang, J.H. Study on fault diagnosis of rolling bearing based on time-frequency generalized dimension. *Shock Vib.* **2015**, *2015*, 808457. [CrossRef]
4. Deng, W.; Zhao, H.M.; Zou, L.; Li, G.Y.; Yang, X.H.; Wu, D.Q. A novel collaborative optimization algorithm in solving complex optimization problem. *Soft Comput.* **2017**, *21*, 4387–4398. [CrossRef]
5. Lei, Y.G.; Lin, J.; He, Z.J.; Zuo, M.J. A review on empirical mode decomposition in fault diagnosis of rotating machinery. *Mech. Syst. Signal Proc.* **2013**, *35*, 108–126. [CrossRef]
6. Zhang, C.; Li, B.; Chen, B.; Cao, H.; Zi, Y.; He, Z. Weak fault signature extraction of rotating machinery using flexible analytic wavelet transform. *Mech. Syst. Signal Proc.* **2015**, *64–65*, 162–187. [CrossRef]
7. Qiao, Z.J.; Pan, Z. SVD principle analysis and fault diagnosis for bearings based on the correlation coefficient. *Meas. Sci. Technol.* **2015**, *26*, 085014. [CrossRef]
8. Wang, Y.; Liu, F.; Jiang, Z.; He, S.; Mo, Q. Complex variational mode decomposition for signal processing application. *Mech. Syst. Signal Proc.* **2017**, *86*, 75–85. [CrossRef]

9. Benzi, R.; Sutera, A.; Vulpian, A. The mechanism of stochastic resonanc. *J. Phys A Math. Gen.* **1981**, *14*, 453–457. [CrossRef]
10. Leng, Y.G.; Leng, Y.S.; Wang, T.Y.; Guo, Y. Numerical analysis and engineering application of large parameter stochastic resonance. *J. Sound Vib.* **2006**, *292*, 788–801. [CrossRef]
11. Tan, J.Y.; Chen, X.F.; Wang, J.; Chen, H.; Cao, H.; Zi, Y.; He, Z. Study of frequency-shifted and re-scaling stochastic resonance and its application to fault diagnosis. *Mech. Syst. Signal Proc.* **2009**, *23*, 811–822. [CrossRef]
12. He, Q.B.; Wang, J.; Liu, Y.B.; Dai, D.Y.; Kong, F. Multiscale noise tuning of stochastic resonance for enhanced fault diagnosis in rotating machine. *Mech. Syst. Signal Proc.* **2012**, *28*, 443–457. [CrossRef]
13. Zhang, Z.H.; Wang, D.; Wang, T.Y.; Lin, J.Z.; Jiang, Y.X. Self-adaptive step-changed stochastic resonance using particle swarm optimization. *J. Vib. Shock* **2013**, *32*, 125–130. (In Chinese)
14. Qiao, Z.; Lei, Y.; Lin, J.; Jia, F. An adaptive unsaturated bistable stochastic resonance method and its application in mechanical fault diagnosis. *Mech. Syst. Signal Proc.* **2017**, *84*, 731–746. [CrossRef]
15. Zhang, H.B.; Zheng, Y.; Kong, F. Weak impulsive signals detection based on step varying asymmetric stochastic resonance. *J. Mech. Eng. Sci.* **2017**, *231*, 242–262. [CrossRef]
16. Zhang, H.B.; He, Q.; Lu, S.L.; Kong, F. Stochastic resonance with a joint woods-saxon and gaussian potential for bearing fault diagnosis. *Math. Probl. Eng.* **2014**, *2014*, 17. [CrossRef]
17. Li, J.M.; Chen, X.F.; He, Z.J. Multi-stable stochastic resonance and its application research on mechanical fault diagnosis. *J. Sound Vib.* **2013**, *332*, 5999–6015. [CrossRef]
18. Han, D.Y.; An, S.J.; Shi, P.M. Multi-frequency weak signal detection based on wavelet transform and parameter compensation band-pass multi-stable stochastic resonance. *Mech. Syst. Signal Proc.* **2016**, *70*, 995–1010. [CrossRef]
19. Liu, X.L.; Liu, H.G.; Yang, J.; Litak, G.; Han, S. Improving the bearing fault diagnosis efficiency by the adaptive stochastic resonance in a new nonlinear system. *Mech. Syst. Signal Proc.* **2017**, *96*, 58–76. [CrossRef]
20. Deng, W.; Zhang, S.J.; Zhao, H.M.; Yang, X.H. A novel fault diagnosis method based on integrating empirical wavelet transform and fuzzy entropy for motor bearing. *IEEE Access* **2018**, *6*, 35042–35056. [CrossRef]
21. Guo, S.K.; Chen, R.; Li, H.; Zhang, T.L.; Liu, Y.Q. Identify severity bug report with distribution imbalance by CR-SMOTE and ELM. *Int. J. Softw. Eng. Knowl. Eng.* **2019**, *29*, 139–175. [CrossRef]
22. Deng, W.; Zhao, H.M.; Yang, X.H.; Xiong, J.X.; Sun, M.; Li, B. Study on an improved adaptive PSO algorithm for solving multi-objective gate assignment. *Appl. Soft Comput.* **2017**, *59*, 288–302. [CrossRef]
23. Lu, S.L.; He, Q.B.; Kong, F. Effects of underdamped step-varying second-order stochastic resonance for weak signal detection. *Digit. Signal Proc.* **2015**, *36*, 93–103. [CrossRef]
24. López, C.; Zhong, W.; Lu, S.; Cong, F.Y.; Cortese, I. Stochastic resonance in an underdamped system with FitzHug-Nagumo potential for weak signal detection. *J. Sound Vib.* **2017**, *411*, 34–46. [CrossRef]
25. Lei, Y.G.; Qiao, Z.J.; Xu, X.F.; Lin, J.; Niu, S.T. An underdamped stochastic resonance method with stable-state matching for incipient fault diagnosis of rolling element bearing. *Mech. Syst. Signal Proc.* **2017**, *94*, 148–164. [CrossRef]
26. Xia, P.; Xu, H.; Lei, M.H.; Ma, Z. An improved stochastic resonance method with arbitrary stable-state matching in underdamped nonlinear systems with a periodic potential for incipient bearing fault diagnosis. *Meas. Sci. Technol.* **2018**, *29*, 085002. [CrossRef]
27. Liu, Y.B.; Dai, Z.J.; Lu, S.L.; Liu, F.; Zhao, J.W.; Shen, J.L. Enhanced bearing fault detection using step-varying vibrational resonance based on duffing oscillator nonlinear system. *Shock Vib.* **2017**, *2017*, 14. [CrossRef]
28. Woods, D.; Saxon, D.S. Diffuse surface optical model for nucleon-nuclei scattering. *Phys. Rev.* **1954**, *95*, 577–584. [CrossRef]
29. Dai, H.; Chen, W.; Song, Y.H.; Zhu, Y.H. Seeker optimization algorithm: A novel stochastic search algorithm for global numerical optimization. *J. Syst. Eng. Electron.* **2011**, *21*, 300–311. [CrossRef]
30. Deng, W.; Xu, J.J.; Zhao, H.M. An improved ant colony optimization algorithm based on hybrid strategies for scheduling problem. *IEEE Access* **2019**, *7*, 20281–20292. [CrossRef]
31. Zhao, H.M.; Zheng, J.J.; Xu, J.J.; Deng, W. Fault diagnosis method based on principal component analysis and broad learning system. *IEEE Access* **2019**. [CrossRef]
32. Liu, Y.Q.; Wang, X.X.; Zhai, Z.G.; Chen, R.; Zhang, B.; Jiang, Y. Timely daily activity recognition from headmost sensor events. *ISA Trans.* **2019**. [CrossRef]

33. Guo, S.K.; Liu, Y.Q.; Chen, R.; Sun, X.; Wang, X.X. Using an improved SMOTE algorithm to deal imbalanced activity classes in smart home. *Neural Process. Lett.* **2019**. [CrossRef]
34. Case Western Reserve University Bearing Data Center. Available online: https://csegroupcasedu/bearingdatacenter/pages/welcome-case-western-reserve-university-bearing-data-center-websit (accessed on 9 March 2018).

Article

Research of the Equipment Self-Calibration Methods for Different Shape Fertilizers Particles Distribution by Size Using Image Processing Measurement Method

Andrius Laucka, Vaida Adaskeviciute and Darius Andriukaitis *

Department of Electronics Engineering, Kaunas University of Technology, Studentu St.50-438,
LT-51368 Kaunas, Lithuania
* Correspondence: darius.andriukaitis@ktu.lt

Received: 4 June 2019; Accepted: 24 June 2019; Published: 27 June 2019

Abstract: The fourth digital revolution of industry makes substantive changes to the rate and methodology of work performance. Machines and robots do the majority of work in robotized and automated factories, while people only supervise them. After an increase of production efficiency, quality control became a critical point. Therefore, quality control systems of computer visions are increasingly installed. The branch of chemical industry requires measurements of quality at as great a frequency as possible. Consequently, indirect measurements are effectively used at this point. This research presents the method of indirect particle measurement. Particles are measured using digital image processing. The algorithm is used for particle measurement to automatically adjust the measurement results. Numerical intelligence is added to the algorithm to increase the accuracy of correction results. The research deals with the problem of matching the results of indirect measurements and the results of the control equipment. For data analysis, fertilizer diameter, mean diameter, aspect ratio, symmetry, sphericity, convexity and some other parameters are used. The mismatch of the artificial neural network results with the control equipment results is slightly higher than 1%.

Keywords: image processing; particle; distribution

---

1. Introduction

The fourth industrial revolution has already become the present-day reality; the development is greatly influenced by the concept of the internet of things. As technologies interconnect physical and digital worlds together, changes are observed in all living spheres of the state. The largest changes occur in those industries where smart factories are established. Robots, which are replacing employees in production, do not get tired, which allows them to fulfill larger orders. It is essential to ensure required qualitative parameters of production together with productivity growth. It is not enough to provide production in series to quality control anymore. Volumes of quality production increase only by using continuous control and by checking every product [1].

The branch of chemical industry, fertilizer production, requires performance of quality checks as frequently as possible because production capacity amounts to hundreds and thousands of tons per hour. Decreasing utilized agricultural areas and the increase of population determines an increasing demand for fertilizers. Hence the growth of larger and more qualitative harvest requires balanced fertilization of plants. The most optimal result of fertilization is facilitated not only by chemical substances but also their physical parameters—the shape, size, and density of fertilizer granules. Therefore, one of the most significant indicators in the process of fertilizer production is distribution of granules in accordance with their diameter.

Every product, whatever industrial branch it belongs to, is characterized by tens or even hundreds of parameters. Due to increasing capacity of production equipment, human resources are not enough for the periodic inspection of production. Parameters of quality are checked by using indirect measurements, for which measurement errors usually are a typical deviation of the measurement result from the fair value.

Direct measurements of fertilizer granules, which are still widely used, are not effective due to the interval of their performance, which is very long. Measurement intervals are reduced by involving indirect measurements carried out alongside the production line. Different indirect methods are used for the measurement of granule distribution: laser diffraction [2–4], acoustic spectroscopy [5–7], infrared spectroscopy [8], dimensional filtering [9,10], radar [11], technique of sedimentation [12], transmission electron microscopy [13,14], X-rays diffractive spectrum [15], and digital image processing (one of the modern methods based on multistage entropy [16]). The processing of visual information provided additional details about particles. It also assesses the roundness of the product; its distribution is in accordance with its quantity and volume. Not to mention that certain information about the color [17,18] or porosity [19] of the product is received by applying an appropriate system of lighting. Systems of computer vision are applied to establish the distribution of particles in liquids [20] by using microscopes (SEM and TEM) [21,22] applying a three-dimensional technique of measurement by analyzing the surface of the granule sample [23,24] or approximating the segmented image of participles [25–27]. The selection of measurement methodology is influenced by the size of an object. The analysis of digital images is applied to assess participles within the interval of 0.01 to 20 mm. Quality control systems of computer vision are increasingly installed in factories due to their large capacity, reliability, mobility, and ability to process additional information. Of course, image processing results strongly depends on granules roundness. In most cases it is needed to recalculate results.

The results of indirect measurements are strongly influenced by the composition of analyzed sample. Different measurement methods are used to obtain reliable results in comparison with control equipment data. Results of the received distribution could be adjusted according to the distribution average coefficient [28] or compared with several other measurements [29,30]. The light reflection is used for the evaluation of larger particles [31]. The surface shape of particles is evaluated more accurately by three-dimensional images obtained with two cameras [32] or the reflection in mirrors [33]. A laser is often used to obtain the optimal result in conjunction with the image system [34,35].

Most methodologies of particle measurement provided in scientific publications make corrections to the measurement of a specific material alongside different production lines. The results are influenced by the temperature and humidity of a granule mix, including some other factors as well. Granules forming equipment is clogged during production. This affects the size and shape of the granules. Following the changes in the shape of the particles, these processes can be precisely identified, and the line stopped for equipment washing. Correction parameters of the results are established one time by carrying out equipment calibration. A significant change in the production process causes a problem—the noncompliance of measurement results exceeds the allowable errors determined by the international standards. An automated algorithm for result correction can make improvements by taking geometric changes in particles and the results of previous measurements into consideration. The universal algorithm may be applied to a wider spectrum of products. Consequently, the key goal of the research analyzed in the article is to develop an automated methodology of result correction to prepare assessment equipment for granule distribution in accordance with their size. The main aim of this research is to achieve a better correlation results between indirect measurements and control equipment. The goal is achieved by improving computer vision system.

2. Materials and Methods

One of the essential features of the fourth industrial revolution is the installation of computer vision systems for the inspection of industrial production quality parameters. The shape of fertilizer granules in the chemical industry has more advantages over the powder shape. The transportation and storage of granules are more convenient, segregation of the product is avoided, and production of homogenous mixes is more effective, as well as the process of fertilization. The measuring of granules by using a computer vision system is based on the geometric shape of particles, which, in an ideal situation, is a sphere.

Digital image processing within an indirect measurement is the primary source of information input. By using a video camera, morphological operators set the outlines of particles in the captured images of granule shadows (part of the researchers use image segmentation [36] by solving the problem of particle overlaying [37]). A precisely accurate result of the measurement is received by using well-calibrated equipment (hardware). To optimize the work of the system, it is important to optimize the computer resources for parallel image processing [38]. Optimal exposure duration ensures the identity of the distinguished granule outline with its real outline. The outlines of particles depend on the binarization of the images [39].

The outlines of granule shadows are approximated by an ellipse [40] because this geometric form most precisely assesses roundness of granules. The outline of an ellipse provides certain information about longitudinal and transverse dimensions of a particle, the ratio of which is the meaning of roundness. The results of particle distribution in accordance with their quantity correspond to the results attained by using the control equipment only in the events when the shape of granules is close to the ideal circle (an interval of roundness of 0.95 to 1.00). That depends on the product type, e.g., ammonium nitrate or carbamide constitutes solid round granules. However, such a product as monocalcium phosphate is small, frail and the shape of its particle is distinguished as anisotropic. Consequently, the image processing algorithm must assess an uneven surface of the granule.

The correlation of the results of the digital image analysis with the results of the control equipment is obtained by evaluating the volume of particle. The latter distribution corresponds to the mass distribution. In the production process, the density of the production in the interval of 4–6 hours varies by 5–7%. However, in the sample for the analysis, this difference is abnormally low (<0.1%), so the distribution must be calculated on the total volume of all particles.

The quality of physical features of fertilizer granules is evaluated by the following factors; an average diameter of granules d50, Uniformity Index, SGN (Size Guide Number), and an average square deviation of the granule size. After the sieve correction factor has been applied for the correction of d50 index, the diameter of every analyzed particle is divided by the latter factor. A more effective correction result is obtained by using the advantage of visual particle information, i.e., distribution of roundness. Then, the sieve correction factor is applied to the particles of an irregular shape only because the diameter of the latter ones depends on the position of falling particles in respect of the camera. The diameter of the granules, which is close to the ideal circle, does not require any correction.

By analyzing the visual information of the granules, it was found that the cumulative particle volume distribution curve accurately reproduces the results of the distribution obtained by the control equipment. Only the shadow of the granules close to the ideal circle was evaluated (an example of an image captured by the camera is provided in Figure 1). Such coincidence of the results depends on the roundness of the analyzed particles (higher rounding index), their mechanical resistance, and the number of granules.

Figure 1. Shadows of granules digital images were captured with a linear video camera.

The regression model functions are used for the correction of the cumulative curve shape of granule distribution. The essence of this new approach is in that the automatic understanding methods were used to develop new generation system [41]. The establishment of the factors of these functions uses samples measured by the control equipment. Depending on the scope of production process variations, the regression model cannot always recalculate the results accurately. That is related to the changes in distribution of particle roundness. If the roundness average has shifted below the 0.9 ratio of the minimum and maximum diameters, the quantity of round particles is not enough to establish accurate results. The assessment of the remaining particles can be carried out by using the results of previous measurements and the results of the most circular particle distribution of a new sample in accordance with their volume. After carrying out an analysis of granule images, the results, together with the data of previous measurements, are used to establish variables of the regressive function. This operation is performed after the analysis of every sample. The algorithm of the methodology is submitted in Figure 2. The cumulative curve of particle distribution describes the function

$$Q(q,c) = \int_{d_u}^{d_o} \frac{4}{3}\pi \cdot r_{max}(q,c) \cdot r_{min}^2(q,c) \tag{1}$$

where r_{max} is defined in

$$r_{max}(q,c) \cdot r_{min}^2(q,c) = \begin{cases} \frac{r_{min}(q)}{r_{max}(q)} > c : r_{max} \times r_{min}^2 \\ 0 \end{cases} \tag{2}$$

where q—analyzed particle; r_{min}—the minimum approximation ellipse radius of analyzed particle [mm]; r_{max}—the maximum approximation ellipse radius of analyzed particle [mm]; d_u and d_o—interval of used sieves [mm]; and c—particle round index (ratio of minimum and maximum radius).

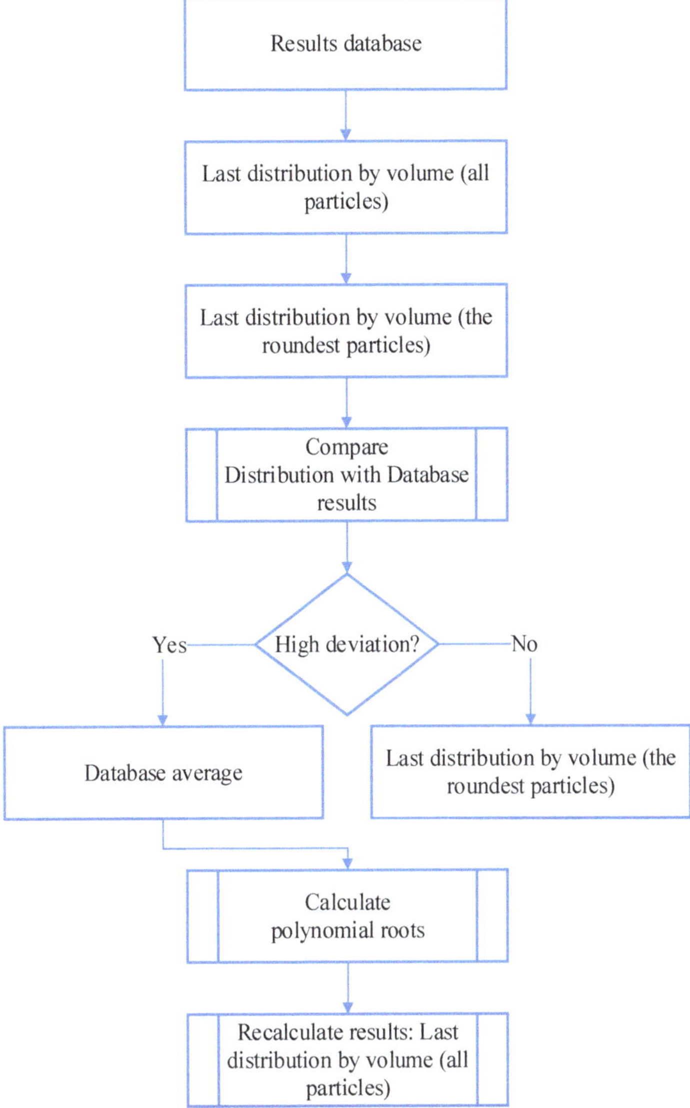

Figure 2. The algorithm to results correction using the function of polynomial.

An artificial neural network can be used to increase the efficiency of the particle correction algorithm. In scientific articles about particle measurement using image processing, authors use the neural network [42] for approximation of individual particles [43]. However, to achieve that, more resources are needed for calculations. Also, supporting vector machines (SVM) are used for particle measurement for the characterization of digital image classification [44]. In this research an artificial neural network is used to correct the cumulative curve. The general formula of the artificial neural network results in the function of activation:

$$y = \varphi(\sum_{j=0}^{n} w_j x_j) \tag{3}$$

where w_j—input weights; x_j—inputs; and y—output [45].

The results of particle image analysis together with the control equipment data of the previous measurement are used for the formation of the regression model function.

Linear video camera was used for particle measurement in the research. Due to their high-speed these cameras can capture real-time shadows of the particles falling from the conveyor, because a new image can be captured parallel with sending the image captured before. A frame of practically unrestricted height is attained. A linear video came typically has smaller geometric distortions; that is why they are more easily corrected than the ones of a matrix camera, because only one-dimensional correction of image distortions is necessary. It captures an image of higher resolution and a larger dynamic range. Moreover, the linear video camera has been selected because of better repeatability of the results. The equipment of experiment is submitted in Figure 3.

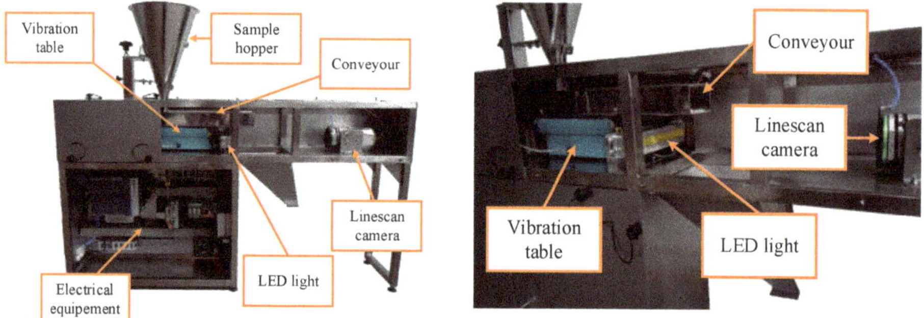

Figure 3. The equipment of the experiment.

By using visual information for the establishment of granule distribution, the minimal diameter of every particle of the sample is assessed. Such classification is an alternative to sieves because it is accepted that a particle can pass through sieves by its narrow par. When carrying out measurements, the volume of every particle is added to the volume of the particles with the same diameter. The intervals of diameters of added particles correspond with sieves. In the end of the analysis, the percentage of the distribution of all granules is attained.

3. Results

In this section, we used different methods for recalculation of the granules cumulative curve. The results of image processing were not accurate because of the anisotropic shape of granules.

3.1. Measurements Repeatability

The investigation of the indirect measurement equipment was completed by repeatedly obtaining the same results to evaluate the reliability of the experimental measurements. The roundness of the particles is related to their mechanical strength, which was also included into considered of the repeatability of the results. In the research, two different material (monocalcium phosphate and ammonium nitrate) samples were used. Each sample was measured three times (Figure 4).

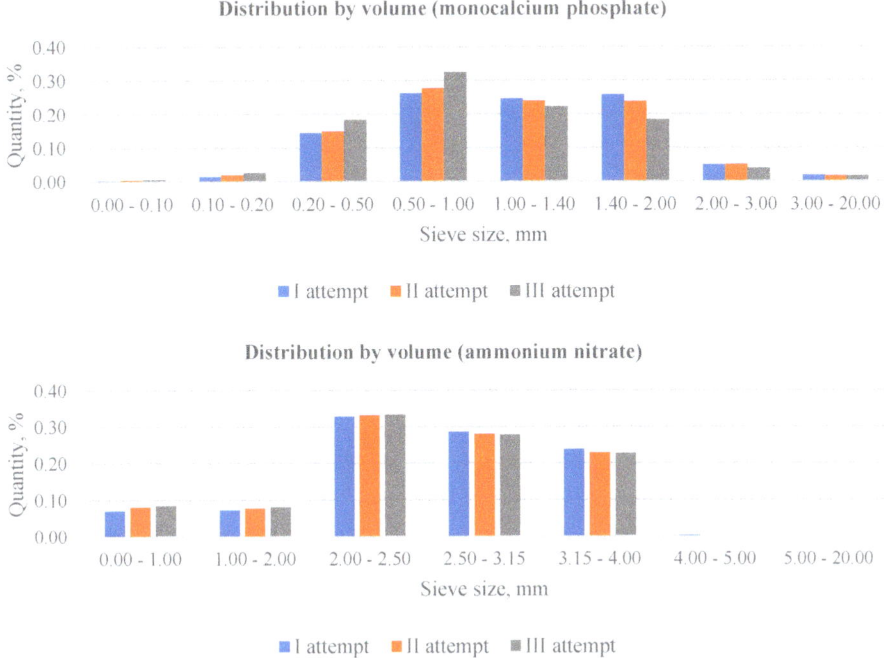

Figure 4. Repeatability of indirect measurement results (particle distribution is estimated by volume).

Additionally, the use of impurities in the production of ammonium nitrate gives better mechanical strength of particles, and therefore they have a better repeatability. Monocalcium phosphate particles crystals are irregular and brittle. This causes the sample deviation after each measurement. A correction of the results is necessary due to the different composition of the particles measurements.

3.2. Cumulative Curve Correction

The coefficient of sieve correction affects the indicator d50 of the average of particle diameter. The diameters of particles are magnified or reduced by moving the cumulative curve. The coefficient of the correction does not affect the shape of cumulative curve. The results of measurement corrections are provided in Figure 5. The latter method is not reliable, especially with small diameter (<1 mm) particle measurements.

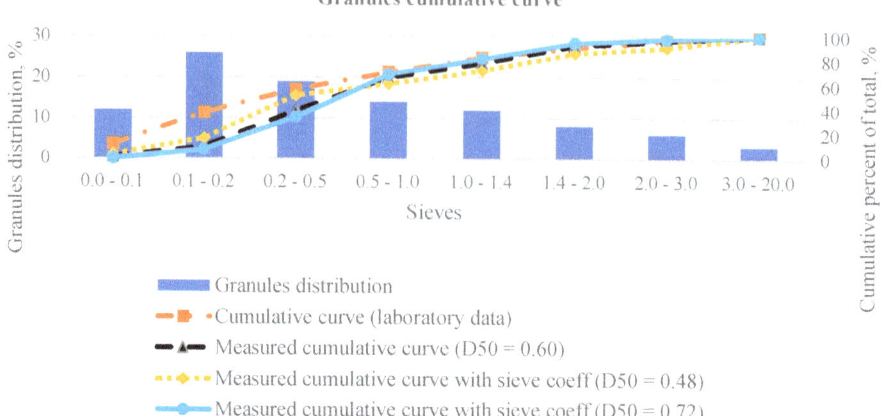

Figure 5. The cumulative curve of particle distribution is changed by applying a sieve correction coefficient.

Two samples were analyzed to evaluate of correlation between indirect measurement results and the control equipment results. Ammonium nitrate material was used for the analysis (two samples of 200 g). The results of the analysis showed that, when evaluating only the roundness of the particles, the received values correlate with the standard particle distribution. However, the amount of analyzed particles also affects the result (see Tables 1 and 2 of the results of measurements and Figure 6).

Table 1. Circularity evaluation results (sample no. 1).

Sample No. 1	Sieve Size					
	0.0–1.0	1.0–2.0	2.0–2.8	2.8–3.15	3.15–4.0	4.0–20.0
Actual weight (g)	0.4	9.0	49.2	59.6	66.8	15
Reference distribution (%)	0.2	4.5	24.6	29.8	33.4	7.5
Cumulative curve	0.2	4.7	29.3	59.1	92.5	100
Particles diagonal ratio (min/max)	**Indirect measurement (distribution by volume)**					
0–100%	0.03	7.01	30.03	25.72	32.24	4.97
90–100%	0.29	5.92	26.94	30.15	32.78	3.92
95–100%	0.14	4.72	25.10	29.60	33.80	6.64
98–100%	0.23	5.43	27.03	31.20	30.94	5.17
Circularity distribution						
Intervals (%)	0–92	92–94	94–96	96–98	98–100	
Quantity (%)	38.25	23.16	15.38	10.79	12.42	

Table 2. Circularity evaluation results (sample no. 2).

Sample No. 2	Sieve Size					
	0.0–1.0	1.0–2.0	2.0–2.8	2.8–3.15	3.15–4.0	4.0–20.0
Actual weight (g)	1.0	11.8	57.6	60.6	57.4	11.6
Reference distribution (%)	0.5	5.9	28.8	30.3	28.7	5.8
Cumulative curve	0.5	6.4	35.2	65.5	94.2	100
Particles diagonal ratio (min/max)	**Indirect measurement (distribution by volume)**					
0–100%	0.68	5.72	23.17	24.06	39.70	6.67
90–100%	0.74	6.14	23.64	26.30	34.50	8.68
95–100%	0.86	5.90	29.06	29.57	29.09	5.52
98–100%	0.79	6.52	30.21	26.14	30.16	6.18
Circularity distribution						
Intervals (%)	0–92	92–94	94–96	96–98	98–100	
Quantity (%)	35.59	21.62	10.84	17.22	14.73	

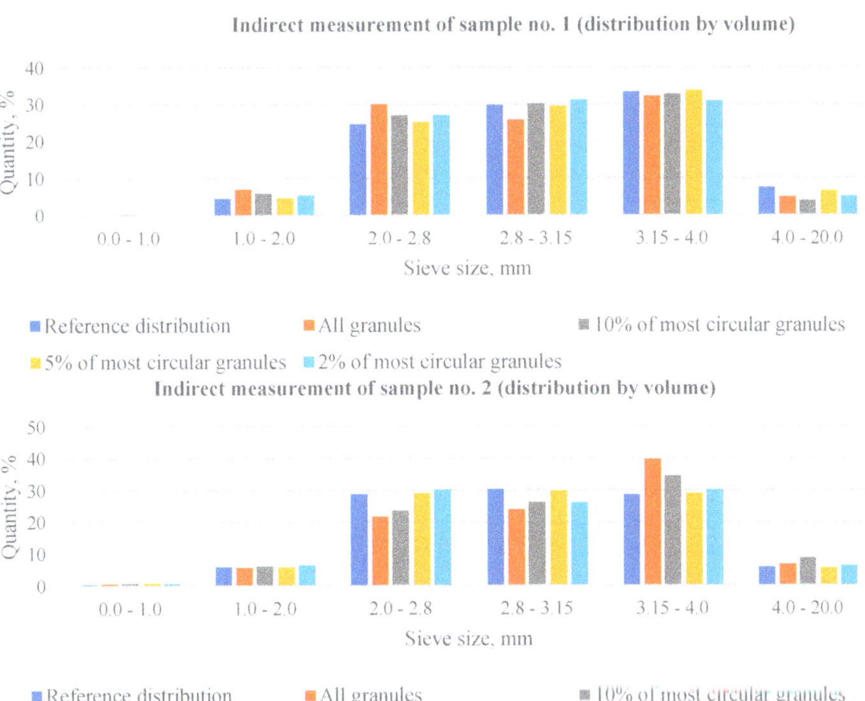

Figure 6. Comparison of indirect measurement results by evaluating only roundness particles.

For the analysis of spherical granules, it is enough to evaluate only part of the granules. If at least 75% of the granules fall within the 0.9 to 1.0 range according to the roundness distribution, it is enough to measure the volume of the granules in the interval 0.95 to 1.00 by roundness. This interval includes approximately 30% of all granules. That is directly related to volume calculation. By analyzing a two-dimensional image the smallest and the largest diameters of granules differ by at least 10%; therefore, it may be assumed that the third diameter (oriented towards the depth of an image) will also be different not more than by 10%. However, this tendency is noticed only with round granules. The more particles of an irregular shape there are in the sample, the worse correspondence to the results is attained by the control equipment. The relative error of the first sample cumulative curve between the control equipment and 5% of round particles is

$$\delta_1 = 0.58\% \tag{4}$$

The relative error of the second sample between cumulative curves:

$$\delta_2 = 0.57\% \tag{5}$$

The first sample (measurement results is provided in Table 1) is consisted of 29322 granules, of which the circular ones were ranked in the interval 0.95 to 1.00 according to the ratio of the smallest and the largest diameters. The number of particles used to find the distribution:

$$Quantity_1 = 29322 \times \frac{12.42 + 10.79 + 10.16}{100} = 9784 \tag{6}$$

The second sample (measurement results is provided in Table 2) is consisted of 30054 particles. The roundest particles from this sample were used to find the distribution by volume:

$$Quantity_2 = 30054 \times \frac{14.73 + 17.22 + 3.19}{100} = 10560 \tag{7}$$

The unplanned changes in production conditions determined problems in the measurement. At that time, it is not possible to rely only on the results of the distribution of a part of the most rounded particles, because the average of the circularity falls below the 0.9 limit. In this case, a correction of results is necessary. One of the most widely used methods is the use of a linear or nonlinear regression model.

The choice of the regression functions used to correct the particle measurement results is influenced by the accuracy which is related to the particle circularity. The experiment determined that the best results are achieved with the linear regression polynomial function (calculation results are presented in Table 3). The test was performed using two samples of monoammonium phosphate, for which four different functions in the MATLAB software environment was adapted.

Table 3. Regression analysis (2 samples).

Sieves	Measurements						Cumulative Curve Recalculation					
	Sample Weight		Indirect Measurements		Polynomial (5th)		Exponential (Half-Life)		Power Curve		Michaelis-Menten (Rectangular Hyperbola)	
	No. 1	No. 2	No. 1	No. 2	No. 1	No. 2	No. 1	No. 2	No. 1	No. 2	No. 1	No. 2
0.0–1.0	0.2	0.3	0.05	0.18	0.19	0.33	−0.14	−0.12	0.00	0.00	0.04	0.16
1.0–2.0	4.7	6.3	5.01	7.03	4.72	6.35	4.30	6.14	3.82	5.69	4.43	6.34
2.0–2.8	29.3	34.2	34.28	38.85	29.38	34.29	31.49	36.28	31.10	35.99	31.45	36.23
2.8–3.15	59.1	64.4	60.07	64.57	59.22	64.35	57.04	62.03	57.33	62.24	56.97	61.93
3.15–4.0	92.5	93.9	93.43	95.36	92.52	94.51	92.64	94.70	92.79	94.77	92.64	94.70
4.0–5.0	98.8	99.1	99.32	99.62	99.01	99.37	99.26	99.39	99.18	99.35	99.28	99.42
5.0–20.0	100	100	100	100	100	100	100	100	100	100	100	100
Goodness measure												
SSE			26.99	24.60	0.063	0.147	9.547	10.87	7.416	9.151	9.508	10.99
RMSE			5.196	4.959	0.251	0.383	3.089	3.297	2.723	3.025	3.083	3.315

The degree of polynomial function was evaluated in the research with the MATLAB software tools. Depending on the amount of sieve used, the best results were recorded using the 5-grade polynomial function (data of experiment is provided in Table 4). The radical of polynomial is found in the evaluation of round particles and all particles distributions.

Table 4. The calculation of polynomial degree (2nd sample).

Sample No. 2	Sieve Size					
	0.0–1.0	1.0–2.0	2.0–2.8	2.8–3.15	3.15–4.0	4.0–20.0
Actual weight (g)	1.0	11.8	57.6	60.6	57.4	11.6
Reference distribution (%)	0.5	5.9	28.8	30.3	28.7	5.8
Cumulative curve	0.5	6.4	35.2	65.5	94.2	100
Particles diagonal ratio (min/max)	**Indirect measurement (distribution by volume) cumulative curve**					
95–100%	0.86	6.76	35.82	65.39	94.48	100
Conversion of results with 3rd degree polynomial						
Recalculated cumulative curve	0.28	7.39	36.50	64.70	96.70	100
SSE [1]			6.6351			
Conversion of results with 4th degree polynomial						
Recalculated cumulative curve	1.08	6.41	36.10	65.20	94.60	100
SSE [1]			0.2919			
Conversion of results with 5th degree polynomial						
Recalculated cumulative curve	0.86	6.75	35.80	65.30	94.30	100
SSE [1]			1.739×10^{-27}			

[1] SSE—sum of squared errors of predictions.

Recalculated results of all sample particles and the results of control equipment correlate well. The relative error between these measurements is

$$\delta = 0.53\% \tag{8}$$

The 5-degree polynomial function results and those of the control equipment correlate well. The changes in the production can lead to a change of results, but not the precise mismatch of it (there may be significant differences between the results and results of the control equipment).

Changed conditions of the production highlight the shortages of the polynomial function. When the standard deviation of individual measurement results changes more than 5% together with the decrease in the average of roundness, the results recalculated in accordance with the polynomial function do not correspond to the standard measurement results. Interconnecting measurements of the particles having the largest roundness with the remaining ones, it is possible to carry out the recalculation of polynomial function radical after every analysis (algorithm is shown in Figure 2 and measurement results are presented in Table 5).

Table 5. The results according to the proposed algorithm.

Sieves (mm)	Sample Measurements Result			Indirect Measurement						
	Weight (g)	Distribution (%)	Cumulative Curve	Last 3 Results from DB Cumulative Curve				Last Meas. Cumulative Curve	10% of most Circular Granules	Polynomial Recalc.
				No. 1	No. 2	No. 3	AVG			
0.0–1.0	8	4	4	3.19	3.47	4.06	3.57	1.48	2.41	3.52
1.0–2.0	14	7	11	10.52	12.27	11.45	11.41	2.53	9.88	11.47
2.0–2.8	32	16	27	29.74	27.82	27.94	28.50	18.44	25.95	28.54
2.8–3.15	44	22	49	51.20	50.77	51.08	51.02	47.29	48.18	50.85
3.15–4.0	52	26	73	73.63	73.19	74.26	73.69	70.97	74.46	72.69
4.0–5.0	32	16	91	93.32	93.15	92.45	92.97	92.68	92.17	90.36
5.0–20.0	18	9	100	100	100	100	100	100	100	99.63

In the experiment an average of distribution results of last three measurements was used. According to the current measurement data, distribution of the particles with the largest roundness is assessed. When carrying out the experiment, 10% of the granules with the largest roundness were assessed. Distribution of the last measurement of all granules is used to establish polynomial radicals together with an average of the measurements from the database. The recalculated result is compared with the distribution of the granules having the largest roundness and with the database average. The distribution of the best result is obtained with the lowest relative error:

$$\delta_{recalc} = 1.22\% \tag{9}$$

$$\delta_{10\%\_circ} = 2.73\% \tag{10}$$

After calculating the estimated errors, the selected particle distribution is recalculated with the polynomial function (calculation results are presented in Table 5 and Figure 7). The received results and the results of control equipment correlate well (distribution by weight).

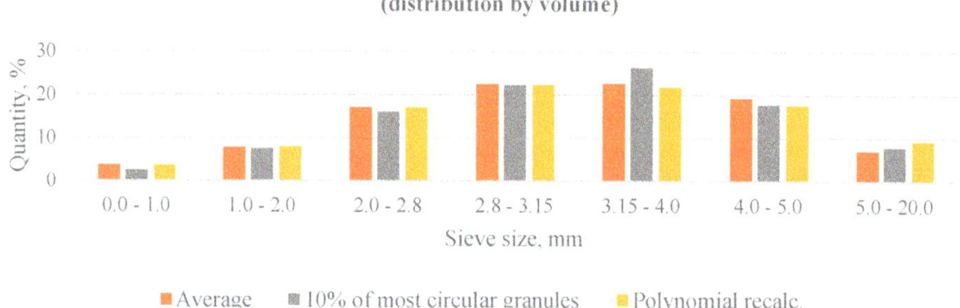

Figure 7. Recalculated indirect measurement results by the average from last 3 measures from database.

3.3. Neural Network Simulation

The Artificial intelligent method was applied to the development of intelligent system [46]. A simulation of the artificial neuron network has been performed using the software MATLAB (structure is provided in Figure 8). The two-layer feed-forward network with sigmoid hidden neurons and linear output neurons was used in the research. The ANN was trained with Levenberg–Marquardt backpropagation algorithm. The results of optical measurement have been recalculated using the samples analyzed by the control equipment in advance. The first sample provided in the table is an average of the last three measurements from the database, which is acknowledged as the standard. In accordance to these measurements, polynomial radicals are calculated and the neuronal network is trained (measurement results are provided in Table 6). Other two measurements use the algorithm provided in Table 5. Distribution of the granules with the largest roundness is presented to the neuron network as well as to the polynomial function. After significant noncompliance has been established between the curves, the cumulative curve of distribution of all particles is presented to the artificial neuron network. The attained results correlate with the mass distribution of the sample.

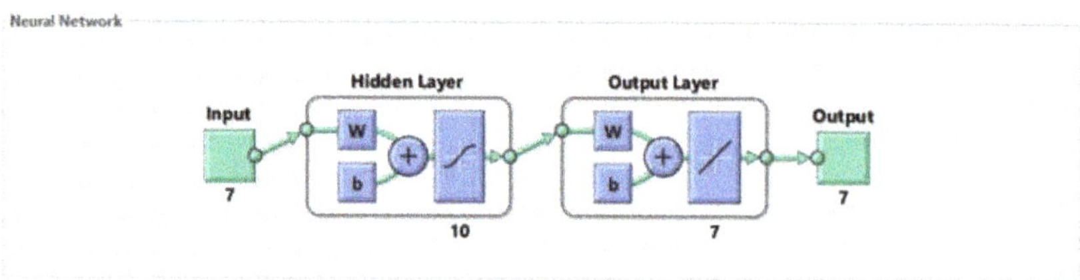

Figure 8. The artificial neural network for the recalculation of fertilizer particles distribution.

Table 6. Artificial neural network comparison with polynomial.

| Sieves | Measurements | | | | Cumulative Curve Recalculation | | | |
| | Weight | Indirect Measurement | | | Polynomial (6th) | | Neural Network | |
		No. 1 (avg)	No. 2	No. 3	No. 2	No. 3	No. 2	No. 3
0.0–1.0	0.2	0.05	0.18	0.14	0.27	0.32	0.14	0.12
1.0–2.0	4.7	5.01	6.17	7.12	6.12	7.32	4.70	4.82
2.0–2.8	29.3	38.37	38.21	39.56	29.08	29.69	29.17	29.26
2.8–3.15	59.1	64.16	63.97	65.61	57.85	60.58	59.22	59.83
3.15–4.0	92.5	93.43	95.08	95.29	89.31	89.45	95.74	96.12
4.0–5.0	98.8	99.32	99.54	99.68	93.83	94.05	98.03	98.07
5.0–20.0	100	100	100	100	94.56	94.55	99.94	99.97
Goodness measure								
SSE		109.1	112.5	162.1	68.1	70.8	11.13	14.19
RMSE		10.45	10.61	12.73	8.25	8.41	3.34	3.77

Shortages of the linear regression model show up by analyzing more samples of the composition, which changes during the production (longer measurement intervals). An artificial neuron network is distinguished for more accurate results of distribution in this situation (results in Table 6). The identity of ANN results with mass distribution of the sample is determined by the quantity of data for network training.

4. Conclusions

In this paper, the method of correction of optical measurement results of fertilizer particles (indirect method) is presented. The results obtained by the digital image processing method depend on the geometric and mechanical properties of the particles. Better repeatability of data is achieved by analyzing mechanically stronger particles, which also have rounder geometric shapes.

The equalization of the obtained results and the control equipment results is converted using a linear regression model. This method identifies the changes in production, but not the size of it. As a result, the obtained results may vary by more than 8% depending on the production conditions and the composition of the material.

The main disadvantage of this method is the dependence of reliability on the sphericity of particles. The best results are obtained when at least 75% of the fertilizer amount falls within the range of 0.9 to 1.0 of distribution by roundness. However, the algorithm can be applied not only to fertilizers, but also to the measurement of other round particles.

While analyzing the production, in which the granules are characterized by large scale roundness, it is enough to assess granules with the largest roundness because the noncompliance reaches up to 0.6%. To increase the effectiveness of the results when the shape of the particle surface varies, it is appropriate to extend the model of linear regression, which is based on the previous results from the database. To achieve result correction the latter algorithm uses not only the results of the distribution of the granules having the largest roundness but also distributions of the previous samples. The attained results precisely correlate with the results obtained by the control equipment even when production conditions significantly change. The noncompliance of the cumulative curve of particles recalculated in the experiment was 1.22%, while the 10% noncompliance of the cumulative curve of the particles with the largest roundness, in comparison with the mass distribution attained by the control equipment, which was 2.73%.

The system was installed in a factory in real production line. It brings real benefits to the operators managing the production process. Also, the fertilizer manufacturer gets more information about their production. One of the most important advantages of the system is that measurements are made very often, which is every 5–7 min.

The most accurate results of indirect measurements for work with many sieves are attained for correction by using an artificial neuron network. The accuracy of the latter results depends on the quantity of the data accumulated for training. Carrying out more accurate measurements requires an assessment of more variables of the production process. In the further investigation, the involvement of humidity and temperature measurements in the artificial neuron network is planned.

Author Contributions: A.L. and V.A. wrote the paper; A.L. and D.A. gave guidance in experiments and data analysis.

Funding: This research received no external funding.

Acknowledgments: The system was created and all researches were made in JSC "Elinta". The owner of intellectual property is JSC "Elinta". We thank JSC "Elinta" and its director Vytautas Jokuzis for the opportunity to publish this research.

Conflicts of Interest: The authors declare no conflict of interest.

References

1. Laucka, A.; Andriukaitis, D. Research of the Defects in Anesthetic Masks. *Radioengineering* **2015**, *24*, 1033–1043. [CrossRef]
2. Stojanovič, Z.; Markovič, S. Determination of Particle Size Distributions by Laser Diffraction. *Tech. New Matter* **2012**, *67*, 11–20.
3. Witt, W.; Heur, M.; Schaller, M. In-line particle sizing for process control in new dimensions. *China Particuol.* **2004**, *2*, 185–188. [CrossRef]
4. Cornillault, J. Particle size analyser. *Appl. Opt.* **1972**, *11*, 265–268. [CrossRef] [PubMed]
5. Coghill, P.J.; Millen, M.J.; Sowerby, B.D. On-line measurement of particle size in mineral slurries. *Miner. Eng.* **2002**, *15*, 83–90. [CrossRef]
6. McClements, D.J. *Ultrasonic Measurements in Particle Size Analysis. Encyclopedia of Analytical Chemistry: Applications, Theory and Instrumentation*; John Wiley Sons: Hoboken, NJ, USA, 2006; pp. 88–127. Available online: 10.1002/9780470027318.a1518 (accessed on 3 June 2019).
7. Wan, Q.; Jiang, W.K. Near field acoustic holography (NAH) theory for cyclostationary sound field and its application. *J. Sound Vib.* **2006**, *290*, 956–967. [CrossRef]
8. Findlay, W.P.; Peck, G.R.; Morris, K.R. Determination of fluidized bed granulation end point using near-infrared spectroscopy and phenomenological analysis. *J. Pharm. Sci.* **2005**, *94*, 604–612. [CrossRef]
9. Petrak, D. Simultaneous measurements of particle size and particle velocity by the spatial filtering. *Part. Part. Syst. Charact.* **2002**, *19*, 391–400. [CrossRef]
10. Dieter, P.; Stefan, D.; Gunter, K. In-line particle sizing for real-time process control by fibre-optical spatial filtering technique (SFT). *Adv. Powder Technol.* **2011**, *22*, 203–208. [CrossRef]
11. Shiina, T.; Muramoto, K. Z-R relation for snowfall using two small doppler radars and snow particle images. In Proceedings of the 2010 IEEE International Geoscience and Remote Sensing Symposium, Honolulu, HI, USA, 25–30 July 2010; pp. 4122–4125. [CrossRef]
12. Kammermeyer, K.; Binder, J. Particle Size Determination by Sedimentation. *Ind. Eng. Chem. Anal. Ed.* **1941**, *13*, 335–337. [CrossRef]
13. Hossain, M.A.; Mori, S. Determination of Particle Size Distribution of Used Black Tea Leaves by Scanning Electron Microscope. *Dhaka Univ. J. Sci.* **2013**, *61*, 111–115. [CrossRef]
14. Ray, O.; Banik, B.; Pani, C. Computational Size Measurement & Study of Nanoparticle Using Transmission Electron Microscopy Data by Image-Processing. In Proceedings of the 2017 IEEE 7th International Advance Computing Conference (IACC), Hyderabad, India, 5–7 January 2017; pp. 656–658. [CrossRef]
15. Gontard, L.C.; Ozkaya, D.; Dunin-Borkowski, R.E. A Simple Algorithm for Measuring Particle Size Distributions on an Uneven Background from TEM Images. *Ultramicroscopy* **2011**, *111*, 101–106. [CrossRef] [PubMed]
16. Yantong, Z.; Guoying, Z.; Yu, G. Particle Size Measurement Based on Image Multivariate Multiscale Entropy. In Proceedings of the 2017 IEEE Trustcom/BigDataSE/ICESS, Sydney, Australia, 1–4 August 2017; pp. 973–977. [CrossRef]

17. Jorgensen, T.; Reinholt, F.; Johnsen, O.M. Automatic Particle Analyzing System. US Patent 7154600, 2001. Available online: https://www.google.lt/patents/US7154600 (accessed on 3 June 2019).

18. Freiherr von Hodenberg, M. Device for Determining Parameters of a Bulk Material Particle Flow. WO Patent 2008046914, 2008. Available online: https://www.google.lt/patents/WO2008046914A1?cl=en (accessed on 3 June 2019).

19. Jorgensen, T.K. Online Sampling Apparatus and Method for Online Sampling. WO Patent 2012083966, 2012. Available online: https://www.google.lt/patents/WO2012083966A1?cl=en (accessed on 3 June 2019).

20. Maaß, S.; Rojahn, J.; Emmerich, J.; Kraume, M. On-line monitoring of fluid particle size distributions in agitated vessels using automated image analysis. In Proceedings of the 14th European Conf. on Mixing, Warsaw, Poland, 10–13 September 2012; pp. 257–262. Available online: http://mixing14.eu/p/mixing14eu_39.pdf (accessed on 3 June 2019).

21. Lu, Z.; Hu, X.; Lu, Y. Particle Morphology Analysis of Biomass Material Based on Improved Image Processing Method. *Int. J. Anal. Chem.* **2017**, *2017*, 1–9. [CrossRef] [PubMed]

22. Hijazi, B.; Cool, S.; Vangeyte, J.; Mertens, K.; Cointault, F.; Paindavoine, M.; Pieters, J. High speed stereovision setup for position and motion estimation of fertilizer particles leaving a centrifugal spreader. *Sensors* **2014**, *14*, 21466–21482. [CrossRef] [PubMed]

23. Närvänen, T.; Seppälä, K.; Antikainen, O.; Yliruusi, J. A New Rapid On-Line Imaging Method to Determine Particle Size Distribution of Granules. *AAPS PharmSciTec.* **2008**, *9*, 282–287. [CrossRef] [PubMed]

24. Kumari, R.; Rana, N. Particle size and shape analysis using Imagej with customized tools for segmentation of particles. *Int. J. Eng. Res. Technol.* **2015**, *4*, 247–250. Available online: https://www.ijert.org/download/14339/particle-size-and-shape-analysis-using-imagej-with-customized-tools-for-segmentation-of-particles (accessed on 3 June 2019).

25. Watano, S.; Numa, T.; Miyanami, K. On-line Monitoring of Granule Growth in High Shear Granulation by an Image Processing System. *Int. J. Chem. Pharm. Bull.* **2000**, *48*, 1154–1159. [CrossRef]

26. Mazzoli, A.; Favoni, O. Particle size, size distribution and morphological evaluation of airbone dust particles of diverse woods by Scanning Electron Microscopy and image processing program. *Int. J. Powder Technol.* **2012**, *225*, 65–71. [CrossRef]

27. Watano, S.; Numa, T.; Miyanami, K. A fuzzy control system of high shear granulation using image processing. *Int. J. Powder Technol.* **2001**, *115*, 124–130. [CrossRef]

28. Sakamoto, Y.; Tamura, Y.; Kawaguchi, K. Measuring Particle Size Distribution. U.S. Patent 4288162, 1981. Available online: https://www.google.lt/patents/US4288162 (accessed on 3 June 2019).

29. Penumadu, D.; Zhao, R.; Steadman, E.F. Particle Size and Shape Distribution Analyser. U.S. Patent 6960756, 1981. Available online: https://www.google.lt/patents/US6960756 (accessed on 3 June 2019).

30. Lieber, K.J.; Browne, I.B.; Tuttle, J. Control Feedback System and Method for Bulk Material Industrial Processes Using Automated Object or Particle. U.S. Patent 6885904, 2005. Available online: https://www.google.lt/patents/US6885904 (accessed on 3 June 2019).

31. Nase, T.F.; Vourisalo, Y.R. Procedure and Apparatus for Determining Size and/or Shape Distribution. WO Patent 1990012310, 1990. Available online: https://www.google.lt/patents/WO1990012310A1?cl=en (accessed on 3 June 2019).

32. Ettmuller, J.; Reindel, K.; Schafer, M. Sample's Particle Individual, Three Dimensional form e.g. Powder Form, Automated Determining Method, Involves Observing Particles from Two Observation Directions, Where Particle Axis is Aligned along Line Transverse to Observation Direction. DE Patent 102005055825, 2007. Available online: https://www.google.lt/patents/DE102005055825A1?cl=en (accessed on 3 June 2019).

33. Canty, T.M.; O'Brien, P.J.; Marks, C.P. Granular Product Inspection Device. U.S. Patent 7009703, 2006. Available online: https://www.google.lt/patents/US7009703 (accessed on 3 June 2019).

34. Niwa, T. Particle Size Measuring Device. U.S. Patent 5379113, 1995. Available online: https://www.google.lt/patents/US5379113 (accessed on 3 June 2019).

35. Schumann, M. Procedure for the Determination of Particle Size Distribution in Particle Mixtures. U.S. Patent 5309215, 1994. Available online: https://www.google.lt/patents/US5309215 (accessed on 3 June 2019).

36. Wang, Y.; Lin, C.L.; Miller, J.D. 3D image segmentation for analysis of multisize particles in a packed particle bed. *Powder Technol.* **2016**, *301*, 160–168. [CrossRef]

37. Priadarsini, S.; Ganesan, S.; Shanthi, C.; Pappa, N. Model based object recognition for particle size distribution. In Proceedings of the 2014 IEEE International Conference on Advanced Communications, Control and Computing Technologies, Ramanathapuram, India, 8–10 May 2014; pp. 1365–1368. [CrossRef]

38. Atasoy, H.; Yildirim, E.; Yildirim, S.; Kutlu, Y. A Real-Time Parallel Image Processing Approach on Regular PCs with Multi-Core CPUs. *Elektronika ir Elektrotechnika* **2017**, *23*, 64–71. [CrossRef]

39. Lech, P.; Okarma, K. Prediction of the Optical Character Recognition Accuracy based on the Combined Assessment of Image Binarization Results. *Elektronika ir Elektrotechnika* **2015**, *21*, 62–65. [CrossRef]

40. Kumara, J.J.; Hayano, K.; Ogiwara, K. Fundamental Study on Particle Size Distribution of Coarse Materials by Image Analysis. Available online: https://www.researchgate.net/publication/301692754_Fundamental_study_on_particle_size_distribution_of_coarse_materials_by_image_analysis (accessed on 3 June 2019).

41. Tadeusiewicz, R.; Ogiela, L.; Ogiela, M.R. Cognitive Analysis Techniques in Business Planning and Decision Support Systems. In Proceedings of the Artificial Intelligence and Soft Computing, Zakopane, Poland, 12–16 June 2006; pp. 1027–1039. [CrossRef]

42. Vrbancic, G.; Podgorelec, V. Automatic Classification of Motor Impairment Neural Disorders from EEG Signals Using Deep Convolutional Neural Networks. *Elektronika ir Elektrotechnika* **2018**, *24*, 3–7. [CrossRef]

43. Ferrari, S.; Piuri, V.; Scotti, F. Image processing for granulometry analysis via neural networks. In Proceedings of the 2008 IEEE International Conference on Computational Intelligence for Measurement Systems and Applications, Instanbul, Turkey, 14–16 July 2008; pp. 28–32. [CrossRef]

44. Zhang, Z. An estimation of coal density distributions by weight based on image analysis and MIV-SVM. In Proceedings of the 2015 IEEE Advanced Information Technology, Electronic and Automation Control Conference (IAEAC), Chongqing, China, 19–20 December 2015; pp. 1110–1113. [CrossRef]

45. White, H. *Artificial Neural Networks: Approximation and Learning Theory*; Blackwell Publishers, Inc.: Cambridge, MA, USA, 1992.

46. Ogiela, L.; Tadeusiewicz, R.; Ogiela, M.R. Cognitive Analysis in Diagnostic DSS-type IT Systems. In Proceedings of the Artificial Intelligence and Soft Computing, Zakopane, Poland, 12–16 June 2006; pp. 962–971. [CrossRef]

 *symmetry*

Article

Data-Driven Adaptive Iterative Learning Method for Active Vibration Control Based on Imprecise Probability

Liang Bai [1,*], Yun-Wen Feng [1], Ning Li [2], Xiao-Feng Xue [1] and Yong Cao [1]

[1] School of Aeronautics, Northwestern Polytechnical University, Western Youyi Street 127, Xi'an 710072, China; fengyunwen033616@163.com (Y.-W.F.); xuexiaofeng033616@163.com (X.-F.X.); caoyong033616@163.com (Y.C.)

[2] College of Sciences, Northeastern University, Shenyang 110819, China; liling80@163.com

[*] Correspondence: bailiangcom@mail.nwpu.edu.cn; Tel.: +86-18392055207

Received: 25 April 2019; Accepted: 29 May 2019; Published: 2 June 2019

Abstract: A data-driven adaptive iterative learning (IL) method is proposed for the active control of structural vibration. Considering the repeatability of structural dynamic responses in the vibration process, the time-varying proportional-type iterative learning (P-type IL) method was applied for the design of feedback controllers. The model-free adaptive (MFA) control, a data-driven method, was used to self-tune the time-varying learning gains of the P-type IL method for improving the control precision of the system and the learning speed of the controllers. By using multi-source information, the state of the controlled system was detected and identified. The square root values of feedback gains can be considered as characteristic parameters and the theory of imprecise probability was investigated as a tool for designing the stopping criteria. The motion equation was driven from dynamic finite element (FE) formulation of piezoelectric material, and then was linearized and transformed properly to design the MFA controller. The proposed method was numerically and experimentally tested for a piezoelectric cantilever plate. The results demonstrate that the proposed method performs excellent in vibration suppression and the controllers had fast learning speeds.

Keywords: time-varying P-type IL method; MFA control; imprecise probability; active control; piezoelectric cantilever plate

---

1. Introduction

Many industrial systems accomplish tasks in a limited period of time and repeat control processes continuously. In these systems, it is attractive to improve the system performance by repeating the control process, which draws attention to intelligent control strategy, named the iterative learning (IL) method. The IL method is applicable to controlled systems with repetitive motion properties. The fundamental IL method is a learning process based on output errors and learning gains. To obtain better control performances, the upgraded system inputs will be generated at the next repetitive processes by the latest tracking errors [1]. In practical industrial processes, the IL method is an effective approach to produce the control inputs, so the system outputs are as close as possible to the desired system outputs, such as: control trajectory tracking for lower limb rehabilitation [2], design of a shaping method for the residual vibration control of industrial robots [3], compensation for aerodynamic disturbance of the aerial refueling system [4], design of the controller for homing guidance of missiles [5]. Considering the repeatability of structural dynamic responses in the vibration process, several research groups have applied the proportional-type iterative learning (P-type IL) method with fixed gain to suppress the vibrations of piezoelectric laminated composite structures. The P-type IL method was first applied to vibration active control of piezoelectric structures by Zhu et al. [6] and Tavakolpour et al. [7], and its control performance has also been studied. Zhang et al. [8] designed a controller to compensate the disturbance estimation and disturbance rejection, where the P-type IL method was used to estimate the system position.

In the P-type IL method, thousands of iterations are necessary to obtain the desired control precision, which may reduce the convergence speed [7,9]. It is worth pointing out that the systems will spillover or even become unstable in practice applications, as the fixed-learning gain are selected unreasonably [10]. For improving the convergence speed and the system stability, the adaptive control is necessary to design the optimal learning gain in each iteration. To design vibration active control systems, both control strategies and structure characteristics need to be seriously considered. Piezoelectric flexible structures, such as plates and shells, challenge researchers in vibration active control for their complexity in the vibration modes [11–13]. In addition, using one single actuator–sensor pair bonded on a plate is obviously unreasonable to suppress all vibration modes and to identify dynamic responses for all locations on the plate. A distributed method that piezoelectric actuator-sensor pairs at discrete locations were bonded on the plate was constructed [14]. Piezoelectric laminated composite structures have low weight and cannot achieve great changes in the dynamic characteristics of the plate. A flexible plate with several piezoelectric actuator–sensor pairs, thus, is a multi-input–multi-output (MIMO) system. If a piezoelectric actuator is not able to satisfactorily perform the given task, the actuators nearby will be influenced negatively. The interaction between actuators will exist in the entire operating process of the system, and the information of the interaction will be uncertain. The obstacle in dealing with the system uncertainty challenges the model-based methods.

A model-free adaptive (MFA) control, a data-driven adaptive control method, can be operated using only input and output (I/O) data from the system [15] and is suitable to deal with uncertainties [16]. Such a method can realize the adaptive adjust in parametric as well as structural manners, and have been successfully incorporated into the IL method for different industrial applications, such as particle quality control for spray fluidized-bed granulation [17], formation control of multi-agent systems [18], freeway traffic iterative learning control [19], and vibration suppression [20]. In this paper, the MFA control was applied to tune the learning gains of the P-type IL method by the system's dynamic behavior. The P-type IL method with time-varying learning gains was used to design the feedback gains, which can accelerate the convergence speed.

To avoid system spillover, the feedback gains must be converged to appropriate values after a period of time. The convergence properties of the controlled system should be seriously considered. The conventional P-type IL method has two domains, namely, time domain and iteration domain. The task of a conventional P-type IL method is implemented in a finite-time interval. Thus, the convergence of the control system in the time domain is not considered, while the convergence of the control system in the iteration domain is focused on. In this paper, the proposed method was different from the conventional P-type IL method, which has only a single domain. In other words, the time domain and iteration domain of the proposed method were overlapping. The convergence properties of the proposed method in overlapping domain need to be considered seriously; however, the convergence analysis in the overlapping domain is still an open problem. Bai et al. [20] designed stopping criteria to guarantee that feedback gains can converge to the appropriate values using the evidence theory. Based on the evidence theory, decision-makings on system states can be carried out by comparing threshold values. However, the results of decision-making are counterintuitive when the given evidence is conflicting [21,22]. In addition, dealing with conflict in evidence theory is still an open question. To make the learning process of feedback gains smoother and to reduce conflicting evidence caused by external noise, sliding mode control (SMC) is used to compensate learning gain in real time [20]. However, the introduction of SMC brings a great computational burden, which results in a great time delay. Generally speaking, the more actuator–sensor pairs that are bonded on the plate, the more the vibration modes of the plate which can be controlled. The time delay also increases with the increase in actuator–sensor pairs, which may limit the applications of the robust MFA-IL control.

In contrast, the theory of imprecise probability can work as a more general model to deal with uncertainties. The probability is represented by intervals, which interprets uncertainty from the perspective of behavior and achieves good results in the application of state diagnosis. In addition, the sliding model controller is removed, which can reduce the computational burden. The theory of imprecise probability provides a formal framework to determine an optimal decision under uncertainties of the state of system, which makes it suitable for a wide range of application areas [23,24]. In this paper, the theory of imprecise probability was used to design the stopping criteria. To deal with the problems of the multicriteria and multiobjectives in vibration control systems, Dempster's combination rule was used to fuse multi-source information. Based on the imprecise probability theory and the combination rule, the learning processes of all controllers can be monitored and diagnosed in real-time.

In this paper, by combining the time-varying P-type IL method with the MFA method, a data-driven adaptive IL method is presented for the vibration active control of piezoelectric laminated composite structures. Considering the system uncertainty in practical applications, the MFA control was incorporated into the time-varying P-type IL method to tune in real-time the learning gains. The square root values of feedback gains were regard as characteristic parameters. Based on the imprecise probability theory, a multi-source information diagnosis technology was presented for the design of the stopping criteria. Decisions made under the imprecise probability theory were used to decide whether the learning processes should be terminated. Numerical simulations and experimental studies were carried out, and the results were analyzed and discussed.

In the rest of the paper, the state–space model of the system is established for the controller design. The motion equation of the piezoelectric structure also driven by the P-type IL method is shown in Section 2. Section 3 introduces the dynamic linearization technique for the state–space system, and the MFA controller is given. The stopping criteria based on the imprecise probability theory and Dempster's combination rule is proposed in Section 4. The proposed method is summarized in Section 5. In Section 6, numerical simulation results are presented for verifying the effectiveness of the proposed method. A complete vibration control system is established, and the results are discussed in Section 7. The conclusions and future outlooks are given in Section 8.

2. State–Space Model and P-Type IL Method

A finite element (FE) formulation for the dynamic response of piezoelectric material has been given as [14]:

$$
\begin{bmatrix} M_{uu} & 0 \\ 0 & 0 \end{bmatrix} \begin{Bmatrix} \ddot{q} \\ \ddot{\phi} \end{Bmatrix} + \begin{bmatrix} C_{uu} & 0 \\ 0 & 0 \end{bmatrix} \begin{Bmatrix} \dot{q} \\ \dot{\phi} \end{Bmatrix} + \begin{bmatrix} K_{uu} & K_{u\phi} \\ K_{\phi u} & K_{\phi\phi} \end{bmatrix} \begin{Bmatrix} q \\ \phi \end{Bmatrix} = \begin{Bmatrix} F_{ue} \\ F_{\phi} \end{Bmatrix}
\tag{1}
$$

where M_{uu} and C_{uu} are the mass matrix and the damping matrix; K_{uu}, $K_{u\phi}$, $K_{\phi u}$, and $K_{\phi\phi}$ represent the stiffness matrix, the piezoelectric coupled matrix, the coupled capacity matrix, and piezoelectric capacity matrix, respectively; F_{ue} and F_{ϕ} are the external force vector and the electric load vector; q and ϕ are the nodal displacement vector and the voltage vector. $\dot{\square}$ and $\ddot{\square}$ denote the first and second derivatives versus time.

The damping matrix is usually linear with respect to the mass matrix and stiffness matrix using the Rayleigh damping coefficients α and β:

$$
C_{uu} = \alpha M_{uu} + \beta K_{uu}
\tag{2}
$$

Equation (1) can be uncoupled into the electric potential:

$$
\phi = K_{\phi\phi}^{-1} F_{\phi} - K_{\phi\phi}^{-1} K_{\phi u} q
\tag{3}
$$

and the reduced motion equation:

$$M_{uu}\ddot{q} + C_{uu}\dot{q} + K_{*}q + K_{u\phi}K_{\phi\phi}^{-1}F_{\phi} = F_{ue} \tag{4}$$

where $K_{*} = K_{uu} - K_{u\phi}K_{\phi\phi}^{-1}K_{\phi u}$.

The electric load vector is usually equal to zero in the sensor. Again using Equation (3), the sensor electric potential is given as $\phi_s = -K_{\phi\phi}^{-1}K_{\phi u}q$. The first-time derivatives of ϕ_s can be given as:

$$\dot{\phi}_s = -K_{\phi\phi}^{-1}K_{\phi u}\dot{q} \tag{5}$$

The system output error can be defined as:

$$e = y_d - y \tag{6}$$

where y_d and y are the desired system output signal and the measured system output signal, respectively.

The desired system output signal y_d is always zero. The measured system output signal y is equal to the first-time derivatives of the sensor electric potential $\dot{\phi}_s$ in Equation (5). The system output error at the kth moment as a discrete-time system is given as:

$$e(k) = 0 - \dot{\phi}_s(k) = K_{\phi\phi}^{-1}K_{\phi u}\dot{q}(k) \tag{7}$$

According to the P-type IL method [7], the feedback gain can be expressed in the iteration form:

$$G(k) = G(k-1) + \delta e(k-1) \tag{8}$$

where δ is the proportional learning gains matrix.

The actuation voltage can be written as:

$$V_a(k) = G(k)e(k) = -G(k)\dot{\phi}_s(k) \tag{9}$$

The electric load vector at kth moment is given as:

$$F_{\phi}(k) = C_a V_a(k) \tag{10}$$

where C_a represents the capacitance constant of the piezoelectric material.

By combining Equations (5), (9), and (10), motion Equation (4) can be approximated as follows:

$$M_{uu}\ddot{q}(k) + [C_{uu} + K_{u\phi}K_{\phi\phi}^{-1}C_a G(k)K_{\phi\phi}^{-1}K_{\phi u}]\dot{q}(k) + K_{*}q(k) = F_{ue}(k) \tag{11}$$

3. Dynamic Linearization and MFA Controller Design

Combining Equations (5) and (7), the state form of system (11) can be rewritten as:

$$\dot{y}(k) = -(K_M C_{uu}K_{\phi u}^{-1}K_{\phi\phi})y(k) - [K_M K_C G(k)]y(k) + K_M K_{*}q(k) - K_M F_{ue}(k) \tag{12}$$

where $K_M = K_{\phi\phi}^{-1}K_{\phi u}M_{uu}^{-1}$, $K_C = C_a K_{u\phi}K_{\phi\phi}^{-1}$.

In the time-varying P-type IL version, the updated rule is given as [25]:

$$G(k) = G(k-1) + \delta(k-1)e(k-1) \tag{13}$$

where $\delta(k-1)$ is the time-varying learning gains matrix.

For the sample period T, we have $\dot{y}(k) = [y(k+1) - y(k)]/T$, and the discrete-time of system (12) can be transformed as follows using Equation (13):

$$y(k+1) = -T[K_M C_{uu} K_{\phi u}^{-1} K_{\phi\phi} + K_M K_C G(k-1) + K_M K_C \delta(k-1)e(k-1) - \frac{1}{T}]y(k) + TK_M F_*(k) \quad (14)$$

where $F_*(k) = K_* q(k) - F_{ue}(k)$, $\delta(k-1)$, and $y(k)$ are the system input and output, respectively.

Lemma 1: *Defining the system output change at an adjacent sampling moment as $\Delta y(k) = y(k) - y(k-1)$, $\Delta\delta(k) = \delta(k) - \delta(k-1)$ is the system input change at an adjacent sampling moment. From Equation (14), the partial derivatives of $y(k+1)$ versus the measured signal$y(k)$ and learning gains $\delta(k-1)$ are continuous. For each moment k, $\|[\Delta y(k), \Delta\delta(k-1)]^T\| \neq 0$, there must exist a pseudo-partial derivative (PPD) matrix$\sigma(k)$, and system (14) can be transformed to a full form dynamic linearization (FFDL) description:*

$$\Delta y(k+1) = \sigma(k)[\Delta y(k), \Delta\delta(k-1)]^T \quad (15)$$

where $\sigma(k) = [\rho_1(k), \rho_2(k)]$, $\rho_1(k), \rho_2(k) \in R^{N \times N}$ and $\|\sigma(k)\| < b$, and b is a positive constant.

The proofs of Lemma 1 can be obtained by similar steps (see Reference [20]) and are omitted. Based on Equation (15), the following dynamic linearization from can be obtained:

$$y(k+1) = \rho_1(k)\Delta y(k) + \rho_2(k)\Delta\delta(k-1) + y(k) \quad (16)$$

where $\varphi_1(k)$ and $\varphi_2(k)$ are dynamically changed.

The MFA controller for calculating the learning gains δ is given as follows [20]:

$$\delta(k-1) = \delta(k-2) + \frac{\varphi\rho_2(k)^T[y_d(k+1) - y(k) - \rho_1(k)\Delta y(k)]}{\|\rho_2(k)\|^2 + \gamma} \quad (17)$$

where a step size constant $\varphi \in (0,1]$ is added to make Equation (17) general.

The parameters of the PPD matrix are estimated as follows [20]:

$$\begin{aligned}
\hat{\rho}_1(k) &= \hat{\rho}_1(k-1) + \frac{\eta[\Delta y(k) - (\hat{\rho}_1(k-1), \hat{\rho}_2(k-1))(\Delta y(k-1), \Delta\delta(k-2))^T]\Delta y(k-1)^T}{\mu + \|\Delta y(k-1)\|^2 + \|\Delta\delta(k-2)\|^2} \\
\hat{\rho}_2(k) &= \hat{\rho}_2(k-1) + \frac{\eta[\Delta y(k) - (\hat{\rho}_1(k-1), \hat{\rho}_2(k-1))(\Delta y(k-1), \Delta\delta(k-2))^T]\Delta\delta(k-2)^T}{\mu + \|\Delta y(k-1)\|^2 + \|\Delta\delta(k-2)\|^2}
\end{aligned} \quad (18)$$

where a step-size constant $\eta \in (0,1]$ is added to make Equation (18) general.

4. The Stopping Criteria Design

4.1. Preliminary Notion of Imprecise Probability

In the theory of imprecise probability, many decision criteria are developed [26]. The Γ–maximin criterion was applied in this paper to design the stopping criteria.

Assuming that a decision d induces a real-value gain J_d, and the set of all available decisions is D, $d \in D$. Our purpose was to identify the optimal decision d in D, and the solution is given as follows:

$$opt(D) = \underset{d \in D}{\arg\max}(J_d) \quad (19)$$

The variables whose values are uncertain can influence the gain J_d. According to the expected utility of its gain, the decision can be ranked reasonably, and the expected utility should be maximized.

$$opt_{E_\mu}(D) = \arg \underset{d \in opt \geqslant (D)}{\max} E_\mu(J_d) \quad (20)$$

where $E_\mu(J_d)$ is the expected utility of the gain J_d, and μ is the probability measure.

As a simplified form for Equation (20), the $\underline{E}_p$ can be seen as the replacement of E_μ, and the Γ−maximin criterion can be written as

$$opt_{\underline{E}_p}(D) = \arg \max_{d \in opt_{\geqslant}(D)} \underline{E}_P(J_d) \tag{21}$$

where $\underline{E}_p$ is a lower expected utility by minimizing E_μ. The Γ−maximin criterion can be understood as a worst-case optimization, and a decision can be made by maximizing the worst expected gain.

4.2. The Diagnosis Method Design

4.2.1. Fault Reliability

The real-time feedback gains of controllers are obtained for diagnosing the system state. Assume that there are N sensors glued on the laminated composite plate in the vibration control system. When the jth sensor works, there are L_j characteristic parameters to represent the state types of the system. For the sake of simplicity, suppose that all state types are independent of each another. Only one state can occur at any given time. Let S_j represents the characteristic parameter vector obtained from the jth sensor:

$$S_j = [s_{j1}s_{j2}\cdots s_{jL_j}]; j = 1, 2, \ldots, N \tag{22}$$

where s_{ji} is the ith element of S_j, the characteristic parameter s_{ji} obtained from jth sensor can be used to identify the certain state i in current circumstances, $i = 1, 2, \ldots, L_j$, L_j is the number of the characteristic parameters provided by the jth sensor.

Considering an exponential function form as the evidence generating function, a basic fault reliability assignment can be defined as:

$$m_{ji} = r_i\left(1 - a_i^{\alpha_i s_{ji}}\right) \tag{23}$$

where r_i, a_i, and α_i are constants, which can be directly determined by expert experience or prior knowledge. In state diagnostics, m_{ji} can be considered as the degree of reliability in the certain state i by evaluating the measurements obtained from the jth sensor.

The fault reliability for the jth sensor can be calculated as follows:

$$m_j = \frac{1}{L_j}\sum_{i=1}^{L_j} m_{ji} \tag{24}$$

4.2.2. Establish Fault Probability Interval

The fault reliabilities obtained from N controllers can be expressed in vector form $M = \{m_1, m_2, \cdots, m_N\}$. After sorting the elements from small to large, a new fault reliability vector can be obtained $M' = \{m'_1, m'_2, \cdots, m'_N\}$, which are divided into two groups, namely, a low fault reliability group $M_{\min}$ and a high fault reliability group $M_{\max}$.

$$\begin{cases} M_{\min} = \{m'_1, m'_2, \cdots, m'_l\}, M_{\max} = \{m'_{l+1}, m'_{l+2}, \cdots, m'_N\}, & N = 2l \\ M_{\min} = \{m'_1, m'_2, \cdots, m'_{l+1}\}, M_{\max} = \{m'_{l+1}, m'_{l+2}, \cdots, m'_N\}, & N = 2l+1 \end{cases} \tag{25}$$

where l is a natural number. The fault reliability conflicts of the $M_{\min}$ and $M_{\max}$ are smaller than that of the M', thus better fused results can be obtained using Dempster's combination rule.

Suppose m'_Π and m'_Λ are two fault reliabilities obtained from the same fault reliability group. The degree of conflict among these two fault reliabilities is shown by the conflict coefficient K as follows [27,28]. The larger the value of K, the more conflicting the two fault reliabilities are:

$$K = \sum_{\Pi \cap \Lambda = \varnothing} m'_\Pi m'_\Lambda \tag{26}$$

where $\varnothing$ is an empty set.

In the same fault reliability group, Dempster's combination rule is given as follows [28,29]:

$$\begin{cases} \frac{1}{1-K} \sum\limits_{\Pi \cap \Lambda = \Omega} m'_\Pi m'_\Lambda & \Omega \neq \varnothing \\ 0 & \Omega = \varnothing \end{cases} \tag{27}$$

Dempster's combination rule in Equation (27) is applied to fuse the fault reliabilities in groups M_{min} and M_{max}, and the two fused fault reliabilities are denoted as m_{min} and m_{max}. Based on the Pignistic probability transformation (PPT), the fused fault reliabilities will be the fault probability, namely, $P_{min} = m_{min}$ and $P_{max} = m_{max}$, when there is only one element in the fault reliability vector. The fault probability interval can be established as $[P_{min}, P_{max}]$.

The system in this paper consists of two types of states: learning termination and normal learning. The fault probability interval mentioned above is the prediction of the following fault indication function.

$$g(\omega) = \begin{cases} 1, & \omega = \omega_1 \\ 0, & \omega = \omega_2 \end{cases} \tag{28}$$

4.2.3. Diagnosis Cost Functions and Decision-Making

The diagnosis cost functions can be designed as follows:

$$f_1 = \begin{cases} a(e), & \omega = \omega_1 \\ b(e), & \omega = \omega_2 \end{cases} ; f_2 = \begin{cases} c(e), & \omega = \omega_1 \\ d(e), & \omega = \omega_2 \end{cases} \tag{29}$$

where f_1 expresses that the fault is occurrence, $a(e)$ is the gain when the fault actually occurs and the state is diagnosed correctly, $b(e)$ is the gain when the fault does not occur and the state is diagnosed incorrectly; f_2 denotes that the fault is not occurrence, $c(e)$ is the gain when the fault actually occurs and the state is diagnosed incorrectly, $d(e)$ is the gain when the fault does not actually occur and the state is diagnosed correctly, and satisfying $a(e) > b(e)$ and $d(e) > c(e)$, and e is the parameter, which can be directly determined by expert experience or prior knowledge.

Considering the fault indication function Equation (28) is predicted, the fault diagnosis problem is transformed into the process of decision making for the expected intervals of f_1 and f_2. In the Γ−maximin criterion, the decision is made by comparing the lower expected utility of f_1 and f_2. The expected intervals of f_1 and f_2 can be calculated as follows:

$$\begin{cases} \underline{E}(f_1) = a(e)P_{min} + b(e)(1 - P_{min}) \\ \overline{E}(f_1) = a(e)P_{max} + b(e)(1 - P_{max}) \\ \underline{E}(f_2) = c(e)P_{max} + d(e)(1 - P_{max}) \\ \overline{E}(f_2) = c(e)P_{min} + d(e)(1 - P_{min}) \end{cases} \tag{30}$$

The square root values of feedback gains are regarded as the characteristic parameters. The basic fault reliability assignment value can be calculated using the exponential function. The fault reliability vectors are divided into two groups, including the high fault reliability group and low fault reliability group. By Dempster's combination rule, the fused results of the two groups above are used to establish the fault probability interval. The Γ−maximin criterion in the theory of imprecise probability is adopted for state diagnosis. The threshold value is predefined to serve as the stopping criteria. Based on the diagnosis results of the Γ−maximin criterion, decision-making can be fulfilled by comparison with the threshold value.

5. The Summary of the Proposed Method

In summary, the flow chart is shown in Figure 1 and detailed as follows:

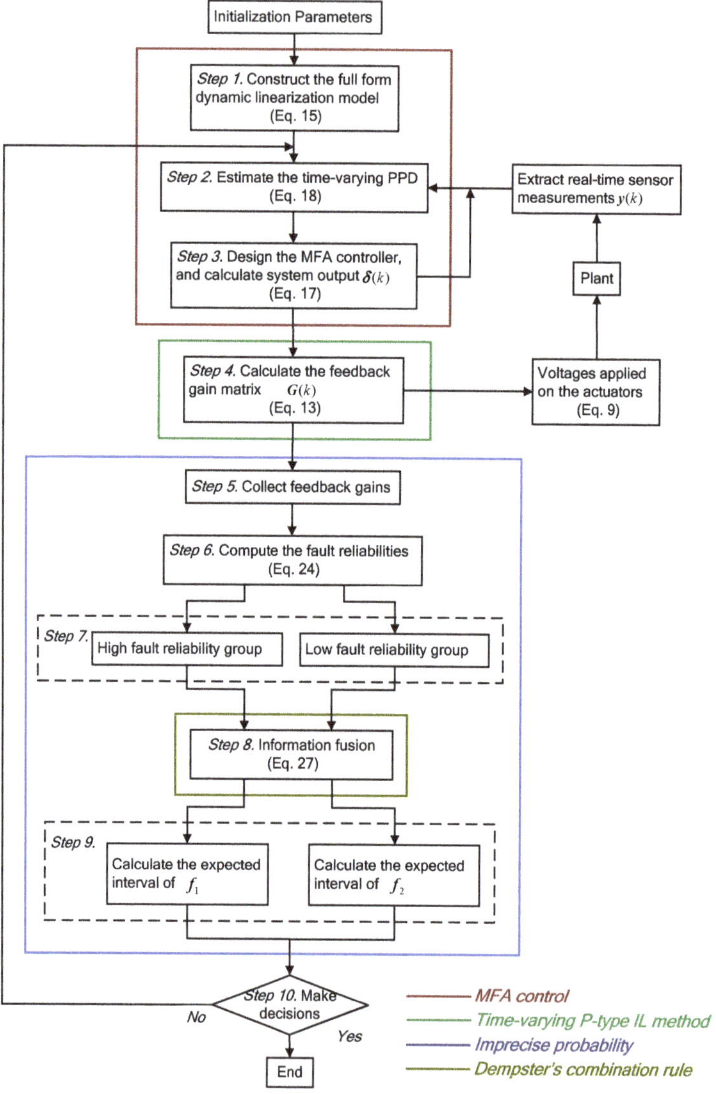

Figure 1. Flow chart of the proposed method.

Step 1. Construct the full-form dynamic linearization model in Equation (15).

Step 2. Predict the time-varying PPD values in Equation (18) merely using the on-line system input $\delta(k)$ and output $y(k)$ data.

Step 3. Design the MFA controller in Equation (17).

Step 4. Calculate the feedback gain matrix $G(k)$ in Equation (13).

Step 5. Extract real-time feedback gains used to transfer the characteristic parameters S_k.

Step 6. Calculate the fault reliabilities in Equation (24) for all sensors.

Step 7. Divide the fault reliability value vector into two groups: the high fault reliability group and the low fault reliability group Equation (25).

Step 8. Respectively fuse the elements of the two groups above using Dempster's combination rule Equation (27).

Step 9. Establish the fault probability interval by the fused results and calculate the expected interval Equation (30) of the diagnoses cost function Equation (29).

Step 10. Make decisions based on the lower expected utility and stopping criteria.

6. Numerical Simulations

6.1. FE Modeling and Setting of Controller Parameters

The numerical simulations were carried out via vibration active control on the cantilevered plate with piezoelectric patches. The piezoelectric cantilevered plate comprise done laminated composite plate (414 mm × 120 mm × 1 mm), on which six piezoelectric patches (60 mm × 24 mm × 1 mm) were bonded in pairs at the plate, as shown in Figure 2. The laminated composite plate was made of graphite-epoxy (GE, carbon-fiber reinforced) composite material, which included five substrate layers. Its total thicknesswas1mm with the angle-ply (0/90/0/90/0), and the thickness of each substrate was 0.2 mm. The upper piezoelectric patches were actuators and the lowers ones worked as sensors. We distinguished the three actuator–sensor pairs as *a*, *b* and *c*, respectively. The positions of the piezoelectric patches were chosen via Reference [29]. The locations of Point A, Point B and Point C are given in Figure 2. The root of the laminated composite plate was clamped. The properties of the laminated composite plate and piezoelectric material are listed in Table 1.

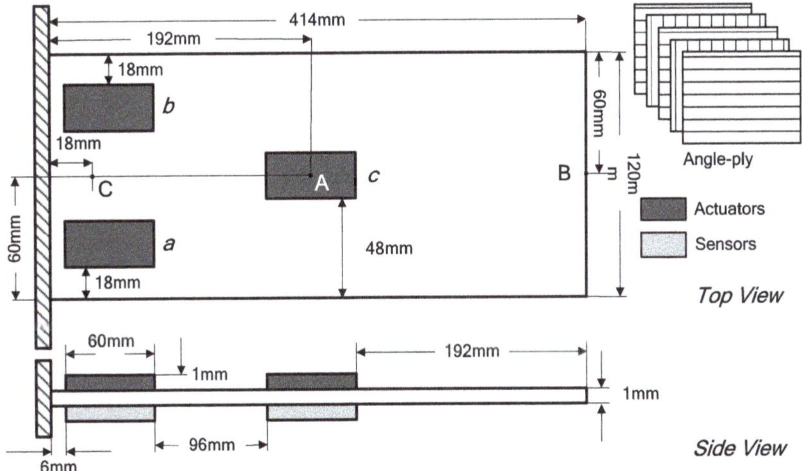

Figure 2. The piezoelectric cantilevered plate.

Table 1. Material properties.

Graphite-Epoxy (GE)	Piezoelectric Material
Yong's modulus (GPa)	Elastic stiffness (GPa)
$E_{11} = 40.51$	$C_{11} = 126$
	$C_{12} = 79.5$
$E_{22} = E_{33} = 13.96$	$C_{13} = 84.1$
	$C_{33} = 117$
Shear modulus (GPa)	$C_{44} = 23.3$
	$C_{66} = 23$
$G_{12} = G_{13} = 3.1$	Piezoelectric stain (C/m^2)
$G_{23} = 1.55$	$e_{16} = e_{25} = 17$
Poisson's ratio	$e_{31} = e_{32} = -6.5$
$v_{12} = v_{13} = 0.22$	$e_{33} = 23.3$
$v_{23} = 0.11$	Permittivity (F/m)
Density (kg/m^3)	$\varepsilon_{11} = \varepsilon_{22} = 1.503 \times 10^{-8}$
$\rho = 1830$	$\varepsilon_{33} = 1.3 \times 10^{-8}$
	Density (kg/m^3)
	$\rho = 7500$

In this paper, the dynamic FE model for simulating the piezoelectric cantilevered plate was constructed using ANSYS. The laminated composite plate and piezoelectric patches were modeled by SOLID46 elements and SOLID5 elements, respectively. The laminated composite plate was meshed with $69 \times 20 \times 1$ elements, and each piezoelectric patch was meshed with $10 \times 4 \times 1$ elements. For the degree of electric freedom, the nodes at the surface of piezoelectric patches were coupled by command CP. Modal analysis was carried out to identify the natural frequencies of the piezoelectric cantilevered plate and to design the sampling period for the numerical simulations [30]. The first three natural frequencies of the piezoelectric cantilevered plate were calculated, which also implied good agreement with the comparison between the numerical results and experimental results in Table 2. The largest error percentage, 13.9%, arose in the second modal frequency. Since the numerical results of the modal frequencies were used to get approximate values to verify the dynamic FE model, the difference in the modal frequencies between the numerical and experimental results were acceptable. The sampling period was taken as $T = 1/(20\omega_1)$, where ω_1 represented the first natural frequency of the piezoelectric cantilevered plate. $\alpha = 2\beta = 0.003$ were the Rayleigh damping coefficients.

Table 2. The results of natural frequencies.

Mode	Numerical (Hz)	Experimental (Hz)	Error Percentage
1	5.4377	5.326	2.1%
2	24.217	21.259	13.9%
3	28.683	31.593	−9.2%

The constants of the MFA controllers were given as: $\gamma = 1$, $\varphi = 1$, $\mu = 1$, and $\eta = 1$. The fault reliability can be calculated in Equation (24). The system in this paper consisted of two types of states: learning termination and normal learning. The square root values of the feedback gains were regarded as the characteristic parameters. The constants for the calculation of the basic fault reliability assignment Equation (23) are given as: $r = 1$, $a = 2.924$, and $\alpha = -1$. The constants of the diagnosis cost functions are defined as: $a(e) = 1$, $b(e) = 1$, $c(e) = 1$, and $d(e) = 1$. The threshold value is predefined to serve as the stopping criteria, and decision-making can be fulfilled by comparing with the threshold value. The controllers connected with different sensors may have distinct convergence speeds. To make all controllers sufficiently learn, two threshold values were defined as: for the lower expected utility of fused fault reliabilities, the threshold value was specified at 0.9323; for the single fault reliability, the threshold value was specified at 0.7978. The learning process should be terminated as long as one of the two threshold values above was met. Otherwise, the learning process should be continued. In the

P-type IL method, the maximum iteration number was defined as 500, and the fixed learning gains were given as $\delta_1 = 0.078$ and $\delta_2 = 0.068$ for various simulations.

6.2. Harmonic Excitation

The vibration active control for the first mode of the piezoelectric cantilevered plate was investigated in this case. Considering the harmonic excitation generated by the function $f(t) = 5\cos(\omega_1 t)N$, the plate was driven at Point C, the constant $\omega_1 = 17.083rad/s(5.4377\text{Hz})$ was the first natural frequency. All numerical results corresponding to the robust MFA-IL control are also given in this section.

In Figure 3a,b, the time-history dynamic responses at Point A and Point B are, respectively, given, and the figure illustrates that the first mode vibration was suppressed effectively by the proposed method, the P-type IL method and robust MFA-IL control. The control performance of the piezoelectric actuators was not able at the places with (e.g., Point A) or without piezoelectric sensors (e.g., Point B). Nevertheless, it is worth noting that the conclusions obtained above were distinct from Saleh's [31]. Saleh pointed out that the P-type IL method cannot effectively control the unwanted vibration at locations of the observation points. Furthermore, it was also noteworthy that the P-type IL method could not obviously control the first mode vibration of the piezoelectric structures.

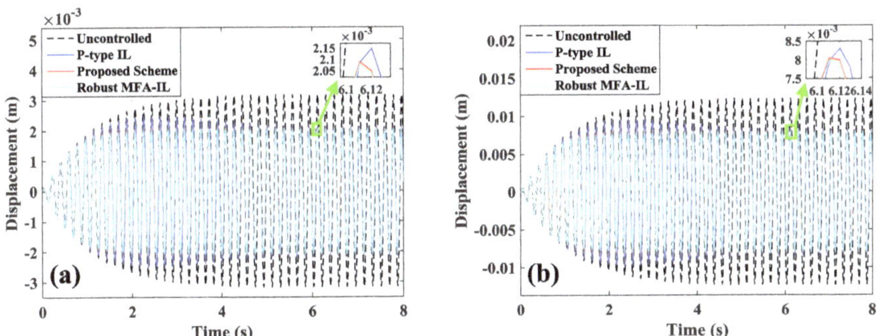

Figure 3. Displacement responses to harmonic excitation: (**a**) Point A, (**b**) Point B.

As an effective vibration active control system, it is possible to reduce the amplitude of the overall structure not merely the parts of the structure. Before designing the vibration active control system, the control rule needs to be considered carefully for achieving satisfactory control results. The control performance of the system also relates to the positions and sizes of the piezoelectric patches [32]. The best positions for the piezoelectric patches were always chosen at the places where the mechanical strain was the largest. To generate satisfactory control forces, the dimensions of the piezoelectric actuators should be investigated and designed. The dimensions of the piezoelectric sensors should be selected appropriately, and then precise information on the structural vibration can be acquired. A misreading of sensor measurement signals may generate unreasonable control force, and the dynamic performance of the system may seriously deteriorate. As long as the positions and dimensions of the piezoelectric patches were chosen appropriately, the P-type IL method presented good performance on the first mode vibration control. Besides, the controllability of structural vibration was notable at the locations with sensors and without sensors.

The actuator time-history voltages are presented in Figure 4a,b, and the actuator voltages changed suddenly at 4.4s while the system was controlled by the P-type IL method. After the learning processes were terminated, the amplitudes of the actuator voltages reconstructed smoothness. The controllers connected with distinct sensors had different convergence speeds in the learning processes, which may cause the control force to mismatch among each other. If a piezoelectric actuator cannot perform as desired, the adjacent piezoelectric actuators will be negatively affected. To avoid this phenomenon, more iterations are needed to improve the control stability. Less iteration numbers may directly lead to system spillover. The instability phenomenon did not occur when the system was controlled by the proposed method and the robust MFA-IL control. The measurement signals from sensor a/b and sensor c are shown in Figure 4c,d. In comparison with the P-type IL method, smaller amplitudes were obtained as long as the piezoelectric cantilevered plate was controlled by the robust MFA-IL control and the proposed method. The root mean square (RMS) values of the dynamic responses and measurement signals were used to quantitatively analyze the performance of the P-type IL method, the robust MFA-IL control, and the proposed method, which are listed in Table 3. From Table 3, both the robust MFA-IL control and the proposed method had better control performance by comparing the P-type IL method. The vibration amplitude was reduced 41.22% under the control of the proposed method, and the vibration amplitudes reduced 40.36% under the control of the robust MFA-IL control. The proposed method and the robust MFA-IL had similar control precision.

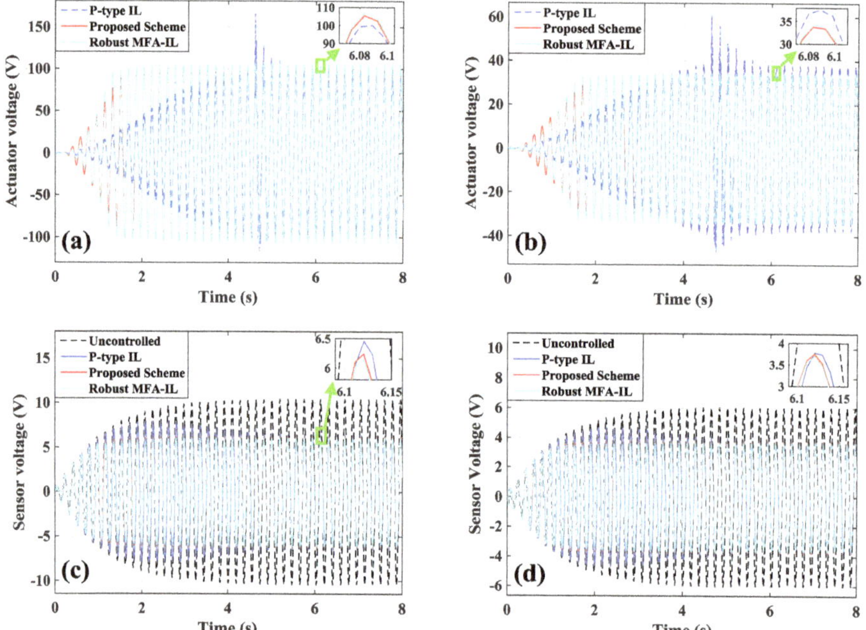

Figure 4. The time-domain measurement signals of the actuators/sensors: (**a**) actuator a/b; (**b**) actuator c; (**c**) sensor a/b; (**d**) sensor c.

Table 3. Root mean square (RMS) results.

Algorithm	Case 1				Case 2				Experiment	
	Point A	Point B	Sensor a/b	Sensor c	Point A	Point B	Sensor a/b	Sensor c	Sensor a/b	Sensor c
Uncontrolled	2.262×10^{-3}	8.805×10^{-3}	7.420	4.314	3.430×10^{-3}	12.793×10^{-3}	9.934	5.249	3.764	2.167
P-type IL	1.526×10^{-3}	5.909×10^{-3}	4.590	2.706	2.238×10^{-3}	8.169×10^{-3}	6.323	3.356	2.563	1.488
Robust MFA-IL	1.479×10^{-3}	5.719×10^{-3}	4.348	2.561	-	-	-	-	-	-
Proposed method	1.488×10^{-3}	5.763×10^{-3}	4.413	2.598	2.080×10^{-3}	7.590×10^{-3}	5.800	3.065	2.480	1.444

The computational time for the various algorithms is shown in Figure 5, including the time for running each iteration and the time for convergence of the feedback gains. From Figure 5, both the robust MFA-IL and the proposed method have fast convergence speed, which makes them overcome the inherent shortcoming of the P-type IL method. By comparing with the proposed method, the computational burden of the robust MFA-IL control was higher when the controller implemented each iteration. The extension of computational time resulted in increasing the time delay. Large time delays will bring uncontrollability to the vibration suppression system. In the proposed method, the main computational cost focused on the part of the MFA control, which was implemented by iterative computation for determining the learning gains in real time. Apart from the MFA control, the SMC was also integrated into the robust MFA-IL control. The introduction of SMC brings great computational burden, and results in great time delays. Generally speaking, the more actuator–sensor pairs bonded on the plate, the more the vibration modes of the plate which can be controlled. Therefore, the time delay also increases with the increase in pairs of actuators–sensors, which may limit the application are as for the robust MFA-IL control. To obtain a slight improvement in control precision, the robust MFA-IL control brings a larger time delay. In practical applications, the design of the vibration control system should be composed of control precision and realization of vibration suppression. The proposed method can be based on a compromise.

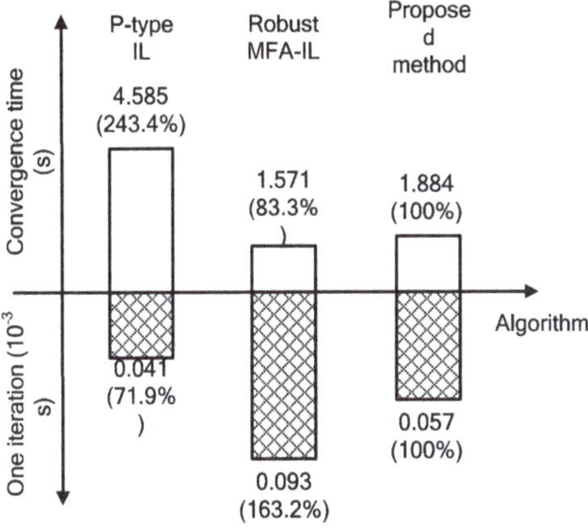

Figure 5. The computational time for the various algorithms.

The learning processes of feedback gains are depicted in Figure 6a,b. From Figures 4 and 6, the proposed method and the robust MFA-IL had fast learning speed and maintained a good control performance and system stability.

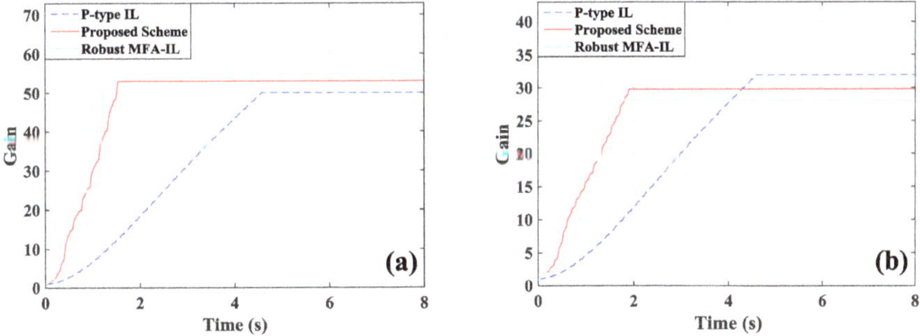

Figure 6. The learning processes of feedback gains: (**a**) actuator *a/b*; (**b**) actuator *c*.

The real-time diagnosis results for the fused information and single information source are given in Figure 7. From Figure 7, the controllers connected with actuator *a/b* had faster convergence speed than that connected with actuator *c*. Based on the theory of imprecise probability, all controllers could learn sufficiently, and satisfactory control performance could be achieved.

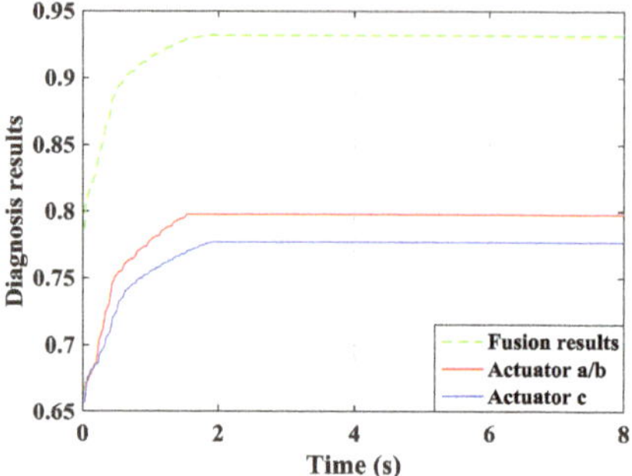

Figure 7. The curves of diagnosis results.

To verify the stability of the controllers, an instability test is carried out in this section. The noise signals are shown in Figure 8 are added to excite the piezoelectric cantilevered plate at Point A. The noise signals start at 6s and lasts only one second. The parameters of controllers are set up the same as mentioned above. The time-history dynamic responses at Point A and Point B are given in Figure 9a, and the measurement signals from sensor *a/b* and sensor *c* are shown in Figure 9b. When noise signals begin to excite, the dynamic responses of the plate and measurement signals from sensors change greatly; however, the divergence phenomenon was not found. After stopping the excitation of the noise signals, the vibration control system was restored to the stability state by the proposed method.

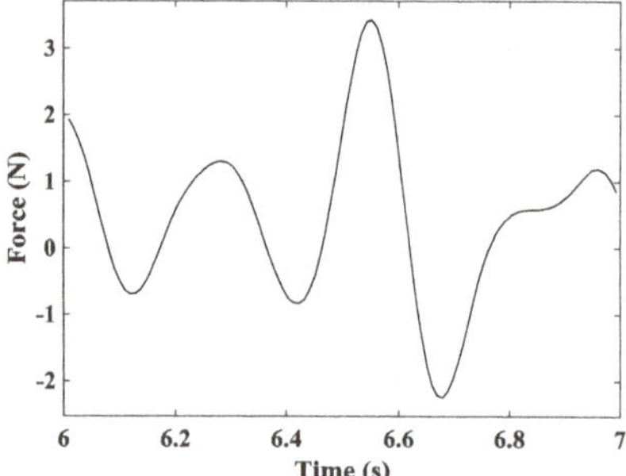

Figure 8. The noise signals.

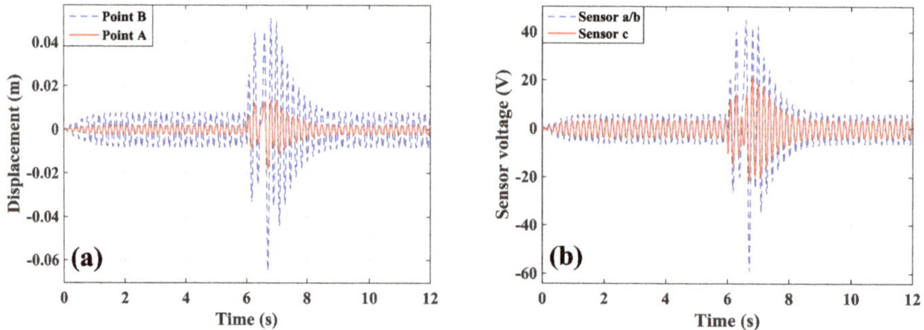

Figure 9. Verification curves with the stability of the proposed method: (**a**) displacement responses and (**b**) measurement signals of sensors.

6.3. Random Excitation

In this case, the plate was driven at Point C by the random force as follows in Figure 10.

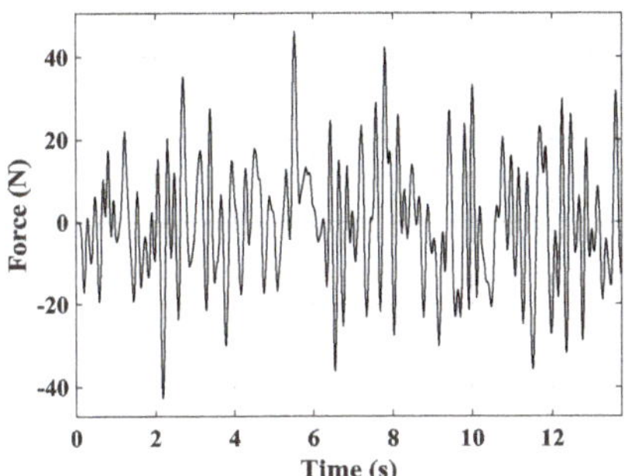

Figure 10. The random excitation.

The time-history dynamic responses at Point A and Point B are shown in Figure 11. The control voltages of actuator *a/b* and actuator *c* are presented in Figure 12a,b. The measurement signals from sensor *a/b* and sensor *c* are displayed in Figure 12c,d. The feedback gains are depicted in Figure 13.

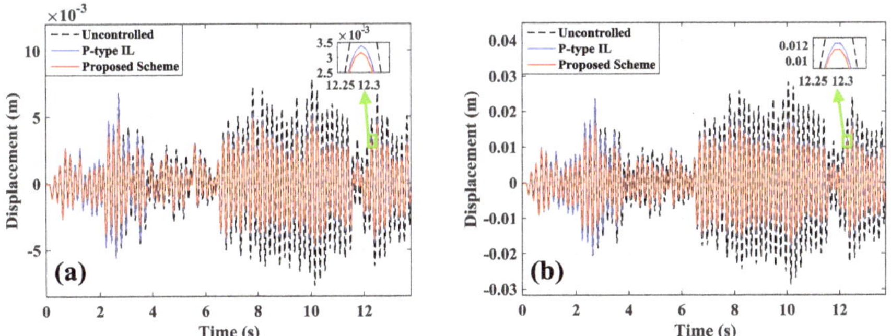

Figure 11. Displacement responses to random excitation: (**a**) Point A; (**b**) Point B.

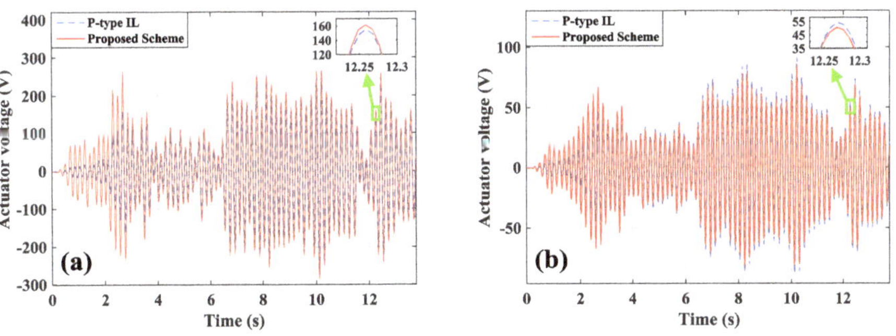

Figure 12. *Cont.*

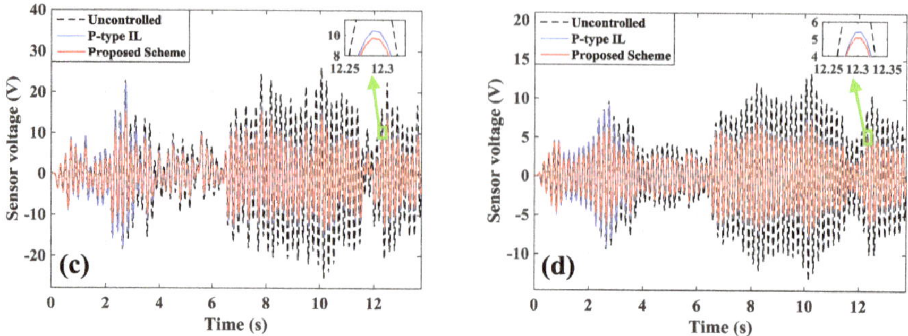

Figure 12. The time-domain measurement signals of actuators/sensors: (**a**) actuator *a/b*; (**b**) actuator *c*; (**c**) sensor *a/b*; (**d**) sensor *c*.

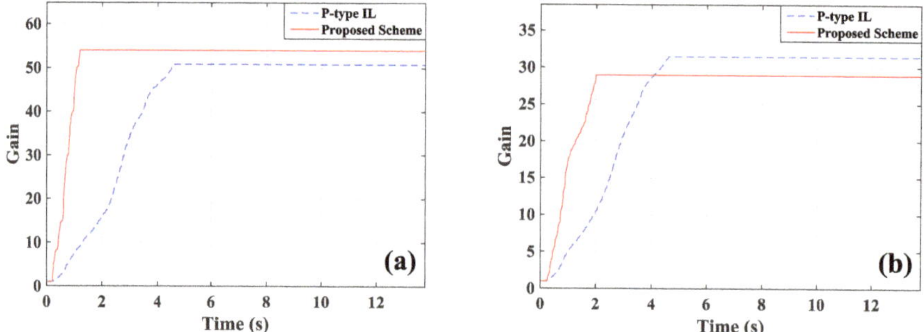

Figure 13. The learning processes of feedback gains: (**a**) actuator *a/b*; (**b**) actuator *c*.

From the results above, the proposed method makes the system have smaller amplitudes of dynamic responses and faster convergence speed. The learning gain δ in P-type IL method is the fixed constant, which is selected based on the practical experience of researchers. A larger learning gain can lead to system instability and robustness reduction [10], thus, a smaller learning gain is necessary to improve the system control precision. However, the smaller the learning gain selected, the more iterations are needed, thus the learning speeds of controllers slow down [7,9]. In the proposed method, the learning gain can be self-tuned by the system's dynamic behavior. The convergence speeds of the controllers are improved, and the high control precision can also be obtained.

In this case, the RMS values for evaluating the proposed method and the P-type IL method are listed in Table 3. The real-time diagnosis results of the system states are given in Figure 14.

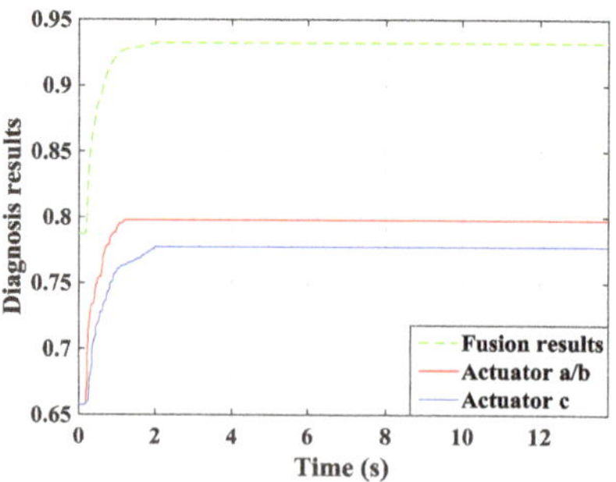

Figure 14. The curves of diagnosis results.

7. Experiments

7.1. ExperimentSetup

To validate the feasibility and control effect of the proposed method, an experimental system to control the vibration of the piezoelectric cantilevered plate was developed, as shown in Figure 15. Experiments on the first mode vibration control were conducted. The experiment setup consisted of a piezoelectric cantilevered plate with one laminated composite plate and six piezoelectric patches, the vibration excitation system, the data acquisition system, and the vibration active control system. The laminated composite plate was made up of GE composite material. The dimensions of the piezoelectric cantilevered plate are given in Section 6.1. The excitation position Point C was replaced by a metal patch.

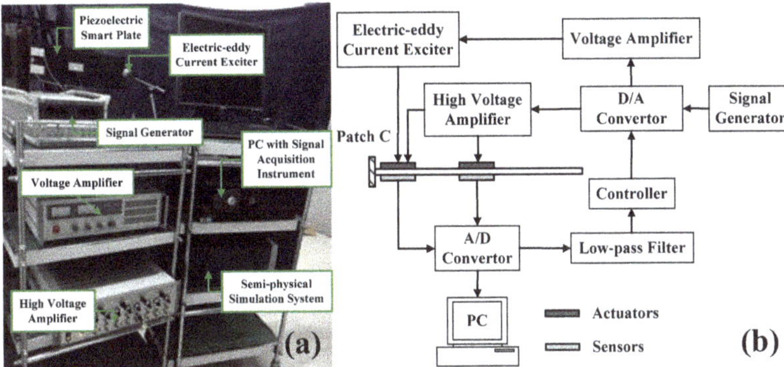

Figure 15. Experimental setup: (**a**) experimental setup and (**b**) experimental principle.

The signal generator (DH1301, Taizhou, China) was used to generate the external excitation signals. After digital to analog (D/A) conversion, the excitation signals were amplified by the voltage amplifier (YE5872A, PA, USA) and then were used to drive the piezoelectric cantilevered plate by the electric-eddy current exciter (JZF-1, Beijing, China). The electric signals were transformed into a mechanical force. Three piezoelectric sensors were applied to detect the vibration information, and their measurement signals were selected as the feedback signals. After analog-to-digital (A/D) conversion, all measurement data were acquired and stored in the PC. Since the control target of the piezoelectric cantilevered plate was the first mode, a low-pass filter was applied to eliminate the high-frequency noise. The controllers implemented the signal processing and calculation in the real-time semi-physical simulation system (Quarc, Toronto, Canada). Running the proposed method, the controllers generated the control signals. After D/A conversion, the control outputs were sent to the high-voltage amplifier (E70, Harbin, China) and then were applied to piezoelectric actuators for vibration suppression. The experimental sample period was chosen as 3 ms.

7.2. Modal Analysis

A swept sine (chirp) signal with an amplitude of 100 V was used to identify the modal frequencies of the system and excite actuator *a*. The initial frequency was 0.5 Hz, and the terminal frequency was 50 Hz.

Fourth-order Butterworth filters were utilized to eliminate high-frequency noises. The cutoff frequency of low-pass filters was specified at 30 Hz in the modal identification, and the cutoff frequency was 14 Hz in the first mode control. After filtering, the time-domain response signal measured by sensor *a* was stored and shown in Figure 16a. The fast Fourier transform (FFT) of the time-domain response data was computed to depict the frequency response of the system in Figure 16b. From Figure 16b, the first three modal frequencies were obtained and are listed in Table 2.

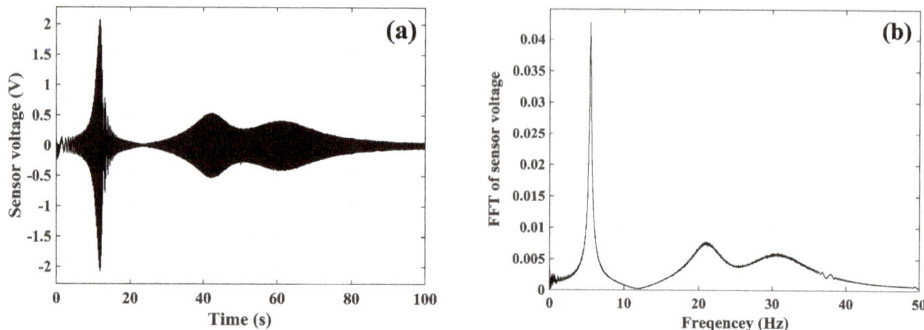

Figure 16. Vibration responses excited by actuator *a*: (**a**) the time-domain measurement signals of the sensor *a*; (**b**) frequency response of the sensor *a*.

7.3. ExperimentResults

The proposed method and the P-type IL method were investigated for vibration active control of the flexible plate during the experiments. In this section, the plate was driven at 5.326 Hz for the first mode control. In the P-type IL method, the number of iterations was predefined as 1500 to improve the system stability and control precision, and the value of learning gain was specified as $\Phi_1 = 0.54$. In the MFA controller, the parameters were selected as $\gamma = 1$, $\varphi = 0.1$, $\mu = 0.1$, and $\eta = 1$. The stopping criteria in this section were the same with those above in Section 6.1. The measurement signals of the sensors (shown in Figure 17a,b) were moved forward for the phase delay compensation due to the hardware factors. The P-type IL method and the proposed method were performed at 5 s after the harmonic excitation started.

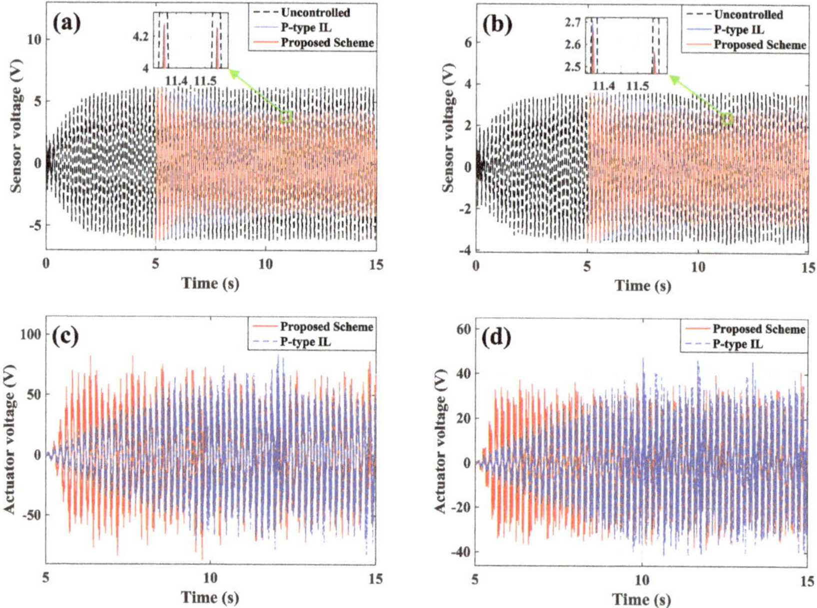

Figure 17. The time-domain measurement signals of actuators/sensors: (**a**) sensor *a/b*; (**b**) sensor *c*; (**c**) actuator *a/b*; (**d**) actuator *c*.

The measurement signals from sensor *a/b* and sensor *c* are presented in Figure 17a,b, and the control voltages of actuator *a/b* and actuator *c* are presented in Figure 17c,d. The data used to calculate the RMSs are recorded after learning termination, and the RMSs are given in Table 3. From Table 3, the proposed method reached the comparatively ideal control performance: the vibration amplitudes were reduced 33.9% under the control of the proposed method, and the vibration amplitudes were reduced 31.8% under the control of the P-type IL method. This excellent performance was obtained by integrating the MFA method into time-varying P-type IL method. The learning processes of the feedback gains are depicted in Figure 18. The proposed method is feasible to simultaneous maintain the control performance and damp down quickly for structural vibration. Under the control of different methods, the feedback gains obtained from the same controllers result in distinct values. The real-time diagnosis curves for the fused information and single information sources are given in Figure 19. By the theory of imprecise probability, the learning processes of feedback gains can be diagnosed in real time. The decisions made based on the designed stopping criteria causes all controllers to learn sufficiently, and excellent control performance was obtained.

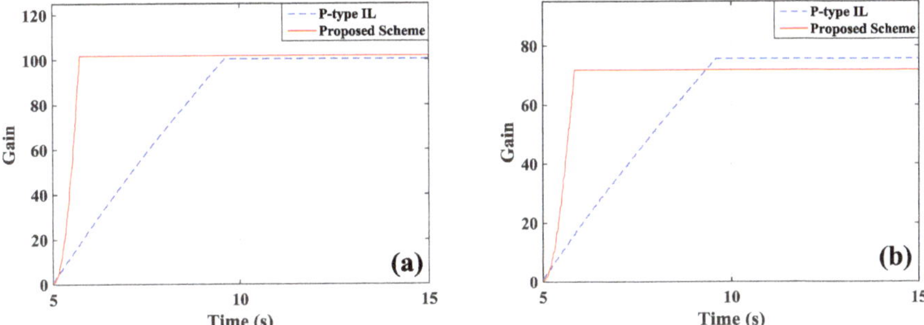

Figure 18. The learning processes of feedback gains: (**a**) actuator *a/b*; (**b**) actuator *c*.

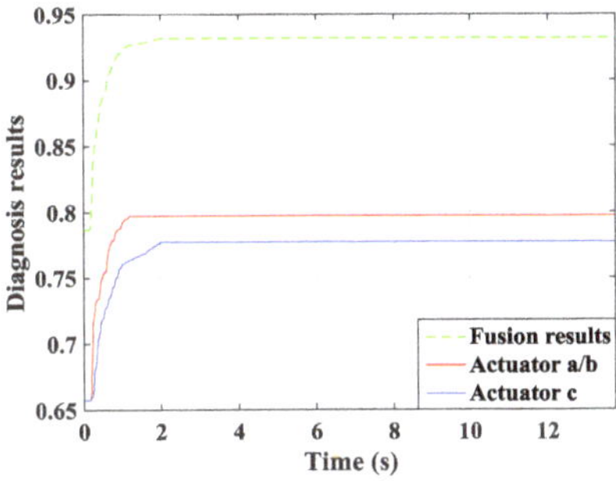

Figure 19. The curves of the diagnosis results.

8. Conclusions and Outlooks

A data-driven adaptive IL method was proposed for vibration active control of piezoelectric laminated composite structures. Based on the P-type IL method, the motion equation of the piezoelectric cantilevered plate was derived by the dynamic FE equations. The PPD matrix is estimated by the modified projection algorithm for dynamically linearizing the motion equation. Considering the uncertain non-linear dynamic processes, the MFA controller was designed. The MFA method was applied to self-tune the learning gains of the time-vary P-type IL method for accelerating the learning speed. The square root values of the feedback gains were regarded as characteristic parameters to diagnose the state of the vibration control system. Based on the theory of imprecise probability, the stopping criteria were designed. On this basis, the decisions were carried out to avoid the over-learning of controllers.

When the positions and dimensions of the piezoelectric patches are chosen appropriately, the P-type IL method shows good effectiveness on the first mode control of the piezoelectric cantilevered plate. Besides, the good controllability of the structural vibration was notable at the locations with sensors or without sensors. The conclusions obtained above in this paper were different other published studies.

Considering the system uncertainties, the MFA method was applied to self-tune in real-time the learning gains by the system's dynamic behavior. The introduction of the MFA method accelerated the convergence speed of the controller and improved the system's control precision. The stopping criteria based on the theory of imprecise probability allowed all controllers to learn sufficiently, and satisfactory control performance was achieved. The proposed method overcomes the shortcomings of the P-type IL method to achieve the expected control performance. The robust MFA-IL control improved control precision at the expense of great time delay, while the proposed method reduced the computational burden and misdiagnosis for system states at the expense of a slight decrease in control precision. The proposed method in this paper can be a compromise.

The proposed method has an open scheme, which can be integrated with other methods, including a model-based method. The data-driven method and model-based method can be complementary and cooperative in the design of a control system. The more precise information and model of the system obtained, the better the control performance of the system can be expected. In the future, the proposed method can be integrated with a model identification method to handle more complex problems in practical applications, which will simplify the structure of controllers and obtain a satisfactory control precision.

Author Contributions: Conceptualization, L.B.; Methodology, L.B.; Software, L.B.; Validation, X.-F.X., N.L. and Y.C.; Formal Analysis, L.B.; Investigation, L.B. and Y.C.; Resources, Y.-W.F.; Data Curation, L.B. and X.-F.X.

Funding: The research was funded by National Natural Science Foundation of China under Grant No. 5187051969.

Acknowledgments: All authors thank the editors, referees and officers in Symmetry for their valuable suggestions and help.

Conflicts of Interest: The authors declare no conflicts of interest.

References

1. Ruan, X.; Bien, Z.Z.; Wang, Q. Convergence characteristic of proportional-type iterative learning control in the sense of Lebesgue-p norm. *IET Control Theory A* **2012**, *6*, 707–714. [CrossRef]
2. Ajjanaromvat, N.; Parnichkun, M. Trajectory tracking using online learning LQR with adaptive learning control of a leg-exoskeleton for disorder gait rehabilitation. *Mechatronics* **2018**, *51*, 85–96. [CrossRef]
3. Xu, J.X.; Huang, D.Q.; Venkatakrisham, V.; Tuong, H. Research on flexible dynamics of a 6-DOF industrial robot and residual vibration control with a pre-adaptive input shaper. *J. Mech. Sci. Tecnol.* **2019**, *33*, 1875–1889.
4. Cao, Z.X.; Zhang, R.D.; Yang, Y.; Lu, J.Y.; Gao, F.R. Iterative learning control and initial value estimation for probe-drogue autonomous aerial refueling of UAVs. *Aerosp. Sci. Technol.* **2018**, *82–83*, 583–593.
5. Freeman, C.T.; Tan, Y. Iterative learning control with mixed constraints for point-to-point tracking. *Defence Technol.* **2017**, *13*, 360–366.
6. Zhu, X.J.; Gao, Z.Y.; Huang, Q.Z.; Yi, J.C. ILC based active vibration control of smart structures. *IEEE Int. Conf.Intell. Comput.Intell. Syst.* **2009**, *2*, 236–240.
7. Tavakolpour, A.R.; Mailah, M.; Intan, Z.; Darus, M.; Tokhi, O. Self-learning active vibration control of a flexible plate structure with piezoelectric actuator. *Simul. Model Pract. Theory* **2010**, *18*, 516–532. [CrossRef]
8. Zhang, Y.L.; Xu, Q.S. Adaptive iterative learning control combined with discrete-time sliding mode control for piezoelectric nanopositioning. In Proceedings of the 35th Chinese Control Conference, Chengdu, China, 27–29 July 2016; pp. 6080–6085.

9. Xu, J.X.; Wang, W.; Huang, D.Q. Iterative learning in ballistic control. In Proceedings of the 2007 American Control Conference, New York, NY, USA, 11–13 July 2007; pp. 1293–1298.

10. Tan, Y.; Dai, H.H.; Huang, D.Q.; Xu, J.X. Unified iterative learning control schemes for nonlinear dynamic systems with nonlinear input uncertainties. *Automatica* **2012**, *48*, 3173–3182. [CrossRef]

11. Taher, A.; Marwan, A.; Binish, J. Wavelets approach for the optimal control of vibrating plates by piezoelectric patches. *J. Vib. Control* **2018**, *24*, 1101–1108.

12. Qiu, Z.C.; Zhang, X.T.; Zhang, X.M.; Han, J.D. A vision-based vibration sensing and active control for a piezoelectric flexible cantilever plate. *J. Vib. Control* **2016**, *22*, 1320–1337. [CrossRef]

13. Li, S.Q.; Li, J.; Mo, Y.P. Piezoelectric multimode vibration control for stiffened plate using ADRC-based acceleration compensation. *IEEE Trans. Ind. Electron.* **2014**, *61*, 6892–6902. [CrossRef]

14. Lim, Y.H. Finite-element simulation of closed loop vibration control of a smart plate under transient loading. *Smart Mater. Struct.* **2003**, *12*, 272–286. [CrossRef]

15. Hou, Z.S.; Wang, Z. From model-based control to data-driven control: Survey, classification and perspective. *Inform. Sci.* **2013**, *235*, 3–35. [CrossRef]

16. Hou, Z.S.; Chi, R.H.; Gao, H.J. An overview of dynamic-linearization-based data-driven control applications. *IEEE Trans. Ind. Electron.* **2017**, *64*, 4076–4089. [CrossRef]

17. Wang, Z.S.; He, D.K.; Zhu, X.; Luo, J.H.; Liang, Y.; Wang, X. Data-driven model-free adaptive control of particle quality in drug development phase of spray fluidized-bed granulation process. *Complexity* **2017**, 4960106. [CrossRef]

18. Bu, X.H.; Cui, L.Z.; Hou, Z.S.; Qian, W. Formation control for a class of nonlinear multiagent systems using model-free adaptive iterative learning. *Int. J. Robust. Nonlinear Control* **2018**, *28*, 1402–1412. [CrossRef]

19. Chi, R.H.; Hou, Z.S. Dual stage optimal iterative learning control for nonlinear non-affine discrete-time system. *ActaAutom. Sin.* **2007**, *33*, 1061–1065. [CrossRef]

20. Bai, L.; Feng, Y.W.; Li, N.; Xue, X.F. Robust model-free adaptive iterative learning control for vibration suppression based on evidential reasoning. *Micromachines* **2019**, *10*, 196. [CrossRef]

21. Jiang, W. A correlation coefficient for belief functions. *Int. J Approx. Reason.* **2018**, *103*, 94–106. [CrossRef]

22. Lu, C.Q.; Wang, S.P.; Wang, X.J. A multi-source information fusion fault diagnosis for aviation hydraulic pump based on the new evidence similarity distance. *Aerosp. Sci. Technol.* **2017**, *71*, 392–401. [CrossRef]

23. Hable, R. Data-based decisions under imprecise probability and least favorable models. *Int. J. Approx. Reason.* **2009**, *50*, 642–654. [CrossRef]

24. Destercke, S. A k-nearest neighbors method based on imprecise probability. *Soft Comput.* **2012**, *16*, 833–844. [CrossRef]

25. Tien, S.C.; Zou, Q.Z.; Devasia, S. Iterative control of dynamics-coupling-caused errors in piezoscanners during high-speed AFM operation. *IEEE T. Contr. Syst. T.* **2005**, *13*, 921–931. [CrossRef]

26. Matthias, C.M.T. Decision making under uncertainty using imprecise probability. *Int. J. Approx. Reason.* **2007**, *47*, 17–29.

27. Razi, S.; Mollaei, M.R.K.M.; Ghasemi, J. A novel method for classification of BCI multi-class motor imagery task based on Dempster-Shafer theory. *Inform. Sci.* **2019**, *484*, 14–26. [CrossRef]

28. Dong, X.J.; Peng, Z.K.; Ye, L.; Hua, H.X.; Meng, G. Performance evaluation of vibration controller for piezoelectric smart structures in finite element environment. *J.Vib. Control.* **2014**, *20*, 2146–2161. [CrossRef]

29. An, J.Y.; Hu, M.; Fu, L.; Zhan, J.W. A novel fuzzy approach for combining uncertain conflict evidences in the Dempster-Shafer theory. *IEEE Access.* **2018**, *7*, 7481–7501. [CrossRef]

30. Malgaca, L. Integration of active vibration control methods with finite element models of smart laminated composite structures. *Compos.Struct.* **2010**, *92*, 1651–1663. [CrossRef]

31. Saleh, A.R.T.; Mailah, M. Control of resonance phenomenon in flexible structures via active support. *J Sound Vib.* **2012**, *331*, 3451–3465. [CrossRef]

32. Nemanja, D.Z.; Aleksandar, M.S.; Zoran, S.M.; Slobodan, N.S. Optimal vibration control of smart composite beams with optimal size and location of piezoelectric sensing and actuation. *J. Intell. Mat. Syst. Struct.* **2012**, *24*, 499–526.

Article

A Method to Determine Core Design Problems and a Corresponding Solution Strategy

Yuanming Xie [1], Wenqiang Li [1,*], Yin Luo [2], Yan Li [2] and Song Li [1]

[1] School of Manufacturing Science & Engineering, Sichuan University, Chengdu 610065, China; xieyuanming999@163.com (Y.X.); SCULS2011@163.com (S.L.)
[2] Unclear Power Institute of China, Chengdu 610065, China; npicluo@sina.com (Y.L.); liyan-npic@sohu.com (Y.L.)
*   Correspondence: liwenqiang@scu.edu.cn; Tel.: +86-028-8540-3211; Fax: +86-028-8540-6988

Received: 6 March 2019; Accepted: 16 April 2019; Published: 19 April 2019

Abstract: The lack of information on the correlation between root causes and corresponding control criteria in the importance calculation of root causes of design problems results in less accurate determinations of core problems. Based on the interaction between customer needs, bad product parameters, and root causes, a hierarchical representation model of the design problem is established in this paper. A network layer of bad parameters, including various types of correlations, and a control layer, including technical feasibility and cost, are constructed. Then, a method based on the network analytic hierarchy process is proposed to rank the importance of root causes of the design problem and determine the core problems. Finally, a product design process based on the core problem solving is established to assist designers with improving design quality and efficiency. The design for the coolant flow distribution device in the lower chamber of a third-generation pressurized water reactor is employed as an example to demonstrate the effectiveness of the proposed method.

Keywords: customer needs; bad parameters; core problems; design process; conceptual design

---

1. Introduction

In the manufacturing industry, most research and development of products is based on previous products. This kind of design can be regarded as incremental innovation [1–3]. The purpose of incremental product innovation design is to improve product performance without changing the working principle of the product. In other words, incremental innovation is aimed at solving various problems that hinder designers from realizing the desired functions, through which the performance of a previous product system will be further improved [4,5].

The key to incremental innovation is to analyze and determine the core problems in the previous product system, and to develop suitable strategies to generate better design solutions. At present, many researchers have integrated innovative methods with complementary advantages to assist designers with improving design quality and efficiency [6,7]. The theory of inventive problem solving (TRIZ) was adopted by Li et al. [8] to define conflicts in the product system and solve these conflicts with effective tools (i.e., a contradiction matrix). The environmentally conscious quality function deployment (ECQFD) approach was proposed by Vinodh et al. [9] to analyze and identify the problem. In this approach, customer needs are converted into corresponding technical feature conflicts, then the conflicts are solved with TRIZ. A process combining the TRIZ method with the design for manufacture and assembly (DFMA) method was proposed by Lucchetta et al. [10] to solve various problems occurring in the simplification process of a product system. Caligiana integrated the QFD and TRIZ methods to define and overcome some core problems that can affect the development of a product [11]. The morphological matrix (MM) method was adopted by Liu et al. [12] to identify various conflicts in the deformation process of product system components, and to solve those conflicts with the TRIZ method. An integrated process was proposed by Li et al. [13] to solve key problems with a process trimming-based TRIZ, and it also helped to identify and inventively solve key problems. The axiomatic design (AD) was adopted by Ko et al. [14] to analyze the problems existing in the product system and solve them with the TRIZ method. The theory of constraints (TOC) method was developed to analyze the root causes of various problems in a product system through current reality trees (CRTs), to identify the conflicts with a conflict resolution diagram (CRD), and to solve these conflicts with the TRIZ method [15–19]. Although these methods provide effective strategies for analyzing various problems and solving conflicts that exist in a product system, there still seems to be some difficulty in identifying the core problems.

The problems that occur in incremental innovation are often not independent. A solution to one problem may affect the solution to others [20]. Hence, it is necessary to determine the solving order of the root causes of problems. The fault tree analysis (FTA) method was adopted by Wei et al. [21] to qualitatively and quantitatively evaluate the importance of root causes. Then, the core problems are identified in the order of importance. A method based on the FTA method was proposed by Shu et al. to calculate the importance of root causes of failure events in a product system [22–24]. Garcia et al. [25] adopted the failure modes and effects analysis (FMEA) method to analyze potential failure problems in a product system and assign weights to them. A method combining FTA and FMEA was proposed by Azadeh et al. to qualitatively and quantitatively calculate the importance of root causes of the fault [26–28]. The fishbone diagram (FD) method was employed by Yazdani et al. [29] to analyze the root causes of the problem and was combined with the analytic hierarchy process (AHP) method to determine the importance of root causes. Although these methods are applied to calculate the importance of root causes, they are based on the premise that root causes are independent of each other. Such an assumption will lead to analysis results deviating from the actual situation. At the same time, there is a lack of corresponding control criteria in analysis process to constrain and evaluate the ranking of the importance of root causes. The above methods will result in less rational and accurate importance order and determination of core problems.

In order to solve the above problems, a hierarchical representation model of the design problem is established in this paper including customer needs, bad parameters, and root causes based on their interaction. The correlations in the bad parameters layer and the control criteria for the ranking of the importance of root causes are taken into consideration. Therefore, a method based on the network analytic hierarchy process is proposed to rank the importance of root causes and determine the core problems. Furthermore, a product design process based on solving the core problem is established to assist designers to improve design quality and efficiency.

2. Establishment and Transformation of Hierarchical Representation Model of Design Problem

The product design process of incremental innovation is a process of converting from problem analysis to concept generation. Before problem analysis, customer needs should be characterized and transformed into technical problems of the product system. Then problem analysis tools are used to analyze the root causes of those problems. Finally, the conflicts of root causes are solved by employing problem solving tools.

Customer needs can be divided into technical targets and economic targets. Technical targets include safety and reliability, functionality, and environmental protection. Economic targets include costs of purchase and use [30,31]. As initial customer needs are vague and involve perceptual understanding of target products, they are characterized by low professionalism and strong subjectivity [32]. Hence, in this paper, customer needs are classified into technical and economic targets to obtain corresponding new product performance targets. These new targets are compared with previous ones, and the performance targets that show poor consistency between the new and previous ones are regarded as the product's bad parameters.

Generally, a product's bad parameters are not independent of each other. A solution to a certain root cause of a bad parameter may lead to the reconstruction of other root causes. Therefore, there is not only an inclusion relationship between the elements in the bad parameters layer and the elements in the root causes layer, but also correlations between bad parameters and root causes.

According to the relationship among customer needs, bad parameters, and root causes, a hierarchical representation model of the design problem is established, as shown in Figure 1. In this model, the customer needs are classified and turned into bad parameters. Then the root causes of the bad parameters are analyzed. Finally, the importance of root causes is analyzed based on the correlation of the bad parameters layer to determine the core problems.

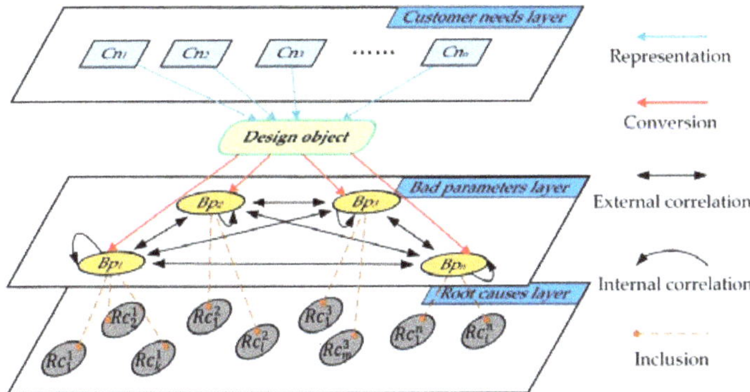

Figure 1. Hierarchical representation model of design problem.

The hierarchical representation model of the design problem that includes customer needs, bad parameters, root causes, and core problems is illuminated as follows:

Customer needs (Cn) layer: An acquisition method is used to obtain various types of customer needs. Then a set of customer needs will be obtained by standardizing and filtering them.

$$Cn = \{Cn_1, Cn_2, Cn_3, \ldots, Cn_n\} \tag{1}$$

Bad parameters (Bp) layer: In this layer, the corresponding new product performance targets are identified and compared with previous ones, then a set of the product's bad parameters is determined.

$$Bp = \{Bp_1, Bp_2, Bp_3, \ldots, Bp_n\} \tag{2}$$

Root causes (Rc) layer: This layer is a set of root causes affecting bad parameters. It includes indirect and direct causes leading to the bad parameters.

$$Rc = \left\{ \left(Rc_1^1, Rc_2^1, \ldots, Rc_k^1\right), \ldots, \left(Rc_1^j, \ldots, Rc_i^j\right), \ldots, \left(Rc_1^n, Rc_m^n\right)\right\}, \tag{3}$$

where Rc_i^j is the ith root cause of the jth bad parameter.

Then, the relationship between the root causes layer and the bad parameters layer can be expressed as

$$Rc = \left\{ \begin{array}{l} \left.\begin{array}{l} Rc_1^1 \\ \dots \\ Rc_k^1 \end{array}\right\} \rightarrow Bp_1 \\ \dots \quad \dots \\ \left.\begin{array}{l} Rc_1^j \\ \dots \\ Rc_l^j \end{array}\right\} \rightarrow Bp_j \\ \dots \quad \dots \\ \left.\begin{array}{l} Rc_1^n \\ \dots \\ Rc_m^n \end{array}\right\} \rightarrow Bp_n \end{array} \right\} = Bp. \tag{4}$$

Core problems (Cp): The core problems in the root causes are the main cause of bad parameters, and their solution may affect the solution to other root causes.

$$Cp_j \in \left(Rc_1^j, Rc_2^j, \dots, Rc_l^j\right), \tag{5}$$

where Cp_j is the core problem that causes the jth bad parameter.

3. Method to Determine Core Problems Based on Analytic Network Process

In order to assist designers with obtaining the importance order of core problems from a large number of problems that affect the product system performance, a determination method of core problems based on ANP is proposed in this paper.

3.1. Construction of Network Layer and Control Layer for Design Problem

In the hierarchical representation model of the design problem, there is not only an inclusion relationship between the elements in the bad parameters layer and the elements in the root causes layer, but also correlations between bad parameters and root causes. These correlations are characterized by vertical clustering and horizontal correlation. In this paper, a network structure of cluster elements [33] is adopted to construct the network layer of bad parameters. In order to analyze the correlations among the clusters of different design problems and obtain the core problems that affect the bad parameters, a control layer, including technical feasibility and cost, is constructed to constrain and evaluate the prospect of solving different design problems.

Analytic network process (ANP) is a multicriteria theoretical measuring method. It obtains the relative priority of an indicator by comparing the relative influence of two elements on a third element under a certain potential criterion [33]. On the one hand, ANP method can be used to consider the correlation of same-level elements with their network structure. On the other hand, it also can be used to constrain and evaluate the ranking of the importance by adopting the control criteria [34,35]. According to the characteristics of ANP method and considering network layer and control layer, the ANP method is adopted in this paper to determine the importance order of root causes.

In this paper, every bad parameter (Bp_i) in the bad parameters layer is called a cluster, and the root cause (Rc_j^i) is regarded as an element in the bad parameter cluster (Bp_i). If an element in a cluster affects or is affected by at least one element in another cluster, it is called an externally dependent correlation between two clusters. If an element in a cluster affects or is affected by at least one element in the same cluster, it is called an internally dependent correlation. On this basis, a product hierarchy model is built, as shown in Figure 2.

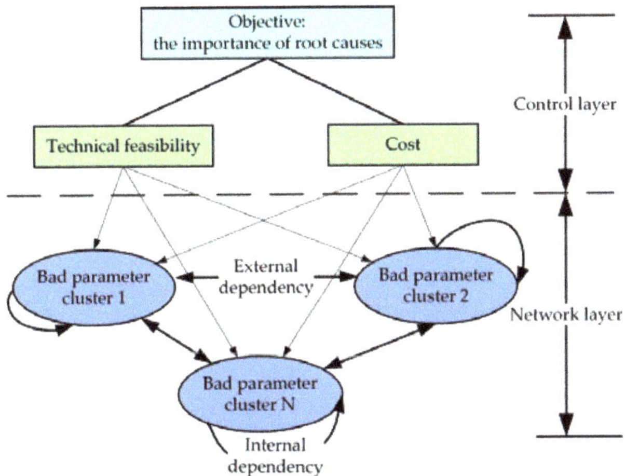

Figure 2. Product hierarchy model.

3.2. Calculation of Importance of Root Causes and Determination of Core Problems

The steps to calculate the importance of root causes and determine the core problems with ANP method are as follows:

Step 1: Construct the judgment matrix (A_m^{ij}) and calculate the weight vector $(w_m^{(ij)})$. If an element (i_j) in a cluster (i) has an external dependency on an element in another cluster (m), or has an internal dependency on an element inside its cluster, this element (i_j) is taken as a subcriterion. The influence degree of the elements in another cluster (m) to this element (i_j) is compared in pairs. In this way, the judgment matrix (A_m^{ij}) is constructed to calculate the corresponding weight vector (w_m^{ij}). When comparing elements in pairs, an industry expert must determine which element has the greater influence on the third party element (i_j) and by how much. The comparison value E of the two elements is assigned on a scale of 0–9, and the ratio scale as shown in Table 1. The constructed judgment matrix (A_m^{ij}) is a positive reciprocal matrix. Taking the above demonstration as an example, the judgment matrix (A_m^{ij}) and weight vector $(w_m^{(ij)} = \left[w_{m_1}^{(ij)}, w_{m_2}^{(ij)}, \ldots, w_{m_n}^{(ij)} \right]^T)$ are obtained by pairwise comparison, as shown in Table 2.

Table 1. The ratio scale [32].

Ratio Scale	Meaning (If: Element 1 Is A; Element 2 Is B; The Third Party Element Is C)
1	A and B have the same effect on C
3	The effect of A on C is slightly larger than the effect of B on C
5	The effect of A on C is medially larger than the effect of B on C
7	The effect of A on C is highly larger than the effect of B on C
9	The effect of A on C is extremely larger than the effect of B on C
2,4,6,8	Median value of the above adjacent scale
Reciprocal of the above scale	The ratio of the effect of A on C to the effect of B on C is E_{AB}^C; The ratio of the effect of B on C to the effect of A on C is $1/E_{AB}^C$

Table 2. Judgment matrix $(A_m^{i_j})$ of elements in cluster (m) based on element (i_j).

i_j	m_1	m_2	$\cdots$	m_n	Weight Vector $w_m^{(i_j)}$
m_1	e_{11}	e_{12}	$\cdots$	e_{1n}	$w_{m_1}^{(i_j)}$
m_2	e_{21}	e_{22}	$\cdots$	e_{2n}	$w_{m_2}^{(i_j)}$
$\vdots$	$\vdots$	$\vdots$	$\ddots$	$\vdots$	$\vdots$
m_n	e_{n1}	e_{n2}	$\cdots$	e_{nn}	$w_{m_n}^{(i_j)}$

Step 2: Construct supermatrix (W_p). The obtained weight vector $(w_m^{i_j})$ in Step 1 is sequentially written into every column of the matrix in Equation (6). In this way, the submatrix (W_{im}) of the supermatrix is constructed. As shown in Equation (7), all clusters are arranged in numerical order. Then the elements in every cluster are placed vertically and horizontally on the left and top of the supermatrix, respectively. Finally, the submatrix (W_{im}) calculated by Equation (6) is filled into the corresponding position of the supermatrix (W_p). (W_p) is constructed with p as a control criterion.

$$
W_{im} = \begin{bmatrix}
w_{m_1}^{(i_1)} & w_{m_1}^{(i_2)} & \cdots & w_{m_1}^{(i_j)} \\
w_{m_2}^{(i_1)} & w_{m_2}^{(i_2)} & \cdots & w_{m_2}^{(i_j)} \\
\vdots & \vdots & \ddots & \vdots \\
w_{m_n}^{(i_1)} & w_{m_n}^{(i_2)} & \cdots & w_{m_n}^{(i_j)}
\end{bmatrix}
\tag{6}
$$

$$
W_p = \begin{array}{cc}
 & \begin{array}{cccc} Bp_1 & Bp_2 & \cdots & Bp_N \\ Rc_1^1 \ldots Rc_{n_1}^1 & Rc_1^2 \ldots Rc_{n_2}^2 & \cdots & Rc_1^N \ldots Rc_{n_N}^N \end{array} \\
\begin{array}{cc}
Bp_1 & \begin{array}{c} Rc_1^1 \\ \cdots \\ Rc_{n_1}^1 \end{array} \\
Bp_2 & \begin{array}{c} Rc_1^2 \\ \cdots \\ Rc_{n_2}^2 \end{array} \\
\vdots & \vdots \\
Bp_N & \begin{array}{c} Rc_1^N \\ \cdots \\ Rc_{n_N}^N \end{array}
\end{array} &
\begin{bmatrix}
W_{11} & W_{12} & \cdots & W_{1N} \\
W_{21} & W_{22} & \cdots & W_{2N} \\
\vdots & & & \vdots \\
W_{N1} & W_{N2} & \cdots & W_{NN}
\end{bmatrix}
\end{array}
\tag{7}
$$

Step 3: Construct a weighted supermatrix and limit weighted supermatrix. A certain cluster is taken as a subcriterion; the relative importance of the two clusters connected with it is compared to obtain the judgment matrix and weight vector. For example, with cluster (Bp_1) as a subcriterion, the judgment matrix is obtained by comparing the connected clusters in pairs, as shown in Table 3.

Table 3. Judgment matrix between other clusters based on cluster (Bp_1).

Bp_1	Bp_1	Bp_2	$\cdots$	Bp_N	Weight Vector
Bp_1	e_{11}	e_{12}	$\cdots$	e_{1n}	a_{11}
Bp_2	e_{21}	e_{22}	$\cdots$	e_{2n}	a_{12}
$\vdots$	$\vdots$	$\vdots$	$\ddots$	$\vdots$	$\vdots$
Bp_N	e_{n1}	e_{n2}	$\cdots$	e_{nn}	a_{1N}

The obtained weight vectors are sequentially written into the respective columns of the matrix in Equation (8) to obtain the weighted matrix A:

$$A = \begin{bmatrix} a_{11} & a_{21} & \cdots & a_{N1} \\ a_{12} & a_{22} & \cdots & a_{N2} \\ \vdots & \vdots & \ddots & \vdots \\ a_{1N} & a_{2N} & \cdots & a_{NN} \end{bmatrix}. \tag{8}$$

Every element $(\overline{W}_{im})$ of the weighted supermatrix $(\overline{W}_p)$ is equal to the corresponding weight (a_{im}) in the weighted matrix A multiplied by the submatrix (W_{im}) in the supermatrix, as shown in Equation (9). Thus, the weighted supermatrix $(\overline{W}_p)$ is obtained:

$$\overline{W}_{im} = a_{im} \times W_{im} \tag{9}$$

$$\overline{W}_p = \begin{bmatrix} a_{11}W_{11} & a_{21}W_{21} & \cdots & a_{N1}W_{N1} \\ a_{12}W_{12} & a_{22}W_{22} & \cdots & a_{N2}W_{N2} \\ \vdots & \vdots & \ddots & \vdots \\ a_{1N}W_{1N} & a_{2N}W_{2N} & \cdots & a_{NN}W_{NN} \end{bmatrix}. \tag{10}$$

In order to make the weighted supermatrix $(\overline{W}_p)$ accurate, it is necessary to perform a stability treatment on $(\overline{W}_p)$ by calculating its nth power, and letting n tend to infinity. As shown in Equation (11), the limit weighted supermatrix (W_∞) is obtained when the value of the columns of the weighted supermatrix $(\overline{W}_p)$ does not change. The value of each row of the limit weighted supermatrix (W_∞) is the importance (w_p) of the corresponding element:

$$W_\infty = \lim_{n \to \infty} \overline{W}_p{}^n. \tag{11}$$

Step 4: Calculate the importance of root causes. Technical feasibility (C_T) and cost (C_C) are taken as control criteria. So importance (w_T) and (w_C) are obtained by calculating the corresponding limit weighted supermatrix according to Steps 1–4. In order to improve the product system performance, two control criteria, technical feasibility (C_T) and cost (C_C), are weighted to obtain the corresponding weights (w_{C_T}) and (w_{C_C}). These weights of the control criteria are weighted for importance (w_T) and (w_C), respectively, as shown in Equation (12), so as to obtain the importance (W) of root causes:

$$W = w_{C_T} \times w_T + w_{C_C} \times w_C. \tag{12}$$

Step 5: Determine the solving order of root causes and core problems based on the importance. (1) The solution of root causes is carried out in the importance order (W). In other words, the higher the importance, the higher the priority of the root causes processing. (2) According to the importance order (W), the first root cause (Rc_i^j) is determined as the core problem (Cp_j) of the bad parameter (Bp_j).

Based on above contents, the quality and effectiveness of design can be effectively guaranteed. On the one hand, product design activities based on the importance order of core problems will improve the direction of the product system and determine the quality of the design. On the other hand, the solution to the core problem that ranks ahead will be beneficial to solve the non-core problems. This will improve the efficiency of the whole design process.

4. Product Design Process Based on Solving the Core Problem

In order to assist designers to determine design core problems and to provide the right design direction for designers to improve design quality and efficiency, this paper establishes a product design process based on solving the core problem, as shown in Figure 3. The specific steps are as follows:

Step 1: Acquire customer needs, then analyze and standardize them. The sorted customer needs are classified by performance under technical and economic targets to obtain corresponding new product performance targets.

Step 2: Compare new product performance targets with previous ones. Performance targets with poor consistency between the new and previous ones are regarded as the product's bad parameters.

Step 3: Employ current reality trees (CRTs) to analyze and determine the root causes of the bad parameters.

Step 4: Use the ANP method to calculate the importance of the root causes and determine the core problems.

Step 5: Identify conflicts of the core problems and employ the TRIZ method to solve the core problems. In addition, design for root cause with lower importance should be based on the design solution for root cause with higher importance.

Step 6: Evaluate the solution to get the best conceptual design scheme.

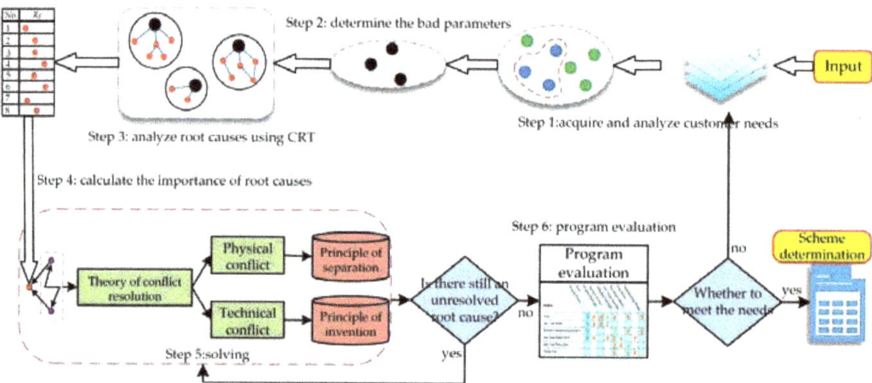

Figure 3. Product design process based on solving the core problem.

5. Case Studies

The coolant flow distribution device in the lower chamber is applied in the third-generation pressurized water reactor, and its main functions include: (1) to mix the coolant in the lower chamber and suppress the generated vortex; and (2) to evenly distribute the coolant from the lower chamber into the core to ensure complete cooling of the core fuel assembly [36]. In addition, the coolant flow distribution device also plays an auxiliary support role in the core, so it has a certain buffering effect on the falling of the core. The following is an incremental design for the coolant flow distribution device. It uses the method proposed in this paper to determine the core design problems and offer solving strategies.

5.1. Importance Analysis of Root Causes of the Coolant Flow Distribution Device

According to the investigation of reactor customers, customer needs for the current coolant flow distribution device [37] include good flow distribution effect (Cn_1), good vortex suppression effect (Cn_2), simple structure (Cn_3), prevention for core falling (Cn_4), sufficient strength (Cn_5), and small vibration (Cn_6). Customer needs are classified by performance under technical and economic targets to achieve corresponding new performance targets of the coolant flow distribution device.

By comparing new performance targets with previous ones, the bad parameters of the coolant flow distribution device are determined as follows: the uniformity for flow distribution is poor (Bp_1), and there are complex structures (Bp_2), and lots of vortices (Bp_3).

According to the bad parameters of the coolant flow distribution device, current reality trees are established to determine the root causes, as shown in Figure 4. Figure 4a–c shows CRT diagrams of the bad parameters: the uniformity for flow distribution is poor (Bp_1), and there are complex structures (Bp_2) and lots of vortices (Bp_3).

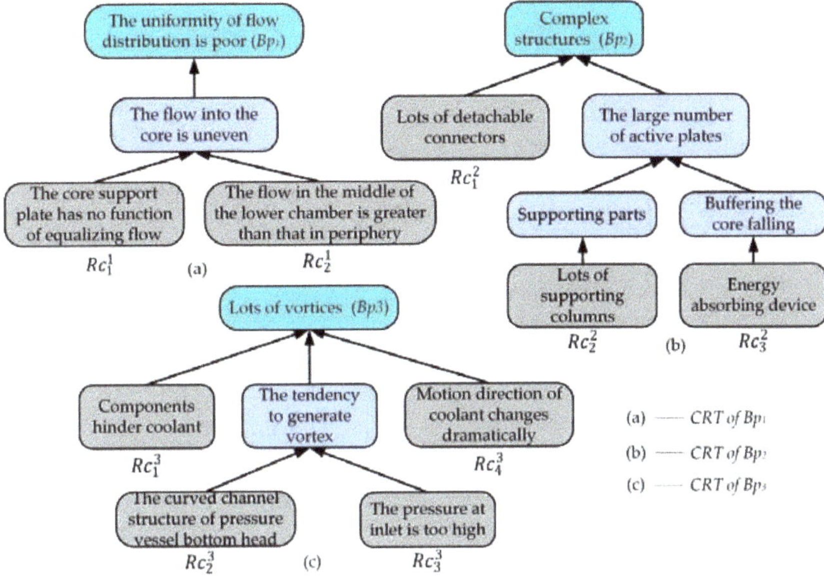

Figure 4. CRT of bad parameters of the coolant flow distribution device.

According to the correlation between bad parameters and internal root causes, the hierarchical structure model of the coolant flow distribution device is set up as shown in Figure 5. The importance of root causes is calculated by using the ANP method.

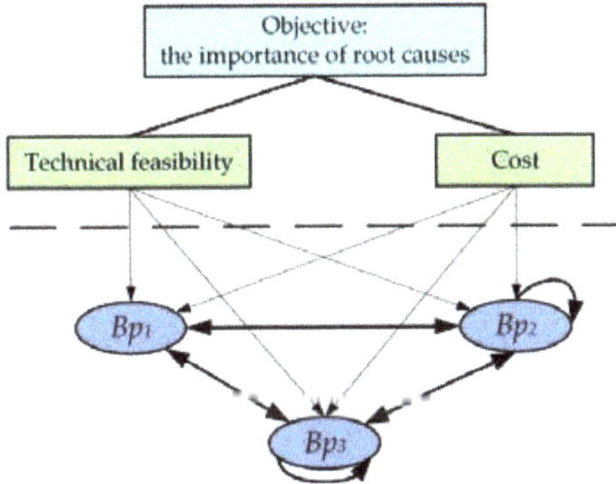

Figure 5. Hierarchical structure model of coolant flow distribution device.

The judgment matrix is constructed according to the influence relationship among root causes. Then the weight vector is calculated. For example, the technical feasibility (C_T) and root cause (Rc_2^1) are taken as the control criterion and subcriterion, respectively. The judgment matrix (A_3^{12}) and weight vector ($w_3^{(12)}$) are obtained by comparing the influence degree between the root causes $\{Rc_1^3, Rc_2^3, Rc_3^3, Rc_4^3\}$ of the bad parameter (Bp_3) cluster, as shown in Table 4.

Table 4. Judgment matrix (A_3^{12}) of elements in cluster (Bp_3) based on element (1_2).

$1_2.$	Rc_1^3	Rc_2^3	Rc_3^3	Rc_4^3	Weight Vector $w_3^{(1_2)}$
Rc_1^3	1	6	3	1/2	0.33393
Rc_2^3	1/6	1	1/2	1/6	0.06607
Rc_3^3	1/3	2	1	1/3	0.13214
Rc_4^3	2	6	3	1	0.46786

Similarly, the judgment matrices (A_m^{ij}) among root causes are constructed, and the weight vectors ($w_m^{(ij)}$) are achieved. The weight vectors are sequentially written into the subcolumn corresponding to the supermatrix (W_T). Then the unweighted supermatrix of the coolant flow distribution device is obtained as shown in Table 5.

Table 5. Unweighted supermatrix (W_T) of coolant flow distribution device.

C_T	Rc_1^1	Rc_2^1	Rc_1^2	Rc_2^2	Rc_3^2	Rc_1^3	Rc_2^3	Rc_3^3	Rc_4^3
Rc_1^1	0.25000	0.33333	0.16667	0.14286	0.12500	0.12500	0.50000	0.16667	0.14286
Rc_2^1	0.75000	0.66667	0.83333	0.85714	0.87500	0.87500	0.50000	0.83333	0.85714
Rc_1^2	1.00000	0.54545	0.59489	0.28571	0.63275	0.30915	1.00000	1.00000	0.28571
Rc_2^2	0.00000	0.27273	0.27661	0.57143	0.19240	0.58126	0.00000	0.00000	0.57143
Rc_3^2	0.00000	0.18182	0.12850	0.14286	0.17485	0.10959	0.00000	0.00000	0.14286
Rc_1^3	0.00000	0.33393	0.33333	0.83333	0.33333	0.15385	0.12500	0.33333	0.54545
Rc_2^3	0.00000	0.06607	0.00000	0.00000	0.00000	0.07692	0.25000	0.33333	0.18182
Rc_3^3	0.00000	0.13214	0.00000	0.00000	0.00000	0.00000	0.00000	0.00000	0.00000
Rc_4^3	1.00000	0.46786	0.66667	0.16667	0.66667	0.76923	0.62500	0.33333	0.27273

According to the mutual influence among the clusters, the relative importance of the two clusters connected with a certain cluster is compared under the subcriterion. Then the judgment matrix and weight vectors are obtained. The weight vectors are sequentially written into the columns of the matrix in Table 6, thereby the weighted matrix (A) of the clusters is obtained.

Table 6. Weighted matrix (A) of each cluster of coolant flow distribution device.

C_T	Bp_1	Bp_2	Bp_3
Bp_1	0.40000	0.26429	0.30915
Bp_2	0.40000	0.64556	0.58126
Bp_3	0.20000	0.09015	0.10959

The corresponding weight (a_{im}) in the weighted matrix A is multiplied by the submatrix (W_{im}) in the supermatrix to obtain the weighted supermatrix ($\overline{W}_T$), as shown in Table 7.

Table 7. Weighted supermatrix ($\overline{W}_T$) of coolant flow distribution device.

C_T	Rc_1^1	Rc_2^1	Rc_1^2	Rc_2^2	Rc_3^2	Rc_1^3	Rc_2^3	Rc_3^3	Rc_4^3
Rc_1^1	0.10000	0.13333	0.04405	0.03776	0.03304	0.03864	0.15458	0.05153	0.04416
Rc_2^1	0.30000	0.26667	0.22024	0.22653	0.23125	0.27051	0.15458	0.25763	0.26499
Rc_1^2	0.40000	0.21818	0.38404	0.18445	0.40848	0.17970	0.58126	0.58126	0.16608
Rc_2^2	0.00000	0.10909	0.17857	0.36889	0.12420	0.33787	0.00000	0.00000	0.33215
Rc_3^2	0.00000	0.07273	0.08296	0.09222	0.11288	0.06370	0.00000	0.00000	0.08304
Rc_1^3	0.00000	0.06679	0.03005	0.07513	0.03005	0.01686	0.01370	0.03653	0.05977
Rc_2^3	0.00000	0.01321	0.00000	0.00000	0.00000	0.00843	0.02740	0.03653	0.01992
Rc_3^3	0.00000	0.02643	0.00000	0.00000	0.00000	0.00000	0.00000	0.00000	0.00000
Rc_4^3	0.20000	0.09357	0.06010	0.01503	0.06010	0.08430	0.06849	0.03653	0.02989

The limit weighted supermatrix (W_∞) with technical feasibility (C_T) as the control criterion is obtained by calculating the limit ($W_\infty = \lim_{n \to \infty} \overline{W}_T{}^n$), as shown in Table 8.

Table 8. Limit weighted supermatrix (W_∞) of coolant flow distribution device.

C_T	Rc_1^1	Rc_2^1	Rc_1^2	Rc_2^2	Rc_3^2	Rc_1^3	Rc_2^3	Rc_3^3	Rc_4^3
Rc_1^1	0.06797	0.06797	0.06797	0.06797	0.06797	0.06797	0.06797	0.06797	0.06797
Rc_2^1	0.24441	0.24441	0.24441	0.24441	0.24441	0.24441	0.24441	0.24441	0.24441
Rc_1^2	0.28464	0.28464	0.28464	0.28464	0.28464	0.28464	0.28464	0.28464	0.28464
Rc_2^2	0.19897	0.19897	0.19897	0.19897	0.19897	0.19897	0.19897	0.19897	0.19897
Rc_3^2	0.07708	0.07708	0.07708	0.07708	0.07708	0.07708	0.07708	0.07708	0.07708
Rc_1^3	0.04730	0.04730	0.04730	0.04730	0.04730	0.04730	0.04730	0.04730	0.04730
Rc_2^3	0.00536	0.00536	0.00536	0.00536	0.00536	0.00536	0.00536	0.00536	0.00536
Rc_3^3	0.00646	0.00646	0.00646	0.00646	0.00646	0.00646	0.00646	0.00646	0.00646
Rc_4^3	0.06781	0.06781	0.06781	0.06781	0.06781	0.06781	0.06781	0.06781	0.06781

When technical feasibility (C_T) is taken as a control criterion, the importance of root causes can be seen (Table 7):

$$w_T = [0.06797,\ 0.24441,\ 0.28464,\ 0.19897,\ 0.07708,\ 0.04730,\ 0.00536,\ 0.00646,\ 0.06781]^T.$$

In the same way, the importance of root causes can be obtained when cost (C_C) is taken as a control criterion:

$$w_C = [0.07361,\ 0.22145,\ 0.21058,\ 0.11387,\ 0.06148,\ 0.12448,\ 0.08122,\ 0.00250,\ 0.11082]^T.$$

The weights of control layer elements given by experience are $w_{C_T} = 0.5$ and $w_{C_C} = 0.5$. According to Equation (12), the importance of every root cause can be obtained as follows:

$$W = [0.07079,\ 0.23293,\ 0.24761,\ 0.15642,\ 0.06828,\ 0.08589,\ 0.04329,\ 0.00448,\ 0.08931]^T.$$

The order of importance of every root cause is $Rc_1^2, Rc_2^1, Rc_2^2, Rc_4^3, Rc_3^2, Rc_1^1, Rc_3^2, Rc_2^3, Rc_3^3$. Therefore, core problems Cp_1, Cp_2, Cp_3 can be determined by the order, which is Rc_2^1, Rc_1^2, Rc_4^3.

5.2. Solving the Root Causes Sequentially Based on Importance

According to the CRD diagram of core problem Cp_2 (a lot of detachable connectors) shown in Figure 6, a conflict between the number of fixed connectors and the number of components exists in the lower chamber. If the number of components is reduced, the number of fixed connectors will increase. The engineering parameters for improvement and deterioration can be defined by TRIZ as energy consumption of stationary objects and productivity, respectively. The vortex suppression plate is designed as a cylindrical shape according to the principle of versatility (No. 6), as shown in the flow distribution cylinder in Figure 7. The cylinder can not only suppress the vortex, but also distribute the flow and effectively support the upper flow equalizing plate. In this way, the number of detachable connectors is reduced. The cylinder has a supporting function. It indirectly reduces the number of supporting columns in the lower chamber and the amount of coolant blocked by the components at a certain extent. Therefore, the solution to root cause (Rc_1^2) also solves root cause (Rc_2^2) (a lot of supporting columns) and root cause (Rc_1^3) (components hinder coolant).

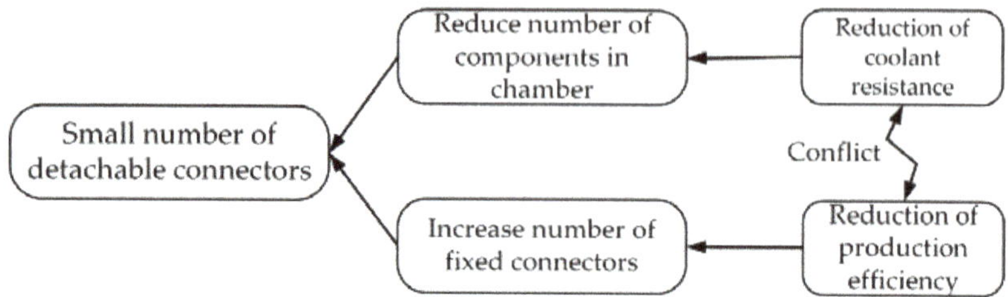

Figure 6. CRD diagram of core problem Cp_2.

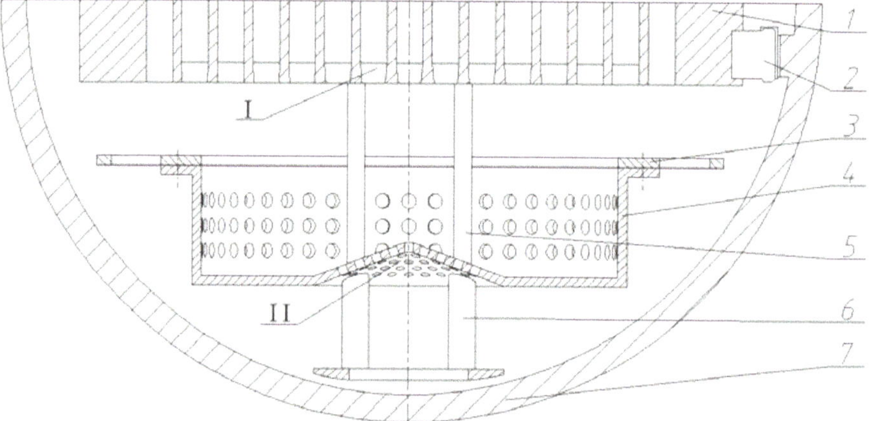

Figure 7. Conceptual design scheme of coolant flow distribution device: (1) core support plate; (2) radial support key; (3) flow equalizing plate; (4) flow distribution cylinder; (5) supporting column; (6) energy-absorbing device; (7) pressure vessel bottom head; (I) inverted cone structure; (II) cap structure.

If the root cause (Rc_2^2) is solved first, the number of supporting columns will be reduced by the trimming method according to TRIZ. However, this solution cannot solve the core problem (Cp_2).

The CRD diagram of core problem Cp_1 (the coolant flow in the middle of the lower chamber is greater than that in the periphery) is set up, as shown in Figure 8. It can be seen that the reduction in the dimension of the active plates requires the setting of special holes in the active plates in the lower chamber. The engineering parameters for improvement and deterioration can be defined by TRIZ as the loss of substance and the quantity of substance or things, respectively. The flow equalizing plate with a cap structure is designed based on the principle of pre-action (No. 10) and the characteristics of flow equalizing plate. It can distribute the intermediate coolant at a certain proportion to the surrounding areas in advance. The design solution of the core problem Cp_2 is flow distribution cylinder. If the cap structure is set on the flow equalizing plate that is installed on the flow distribution cylinder, the cap structure will hinder the flow distribution function of the cylinder. According to above definition: design for root cause with lower importance should be based on the design solution for root cause with higher importance, and in order to eliminate the adverse effect on flow distribution function, the cap structure should be set at the bottom of the flow distribution cylinder, as shown in Figure 7. There is a certain inclination angle on the surface of the cap structure, and its holes are set perpendicular to the surface. When the intermediate fluid passes through the cap structure, the cap structure can distribute the intermediate coolant to the surroundings according to its surface and the holes.

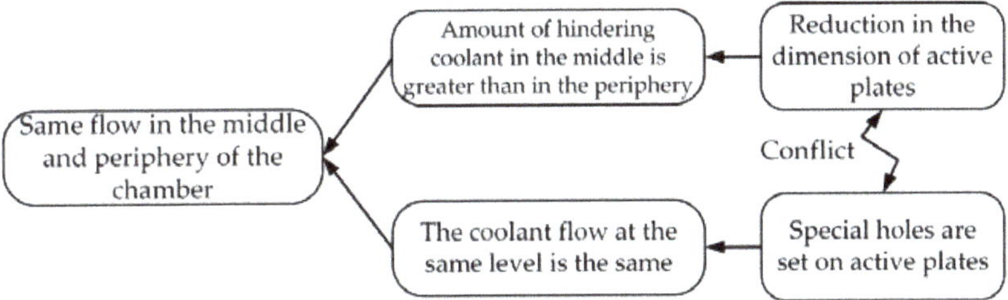

Figure 8. CRD diagram of core problem Cp_1.

If the solution order of the core problem Cp_1 and Cp_2 is exchanged, the flow equalizing plate with a cap structure is designed first, then the flow distribution cylinder is designed. According to above definition, the flow equalizing plate with a cap structure is installed on the cylinder. In this way, the coolant passes through the cylinder and then passes through the cap structure. However, the cylinder has already distributed the coolant. When the coolant passes through the flow equalizing plate, the cap structure will distribute the coolant again. It will cause a negative effect and is not good for the prediction of coolant movement.

The CRD diagram of core problem Cp_3 (motion direction of coolant changes dramatically) is established, as shown in Figure 9. Whether the structure of the flow channel should be changed or not is a physical conflict in the lower chamber. The principle of spatial dimension change (No. 17) in spatial separation is selected to design a stepped flow channel inside the pressure vessel bottom head. In this way, the variation degree of flow direction of coolant is effectively reduced. The structures inside the pressure vessel bottom head is changed by the stepped flow channel. Thereby, it also solves the root cause (Rc_2^3) (curved channel structure of pressure vessel bottom head).

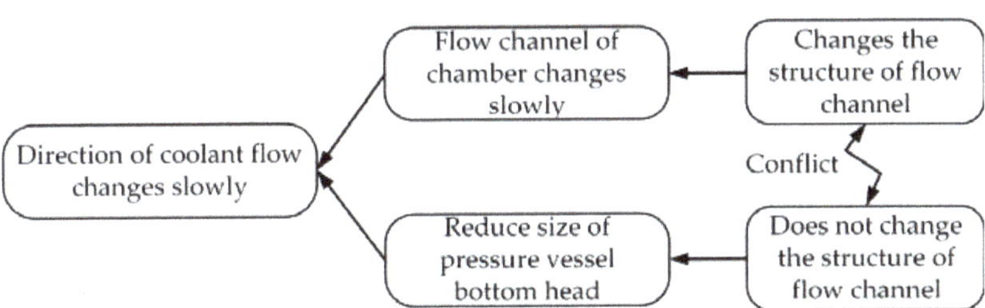

Figure 9. CRD diagram of the core problem Cp_3.

The CRD diagram of the root cause (Rc_1^1) (core support plate has no function of equalizing flow) is established as shown in Figure 10. A pair of physical conflicts can be seen over the unequal or equal dimension of the hole on the core support plate. In the light of the principle of local mass (No. 3) in spatial separation and the continuity equation of hydrodynamics, the cross section of every inlet hole of the core support plate enlarges gradually from the middle to the periphery, and every outlet hole has the same cross section, as shown in the inverted cone structure in Figure 7.

The root causes Rc_3^2 (energy-absorbing device) and Rc_3^2 (the pressure at the inlet is too high) are less important and not the core problems. Moreover, the energy-absorbing device is a measure for preventing the core from falling. The pressure at the inlet is an unchangeable parameter set by the nuclear reactor. Therefore, it is verified that the importance of root causes and core problems determined in this paper is in line with the actual design requirements.

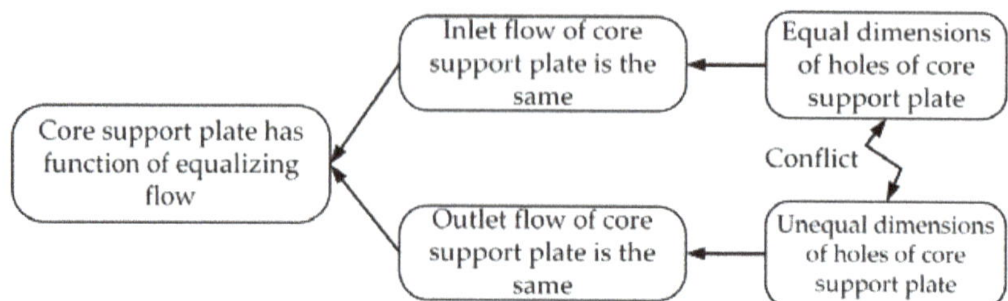

Figure 10. CRD diagram of the root cause Rc_1^1.

5.3. Conceptual Design Scheme Generation

Through the optimization of the scheme, the conceptual design scheme of the coolant flow distribution device was finally obtained, as shown in Figure 7. The hole of the core support plate retains the previous upper straight hole, and the taper hole is set on its bottom. The taper gradually decreases from the middle to the periphery at a certain proportion, and the axial length ratio of the two types of holes and the taper ratio of the taper hole can be adjusted according to the actual requirements. There is not only the function of vortex suppression, but also the function of flow distribution on the flow distribution cylinder. It can effectively support the upper flow equalizing plate, and the bottom of the cylinder is also provided with a cap structure. The surface of the cap structure has a certain inclination angle and is also provided with holes perpendicular to the surface.

5.4. Simulation Analysis

Computational Fluid Dynamics (CFD) software has been widely used in engineering fields [38,39]. Therefore, in order to verify the effectiveness of conceptual design scheme, CFD software is used to analyze the coolant flow characteristics of design scheme. Firstly, a three-dimensional model of the design scheme was built by Solidworks (Dassault Systemes: Massachusetts, MA, USA); secondly, the fluid field was divided into meshes by Meshing (Ansys: Canonsburg, PA, USA); and finally, the flow field was calculated by Fluent (Ansys: Canonsburg, PA, USA) software. The boundary conditions were set as follows:

(1) Turbulent model: K-epsilon (2 eqn)-standard
(2) Fluid material: water (liquid)
(3) Wall: no slip
(4) Inlet (velocity inlet): velocity magnitude: 5.084 m/s; turbulent intensity: 1.755%; hydraulic diameter: 1200 mm
(5) Outlet (pressure outlet): gauge pressure: 0 pa
(6) Convergence absolute criterion: 10^{-3}
(7) Number of meshes: 5.37 million

After calculation, the flow field of the design scheme of the lower chamber is obtained, as shown in Figure 11.

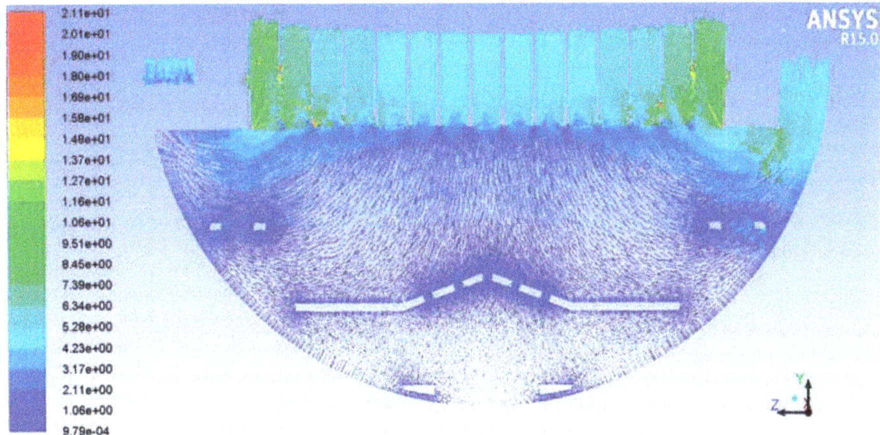

Figure 11. Velocity vectors of cross section of lower chamber. The velocity of the coolant around the upper surface of the core support plate is slightly greater than that in the middle. Although the velocity is not exactly the same on the upper surface, the velocity of the coolant in the middle is basically the same. What is more, the change of velocity consistency from the middle to the periphery is very small. This proves the effectiveness of the flow distribution effect of the design scheme. The direction of coolant flow in the lower chamber is stable, the variation degree of the flow direction is slow, and there is only a few of vortices. Structurally, the structure of the lower chamber is simplified by reducing the number of supporting columns and detachable connectors. Therefore, the design scheme has the characteristics of good flow distribution effect, a few of vortices, and simple structures.

6. Conclusions

In this paper, according to the characteristics of incremental innovation problems, the design core problems and the corresponding solving strategy were obtained based on the hierarchical representation model of the design problem. The main work included the following:

(1) A hierarchical representation model of the design problem was established, including customer needs, bad parameters, and root causes. It is based on the interaction among them, and can help designers to determine design core problems.

(2) This paper took into account the network layer and control layer in the hierarchical representation model of the design problem. Then, a method based on the network analytic hierarchy process was proposed to rank the importance of the root causes of the design problem and determine the core problems.

(3) Based on the core problems and the importance of the root causes of the bad parameters in the product system, an incremental product design process adopting TOC and TRIZ tools is established.

(4) The effectiveness of the proposed method was verified by a specific application in the design for a coolant flow distribution device in the lower chamber of a third-generation pressurized water reactor.

In order to better apply the proposed method to practical design activities, several aspects still need to be addressed: (1) improving the reasoning mechanism of design knowledge in the process of solving design problems and establishing a corresponding design knowledge base; and (2) developing a corresponding computer-aided design system based on the determination of core problems.

Author Contributions: Resources, Y.L. (Yin Luo) and Y.L. (Yan Li); Writing—original draft, Y.X.; Writing—review and editing, W.Q.L. and S.L.

Funding: This work was supported by the National Natural Science Foundation of China (grant number 51435011), the Science and Technology Ministry Innovation Method Program, China (grant number 2017IM040100), and Sichuan university-Luzhou strategic cooperation project (2017CDLZ-G11).

Acknowledgments: The authors would like to thank the anonymous referees and academic editor for their very valuable comments.

Conflicts of Interest: The authors declare no conflict of interest.

References

1. Mugge, R.; Dahl, D.W. Seeking the Ideal Level of Design Newness: Consumer Response to Radical and Incremental Product Design. *J. Prod. Innovat. Mang.* **2013**, *30*, 34–47. [CrossRef]
2. Wood, K.L.; Jensen, D.; Bezdek, J. Reverse engineering and redesign: Courses to incrementally and systematically teach design. *J. Eng. Educ.* **2001**, *90*, 363–374. [CrossRef]
3. Parsons, G. *The Philosophy of Design*, 1st ed.; Polity Press: Cambridge, UK, 2016.
4. Li, Y.; Li, X.L.; Zhao, W. Research on product creative design with cognitive psychology. *Comput. Integr. Manuf. Syst.* **2005**, *9*, 1201–1207.
5. Ko, Y.; Chen, M.; Yang, C. Modelling a contradiction-oriented design approach for innovative product design. *Proc. Inst. Mech. Eng. B-J. Eng.* **2014**, *229*, 199–211. [CrossRef]
6. Li, W.; Li, Y.; Wang, J. The process model to aid innovation of products conceptual design. *Expert Syst. Appl.* **2010**, *37*, 3574–3587. [CrossRef]
7. Cho, H.; Park, J. Cost-effective concept development using functional modeling guidelines. *Robot Comput. Integr. Manuf.* **2019**, *55*, 234–249. [CrossRef]
8. Li, T. RETRACTED ARTICLE: Applying TRIZ and AHP to develop innovative design for automated assembly systems. *Int. J. Adv. Manuf. Technol.* **2010**, *46*, 1–4. [CrossRef]
9. Vinodh, S.; Kamala, V.; Jayakrishna, K. Integration of ECQFD, TRIZ, and AHP for innovative and sustainable product development. *Appl. Math. Model.* **2014**, *38*, 2758–2770. [CrossRef]
10. Lucchetta, G.; Bariani, P.F.; Knight, W.A. Integrated Design Analysis for Product Simplification. *CIRP. Ann.-Manuf Technol.* **2005**, *54*, 147–150. [CrossRef]
11. Caligiana, G.; Liverani, A.; Francia, D. Integrating QFD and TRIZ for innovative design. *J. Adv. Mech. Des. Sys. Manuf.* **2017**, *11*, JAMDSM0015. [CrossRef]
12. Liu, X.Z.; Qi, G.N.; Fu, J.Z. A design process model of integrated morphological matrix and conflict resolving principles. *J. Zhejiang Univ.-SCI.* **2012**, *12*, 2243–2251.
13. Li, M.; Ming, X.; Zheng, M. An integrated TRIZ approach for technological process and product innovation. *Proc. Inst. Mech. Eng. B-J. Eng.* **2015**, *231*, 1062–1077. [CrossRef]
14. Ko, Y. Modeling a hybrid-compact design matrix for new product innovation. *Comput. Ind. Eng.* **2017**, *107*, 345–359. [CrossRef]
15. Stratton, R.; Mann, D. Systematic innovation and the underlying principles behind TRIZ and TOC. *J. Mater. Process Technol.* **2003**, *139*, 120–126. [CrossRef]
16. Nahavandi, N.; Parsaei, Z.; Montazeri, M. Integrated framework for using TRIZ and TOC together: A case study. *Int. J. Bus. Innov. Res.* **2011**, *5*, 309–324. [CrossRef]
17. Huang, W.; Hou, L.; Zhao, N. Product Innovation and Evaluation Based on TOC and TRIZ. *Adv. Mater. Res.* **2012**, *421*, 709–712. [CrossRef]
18. Hua, Z.; Gu, L.; Wang, W. TOC & TRIZ based product design method and its application. *Comput. Integr. Manuf. Syst.* **2006**, 817–822. [CrossRef]
19. Huang, S.; Liu, X.; Ai, H. Research on application of process model for product concept creative design based on TRIZ and TOC. *Int. J. Interact. Des. Manuf.* **2017**, *11*, 957–966. [CrossRef]
20. Zhang, J.; Liang, R.; Hao, B. The Problem Flow Network Building and Solving Process Model for Complex Product. *Chin. J. Mech. Eng.* **2018**, 1–15. [CrossRef]
21. Wei, Z.; Tan, R.; Ma, L. Research on complex problem analysis in TRIZ. In Proceedings of the IEEE 2008 International Conference on Management of Innovation and Technology, Bangkok, Thailand, 21–24 September 2008. [CrossRef]
22. Shu, M.; Cheng, C.; Chang, J. Using intuitionistic fuzzy sets for fault-tree analysis on printed circuit board assembly. *Microelectron. Reliab.* **2006**, *46*, 2139–2148. [CrossRef]
23. Hiraoka, Y.; Murakami, T.; Yamamoto, K. Method of Computer-Aided Fault Tree Analysis for High-Reliable and Safety Design. *IEEE Trans. Reliab.* **2016**, *65*, 687–703. [CrossRef]

24. Morello, M.G.; Cavalca, K.L.; Silveira, Z.D.C. Development and reduction of a fault tree for gearboxes of heavy commercial vehicles based on identification of critical components. *Qual. Reliab. Eng. Int.* **2010**, *24*, 183–198. [CrossRef]
25. Garcia, P.A.A.; Schirru, R.; Frutuoso, E.; Melo, P.F. A fuzzy data envelopment analysis approach for FMEA. *Prog. Nucl. Energy* **2005**, *46*, 359–373. [CrossRef]
26. Azadeh, A.; Sheikhalishahi, M.; Aghsami, A. An integrated FTA-DFMEA approach for reliability analysis and product configuration considering warranty cost. *Prod. Eng. Res. Devel.* **2015**, *9*, 635–646. [CrossRef]
27. Zhai, G.; Zhou, Y.; Ye, X. A method of multi-objective reliability tolerance design for electronic circuits. *Chin. J. Aeronaut.* **2013**, *26*, 161–170. [CrossRef]
28. Mocko, G.M.; Paasch, R. Incorporating Uncertainty in Diagnostic Analysis of Mechanical Systems. In Proceedings of the ASME 2002 International Design Engineering Technical Conferences and Computers and Information in Engineering Conference, Montreal, QC, Canada, 29 September–2 October 2002. [CrossRef]
29. Yazdani, A.; Tavakkoli-Moghaddam, R. Integration of the fish bone diagram, brainstorming, and AHP method for problem solving and decision making—A case study. *Int. J. Adv. Manuf. Technol.* **2012**, *63*, 651–657. [CrossRef]
30. Zhang, X.; Auriol, G.; Eres, H. A prescriptive approach to qualify and quantify customer value for value-based requirements engineering. *Int. J. Comput. Integr. Manuf.* **2013**, *26*, 327–345. [CrossRef]
31. Shimomura, Y.; Nemoto, Y.; Ishii, T. A method for identifying customer orientations and requirements for product–service systems design. *Int. J. Prod. Res.* **2017**, 1–11. [CrossRef]
32. Chen, Y.; Zhao, M.; Xie, Y. A new model of conceptual design based on Scientific Ontology and intentionality theory. Part II: The process model. *Des. Stud.* **2015**, *38*, 139–160. [CrossRef]
33. Thomas, L.S.; Luis, G.V. *Decision Making with the Analytic Network Process*; Springer: Berlin, Germany, 2013.
34. Huang, Y.; Bian, L. A Bayesian network and analytic hierarchy process based personalized recommendations for tourist attractions over the Internet. *Expert Syst. Appl.* **2009**, *36*, 933–943. [CrossRef]
35. Yüksel, İ.; Dag˘deviren, M. Using the analytic network process (ANP) in a SWOT analysis—A case study for a textile firm. *Inform Sci.* **2007**, *177*, 3364–3382. [CrossRef]
36. Jeong, J.H.; Han, B. Coolant flow field in a real geometry of PWR downcomer and lower plenum. *Ann. Nucl. Energy* **2008**, *35*, 610–619. [CrossRef]
37. Zhang, H.; Liu, H.; Fang, C. Optimization Design of Reactor Lower Plenum. *Nucl. Power Eng.* **2014**, 59–63. [CrossRef]
38. Yan, B.H.; Zhang, G.; Gu, H.Y. CFD analysis of the effect of rolling motion on the flow distribution at the core inlet. *Ann. Nucl. Energy* **2012**, *41*, 17–25. [CrossRef]
39. Piancastelli, L.; Gatti, A.; Frizziero, L. CFD analysis of the Zimmerman's V173 stol aircraft. *Asian Res. Publ. Netw. (ARPN) J. Eng. Appl. Sci.* **2015**, *10*, 8063–8070.

Article

A Generalised Bayesian Inference Method for Maritime Surveillance Using Historical Data

Jia Li [1], Xiumin Chu [1], Wei He [2,*], Feng Ma [1], Reza Malekian [3] and Zhixiong Li [4]

[1] Intelligent Transport System Research Center, Wuhan University of Technology, Wuhan 430068, China; my13638682566@163.com (J.L.); chuxm@whut.edu.cn (X.C.); matin7wind@163.com (F.M.)
[2] College of Marine Sciences, Minjiang University, Fuzhou 350108, China
[3] Department of Electrical, Electronic & Computer Engineering, University of Pretoria, Pretoria 0002, South Africa; reza.malekian@ieee.org
[4] School of Mechanical, Materials, Mechatronic and Biomedical Engineering, University of Wollongong, Wollongong, NSW 2522, Australia; zhixiong_li@uow.edu.au
* Correspondence: hewei11@mju.edu.cn; Tel.: +86-186-069-90698

Received: 30 November 2018; Accepted: 31 January 2019; Published: 8 February 2019

Abstract: In practice, maritime monitoring systems rely on manual work to identify the authenticities, risks, behaviours and importance of moving objects, which cannot be obtained directly through sensors, especially from marine radar. This paper proposes a generalised Bayesian inference-based artificial intelligence that is capable of identifying these patterns of moving objects based on their dynamic attributes and historical data. First of all, based on dependable prior data, likelihood information about objects of interest is obtained in terms of dynamic attributes, such as speed, direction and position. Observations on these attributes of a new object can be obtained as pieces of evidence profiled as probability distributions or generally belief distributions if ambiguity appears in the observations. Using likelihood modelling, the observed pieces of evidence are independent of the prior distribution patterns. Subsequently, Dempster's rule is used to combine the pieces of evidence under consideration of their weight and reliability to identify the moving object. A real world case study of maritime radar surveillance is conducted to validate and prove the efficiency of the proposed approach. Overall, this approach is capable of providing a probabilistic and rigorous recognition result for pattern recognition of moving objects, which is suitable for any other actively detecting applications in transportation systems.

Keywords: Dempster's rule; evidence distance; pattern recognition; maritime surveillance

1. Introduction

In diverse transportation systems, the concerns about management efficiency and public safety bring forward high request to sensor networks [1–3]. To obtain more detailed and real-time information of traffic objects, varieties of powerful sensors have been invented and equipped, leading to information explosions [4–6]. For example, IMO (International Marine Organization) demands all the ports and vessels to equip Radar, AIS (Automatic Identification System) and satellite monitoring systems [7]. It can be inferred that the operators of transportation systems have to deal with much information, which requires them to choose, filter and determine useful data [8]. In practice, there are limitations respectively in different kinds of sensors. Meanwhile, it is widely recognised that information from multisources will conflict frequently with each other. Hence, the radar systems are still the most reliable tool for the maritime surveillance in vessel transportation systems [9].

However, radars generally provide multiple objects indiscriminately. Even worse, the sensors often misunderstand the behaviours of targets, leading to management difficulties [10]. To address this issue, sensor manufactures have paid enormous effort to improve the sensitivity and the capability of noise suppression. For example, the late-model marine radar is capable of tracking a 0.5-square metre target at a distance of 5 miles [9]. However, the differences between a drowning person, a clump of sea-grass and a canoe cannot be obtained from radars [11]. In fact, high sensitivity will bring more false and useless information. Therefore, most radar monitoring systems need manual assistant to identify moving targets, such as whether the corresponding object is a moving or anchoring [12]. Nevertheless, in most situations where many nondistinctive false or unimportant targets appear on the screen will distract the attention of supervisors. In the increasingly crowded harbours, the manual identification becomes impractical. For instance, downstream of the Yangtze Rivers, there might be 20,000 vessels passing through during peak hours. It is impossible to ensure that any single radar object is inspected manually [13,14]. Therefore, it is practical to propose an intelligent approach to identify the patterns of objects, which will reduce the burden of supervisors observably and improve the safety.

It is possible to establish the correlations between dynamic attributes and the patterns of moving targets based on historical data [15]. Hence, a probabilistic inference is appropriate for identifying the objects [16]. It is worth noting that the prior distribution of patterns is generally an indispensable part in conventional probabilistic inference process [17]. However, it is very difficult to estimate or observe in practice. For instance, in coastal radar surveillance, blips caused by noises might confuse the supervisors and need further identification on their authenticity. However, the proportion of false targets is erratic and will be affected by weather, blocks and other factors. Therefore, in the former research, the assumptions of prior distribution have to be made before the probabilistic inference process [18]. Actually, supervisors normally do not take such a proportion or prior distribution into consideration, but would like to identify the blips with the correlations between authenticities and dynamic attributes such as speed, course and location. For instance, a vessel is barely allowed to operate at a speed higher than 30 knots in the Yangtze Rivers. As a result, the speed, trajectory or other attributes can be used to build the correlations through data mining. However, very limited work has been done to address this issue by building a relationship between the attributes and patterns of moving objects without prior pattern distributions.

In order to bridge this research gap, a new method is proposed in this work using the Evidential Reasoning (ER) rule [19] to model the correlations the attributes and patterns of moving objects and make the conjunctive inference. Different from conventional Bayesian inference, the ER rule no longer needs a prior pattern distribution for inference as it constitutes a likelihood-based inference process [20]. Using the likelihood-based inference process, multiple pieces of evidence can be generated from multisource-verified samples and human experts. Meanwhile, the ER rule provides a probabilistic process for the conjunctive combination of independent evidence, while the quality of data can be taken into account in the form of evidence reliability. Experimental data acquired from the Yangtze Rivers have been used to demonstrate the effectiveness of the proposed method for maritime surveillance.

The rest of the paper is organised as follows. Section 2 reviews the related work. Section 3 describes the proposed approach. In Section 4, a case study is conducted to validate the proposed approach. Section 5 concludes this paper.

2. Related Work

As discussed, the dynamic attributes, including speed, course and trajectory, are the kernel evidence for manual identification. These attributes are widely studied in pattern recognition for transportation systems. In general, the recognition methods can be divided into two categories: nonprobabilistic inference and probabilistic inference.

2.1. Nonprobabilistic Inference Identification

Nonprobabilistic inference relies on the attributes in vector space or self-adapted classification model, which is not based on probability theory. Generally, it is more widely used. Typical algorithms include fuzzy k-means (FCM) [21], k Nearest Neighbour (KNN) [22], principal component analysis (PCA) [23], support vector machine (SVM) and artificial neural network (ANN) [24].

The k-means is popular for cluster analysis in data mining. It divides the observations into k clusters in which each observation belongs to the cluster with the nearest mean, serving as a prototype of the cluster. In FCM observation can belong to more than one cluster. Ma et al. [10] proposed a method on classification to identify the false targets in maritime radar surveillance using FCM, and the accuracy reached 91.0%. Tang et al. [21] used FCM to identify missing data in traffic perception. However, the problem of FCM recognition is that the model could not give the probabilities to individual groups and there is no clear meaning for distances in vector space. Different from FCM, KNN does not specify the edges of different groups, but use the distance to nearest neighbour points as the basis for classification. Zheng et al. [22] uses KNN-enhanced constrained linearly sewing principle to recognize the traffic pattern and predict the traffic flow. It found that KNN is suitable for datasets with incomplete information.

Generally, more attributes would lead better pattern recognition. However, too many attributes would increase the computation difficulty. Therefore, PCA is proposed to address this problem by using an orthogonal transformation to convert a set of possibly correlated variables into a set of linearly uncorrelated variables. Jin et al. [23] used an improved PCA-based method to identify abnormal traffic flow pattern isolation and loop detector fault detection.

SVM adopts supervised learning to analyse data and recognize patterns. Although SVM and its improved algorithms have proven to be effective in transportation research, the problem is that the classification in vector space is based on morphological characteristics which have no physical meaning to the real world. In addition to SVM, ANN is another nonprobabilistic method for pattern recognition in transportation research. Srinivasana et al. [25] proposed an ANN model based on reduced multivariate polynomial pattern classifier for freeway incident detection. With enough training samples, ANN could solve the best solution and be able to restrain the noise efficiently. Different from SVMs, ANN does not need a vector space for illustration and modelling.

In summary, nonprobabilistic recognition algorithms are efficient for applications with complete prior probability in transportation systems. However, it is difficult to deal with uncertainties using nonprobabilistic recognition algorithms.

2.2. Probabilistic Inference Identification

In traffic object detection, all the dynamic attributes are quantitative data, which can be acquired from various sensor systems. Therefore, the probabilistic inference identification is feasible in transportation systems. The commonly used methods include Gaussian mixture model (GMM), Bayes' rule, Dempster's rule, Evidence Combine rule (ECR), probability box and ER rule. GMM refers to a mixture model that is a probabilistic model for representing the presence of subpopulations within an overall population. With a specific prior distribution discovered from historical data, any observation could be defined into groups according to probability [14]. In the circumstances with incomplete prior data GMM assumes all the random process obeys indeterminate Gaussian distribution. Xia et al. [6] proposed a GMM for vehicle segmentation for traffic flow estimation.

Bayes' rule is suitable for applications with abundant and sufficient prior samples. However, Bayes' rule strictly requires that the elements in discrimination framework must be independent and all the source information should be reliable and equally treated. Furthermore, the prior distribution in the discrimination framework should be stable. Bayesian Network (BN) can overcome these problems and is highly accepted in the uncertain inference [26]. Zhang et al. [16] used BN to reason the block risk in the Yangtze River. The BN provides a conjunctive probabilistic inference, and keep the consistency with Bayes' rule. Besides BN, Dempster's rule is another method for probabilistic inference, which constitutes a conjunctive probabilistic process. ECR is a derivative of Dempster's rule [27] and is used to address the paradoxes and counterintuitive problems, such as the Yager's rule, Dubois and Prade's rule [28]. The literature review indicates that the probabilistic inference methods are suitable for traffic object identification if multisource prior information is provided [29–32]. The combination of Bayesian reference logic and Dempster's rule is appropriate to address the characteristic discrimination problem in traffic objects. To our best knowledge, the Bayesian reference based Dempster's rule method has not been found to model the uncertainty and detect moving objects in transportation systems.

3. The Proposed Approach

3.1. Likelihood Modelling

To satisfy the conceivable dependency of propositions, Dempster's rule is investigated on a power set $P(\Theta)$. Suppose $\Theta = \{\theta_1, \cdots, \theta_N\}$ is the set of mutually exclusive and collectively exhaustive propositions, with $\theta_i \cap \theta_j = \phi$ $(i, j = 1, \cdots, N)$, where ϕ denotes an empty set and Θ refers to a frame of discernment. A Basic Probability Assignment (pba) is a function m: $2^{\Theta} \to [0, 1]$, satisfying

$$m(\phi) = 0, \quad \sum_{\theta \subseteq \Theta} m(\theta) = 1 \tag{1}$$

Let 2^{Θ} or $P(\Theta)$ represent the power set of Θ,

$$P(\Theta) = 2^{\Theta} = \{\phi, \theta_1, \cdots, \theta_N, \{\theta_1, \theta_2\}, \cdots, \{\theta_1, \theta_N\} \cdots, \{\theta_1, \cdots, \theta_{N-1}\}, \Theta\} \tag{2}$$

then $m(\theta)$ denotes a basic probability mass that is assigned exactly to a proposition θ. Therefore, the unions of any propositions have been considered as individual elements in discrimination.

The pattern recognitions on traffic objects include authenticity, importance, risk and behaviours. As discussed in the introduction, in order to imitate the manual work, the initial discrimination framework can be modelled. Let X be a specific characteristic,

$$X = \{x_1, x_2, \ldots, x_n\} \tag{3}$$

The basic concept is to classify the targets into the discrete groups and evaluate the probabilities to individual classifications. Based on the power set, the discrimination framework transforms to

$$P(\Theta) = 2^{\Theta} = \{\phi, x_1, \cdots, x_n, \{x_1, x_2\}, \cdots, \{x_1, x_N\} \cdots, \{x_1, \cdots, x_{N-1}\}, \Theta\} \tag{4}$$

Because of empty set ϕ, most of the unions of elements, $\{x_1, x_2\}, \cdots, \{x_1, x_N\} \cdots, \{x_1, \cdots, x_{N-1}\}$ are meaningless. Therefore, the investigated discrimination framework is

$$P'(\Theta) = \{x_1, \cdots, x_n, \Theta\} \tag{5}$$

In particular, linear identification problems can be discretised, making it compatible with Equation (5). The actively detecting tool in maritime surveillance is able to obtain the target dynamic attributes, including speed, course, location, acceleration, image size and so on. The objective of likelihood modelling is to find out the correlations between such attributes and the probabilities to specific classification of a characteristic. For example, when a target is moving at the speed of 100 km/h in the harbour, the probability of being a real vessel is negligible. Based on this concept, the verified historical data could be represented as Table 1. There are m separated values on an attribute offered by m sensors. Let y_i^j be the frequency of a singleton value i in specific characteristic j, $(i = 1, 2, \ldots, m, j = 1, 2, \ldots, n)$, Q^j represents the sum of sample in classification j.

Table 1. Verified sample on characteristic.

Classification on X	Verified Sample Observation Attribute Value					Total
	Value 1	...	Value i	...	Value m	
x_1	y_1^1	...	y_i^1	...	y_m^1	$Q^1 = \sum_{i=1}^{m} y_i^1$
$\vdots$	$\vdots$	$\ddots$	$\vdots$	$\ddots$	$\vdots$	$\vdots$
x_j	y_1^j	...	y_i^j	...	y_m^n	$Q^j = \sum_{i=1}^{m} y_i^j$
$\vdots$	$\vdots$	$\ddots$	$\vdots$	$\ddots$	$\vdots$	$\vdots$
x_n	y_1^n	...	y_i^n	...	y_m^n	$Q^n = \sum_{i=1}^{m} y_i^n$
$\Theta = \{x_1, \dots, x_n\}$	y_1^{Θ}	...	y_i^{Θ}	...	y_m^{Θ}	$Q^{\Theta} = \sum_{i=1}^{m} y_i^{\Theta}$

It is worth noting that the distributions of attributes in different groups could be observed from multiple sources. In maritime management, AIS uses GPS to get reliable positioning information. In the conventional Bayes' rule-based inference, AIS data is used to model conditional probabilities. However, the conditional probabilities require the prior distributions of $\{x_1, \dots, x_n, \Theta\}$, which might not be appropriate. To this end, a likelihood modelling method is proposed, which is independent of prior distributions of $\{x_1, \dots, x_n, \Theta\}$. Based on different attributes, it is feasible to build up several pieces of evidence to identify the targets. Actually, the dynamic attributes only relate to the inherent features of moving object and regulations. In this light, based on Table 1 the likelihood (probability) of attribute value to specific classification of characteristic can be presented in a new way, as shown in Table 2 and Equations (6) and (7).

$$c_i^j = y_i^j / Q^j; \text{ for } i = 1, 2, \dots, m, \; j = 1, 2, \dots, n, \; \Theta. \tag{6}$$

$$p_i^j = c_i^j / (c_i^{\Theta} + \sum_{k=1}^{n} c_i^k); \text{ for } i = 1, 2, \dots, m, \; j = 1, 2, \dots, n \tag{7}$$

where c_i^j denotes the likelihood to which the value i is expected to occur given that the classification x_j is true and p_i^j denotes the likelihoods to the normalization form among $x_1, \dots, x_n$. The prior distribution of a characteristic can be treated as a normal piece of evidence [20]. Whether to use it in the conjunctive inference depends on its stability.

Table 2. Likelihoods without classification prior distribution.

Classification on X	Verified Sample Observation Attribute Value Likelihood				
	Value 1	...	Value i	...	Value m
x_1	c_1^1	...	c_i^1	...	c_m^1
...	...	...	...	...	...
x_j	c_1^j	...	c_i^j	...	c_m^j
...	...	...	...	...	...
x_n	c_1^n	...	c_i^n	...	c_m^n
$\Theta = \{x_1, \dots, x_n\}$	c_1^{Θ}	...	c_i^{Θ}	...	c_m^{Θ}

3.2. Conjunctive Inference with Dempster's Rule

With proper likelihood model, all the dynamic attributes could be mapped into pieces of evidence. Since all the dynamic attributes of each radar target are varying randomly and almost independently, the pieces of evidence can be regarded as approximately independent component. Hence, in this research, Dempster's rule is applicable in the evidence combination [14].

Assume a target located at the position (x_k, y_k), its probability of being a real moving vessel is estimated based on Section 3.1, in accordance with three pieces of evidence (i.e., speed, course and location). Dempster's rule can be used to combine the three pieces of evidence as

$$m(\theta) = [m_1 \oplus m_2] = \begin{cases} 0 & \theta = \varnothing \\ \frac{\sum_{B \cap C = \theta} m_1(B)m_2(C)}{1 - \sum_{B \cap C = \phi} m_1(B)m_2(C)} & \theta \neq \varnothing \end{cases} \tag{8}$$

4. A Case Study

To validate the method proposed in this paper, a case study on a waterway management is carried out using coastal surveillance radar to identify passing vessels.

4.1. Experimental Platform and Discrimination Framework

In port management, the supervisors have to face crowded vessels and narrow waterways, which put forward a high request for surveillance. Although several monitoring systems have been equipped on vessels requested by IMO, the radar is the most credible tool. However, a non-ignorable part in the radar systems is the effect of multipath, waves and noise, which leads to false probability distribution. In practice, the supervisors generally identify the targets by dynamic attributes measured by radar, especially for the speed, course and trajectory. Another monitoring system is AIS. The supervisors are able to use the AIS directly through database sharing systems. Therefore, it is logical to investigate the correlations in the AIS record and identify the authenticity of targets.

An experiment platform has been built at the Yangtze River, Wuhan, which contains X-band maritime radar with ARPA (Automatic Radar Plotting Aid) function, and three individual AIS receivers to provide historical data, as shown in Figure 1.

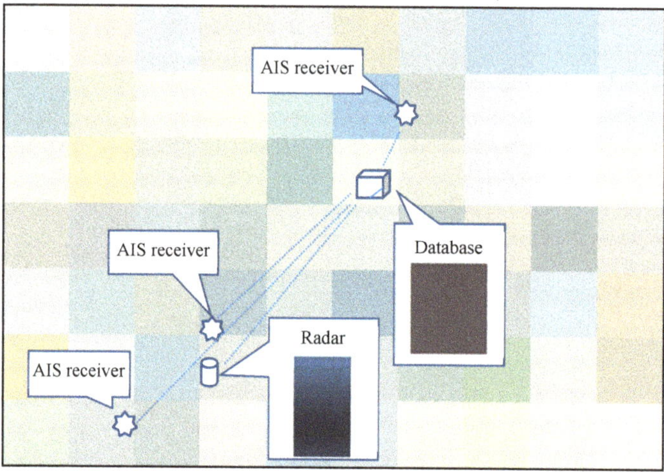

Figure 1. Experiment platform.

Three experiments have been conducted. Experiment ①: The AIS receivers obtained 13,087,777 messages from 2014-04-12 17:11:45 to 2014-06-07 11:31:01 between the lines of latitude 30.1244N and 30.4566N, which is also the capturing area of the ARPA radar. To clarify, during the experiment, the external environment is steady. Since only the real vessels would report AIS messages, all the AIS targets could be considered as verified vessel samples.

Different from real vessels, the noise targets would not report information. Moreover, the noise targets are generally cluttered with real vessels. Therefore, the verified noise ARPA targets are unattainable in normal circumstances. To address such problem, the paper chose a special situation to obtain verified noise targets as followed.

Experiment ②: From 2014-06-07 10:31:00 to 2014-06-07 12:30:59, the measured waterway was temporally forbidden. During this time, the ARPA captured 19,549 records which include 120 individual false targets. As the noise targets normally are caused by the multipath effects and background noise, which do not correlate to the real vessels; an assumption could be made that the 2 h noise sample represents the generic distribution characteristics of noise speed, course.

Experiment ③: A second round of radar testing is running from 13:35 to 16:65 on 10th November 2013, which is proceed for method validation. In total, 55 targets have been captured and verified, which include 22 vessels and 33 noises. Particularly, the verification on targets is very difficult, as the telescope observation is the only way that affected by the visibility and weather. Only if the target was sailing nearby the known buoys was the position conjecturable. Therefore, although more than 500 hundred targets were captured, only 55 targets had been verified.

This paper aims to use Dempster's rule to identify the real targets from the ARPA targets offered by the radar. Therefore, the discrimination framework is {TRUE, FALSE, UNKNOWN}.

4.2. Step 1: Object Attribute Likelihood Modelling

Experiment ① and ② provided the verified samples of vessel and noise targets to model likelihood. Consequently, the speed and course normalised likelihoods to vessel and noise are represented in Figures 2 and 3. The X axis represents the speed and course values and the Y axis represents the normalised likelihoods given by Table 2 and Equation (6).

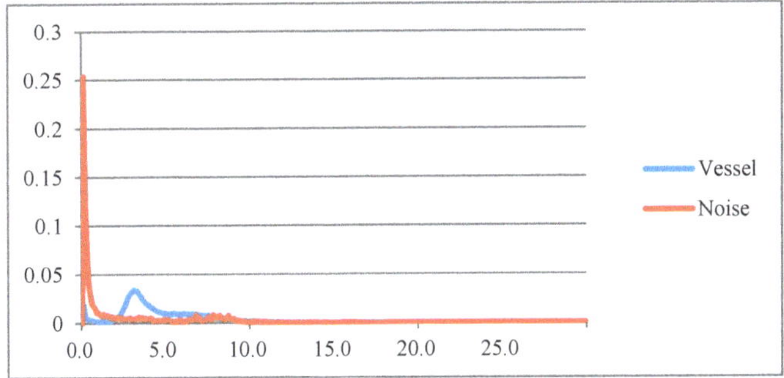

Figure 2. Verified noised and vessel objects' normalised likelihoods on speed.

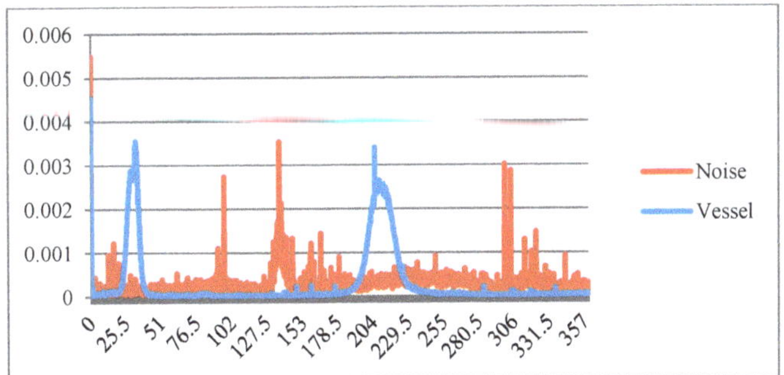

Figure 3. Verified noised and vessel objects' normalised likelihoods on course.

As per Figures 2 and 3, the speed and course distribution of vessels and noise follows certain patterns. To save fuel in the specific area where the experiments were conducted, each vessel normally maintained a stable speed; therefore, the speed distribution conforms to a regular pattern that is determined by the actual operation. However, because false targets are caused by factors such as noise and multipath effects, the speed distribution of noise ARPA targets would conform to a another pattern determined by the ARPA algorithm embedded in radars.

Similarly, such pattern differences could be found in course evidence. In Figure 4, the two dash circles marked two major courses of vessels, which are exactly the upstream and downstream courses. On contrary, there is no particular pattern in noise target courses, and the courses distribute randomly.

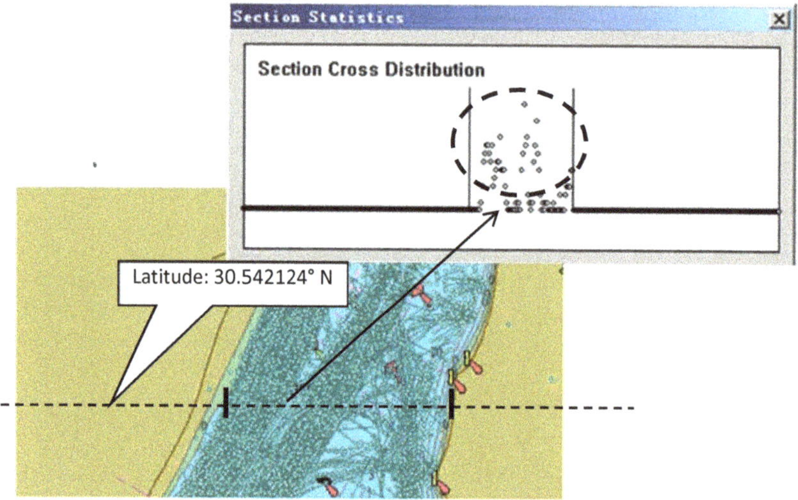

Figure 4. 2,555 Vessels' Paths and Distributions on the some Latitude section.

Different from the speed and course evidence, the location is a two dimensional variable, which contains latitude and longitude. To analyse it, the paper divided the waterway of Figure 4 into cross-sections. Since the studied waterway is south–north direction, using the latitude lines as the cross-sections is a practical option. A typical cross-section of vessel densities is represented in Figure 4. The dash line denotes the studied cross line, and the bold lines marked the waterway boundaries. The vessel distribution on this section is represented in the dialog box, which is marked in the dash circle. Another problem is how to estimate the distribution of noise targets. The experiment ② has not provided enough samples to model distributions in this section, meanwhile the occurrence principles of noise ARPA targets are difficult to estimate. Hence, the distribution of noise on sections is assumed to be normal. Another noteworthy factor is that there are some places in the waterway where no vessel has passed in records, but there is still a chance of some vessels passing through. In this instance, on such locations, a {UNKOWN} proposition is activated.

4.3. Step 2: Dempster's Rule-Based Conjunctive Inference

To validate the methods, software on Windows 7 has been developed and shown in Figure 5. In this dialog, the ARPA targets are marked as green circles, the bold lines denote the directions, the right bottom number denotes the targets number and the right top number denotes the probabilities of being a real vessel provided by Dempster's rule. It is worth noting that there are two values separated by ':'. Taking the samples from experiment ③ as the research objects, the recognition result is presented in Table 3.

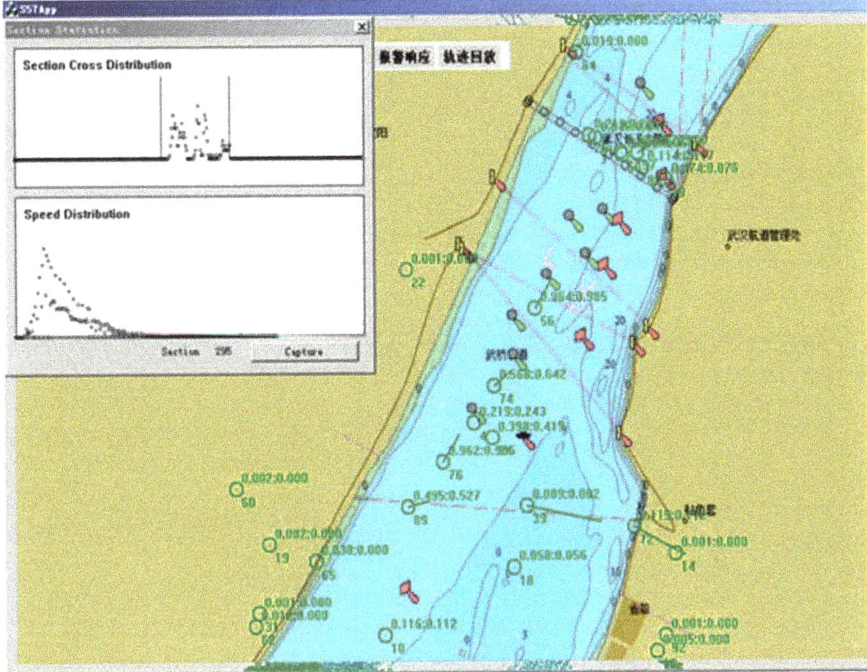

Figure 5. The Experiment.

Table 3. Confusion matrix.

Verified Observations	Vessel Probability $\geq$ 50%	Noise Probability $\geq$ 50%	Accuracy
vessel	139	21	86.88%
noise	118	1355	91.99%

The total accuracy is 91.43%. In similar research, the improved Fuzzy C Means algorithm can recognize the false targets with an accuracy of 82.0% under same sample [10], which is also lower than the accuracy when the method proposed in this work is used. The proposed approach has been proved to be efficient.

5. Conclusions and Discussion

Moving targets captured by actively detecting tools need further identification, which includes risk, authenticity, importance and behaviours. The paper proposes a likelihood modelling and Dempster's rule-based inference approach to identify targets' patterns based on dynamic attribute values. The conclusions are below.

1. In conventional probabilistic inference, the prior distribution of discrimination framework has been incorporated into pieces of evidence. However, such prior distribution is generally erratic or even not existing at all in transportation systems.
2. The likelihoods to propositions of moving object dynamic attributes are steady and stable, and only relates to the inherent characteristics of objects or local regulations. Based on this, the likelihood modelling in this paper is more persuasive.

With the help of Dempster's rule, it is possible to build artificial intelligence to identify the characteristics of moving targets in transportation systems and improve the management. However, there are several problems that should be discussed in the future.

1. The pieces of evidence might have correlations to each other, which does conform to the preconditions of Dempster's rule. For example, when a vessel is passing through a bridge, there is a speed recommended by local supervisors. In this instance, there is an obvious connection between the speed and position. To make the model practical and rigorous, the correlations among the pieces of evidence should be taken into consideration.
2. Furthermore, it is very practical to develop an interactive optimization in response to the feedback from supervisors.

In maritime management, consider an extreme case where an accident occurred in a waterway, leading to a blockage. Suppose that no vessel is sailing, all of the targets are false in radar observation, and the authenticity prior distribution is definitely invalid. The error rate in radar is erratic and affected by many factors, which means the prior distribution of false targets might not exist at all. The assumption on the distribution in certain period could indicate illogical observation. Therefore, the valid evidence for automobile type identification have to be the size, plate colour, which have a steady correlation to automobile type that would not be changed easily; such evidence could be modelled as Table 2 and Equations (6) and (7).

In fact, if you take a deep look into manual judgment, it is easy to find out that the identification on characteristic is irrelevant to certain sample, only to the inherent properties of objects. In inland river waterways it is unauthentic for an object to move at 100 km/h. There are two different views. First, based on a certain sample, few vessels have sailed at this high speed, most high speed targets are noise. The other view is that, based on distribution of vessels, the vessels is not likely to have the ability to sail at this speed. Actually, the second one is more acceptable and generic. The likelihood modelling above is exactly the representation of the second view.

Author Contributions: J.L. conceived and designed the experiments; F.M. performed the experiments and wrote the paper; X.C. and Z.L. revised the paper; and W.H. and R.M. analysed the data.

Acknowledgments: This study is supported by the National Natural Science Foundation of China (Grant No. 51479155), the Fujian Province Natural Science Foundation (No: 2018J01506), the Young Teachers Education Research Project of Department of Education of Fujian Province (No. JK2017038, JAT170439), the 2017th Outstanding Young Scientist Training Program of Colleges in Fujian Province, the Fuzhou Science and Technology Planning Project (No: 2018-G-92, 2018-S-113) and Australia ARC DECRA (No. DE190100931). The authors would like to express their thanks to Wuhan Maritime Bureau for providing AIS data.

Conflicts of Interest: The authors declare no conflicts of interest.

References

1. Tian, Z.; Liu, F.; Li, Z.; Malekian, R.; Xie, Y. The development of key technologies in applications of vessels connected to the internet. *Symmetry* **2017**, *9*, 211. [CrossRef]
2. Ma, F.; Chen, Y.W.; Yan, X.P.; Chu, X.M.; Wang, J. A novel marine radar targets extraction approach based on sequential images and Bayesian Network. *Ocean Eng.* **2016**, *120*, 64–77. [CrossRef]
3. He, W.; Li, Z.; Malekian, R.; Liu, X.; Duan, Z. An internet of things approach for extracting featured data using AIS database: An application based on the viewpoint of connected ships. *Symmetry* **2017**, *9*, 186. [CrossRef]
4. Ma, F.; Chu, X.M.; Yan, X.P. Short message characteristics of AIS base stations. *Jiaotong Yunshu Gongcheng Xuebao* **2012**, *12*, 111–118.
5. Li, Z.; Jiang, Y.; Guo, Q.; Hu, C.; Peng, Z. Multi-dimensional variational mode decomposition for bearing-crack detection in wind turbines with large driving-speed variations. *Renew. Energy* **2018**, *116*, 55–73. [CrossRef]
6. Xia, Y.; Shi, X.; Song, G.; Geng, Q.; Liu, Y. Towards improving quality of video-based vehicle counting method for traffic flow estimation. *Signal Process.* **2016**, *120*, 672–681. [CrossRef]

7. Guerriero, M.; Willett, P.; Coraluppi, S.; Carthel, C. Radar/AIS data fusion and SAR tasking for maritime surveillance. In Proceedings of the International Conference on Information Fusion, Cologne, Germany, 30 June–3 July 2008; pp. 1–5.

8. Li, H.; Shen, Y.; Liu, Y. Estimation of detection threshold in multiple ship target situations with HF ground wave radar. *J. Syst. Eng. Electron.* **2007**, *18*, 739–744.

9. Lee, P.T.W.; Yang, Z. (Eds.) *Multi-Criteria Decision Making in Maritime Studies and Logistics*; Springer: Cham, Switzerland, 2018.

10. Ma, F.; Wu, Q.; Yan, X.; Chu, X.; Zhang, D. Classification of automatic radar plotting aid targets based on improved Fuzzy C-means. *Transp. Res. Part C Emerg. Technol.* **2015**, *51*, 180–195. [CrossRef]

11. Liu, G.P.; Yang, J.B.; Whidborne, J.F. *Multiobjective Optimisation and Control*; Research Studies Press: Baldock, UK, 2003.

12. Lin, B.; Huang, C.H. Comparison between ARPA radar and AIS characteristics for vessel traffic services. *J. Mar. Sci. Technol.* **2006**, *14*, 182–189.

13. Talavera, A.; Aguasca, R.; Galván, B.; Cacereño, A. Application of Dempster—Shafer theory for the quantification and propagation of the uncertainty caused by the use of AIS data. *Reliab. Eng. Syst. Saf.* **2013**, *111*, 95–105. [CrossRef]

14. Ma, F.; Chen, Y.W.; Huang, Z.C.; Yan, X.P.; Wang, J. A novel approach of collision assessment for coastal radar surveillance. *Reliab. Eng. Syst. Saf.* **2016**, *155*, 179–195. [CrossRef]

15. Smarandache, F.; Dezert, J.; Tacnet, J. Fusion of sources of evidence with different importances and reliabilities. In Proceedings of the 2010 13th Conference on Information Fusion (FUSION), Edinburgh, UK, 26–29 July 2010; pp. 1–8.

16. Zhang, D.; Yan, X.P.; Yang, Z.L.; Wall, A.; Wang, J. Incorporation of formal safety assessment and Bayesian network in navigational risk estimation of the Yangtze River. *Reliab. Eng. Syst. Saf.* **2013**, *118*, 93–105. [CrossRef]

17. Shafer, G.; Pearl, J. *Readings in Uncertain Reasoning*; Morgan Kaufmann Publishers Inc.: San Mateo, CA, USA, 1990.

18. Trucco, P.; Cagno, E.; Ruggeri, F.; Grande, O. A Bayesian Belief Network modelling of organisational factors in risk analysis: A case study in maritime transportation. *Reliab. Eng. Syst. Saf.* **2008**, *93*, 845–856. [CrossRef]

19. Yang, J.B.; Xu, D.L. ER rule for evidence combination. *Artif. Intell.* **2013**, *205*, 1–29. [CrossRef]

20. Yang, J.B.; Xu, D.L. A Study on Generalising Bayesian Inference to Evidential Reasoning. In *Belief Functions: Theory and Applications*; Springer: Cham, Switzerland, 2014; pp. 180–189.

21. Tang, J.; Zhang, G.; Wang, Y.; Wang, H.; Liu, F. A hybrid approach to integrate fuzzy C-means based imputation method with genetic algorithm for missing traffic volume data estimation. *Transp. Res. Part C Emerg. Technol.* **2015**, *51*, 29–40. [CrossRef]

22. Zheng, Z.; Su, D. Short-term traffic volume forecasting: A k-nearest neighbor approach enhanced by constrained linearly sewing principle component algorithm. *Transp. Res. Part C Emerg. Technol.* **2014**, *43*, 143–157. [CrossRef]

23. Jin, X.; Zhang, Y.; Li, L.; Hu, J. Robust PCA-based abnormal traffic flow pattern isolation and loop detector fault detection. *Tsinghua Sci. Technol.* **2008**, *13*, 829–835. [CrossRef]

24. Islam, T.; Rico-Ramirez, M.A.; Han, D.; Srivastava, P.K. Artificial intelligence techniques for clutter identification with polarimetric radar signatures. *Atmos. Res.* **2012**, *109*, 95–113. [CrossRef]

25. Srinivasan, D.; Sharma, V.; Toh, K.A. Reduced multivariate polynomial-based neural network for automated traffic incident detection. *Neural Netw.* **2008**, *21*, 484–492. [CrossRef]

26. Hossain, M.; Muromachi, Y. A Bayesian network based framework for real-time crash prediction on the basic freeway segments of urban expressways. *Accid. Anal. Prev.* **2012**, *45*, 373–381. [CrossRef]

27. Dempster, A.P. Upper and lower probabilities induced by a multivalued mapping. *Ann. Math. Stat.* **1967**, *38*, 325–339. [CrossRef]

28. Shafer, G. *A Mathematical Theory of Evidence*; Princeton University Press: Princeton, NJ, USA, 1976; Volume 1.

29. Howells, J. Innovation, consumption and services: Encapsulation and the combinatorial role of services. *Serv. Ind. J.* **2004**, *24*, 19–36. [CrossRef]

30. Kritayakirana, K.; Gerdes, J.C. Autonomous vehicle control at the limits of handling. *Int. J. Veh. Auton. Syst.* **2012**, *10*, 271–296. [CrossRef]

31. Petsios, M.N.; Alivizatos, E.G.; Uzunoglu, N.K. Solving the association problem for a multistatic range-only radar target tracker. *Signal Process.* **2008**, *88*, 2254–2277. [CrossRef]
32. Sun, S.; Fu, G.; Djordjević, S.; Khu, S.-T. Separating aleatory and epistemic uncertainties: Probabilistic sewer flooding evaluation using probability box. *J. Hydrol.* **2012**, *420*, 360–372. [CrossRef]

Article

Matching Model of Dual Mass Flywheel and Power Transmission Based on the Structural Sensitivity Analysis Method

Lei Chen [1] , Xiao Zhang [1], Zhengfeng Yan [2],* and Rong Zeng [3]

[1] Hubei Digital Manufacturing Key Laboratory, School of Mechanical and Electronic Engineering,
    Wuhan University of Technology, Wuhan 430070, China; chenlei811001@163.com (L.C.);
    zxkingxz@outlook.com (X.Z.)
[2] School of Automotive and Transportation Engineering, Hefei University of Technology, Hefei 230009, China
[3] Key Laboratory of Agricultural Equipment in Mid-lower Yangtze River, Ministry of Agriculture, College of
    Engineering, Huazhong Agricultural University, Wuhan 430070, China; zengrong@mail.hzau.edu.cn
* Correspondence: zf.yan@hfut.eud.cn

Received: 6 December 2018; Accepted: 22 January 2019; Published: 7 February 2019

Abstract: As a new torsional vibration absorber, the dual mass flywheel (DMF) contains a symmetric structure in which the damping element is a pair of springs symmetrically distributed along the circumference direction. Through reasonable matching parameters, the DMF functions in isolating torsional vibrations caused by the engine from the transmission system. Our work aims to solve the accuracy of matching models between the DMF and power transmission system. The critical structural parameters of each order modal are treated consecutively by two methods: Absolute sensitivity (e.g., under the idle condition and driving condition), and relative sensitivity. The operation achieves a separation of the parameters and diagnosis of the relationship between these parameters and the natural frequency in the system. In addition, the natural frequency range is determined based upon the area of the resonance speed. As a result, the matching model is established based on the sensitivity analysis method and the natural frequency range, which means the moment of inertia distribution (its coefficient should be used as one structural parameter in relative sensitivity analysis) and the torsional stiffness in multiple stages can be observed under the combined values. The effectiveness of the matching model is verified by experiments of a real vehicle test under the idling condition and driving condition. It is concluded that the analysis study can be applied to solve the parameters matching accuracy among certain multi-degree-of-freedom dynamic models.

Keywords: dual mass flywheel; absolute sensitivity; relative sensitivity; torsional vibration; spring

---

1. Introduction

As the automobile power output and transmission are linked, dynamic characteristics of power transmissions are an important factor in ride safety, fuel economy, and NVH (noise, vibration, and harshness) performance of vehicles. It is recognized that vibrations and noises are the most important indicators to evaluate the vehicle NVH performance [1]. Vehicle vibration noises can be caused by the power source, aerodynamics, tires, transmission system, and uneven loads and so on. Among them, power source vibration noises account for more than one half of vibration noises [2,3]. In fact, torsional vibration is the main source of the vibration noises of power transmission. There are several ways to suppress torsional vibration of the power transmission. The traditional way uses the elastic element to change the natural frequency to avoid the resonance zone and the damping element to attenuate the vibration amplitude [4]. Traditionally, we used a driven plate type clutch torsional vibration damper. However, due to its space structure constrains, the damping performance is not satisfactory.

DMF (Dual Mass Flywheel) is a new kind of vehicle torsional damper, which not only has the function of the single mass flywheel, but also a driven plate type clutch torsional vibration damper [5]. Due to its construction on reasonable inertia distribution and torsional stiffness, DMF can make the resonance rotating speed lower than the idling speed through power transmission, thus attenuating torsional vibrations under driving conditions [6]. Indeed, DMF has been widely used both in traditional ICE (internal combustion engine) vehicles and HEVs (hybrid electric vehicle), providing a more efficient damping performance. Figure 1 shows a schematic diagram of the power transmission with the DMF, which consists of a primary flywheel assembly, and a secondary flywheel assembly and a damper. Figure 2 shows a schematic diagram of the DMF, which consists of a primary flywheel assembly, a secondary flywheel assembly, and a damper. The primary flywheel assembly includes a starting gear ring, a signal ring, a cover, and a primary flywheel. The secondary flywheel assembly comprises a flange, a seal disc, and a secondary flywheel. The damper is composed of springs and damping elements. DMFs can be divided into several types according to the structure and the form of the springs, in which the circumferential arc spring dual mass flywheel (DMF-CS) is the most widely used type. As shown in Figure 1, the primary assembly and the crankshaft are connected by bolts. In addition, the clutch and the AT (automatic transmission) can be connected by the secondary assembly. Thus, power from the engine can be initially transmitted to the primary assembly, and then to the secondary assembly by compression of the flange into the arc springs. In the end, the power reaches the power transmission, leading to the car's driving.

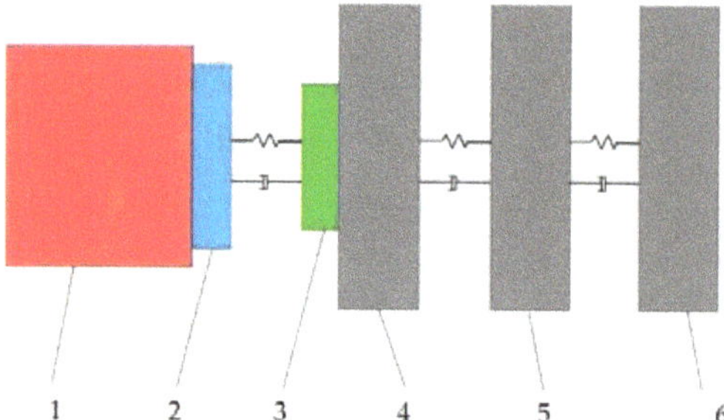

Figure 1. The power transmission with the DMF (Dual Mass Flywheel). 1. Engine; 2. Primary flywheel assembly; 3. Secondary flywheel assembly; 4. Clutch and gear box; 5. Transmission shaft; 6. Vehicle load.

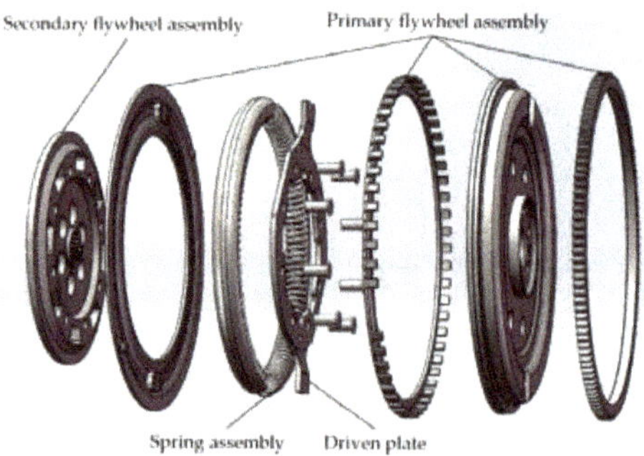

Figure 2. A schematic diagram of the DMF.

Despite recent advances in the DMF's experiments, simulations, structure innovations, and performance comparisons, little is known about the matching method between the DMF and power transmission. Some studies by Hartmut [7], Zeng [8], and Maffiodo [9] suggested that the excellent damping performance of DMF is associated with idling and low speed conditions by simulating characteristics of a power transmission with DMF and reviewing the angular accelerations and displacements between the output of the engine and the input of the gearbox under idling, sliding, and accelerating conditions. Others [10] found that although the angular acceleration of the crankshaft increased with the DMF, the corresponding dynamic load of the crankshaft decreased as the inertia was reduced. Kang, Kauh, and Ha [11] proposed the development of the displacement measuring system for the DMF based on the principle of the linear variable differential transformer (LVDT), which was used for installation in a real vehicle. Liupeng He [12] suggested a method for estimating the instantaneous engine torque of vehicles with conventional combustion engines and the DMF to obtain better control of engines equipped with a DMF. Peter and Robert [13] experimentally studied the dynamic change rules of the torsional stiffness and damping of the DMF. Li [14] and Wang [15] studied the natural torsion characteristic of DMF, and proposed the result that the natural frequency would be the minimum when the inertia ratio of the primary and secondary flywheel was 1:1. Yadav, Birari, et al. created a two-degree freedom dynamic model without a transmission system in the design of a crankshaft torsional vibration damper, and they found that the torsional vibration of the engine was attenuated when the natural frequency of the torsional vibration damper was equal to the first natural frequency of the engine, but this also introduced two new resonating frequencies to the original system [16]. Shangguan, Liu, and Rakheja proposed that the reduction of the torsional stiffness of a clutch was the most effective way to reduce gear rattle, and the torsional stiffness of a clutch at the first stage was determined by considering the excitation frequency of an engine at idle [17].

Structural sensitivity can reflect the gradient of the structural parameters to the response of the system. It is accepted that structural sensitivity will facilitate the optimization of the dynamic characteristics by modifying the structural parameters. Yue, et al. [18] studied the design parameters of a quarter wave tuner through sensitivity analysis by using acoustic simulations of the orifice noise of an intake system. Moreover, others [19] presented an explicit time-domain method for sensitivity analysis of structural responses under non-stationary random excitations and a new and more concise time-domain explicit expression of response sensitivity was derived using the direct differentiation method (DDM) based on time-domain explicit expressions of dynamic responses. A well-defined vibration mode [20,21] was used in the properties of a new micro machined tuning fork gyroscope (TFG) with an anchored diamond coupling mechanism to calculate Eigen sensitivities and establish exact formulae to connect the natural frequency sensitivity with the modal strain or kinetic energy, and determine the sensitivity to all stiffness and inertia parameters by the modal energy distribution.

The literature [7–15] shows that DMF can greatly improve the dynamic load of the crankshaft and can effectively isolate the torsional vibration caused by the engine at idling speed and in the low speed region. Moreover, the inertias of the primary and secondary flywheel assembly and the multi-stage torsional stiffness have a great influence on the characteristics of the power transmission. Therefore, the structural parameters of the DMF may be a decisive factor of the damping effect when the value reasonably matches the power transmission. It can be concluded from the literature [16,17] that the vibration reduction principle and the structure of a crankshaft torsional vibration damper are completely different from that of the DMF, which has only two structural parameters, and a torsional vibration damper of the clutch driven disk has only one structural parameter (torsional stiffness). Therefore, the matching model of the DMF and power transmission is different from that of a crankshaft torsional vibration damper and a torsional vibration damper of a clutch driven disk. Studies have shown that the sensitivity analysis method can be widely used in mechanical dynamical analysis and can also directly reflect the relationship between the structural parameters and the dynamic response of the system [18–21].

The literature shows the data of the main structural parameters of the DMF, including the inertias of the primary and secondary flywheel assemblies as well as the multi-stage torsional stiffness. Additionally, the sum of inertia of the primary and the secondary flywheel assembly is equal to the moment of inertia of the single mass flywheel, indicating the moment of inertia of the dual mass flywheel is a constant for a certain type of engine. The inertias of the primary and secondary flywheel assembly and the multi-stage torsional stiffness need to be reasonably determined in the process of matching, which suggests the matching problem between DMF and the power transmission is actually a multivariable matching problem. The literature recommends that sensitivity analysis is suitable for multivariable matching problems. This paper achieves the matching of DMF and the power transmission by integration of the sensitivity analysis method and the vibration reduction theories. Firstly, we demonstrated the modal analysis of the power transmission with the DMF. According to the analysis results, the absolute sensitivity analysis method was used to determine the main structural parameters, and the relative sensitivity analysis was used for the mathematical relationship between the main structural parameters and the natural frequencies of the system. Through the relative sensitivity analysis, the inertias of the primary and secondary flywheel assemblies can be combined as one structural parameter because of the constraint relation of the moment of inertia of the dual mass flywheel, namely the inertia ratio of them. The parameter should directly reflect the influence of the change of the inertias on the natural frequency of the system. Secondly, the range of the natural frequencies of the system was determined according to the vibration attenuation theories. Finally, the matching data between the DMF and the power transmission were predicted by using the above mathematical relationship and the range of natural frequencies.

2. Structural Sensitivity Analysis Method of Automobile Power Transmission

Sensitivity is widely used with different meanings in different areas. The meanings of sensitivity can be summarized as the gradient of a structural parameter or a variable to the system response or a solution of a function [16]. As a multivariable function, $f(x_i)$, with regard to $x_i (i = 1, 2 \ldots, n)$, the sensitivity of $f(x_i)$ related to x_i can be expressed as:

$$S_{ab}(f/x_i) = \lim_{\Delta x_i \to 0} \frac{\Delta f}{\Delta x_i} = \frac{\partial f}{\partial x_i} \tag{1}$$

where S_{ab} is the absolute sensitivity, of which the value denotes the influence of the variable, x_i, on $f(x_i)$. If we change the numerator and denominator of Equation (1) into the change rates of $f(x_i)$ and x_i, shown in Equation (2), S_{rt} is the relative sensitivity, of which the value denotes the relation between the change rates of $f(x_i)$ and x_i:

$$S_{rt}(f/x_i) = \lim_{\Delta x_i \to 0} \frac{\frac{\Delta f}{f}}{\frac{\Delta x_i}{x_i}} = \frac{\frac{\partial f}{f}}{\frac{\partial x_i}{x_i}} = \frac{x_i}{f} \frac{\partial f}{\partial x_i} \tag{2}$$

The structural sensitivity analysis method can be regarded as the application of the sensitivity analysis method in mechanical dynamics. Using this method, we can evaluate the influence of the change of system structural parameters on the system dynamic response. The dynamic characteristics of the power transmission generally cover amplitude-frequency and phase-frequency characteristics. Normally, DMF change the natural frequency of the power transmission by matching inertias and decreasing stiffness to avoid the resonance zone. Therefore, the structural sensitivity analysis can only involve the gradients of the system's natural frequencies to the inertias and stiffness under free vibration.

With rotational motion, the dynamic model of automobile power transmission is a torsional vibration model. The dynamic equation without damping is given by:

$$\left([K] - \omega_i^2 [J] \right) \{\theta\}_i = 0 \tag{3}$$

where $[K]$ and $[J]$ are the torsional stiffness matrix and inertia matrix, respectively, ω_i is the ith order natural frequency, and $\{\theta\}_i$ is the ith order modal shape. Structural damping and viscous damping still exist in the actual model; however, damping elements have little influence on the natural frequency of the system because of a small damping coefficient [4,22]. Furthermore, viscous friction and coulomb friction can cause a DMF to assume the hysteresis nonlinearity; however, the nonlinear model needs to be identified by the modified Bouc-Wen model combined with experimental data [23]. That is, the nonlinear model must be determined after a DMF is manufactured. Some studies [23] showed that the real natural frequency is approximately equal to the real natural frequency of the system without damping at low rotational speed. Therefore, the dynamic Equation (3) can be used to analyze the model in the process of matching.

2.1. Sensitivity of Natural Frequencies of Torsional Vibration to Torsional Stiffness

Both $[K]$ and $[J]$ are the real symmetric matrix. To simplify the calculation, Equation (3) is pre-multiplied by θ_i^T to obtain Equation (4):

$$\{\theta\}_i^T \left([K] - \omega_i^2 [J]\right) \{\theta\}_i = 0 \tag{4}$$

$$\{\theta\}_i^T [J] \{\theta\}_i = M_i \tag{5}$$

where M_i is the modal mass under the ith order. Let the absolute and relative sensitivities of ω_i to the torsional stiffness of the jth unit be $S_{ab}(\omega_i/K_j)$ and $S_{rt}(\omega_i/K_j)$, respectively. Referring to Equations (1) and (2), the partial derivative with respect to K_j in Equation (4) is operated to obtain Equation (6). Thus, $S_{ab}(\omega_i/K_j)$ and $S_{rt}(\omega_i/K_j)$ can be derived as:

$$\theta_i^T \left(\frac{\partial [K]}{\partial K_j} - 2\omega_i \frac{\partial \omega_i}{\partial K_j}[J] - \frac{\omega_i^2 \partial [J]}{\partial K_j}\right) \theta_i = 0 \tag{6}$$

$$S_{ab}(\omega_i/K_j) = \frac{\partial \omega_i}{\partial K_j} = \frac{\theta_i^T \frac{\partial [K]}{\partial K_j} \theta_i}{2\omega_i M_i} \tag{7}$$

$$S_{rt}(\omega_i/K_j) = \frac{\partial \omega_i/\omega_i}{\partial K_j/K_i} = \frac{\theta_i^T \frac{\partial [K]}{\partial K_j} \theta_i}{2\omega_i M_i} \frac{K_j}{\omega_i} \tag{8}$$

$[K]$ is expressed as Equation (9), so $\frac{\partial [K]}{\partial K_j}$ can be obtained as Equation (10) when $j < n - 1$ and $\frac{\partial [K]}{\partial K_j}$ can be given by Equation (11) when $j = n - 1$. Where n is the degree of freedom of the system.

$$[K] = \begin{bmatrix} K_1 & -K_1 & & & & & & \\ -K_1 & K_1+K_2 & -K_2 & & & 0 & & \\ & -K_2 & K_2+K_3 & -K_3 & & & & \\ & & \ddots & \ddots & \ddots & & & \\ & & & -K_{i-1} & K_{j-1}+K_j & -K_j & & \\ & & & & \ddots & \ddots & \ddots & \\ & 0 & & & & -K_{n-2} & K_{n-2}+K_{n-1} & -K_{n-1} \\ & & & & & & -K_{n-1} & K_{n-1} \end{bmatrix} \tag{9}$$

$$\frac{\partial[K]}{\partial K_j} = \begin{bmatrix} 0 & & & & & & \\ & & & & 0 & & \\ & & \ddots & \ddots & \ddots & & \\ & & & 1 & -1 & & \\ & & & \ddots & \ddots & \ddots & \\ & 0 & & & & & \\ & & & & & & 0 \end{bmatrix} \tag{10}$$

$$\frac{\partial[K]}{\partial K_j} = \begin{bmatrix} 0 & & & & & \\ & & & & 0 & \\ & & \ddots & \ddots & \ddots & \\ & & & & & \\ & & & \ddots & \ddots & \ddots \\ & 0 & & & & \\ & & & & -1 & 1 \end{bmatrix} \tag{11}$$

Combining Equations (7), (8), (10), and (11), $S_{ab}(\omega_i/k_j)$ and $S_{rt}(\omega_i/k_j)$ are given by:

$$S_{ab}(\omega_i/K_j) = \frac{\left[(\theta_i)_j - (\theta_i)_{j+1}\right]^2}{2\omega_i M_i} \tag{12}$$

$$S_{rt}(\omega_i/K_j) = \frac{\left[(\theta_i)_j - (\theta_i)_{j+1}\right]^2}{2\omega_i M_i} \frac{K_j}{\omega_i} \tag{13}$$

2.2. Sensitivity of Natural Frequencies of Torsional Vibration to Inertias

Let the absolute sensitivity and relative sensitivity of the ith natural frequency, ω_i to the torsional stiffness of the jth unit be $S_{ab}(\omega_i/J_j)$ and $S_{rt}(\omega_i/J_j)$. By seeking the partial derivative with respect to J_j in Equation (4), Equation (14) can be obtained as:

$$\theta_i^T \left(\frac{\partial[K]}{\partial J_j} - 2\omega_i \frac{\partial \omega_i}{\partial J_j}[J] - \frac{\omega_i^2 \partial[J]}{\partial J_j} \right) \theta_i = 0 \tag{14}$$

Therefore, the absolute and relative sensitivities can be calculated by:

$$S_{ab}(\omega_i/J_j) = \frac{\partial \omega_i}{\partial J_j} = -\frac{\omega_i \theta_i^T \frac{\partial[J]}{\partial J_j} \theta_i}{2M_i} \tag{15}$$

$$S_{rt}(\omega_i/J_j) = \frac{\partial \omega_i/\omega_i}{\partial J_j/J_j} = -\frac{J_j \theta_i^T \frac{\partial[J]}{\partial J_j} \theta_i}{2M_i} \tag{16}$$

$[J]$ is expressed as Equation (17), thus $\frac{\partial[J]}{\partial J_j}$ is expressed as Equation (18):

$$[J] = \begin{bmatrix} J_1 & & & & & \\ & J_2 & & & 0 & \\ & & \ddots & & & \\ & & & J_j & & \\ & 0 & & & \ddots & \\ & & & & & J_n \end{bmatrix} \tag{17}$$

$$\frac{\partial[J]}{\partial J_j} = \begin{bmatrix} 0 & & & & & \\ & 0 & & & 0 & \\ & & \ddots & & & \\ & & & 1 & & \\ & 0 & & & \ddots & \\ & & & & & 0 \end{bmatrix} \tag{18}$$

Combining Equations (15), (16), and (18), $S_{ab}(\omega_i/J_j)$ and $S_{rt}(\omega_i/J_j)$ are obtained as:

$$S_{ab}(\omega_i/J_j) = -\frac{\omega_i[(\theta_i)_j]^2}{2M_i} \tag{19}$$

$$S_{rt}(\omega_i/J_j) = -\frac{J_j[(\theta_i)_j]^2}{2M_i} \tag{20}$$

3. Matching Model of DMF and the Power Transmission Based on the Structural Sensitivity Analysis Method

Comparative analysis between the DMF and clutch suggested that the DMF can effectively attenuate the torsional vibrations under the idling condition and in the low engine speed zone (1200–3000 r/min) and exhibit a similar damping performance to the clutch in the high engine speed region (above 3000 r/min) [8]. The goals of this study were to describe the reasonable inertia distributions of the primary and secondary flywheels and multi-stage torsional stiffness, and to identify a potential association of a matching DMF and power transmission in terms of avoiding resonances under the idling condition and low speed zone. Only the first order torsional vibration will occur in the power transmission system under the idling condition and the modal vibrations of the system in the low speed zone under driving conditions are usually much more complicated, which are determined by the absolute structural sensitivity analysis method in the low speed zone. The natural frequency ranges of each mode can be established based on the resonance speed zone and then the matching model is created based on the natural frequency ranges and the relative structural sensitivity analysis method. The steps of building a matching model are shown in Figure 3.

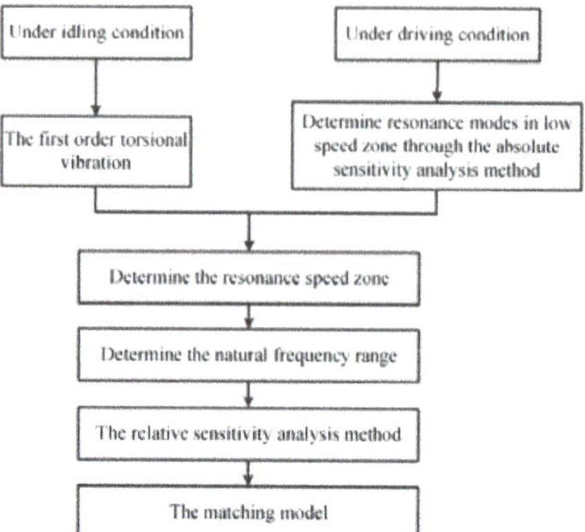

Figure 3. The steps of building a matching model.

3.1. Matching Model of Inertia and Torsional Stiffness of the DMF under the Idling Condition

The rotational speed of the engine under the idling condition is relatively low, usually around 800 r/min. For four stroke engines, the first order modal resonance under the 0.5th, 1st, 1.5th, and 2nd harmonic excitations will occur in this range of speed. In theory, the DMF can reduce the 1st natural frequency of the idling condition to be lower than the frequency corresponding to the idling speed by adjustment the inertias of the primary and secondary flywheels and torsional stiffness. However, in practice, many factors may influence the matching of the inertias of the DMF, such as the installation space, the dynamic load bearing capacity of the engine crankshaft, and the transmission shifting impact. Since the change interval of the inertias is limited, the natural frequency under the idling condition may be higher than the frequency corresponding to the idling speed. Therefore, two situations should be considered:

(1) When the 1st order modal resonance speed of the power transmission is lower than the idling speed, the 0.5th and 1st harmonic resonances should be avoided. In this case, we should compare the vector sums of the relative amplitudes of the 0.5th and the 1st harmonic orders to determine the main harmonic excitations that should be avoided.

(2) When the 1st order modal resonance speed of the power transmission is higher than the idling speed, the 1st, 1.5th, and 2nd harmonic resonances should be avoided. Under the idling condition, since nodes of the 1st order modal shape will not exist in the engine blocks, the main harmonic order will be the 2nd one for four-cylinder engines. In this instance, the 2nd order harmonic torsional vibration should be avoided.

Let the 1st order natural frequency be f, the resonance and the idling speed be n_1 and n_2, respectively, the harmonic order be I, and the resonance speed zone be Z_n, where $n_1 = i \cdot n_2 = 60f$. According to vibration attenuation theories, the resonance speed zone, Z_n, will be from $0.8 \cdot n_1$ to $1.2 \cdot n_1$, that is, $Z_n = (0.8, 1.2) \cdot n_1$ [20]. We will discuss the two cases respectively.

(1) $n_1 < n_2$, $i = 0.5$, 1. In this case, f should meet the following requirements:

$$60f < \frac{in_2}{1.2} \tag{21}$$

$$60f > \frac{in_2}{0.8} \tag{22}$$

For the 0.5th order harmonic excitation, the natural frequency is relatively low. Two situations will occur when using Equation (21) to design the natural frequency. Firstly, the torsional stiffness at the idling condition is so low that the torsional stiffness at driving conditions will be exclusively high.

Thus, resonances will occur under driving conditions. Secondly, the engine starts at an instant speed of about 200 r/min, which will cause start-up resonance and difficulties in starting. Therefore, only Equation (22) will be available.

For the 1st order harmonic excitation, the 1.5th and 2nd order resonances will occur when using Equation (22) to design the natural frequency. Therefore, only Equation (21) will be available in this situation.

In a word, the natural frequency under the idling condition will be:

$$\frac{0.5n_2}{0.8} < 60f < \frac{n_2}{1.2} \tag{23}$$

At the 1st mode, when the vector sum of the relative amplitude of the 0.5th harmonic is larger than that of the 1st harmonic, it is assumed that the resonance speed zone is $Z_n = (x, 1.2) \cdot n_1$. Thus, x should satisfy Equation (24):

$$\frac{0.5n_2}{x} < \frac{n_2}{1.2} \tag{24}$$

According to Equation (24), that is, $0.6 < x < 0.8$, so x can be valued at 0.7. Therefore, f can be calculated by:

$$\frac{0.5n_2}{0.7} < 60f < \frac{n_2}{1.2} \tag{25}$$

At the 1st mode, when the vector sum of the relative amplitude of the 1st harmonic is larger than that of the 0.5th harmonic, it is assumed that the resonance speed zone is $Z_n = (0.8, x) \cdot n_1$. Thus, x should satisfy Equation (26):

$$\frac{0.5n_2}{0.8} < \frac{n_2}{x} \tag{26}$$

According to Equation (26), that is, $1.2 < x < 1.6$, so x can be valued at $\sqrt{2}$. Therefore, x can be calculated by:

$$\frac{0.5n_2}{0.8} < 60f < \frac{n_2}{\sqrt{2}} \tag{27}$$

(2) $n_1 > n_2, i = 1, 1.5, 2$.

Similarly, for each order harmonic excitation, f should also satisfy Equations (21) and (22), which can be expressed as:

$$\begin{cases} \frac{n_2}{0.8} < 60f < \frac{1.5n_2}{1.2} \\ \frac{1.5n_2}{0.8} < 60f < \frac{2n_2}{1.2} \end{cases} \tag{28}$$

In fact, f cannot satisfy Equation (28). Under such a circumstance, resonances under the 1st and 2nd order harmonic excitations can only be considered. Furthermore, since the 2nd order harmonic is the main one for the four-cylinder engine, f should firstly satisfy Equation (29), and then satisfy the Equation (30):

$$\frac{n_2}{0.8} < 60f < \frac{2n_2}{\sqrt{2}} \tag{29}$$

$$\frac{n_2}{0.8} < 60f < \frac{2n_2}{1.2} \tag{30}$$

In summary, the 1st order natural frequency should be lower than the frequency corresponding to the idling speed as much as possible. Otherwise, the 1.5th order resonance will not be avoided. It is assumed that the inertias of the primary and secondary flywheel assembly are J_1 and J_2, respectively, and the inertia of the single mass flywheel matched to the engine is J_3. J_3 is provide by the engine manufacturer, and will usually be within a certain range; that is, $J_3 \in (J_x, J_y)$. Thus, $J_1 + J_2 \in (J_x, J_y)$. Furthermore, the inertia ratio of the primary and secondary flywheel assembly can be obtained by the constraints of the inertias, the masses, and the installation spaces of the primary and secondary flywheel assembly. The inertia ratio can be expressed as λ and $\lambda \in (a, b)$. With initial conditions, the initial values of J_1 and J_2 can be determined by:

$$\lambda = \frac{a+b}{2} \quad J_1 + J_2 = \frac{J_x + J_y}{2} \tag{31}$$

$$J_1 = \frac{\lambda(J_x + J_y)}{2(\lambda+1)} \quad J_2 = \frac{J_x + J_y}{2(\lambda+1)} \tag{32}$$

Let the torsional stiffness of DMF at the idling stage be K_1. Based on the initial conditions of the moment inertias of the primary and secondary flywheel assembly and the torsional stiffness at the idling stage, combined with the value range of the 1st order resonance and the analysis method of structural sensitivity, the matching of J_1, J_2, and K_1 to the power transmission follows the procedure outlined in Figure 4.

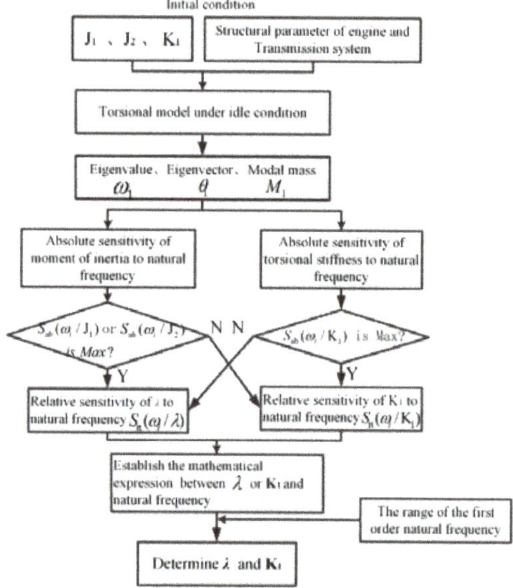

Figure 4. Structure parameters matching method of the dual mass flywheel under the idle condition based on structural sensitivity.

In Figure 4, $S_{ab}(\omega_1/J_1)$ and $S_{ab}(\omega_1/J_2)$ denote the absolute sensitivities of the 1st order natural frequency to the moment of inertias of the primary and secondary flywheel assembly, respectively, and $S_{rt}(\omega_1/\lambda)$ denotes the relative sensitivity of the 1st order natural frequency to the inertia ratio. Meanwhile, $S_{ab}(\omega_1/K_1)$ and $S_{rt}(\omega_1/K_1)$ are the absolute and relative sensitivities of the 1st order natural frequency to the torsional stiffness at the idling stage, respectively. If both $S_{ab}(\omega_1/J_1)$ and $S_{ab}(\omega_1/J_2)$ are not significant, J_1 and J_2 will not be the main structural parameters affecting the 1st order natural frequency. Thus, K_1 should be the crucial structural parameter to tune the natural frequency. On the other hand, if $S_{ab}(\omega_1/K_1)$ is not the largest sensitivity, J_1 and J_2 will be the key structural parameters to adjust the natural frequency. In addition, if all the three sensitivities are significant, K_1, J_1, and J_2 will be the key parameters to adjust the natural frequency. In the matching process, because of the constraints of the primary and secondary flywheel assembly, any change of the moment of inertia of the flywheel assembly will cause the change of the other. Thus, the moment of inertia ratio, λ, should be used instead of J_1 and J_2 as the structural parameter to conduct the calculations when analyzing the gradient relationship between the change of the primary and secondary flywheel assembly and that of the natural frequency. Therefore, Equation (16) involving λ can be rewritten as:

$$S_{rt}(\omega_i/\lambda) = \frac{\partial \omega_i/\omega_i}{\partial \lambda/\lambda} = -\frac{\lambda \theta_i^T \frac{\partial [J]}{\partial \lambda} \theta_i}{2M_i} \tag{33}$$

Let the jth and $(j+1)^{\text{th}}$ units be the primary flywheel assembly and secondary flywheel assembly, respectively, then:

$$[J] = \begin{bmatrix} J_1 & & & & & & \\ & J_2 & & & & 0 & \\ & & \ddots & & & & \\ & & & \frac{\lambda(J_x+J_y)}{2(\lambda+1)} & & & \\ & 0 & & & \frac{J_x+J_y}{2(\lambda+1)} & & \\ & & & & & \ddots & \\ & & & & & & J_n \end{bmatrix} \tag{34}$$

$$\frac{\partial [J]}{\partial J_j} == \begin{bmatrix} 0 & & & & & & \\ & & & & & 0 & \\ & & \ddots & & & & \\ & & & \frac{(J_x+J_y)}{2(\lambda+1)^2} & & & \\ & 0 & & & -\frac{J_x+J_y}{2(\lambda+1)^2} & & \\ & & & & & \ddots & \\ & & & & & & 0 \end{bmatrix} \tag{35}$$

Substituting Equation (35) into Equation (33), it can be rewritten as:

$$S_{rt}(\omega_i/\lambda) = -\frac{\lambda}{2M_i} \frac{J_x+J_y}{2(\lambda+1)^2}\left[(\theta_i)_j^2 - (\theta_i)_{j+1}^2\right] \tag{36}$$

According to $S_{rt}(\omega_1/K_1)$ and $S_{rt}(\omega_1/\lambda)$, the mathematical relationships between the 1st order natural frequency and λ, K_1 can be respectively established as:

$$\Delta\lambda = \frac{\Delta\omega_1/\omega_1}{S_{rt}(\omega_1/\lambda)}\lambda \tag{37}$$

$$\Delta K_1 = \frac{\Delta\omega_1/\omega_1}{S_{rt}(\omega_i/K_1)}K_1 \tag{38}$$

where $\Delta\lambda$ is the variation based on the initial value of λ, $\Delta\omega_1$ is the variation of the 1st order natural frequency, ω_1, based on the initial conditions, and ΔK_1 is the variation based on the initial value of K_1. Thus J_1, J_2, and K_1 are obtained as:

$$J_1 = \frac{(\lambda+\Delta\lambda)(J_x+J_y)}{2(\lambda+\Delta\lambda+1)} \qquad J_2 = \frac{J_x+J_y}{2(\lambda+\Delta\lambda+1)} \tag{39}$$

$$K_1 = K_1 + \Delta K_1 \tag{40}$$

$\Delta\omega_1$ can be determined by the difference between the actual value and the value range of ω_1. Then, the range of λ and K_1 can be determined by Equations (39) and (40). In this process, $Max\left(S_{rt}\left(\frac{\omega_1}{K_1}\right), S_{rt}(\omega_1/\lambda)\right)$ should be the structural parameter to be adjusted firstly. When it cannot meet the requirement, another structural parameter should be adjusted.

3.2. Matching Model of Inertia and Torsional Stiffness of the DMF under the Driving Condition

Under driving conditions, the torsional stiffness of the DMF at the driving stage that is K_2 can both transfer the engine power and adjust the system natural frequency. Let the operating angle of DMF at K_1 and K_2 be θ_1 and θ_2, respectively. Generally, the total torsion angle of the DMF springs being θ is about 65°–70° [5], thus, $\theta_1 + \theta_2 = \theta$. θ_1 can be primarily valued as:

$$\theta_1 = \frac{T_1}{K_1} \tag{41}$$

where T_1 is the moment of inertia of the power transmission under the idling condition, which is related to the inertias of the secondary flywheel assembly, the clutch and the input shaft of the transmission, and the angular accelerations of the starting motor. Accordingly, K_2 can be primarily calculated as:

$$K_2 = \frac{\xi T_{max}}{\theta - \theta_1} \tag{42}$$

where T_{max} is the maximum torque from the engine, and ξ is the torque backup coefficient, which is related to the real car.

Figure 5 shows the matching process of the structural parameters of the DMF using the structural sensitivity analysis method. Taking J_1, J_2, and K_2 as initial conditions, the torsional vibration model of the system can be established firstly. Then, modal analysis will be conducted to determine whether the resonance speed is in the low speed region. If the resonance speed deviates from the low speed region, J_1, J_2, K_1, and K_2 will be the final structural parameters of DMF. Whereas, if the resonance speed is in the low speed region, we should firstly obtain the order set of resonances, which is order_set1. Then, the absolute sensitivities of J_1, J_2, and K_2 to the natural frequency are analyzed for each order in order_set1 to obtain order_set2 associated with J_1, J_2, and K_2. Finally, the structural sensitivities of K_2 and λ are analyzed in order_set2, and their values are matched.

In this process, for the orders of resonance, the ranges of the natural frequency can be determined by Equations (21) and (22). Meanwhile, the relative sensitivities of K_2 and λ to each order natural frequency can be calculated. Then, referring to Equations (39) and (40), the ranges of K_2 and λ can be determined and stored in the K_2_set and λ_set, respectively. After traversing order_set2, the intersection of all values in the λ_ set and K_2_set will be obtained, and the values of K_2 and λ will be determined accordingly. After the above calculations under driving conditions, K_2 will change to be K_3. If $K_3 > K_2$, we value the torsional stiffness of the DMF at the driving stage as K_3; that is $K_2 = K_3$. If $K_3 < K_2$, K_3 cannot meet the requirement of torque transmission. Therefore, in this case, the intersection of the ranges of K_1 and K_3 under the idling condition should be determined firstly. In this intersection, by increasing K_1 and its operating angle, θ_1, K_2 will finally be determined according to Equation (42).

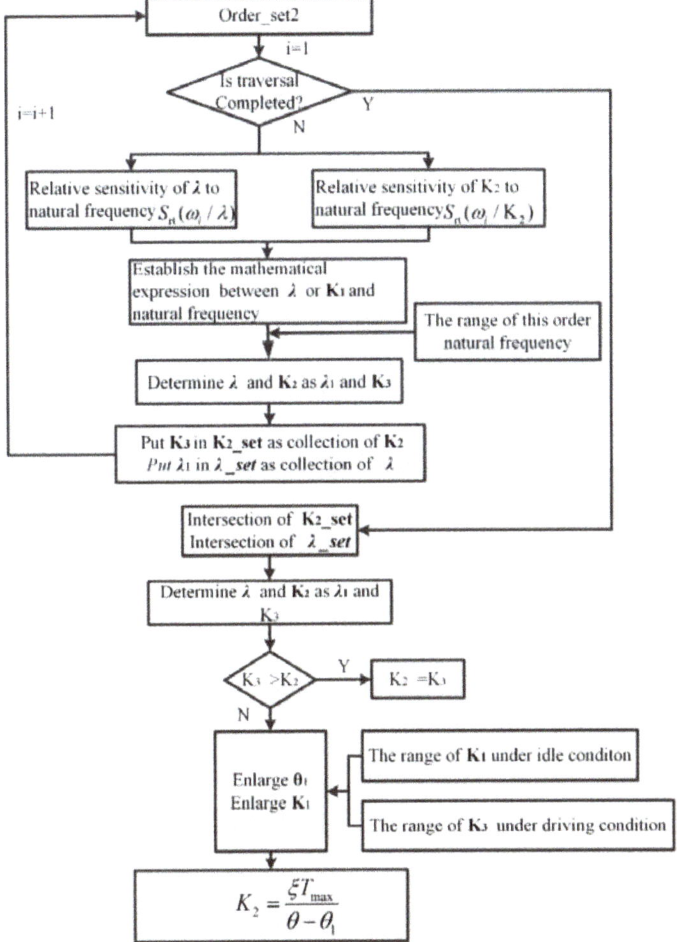

Figure 5. Structure parameters matching method of the dual mass flywheel under the driving condition based on structural sensitivity.

4. Matching Example and Real Vehicle Test of DMF

4.1. Matching Example of the DMF Based on the Structural Sensitivity Analysis Method

Taking a car matching a CVT (continuously variable transmission) as an example, the undamped torsional vibration models under idling and driving conditions are shown in Figures 6 and 7, respectively, where J_i denotes the moment of inertia and K_i denotes the torsional stiffness linking the two lumped masses. The structural parameters of the power transmission are listed in Table 1, where the units of the moment of inertia and torsion stiffness are kg·m^2 and N·m/rad, respectively.

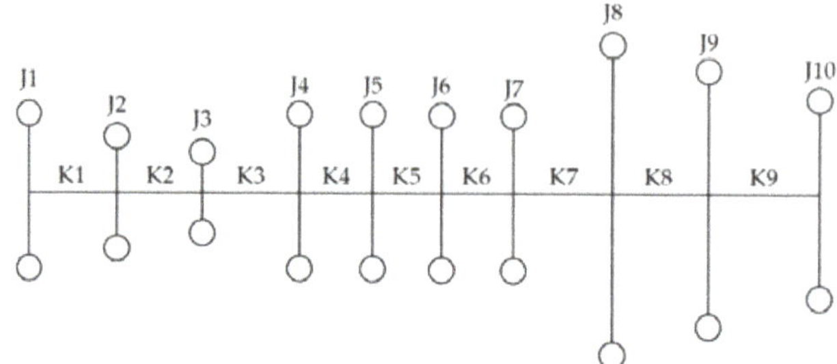

Figure 6. Undamped torsional vibration model under the idling condition.

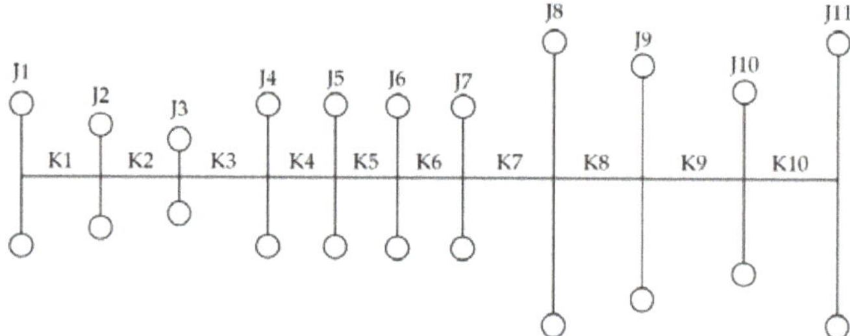

Figure 7. Undamped torsional vibration model under the driving condition.

Table 1. Structural parameters of the power transmission.

Names of Elements	Inertia	Value of Inertia (kg·m^2)	Torsional Stiffness	Value of Torsional Stiffness (N·m/rad)
Driven part of rubber damper	J_1	4.795×10^{-3}	K_1	14,320.0
Driving part of rubber damper	J_2	2.038×10^{-3}	K_2	74,636.1
Accessories	J_3	9.74×10^{-5}	K_3	356,080.5
Cylinder 1	J_4	4.669×10^{-3}	K_4	358,856.9
Cylinder 2	J_5	4.712×10^{-3}	K_5	360,638.8
Cylinder 3	J_6	4.712×10^{-3}	K_6	359,640
Cylinder 4	J_7	4.686×10^{-3}	K_7	1,871,080
Primary flywheel assembly	J_8	0.08	K_8	160.43; 733.39
Secondary flywheel assembly	J_9	0.012	K_9	98,731.62
Input shaft of CVT	J_{10}	7.312×10^{-3}	K_{10}	48,483.78
Driving cone of CVT	J_{11}	0.0268368		

The idling speed and maximum torque of the engine of this car are 750 r/min and 225 N·m, respectively. According to the technical parameters provided by the vehicle factory, the range of the total inertia of the DMF is 0.08–0.11 kg·m^2, where the inertia of the primary flywheel assembly is not less than 0.075 kg·m^2, and the secondary flywheel assembly quality should be less than 5 kg. Furthermore, $\xi = 1.33$, thus $4.2 < \lambda < 9$.

The car has been fitted with a DMF. The inertias of the primary and secondary flywheel assembly are 0.08 kg·m^2 and 0.012 kg·m^2, respectively, and the torsional stiffness at the idling stage and driving stage is 160.43 N·m/rad ($\theta_1 = 45.25°$) and 733.39 N·m/rad ($\theta_2 = 15°$), respectively. In addition, the hollow travel angle is 4.75°, and the total torsional stiffness is 65°, that is, $\theta = 65°$.

Based on the above data, the simulation model is established by using Adams software. Figures 8 and 9 are the simulation models of the automotive power transmission system with the DMF under idle and driving conditions, respectively. The simulation models created by Adams software usually consist of bodies, connectors, and forces, where bodies represent centralized mass units and constraints on each body are implemented by connectors, and the force transfer between bodies is achieved by torsion springs in forces.

Figure 8. The simulation model under the driving condition.

Figure 9. The simulation model under the driving condition.

The modal analysis is carried out according to the above simulation models, and the natural frequencies under idling and driving conditions are listed in Tables 2 and 3.

Table 2. Natural frequencies under the idling condition (Hz).

f_1	f_2	f_3	f_4	f_5	f_6	f_7	f_8	f_9
15.8	239.8	603.8	742	1053	1731	2481	3602	10,667

Table 3. System natural frequency under the driving condition (Hz).

f_1	f_2	f_3	f_4	f_5	f_6	f_7	f_8	f_9	f_{10}
22.5	239.8	300	603.8	821.79	1053	1731	2481	3602	10,667

It is shown that the 1st order natural frequency is 15.8 Hz under the idling condition and the corresponding resonance speed will be 940 r/min under the idling condition, which can meet the requirement of Equation (29). The 1st order natural frequency is 22.5 Hz under the driving condition and the corresponding resonance speed is 1350 r/min under the driving condition, which means that resonance will occur in the low speed region. Therefore, the structural parameters of the original DMF needed to be modified.

The calculation flow of sensitivity is shown in Figure 10, where n is the degree of freedom of the system, and i stands for the ith order and j represents the jth unit. According to this calculation flow, the program of m file is coded by using MATLAB, and the corresponding program code is shown in Appendix A. The absolute sensitivities of the 1st order natural frequency to the inertias and torsional stiffness can be obtained based on the torsional vibration model under the driving condition, as shown in Tables 4 and 5 and Figures 11 and 12.

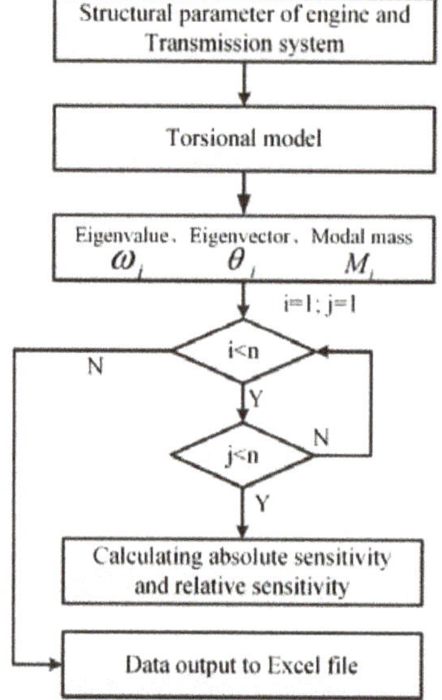

Figure 10. The calculation flow of sensitivity.

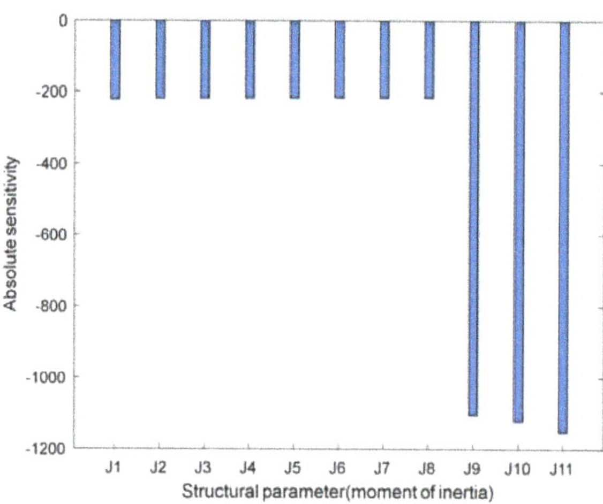

Figure 11. The absolute sensitivities of the 1st order natural frequency to inertias under the driving condition.

Table 4. The absolute sensitivities of the 1st order natural frequency to inertias under the driving condition.

J_1	J_2	J_3	J_4	J_5	J_6	J_7	J_8	J_9	J_{10}	J_{11}
-221.583	-218.24	-217.333	-217.14	-216.821	-216.376	-215.803	-215.668	-1104.44	-1122.1	-1150.73

Table 5. The absolute sensitivities of the 1st order natural frequency to torsional stiffness under the driving condition.

K_1	K_2	K_3	K_4	K_5	K_6	K_7	K_8	K_9	K_{10}
5.62×10^{-7}	4.18×10^{-8}	1.89×10^{-9}	5.18×10^{-9}	1.01×10^{-8}	1.68×10^{-8}	9.29×10^{-10}	0.101543	3.1×10^{-6}	7.97×10^{-6}

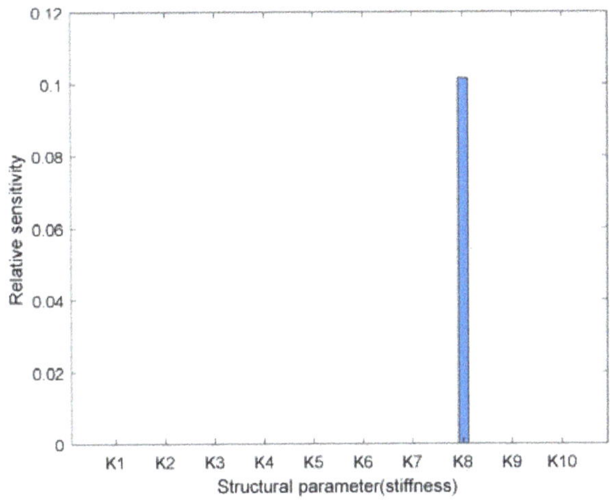

Figure 12. The absolute sensitivity of the 1st order natural frequency to torsional stiffness under the driving condition.

For inertias, the absolute sensitivities show that the inertias of the secondary flywheel assembly and the CVT component have the greatest influence on the 1st order natural frequency under the driving condition and the natural frequency will decrease with the increase of inertias. For the torsional stiffness, the torsional stiffness of the DMF at the driving stage has the greatest influence on the 1st order natural frequency and the natural frequency will increase with the increase of the torsional stiffness. Since the structural parameters of the CVT cannot be modified, the key structural parameters will be the inertia ratio of the primary and secondary flywheel assembly (λ) and the torsional stiffness of the DMF under the driving condition (K_8).

The relative sensitivities of the 1st order natural frequency to the torsional stiffness driving condition are shown in Table 6 and Figure 13, where $S_{rt}(\omega_1/K_8) = 0.495$, and it can be calculated that $S_{rt}(\omega_1/\lambda) = 0.062$ according to Equation (36). Combined with Equations (37) and (38), Figure 4, Figure 5, and the initial conditions, the matching results can be obtained, which are $J_8 = 0.077$ kg·m², $J_9 = 0.018$ kg·m², $K_8 = 257.8$ N·m/rad ($\theta_1 = 54.25$ °) (the torsion stiffness under idling condition), $K_8 = 710.5$ N·m/rad ($\theta_2 = 6$ °) (the torsion stiffness under driving condition), $\theta_0 = 4.75$ ° (the hollow travel angle), and $\theta = 65$ ° (the total torsion angle). The rematching DMF is shown as Figure 14.

Table 6. The relative sensitivities of the 1st order natural frequency to torsional stiffness under the driving condition.

K_1	K_2	K_3	K_4	K_5	K_6	K_7	K_8	K_9	K_{10}
5.35×10^{-5}	2.08×10^{-5}	4.47×10^{-6}	1.24×10^{-5}	2.42×10^{-5}	4.03×10^{-5}	1.16×10^{-5}	0.495229	0.002033	0.00257

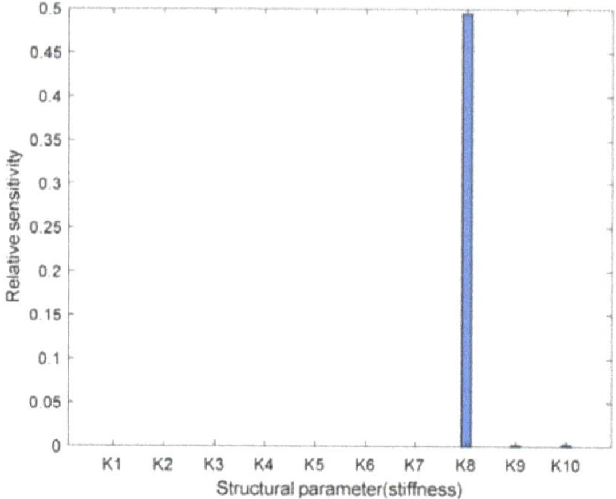

Figure 13. The relative sensitivities of the 1st order natural frequency to torsional stiffness under the driving condition.

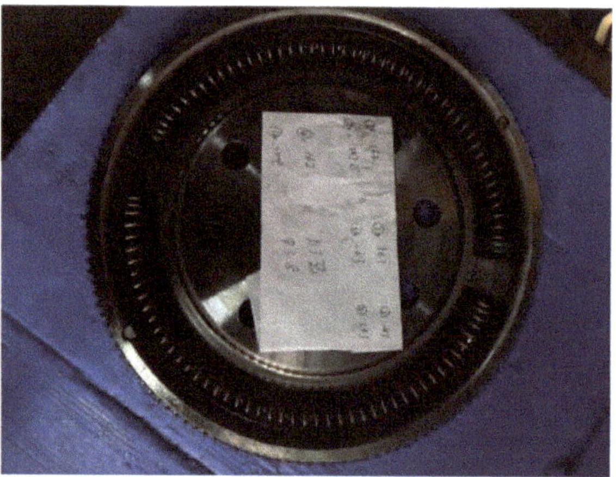

Figure 14. The rematching DMF before assembly.

After rematching, the natural frequencies of the power transmission under idling and driving conditions are obtained as shown in Tables 7 and 8. The 1st order natural frequency is 18.8 Hz and the corresponding resonance speed is 1128 r/min under the idling condition, which can meet the requirement of Equation (29). Furthermore, the 1st order natural frequency under the driving condition is 15 Hz and the corresponding resonance speed is 900 r/min under the driving condition, which can avoid resonances in the low speed region.

Table 7. Natural frequencies under the idling condition (Hz).

f_1	f_2	f_3	f_4	f_5	f_6	f_7	f_8	f_9
18.8	240.1	605.1	713.4	1053.6	1731.9	2481.7	3604.7	10,667.8

Table 8. Natural frequencies under the driving condition (Hz).

f_1	f_2	f_3	f_4	f_5	f_6	f_7	f_8	f_9	f_{10}
15	240.1	282.9	605.1	801.7	1053.6	1731.9	2481.7	3604.7	10,667.8

4.2. Real Vehicle Test

Real vehicle tests were carried out for the power transmissions matching the DMFs before and after optimization, as shown in Figures 15 and 16. As is shown in Figure 16, two electromagnetic speed sensors of which the model number is ONOSOKKI-MP-910 were mounted on the housing of the transmission. Furthermore, the arrangement details of these two sensors are shown in Figure 17, where sensor 1 is pointed to signal gear 1 on the primary flywheel of the DMF and sensor 2 is pointed to signal gear 2 on the input shaft of the transmission. As is shown in Figure 18, with the rotation of the gear, the clearance between the gear and the sensor will also change due to the different distances between the addendum and the dedendum of the gear and the sensor, which will cause a variation of the magnetic flux of the coil in the sensor, thus the output signal of the sensor is similar to a sinusoidal wave. Let the rotating speed of the gear be ω (r/min), the number of teeth of the gear be Z, and the frequency of the signal be f(Hz), so ω will be:

$$\omega = \frac{f}{Z} \times 60 \tag{43}$$

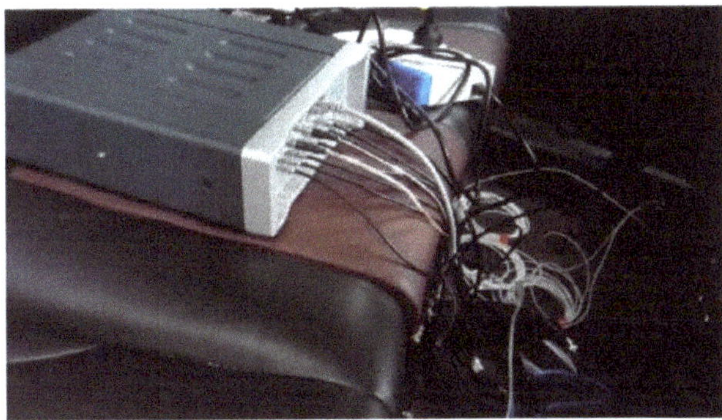

Figure 15. Digital acquisition hardware inside the car.

Figure 16. Electromagnetic rotating speed sensor of the transmission.

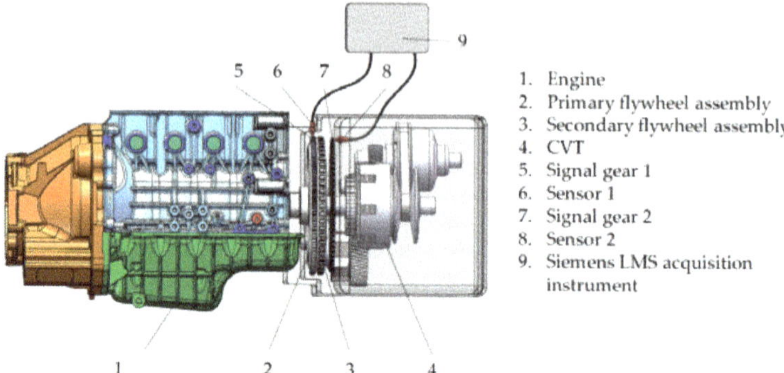

Figure 17. Schematic diagram of the electromagnetic rotating speed sensors' arrangement.

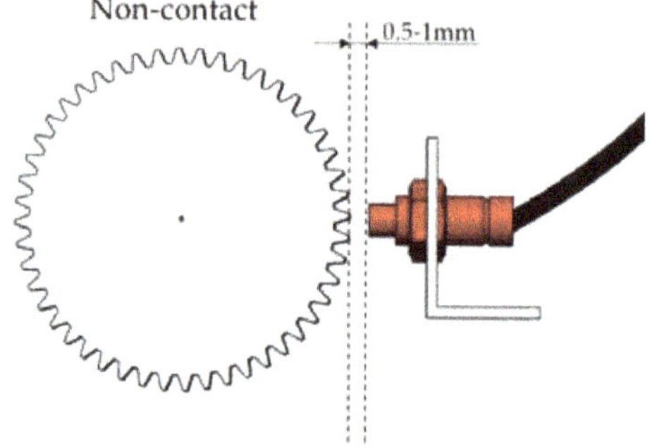

Figure 18. Schematic diagram of the measuring mode of the electromagnetic rotating speed sensor.

In this experiment, the signal wires of these two rotating speed sensors were connected with a Siemens data acquisition instrument, LMS SCADAS302VB. The collected data were processed by Siemens LMS Test.Lab 15A, and the tracking settings for the two signal channels are shown in Figure 19, where the number of teeth of signal gear 1 (as shown in Figure 17) is 133 and the number of teeth of signal gear 2 (as shown in Figure 17) is 60. In Figure 20, tacho1 and tacho2 are the rotating speed signal channel of the primary flywheel and the rotating speed signal channel of the input shaft of the transmission, respectively.

Speed signals were collected under idling and driving conditions and the angular acceleration of the primary flywheel and the transmission input were analyzed. For the engine mounted with the original DMF, under the idling condition, the engine speed starts at around 750 r/min, as shown in Figure 21, in which the red curve represents the engine speed. Figure 22 shows the angular acceleration, in which the red and green curves respectively present the angular acceleration of the primary flywheel and the transmission input, where the maximum angular acceleration of the transmission input is about 83 rad/s^2. Under the driving condition, the range of the engine speed is 750 r/min–1500 r/min, as shown in Figure 23, in which the red curve represents the engine speed in the time-domain. Figure 24 shows the angular acceleration, in which the red and green curves respectively present the angular acceleration of the primary flywheel and the transmission input, where the maximum angular acceleration of the transmission input is about 203 rad/s^2 around 1210 r/min.

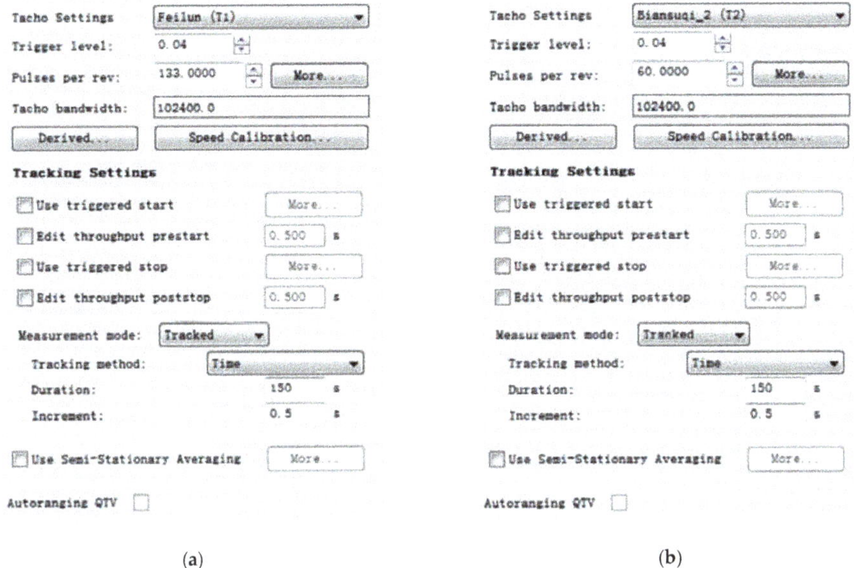

(a)   (b)

Figure 19. Measurement tracking setting: (**a**) tracking setting of the rotating speed signal channel of the primary flywheel and (**b**) tracking setting of the rotating speed signal channel of the input shaft of the transmission.

Figure 20. Measurement channel setting.

For the rematching DMF, the engine speed was around 750 r/min under the idling condition, as shown in Figure 25, in which the red curve represents the engine speed. Figure 26 shows the angular acceleration under the idling condition, in which the red and green curves respectively present the angular acceleration of the primary flywheel and the transmission input, where the maximum angular acceleration of the transmission input is about 25 rad/s^2. Under the driving condition, the range of the engine speed is 750 r/min–2500 r/min, as shown in Figure 27, in which the red curve represents the engine speed in the time-domain. Figure 28 shows the angular acceleration, in which the red and green curves respectively present the angular acceleration of the primary flywheel and the transmission input, where the maximum angular acceleration of the transmission input is 184 rad/s^2 around 836 r/min.

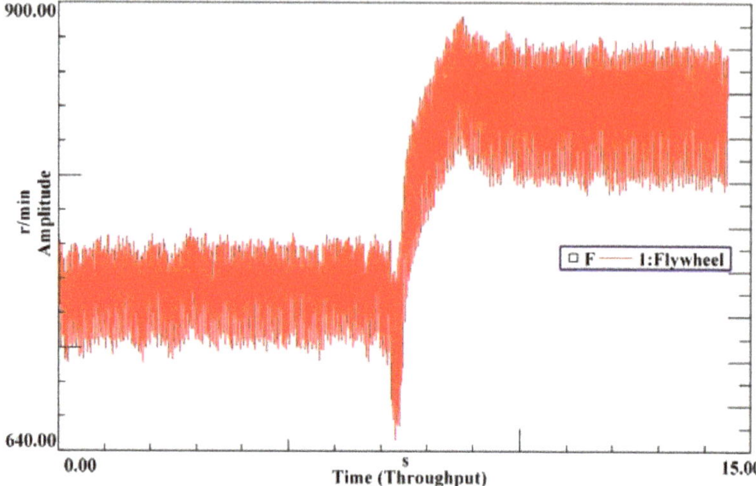

Figure 21. Time-domain speed signal of the engine with the original DMF under the idling condition.

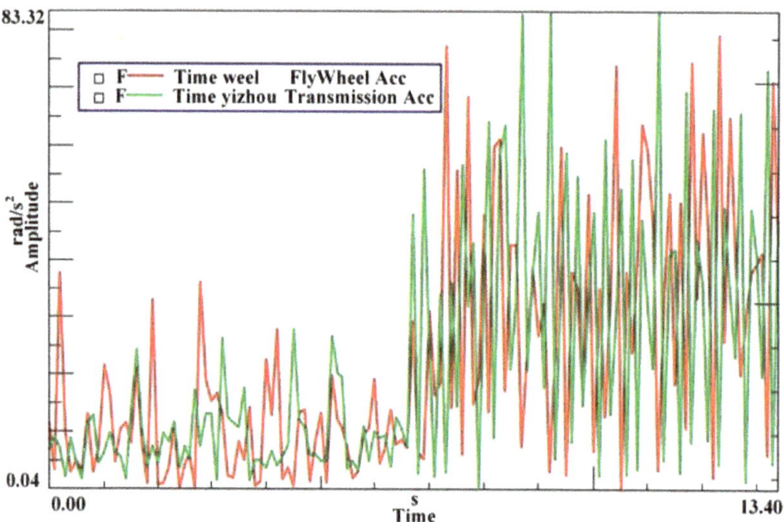

Figure 22. The angular acceleration of the primary flywheel and the transmission input under the idling condition for the engine with the original DMF.

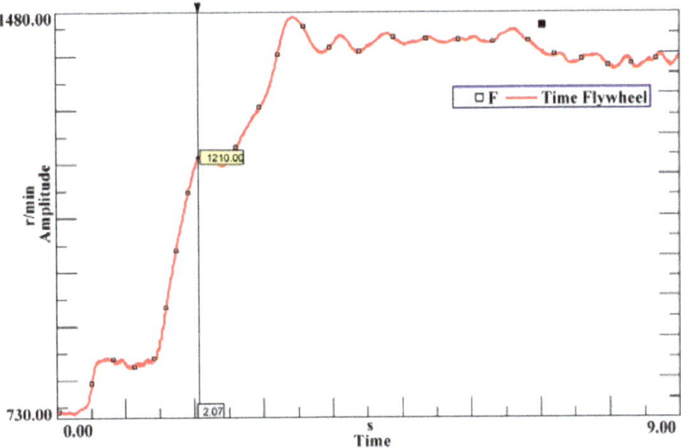

Figure 23. Time-domain speed signal of the engine with the original DMF under the driving condition.

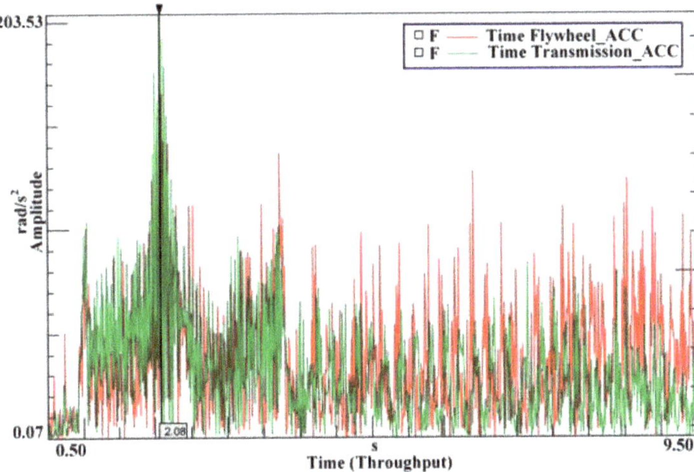

Figure 24. The angular acceleration of the primary flywheel and the transmission input under the driving condition for the engine with the original DMF.

Theoretical calculation results and the experimental results of the engine with the original DMF and rematching DMF are compared as shown in Table 9. Under the idling condition, the theoretical value of the first order resonance speed of the system is 940 r/min with the original DMF and 1128 r/min with the rematching DMF, however, there is no change in terms of the engine speed, remaining around 750 r/min under the idling condition. Thus, the resonance speed appears unmeasured. Figure 25 shows that the resonance speed under the idling condition occurs around 750 r/min; Figure 26 shows that the resonance speed under the idling condition remains around 750 r/min and the maximum angular acceleration of the transmission input is decreased as the torsional stiffness of the DMF at the idling stage rises, suggesting an acceptance of the theoretical calculation result.

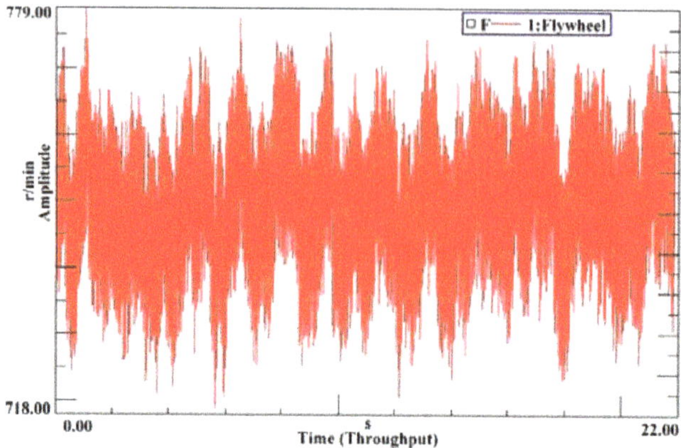

Figure 25. Time-domain speed signal of the engine with the rematching DMF under the idling condition.

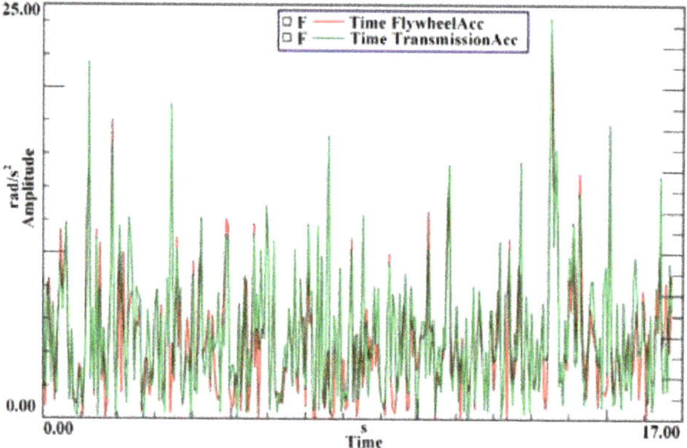

Figure 26. The angular acceleration of the primary flywheel and the transmission input under the idling condition for the engine with the rematching DMF.

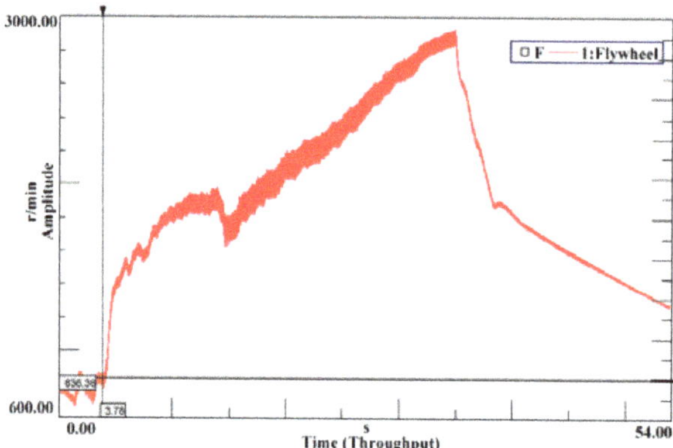

Figure 27. Time-domain speed signal of the engine with the rematching DMF under the driving condition.

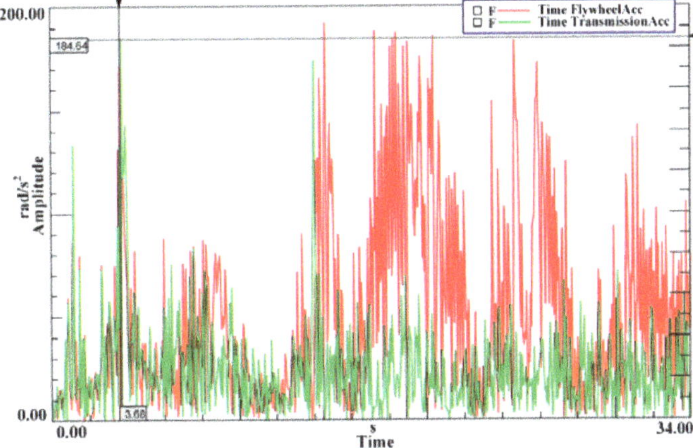

Figure 28. The angular acceleration of the primary flywheel and the transmission input under the driving condition for the engine with the rematching DMF.

Table 9. Comparative analysis of the results.

Items	Under Idling Condition		Under Driving Conditon	
	Original DMF	Rematching DMF	Original DMF	Rematching DMF
The maximum angular acceleration (rad/s^2)	83	25	203	184
Measured resonance speed (r/min)			1210	836
Theoretical resonance speed (r/min)	940	1128	1350	900

Under the driving condition, the actual measurement result of the first order resonance speed of the engine with the original DMF is 1210 r/min, and then the theoretical calculation of the resonance speed is 1350 r/min based on the undamped torsional vibration model; that is, the error is about 11%. Moreover, the actual measurement result of the first order resonance speed of the engine with the rematching DMF is 836 r/min and the theoretical calculation of the resonance speed is 900 r/min based on the undamped torsional vibration model; that is, the error is about 7%, which means the theoretical model can be used to analyze the modal in the process of matching.

Under the idling condition, the angular acceleration of the transmission input is not damped compared with that of the primary flywheel because the resonance speed is higher than the idling speed for the engine with the rematching DMF. Nevertheless, the maximum angular acceleration of the transmission input decreases from 83 rad/s² to 25 rad/s². Under the driving condition, the maximum angular acceleration of the transmission input decreases from 203 rad/s² to 184 rad/s² for the engine with the rematching DMF. Moreover, the resonance speed is reduced from 1210 r/min to 836 r/min for the engine with the rematching DMF under the driving condition, avoiding the resonance at the low speed region. Obviously, the damping performance of the rematching DMF is better than the original DMF, thus these results confirm the validity of the matching model.

5. Conclusions

This study addressed the matching model of the DMF and the power transmission by integration of the sensitivity analysis method and the vibration reduction theories. Based on the mathematical meaning of sensitivity, the structural sensitivity analysis method for an automobile power transmission system was obtained. Furthermore, considering the constraints of the inertia of the primary and secondary flywheel assembly, the inertia ratio of them was taken as one structural parameter completing a successful sensitivity analysis. The main function of the DMF is to attenuate torsional vibration from the engine and hence the resonance speed zones were defined in light of the vibration reduction theories and their working characteristics. The matching model of the DMF and the power transmission was established based on the resonance speed zones and the structural sensitivity analysis method. The matching model was applied to modify the structural parameters of the DMF of a certain vehicle and then the effectiveness of the matching model was verified by experiments of a real vehicle test under the idling condition and driving condition. The main conclusions in this research are summarized as follows:

(1) The absolute structural sensitivity can effectively isolate key structural parameters of the vibration modal at each stage and the resultant parameters can be quantitatively revised by the relative structural sensitivity. Under the driving condition, the inertia of the secondary flywheel assembly and torsional stiffness of DMF have a significant influence on the 1st order natural frequency of automotive power transmission. The inertia of the secondary flywheel assembly is inversely proportional to the 1st order natural frequency. In contrast, the torsional stiffness is positively proportional to the 1st order and the inertia ratio of the primary and secondary assembly is positively associated to the 1st order.

(2) Given that the resonance speed is higher than the idle speed under the idling condition, the 1st order natural frequency of the system should be increased through enhancement of the torsional stiffness of the DMF at the idling stage to decrease the angular acceleration amplitude of the input shaft of the transmission. In contrast, the 1st order natural frequency of the system should be decreased through reduction of the torsional stiffness of the DMF at the idling stage to attenuate the angular acceleration amplitude of the input shaft of the transmission.

(3) Under the driving condition, the 1st order natural frequency of the system should be decreased by reduction of the inertia ratio and the torsional stiffness of the DMF at the driving stage, which appears to protect resonances in low speed zones and attenuates the angular acceleration amplitude of the input shaft of the transmission.

(4) Given that the torsional stiffness at the driving stage cannot meet the requirements of the matching model, the operation range of the torsional stiffness at the idling stage should be enlarged to make it work under driving conditions.

As the matching model proposed in this paper can achieve reasonable matching between structural parameters of the DMF and power transmission system, it is useful to make a transmission system that is insulated from torsional vibrations caused by the engine. The matching model can also be applied in structural parameters matching of other types of shock absorber in vehicle systems and other mechanical systems after adjusting the resonance frequency ranges and vibration model according to the actual operating conditions of the system.

The matching model can be used to preliminarily determine the main structural parameters of the DMF, however, these were not further optimized according to the dynamic non-linear characteristics of the DMF, so in future study, we will combine the dynamic non-linearity model of the DMF with the linear matching model to further optimize the structural parameters in order to obtain a better damping performance.

Author Contributions: Conceptualization, L.C.; Methodology, L.C. and X.Z.; Software, X.Z. and R.Z.; Validation, Z.Y.; Investigation, X.Z. and L.C.; Writing—original draft preparation, L.C.; Writing—review and editing, Z.Y.; Funding acquisition, L.C. and R.Z.

Funding: This research was funded by the National Natural Science Foundation of China (Grant No. 51405355) and the National Natural Science Foundation of China (Grant No. 51605182).

Appendix A

```
function [m1,m2,m3,m4,m5,m6]=Sensibilty()
JZ11=0.004795;
JZ12=0.002038;
JZ13=0.0000974;
JZ14=0.004669;
JZ15=0.004712;
JZ16=0.004712;
JZ17=0.004686;
JZ18=0.08;
JZ19=0.012;
JZ110=0.00731216;
JZ111=0.0268368;
KZ11=14320;
KZ12=74636.1;
KZ13=356080.5;
KZ14=358856.9;
KZ15=360638.8;
KZ16=359640;
KZ17=1871080;
KZ18=733.39;
KZ19=98731.62;
KZ110=48483.78;
J1=[JZ11 JZ12 JZ13 JZ14 JZ15 JZ16 JZ17 JZ18 JZ19 JZ110 JZ111];
K1=[KZ11 KZ12 KZ13 KZ14 KZ15 KZ16 KZ17 KZ18 KZ19 KZ110];
J=diag(J1);
K=[KZ11 -KZ11 0 0 0 0 0 0 0 0 0;
  -KZ11 KZ12+KZ11 -KZ12 0 0 0 0 0 0 0 0;
  0 -KZ12 KZ12+KZ13 -KZ13 0 0 0 0 0 0 0;
  0 0 -KZ13 KZ14+KZ13 -KZ14 0 0 0 0 0 0;
  0 0 0 -KZ14 KZ14+KZ15 -KZ15 0 0 0 0 0;
  0 0 0 0 -KZ15 KZ16+KZ15 -KZ16 0 0 0 0;
  0 0 0 0 0 -KZ16 KZ16+KZ17 -KZ17 0 0 0;
  0 0 0 0 0 0 -KZ17 KZ18+KZ17 -KZ18 0 0;
  0 0 0 0 0 0 0 -KZ18 KZ18+KZ19 -KZ19 0;
  0 0 0 0 0 0 0 0 -KZ19 KZ110+KZ19 -KZ110;
  0 0 0 0 0 0 0 0 0 -KZ110 KZ110;];
J2=inv(J);
[v,d]=eig(J2*K);
d=d^0.5;
m5=abs(diag(d))';
v1=[v;abs(diag(d))']';
v2=sortrows(v1,12);
n=11;
senb_J=zeros(n,n);
senb_K=zeros(n,n-1);
senb_J2=zeros(n,n);
senb_K2=zeros(n,n-1);
modal_mass=zeros(1,n);
Temp_V1=zeros(1,n);
```

```
Temp_V2=0;
Temp_V3=0;
for i=1:n
    for j=1:n
        Temp_V(1,j)=v2(i,j);
    end
    modal_mass(1,i)=Temp_V*J*Temp_V';
end
Temp_V4=0;
y=JZ18/JZ19;
senb_J3=0;
for i=1:n
    for j=1:n
        Temp_V2=-v2(i,j)^2*v2(i,n+1);
        if j<n
            Temp_V3=(v2(i,j)-v2(i,j+1))^2;
            senb_K(i,j)=Temp_V3/(2*v2(i,n+1)*modal_mass(1,i));
            senb_K2(i,j)=senb_K(i,j)*K1(1,j)/v2(i,n+1);
        end
            senb_J(i,j)=Temp_V2/(2*modal_mass(1,i));
            senb_J2(i,j)=senb_J(i,j)*J1(1,j)/v2(i,n+1);
    end
end
m1=senb_K; % The absolute sensitivities of natural frequency to torsional stiffness
m2=senb_K2; % The relative sensitivities of natural frequency to torsional stiffness
m3=senb_J; % The absolute sensitivities of natural frequency to inertias
m4=senb_J2; % The relative sensitivities of natural frequency to inertias
m6=modal_mass; % modal mass
m5=m5/(2*pi);
xlswrite('.\sensibility.xls',senb_K,'A1:J11');
xlswrite('.\sensibility.xls',senb_K2,'A13:J23');
xlswrite('.\sensibility.xls',senb_J,'A25:K35');
xlswrite('.\sensibility.xls',senb_J2,'A37:K47');
```

References

1. Hao, Y.D.; He, Z.C.; Li, G.Y.; Li, E.; Huang, Y.Y. Uncertainty Analysis and Optimization of Automotive Driveline Torsional Vibration with a Driveline and Rear Axle Coupled Model. *Eng. Optim.* **2018**, *50*, 1871–1893. [CrossRef]
2. Sayin, P.; Schoftner, J. Analytical Investigation on the Damping Performance of A torque Converter in an Automotive Driveline Model. In Proceedings of the 2017 International Conference on Mechanical, Aeronautical and Automotive Engineering (ICMAA 2017), Malacca, Malaysia, 25–27 February 2017; pp. 1–7.
3. Ahmed, H.; Eliot, M.; Stephanos, T.; Homer, R.; Patrick, K.; Alexander, V.; Lawrence, A.B.; Donald, M.M. A Study on Torsional Vibration Attenuation in Automotive Drivetrains Using Absorbers with Smooth and Non-smooth Nonlinearities. *Appl. Math. Model.* **2017**, *46*, 674–690.
4. Pavel, N.; Aleš, P.; Martin, Z.; Kamil, Ř. Investigating the Influence of Computational Model Complexity on Noise and Vibration Modeling of Powertrain. *J. Vibroeng.* **2016**, *18*, 378–393.
5. Tang, X.; Hu, X.; Yang, W.; Yu, H. Novel Torsional Vibration Modeling and Assessment of a Power-Split Hybrid Electric Vehicle Equipped with a Dual-Mass Flywheel. *IEEE Trans. Veh. Technol.* **2018**, *67*, 1990–2000. [CrossRef]
6. Szpica, D. Modelling of the Operation of a Dual Mass Flywheel (DMF) for Different Engine-related Distortions. *Math. Comput. Model. Dyn. Syst.* **2018**, *24*, 623–640. [CrossRef]

7.   Hartmut, B.; Wolfgang, K. Simulation of Various Operating Conditions of Powertrains with a Dual Mass Flywheel. *Drive Syst.* **2008**, *22*, 3–16.

8.   Zeng, L.P.; Chen, Q.P.; Yuan, X.X. Study on Nonlinear Vibration of Vehicle Dual Mass Flywheels with Three-stage Piecewise Stiffnesses. *Chin. Mech. Eng.* **2018**, *29*, 2453–2459.

9.   Maffiodo, D.; Sesana, R.; Paolucci, D.; Bertaggia, S. Finite Life Fatigue Design of Spiral Springs of Dual-mass Flywheels: Analytical Estimation and Experimental Results. *Adv. Mech. Eng.* **2018**, *10*, 1–3. [CrossRef]

10.   Duran, E.T.; Sever, A.C. *Dynamic Simulation and Endurance Limit Safety Factor Calculation for Crankshaft-comparison of Single Mass and Dual Mass Flywheel*; SAE Technical Paper 2008-01-2622; SAE: Warrendale, PA, USA, 2008. [CrossRef]

11.   Kang, T.-S.; Kauh, S.-K.; Ha, K.-P. Development of the Displacement Measuring System for a Dual Mass Flywheel in a Vehicle. *J. Automob. Eng.* **2009**, *223*, 1273–1281. [CrossRef]

12.   He, L.; Xia, C.; Chen, S.; Guo, J.; Liu, Y. Parametric Investigation of Dual-Mass Flywheel Based on Driveline Start-Up Torsional Vibration Control. *Shock Vib.* **2019**, *2019*, 1–12. [CrossRef]

13.   Peter, B.; Robert, G. Comparison of Dynamic Properties of Dual Mass Flywheel. *Diagnostyka* **2015**, *16*, 29–33.

14.   Li, Q.S.; Li, P.Z.; Shu, P.Z.; Jian, D.Z.; Hong, E.N. Design and Analysis of a Dual Mass Flywheel with Continuously Variable Stiffness Based on Compensation Principle. *Mech. Mach. Theory* **2014**, *79*, 124–140.

15.   Wang, Y.L.; Qin, X.P.; Huang, S.; Deng, S. Design and Analysis of a Multi-stage Torsional Stiffness Dual Mass Flywheel Based on Vibration Control. *Appl. Acoust.* **2016**, *104*, 172–181. [CrossRef]

16.   Yadav, A.; Birari, M.; Bijwe, V.; Billade, D. *Critique of Torsional Vibration Damper (TVD) Design for Powertrain NVH*; SAE Technical Paper 2017-26-0217; SAE: Warrendale, PA, USA, 2017. [CrossRef]

17.   Shangguan, W.B.; Liu, X.L.; Yin, Y.; Rakheja, S. Modeling of automotive driveline system for reducing gear rattles. *J. Sound Vib.* **2018**, *416*, 136–153. [CrossRef]

18.   Yue, G.P.; Zhang, Y.M. Structure Betterment of Intake System Based on Sensitivity Analysis of Orifice Noise. *J. Mech. Eng.* **2010**, *46*, 77–81. [CrossRef]

19.   Hu, Z.Q.; Su, C.; Chen, T.C.; Ma, H.T. An Explicit Time-domain Approach for Sensitivity Analysis of Non-stationary Random Vibration Problems. *J. Sound Vib.* **2016**, *382*, 122–139. [CrossRef]

20.   Guan, Y.W.; Gao, S.Q.; Liu, H.P.; Jin, L.; Niu, S.H. Design and Vibration Sensitivity Analysis of a MEMS Tuning Fork Gyroscope with an Anchored Diamond Coupling Mechanism. *Sensors* **2016**, *16*, 468. [CrossRef] [PubMed]

21.   Guan, Y.W.; Gao, S.Q.; Jin, L.; Cao, L.M. Design and vibration sensitivity of a MEMS tuning fork gyroscope with anchored coupling mechanism. *Microsyst. Technol.* **2016**, *22*, 247–254. [CrossRef]

22.   Huang, Y.; Li, D. Subjective Discomfort Model of the Micro Commercial Vehicle Vibration over Different Road Conditions. *Appl. Acoust.* **2019**, *145*, 385–392. [CrossRef]

23.   Chen, L.; Zeng, R.; Jiang, Z.F. Nonlinear dynamical model of an automotive dual mass flywheel. *Adv. Mech. Eng.* **2015**, *7*, 1687814015589533. [CrossRef]

Article

On the Identification of Sectional Deformation Modes of Thin-Walled Structures with Doubly Symmetric Cross-Sections Based on the Shell-Like Deformation

Lei Zhang [1,*], Aimin Ji [1,*], Weidong Zhu [2] and Liping Peng [3]

1   College of Mechanical and Electrical Engineering, Hohai University, Changzhou 213022, China
2   Department of Mechanical Engineering, University of Maryland, Baltimore County, MD 21250, USA;
    wzhu@umbc.edu
3   State Key Laboratory of Mineral Processing, Beijing General Research Institute of Mining and Metallurgy,
    Beijing 102600, China; plp_hhu@163.com
*   Correspondence: leizhang@hhu.edu.cn (L.Z.); jam@ustc.edu (A.J.); Tel.: +86-0519-85191840 (L.Z.)

Received: 19 November 2018; Accepted: 11 December 2018; Published: 16 December 2018

Abstract: In this paper, a new approach is proposed to identify sectional deformation modes of the doubly symmetric thin-walled cross-section, which are to be employed in formulating a one-dimensional model of thin-walled structures. The approach considers the three-dimensional displacement field of the structure as the linear superposition of a set of sectional deformation modes. To retrieve these modes, the modal analysis of a thin-walled structure is carried out based on shell/plate theory, with the shell-like deformation shapes extracted. The components of classical modes are removed from these shapes based on a novel criterion, with residual deformation shapes left. By introducing benchmark points, these shapes are further classified into several deformation patterns, and within each pattern, higher-order deformation modes are derived by removing the components of identified ones. Considering the doubly symmetric cross-section, these modes are approximated with shape functions applying the interpolation method. The identified modes are finally used to deduce the governing equations of the thin-walled structure, applying Hamilton's principle. Numerical examples are also presented to validate the accuracy and efficiency of the new model in reproducing three-dimensional behaviors of thin-walled structures.

Keywords: thin-walled structures; higher-order deformation modes; identification; doubly symmetric cross-sections; shell-like deformation

---

1. Introduction

Thin-walled structures are widely used in civil, aeronautical, and mechanical engineering. In the processes of designing and manufacturing them, a mathematical model is essential to predict their structural behaviors. For simplicity and efficiency, one-dimensional (beam) models are more widely used than two-dimensional (plate/shell) and three-dimensional (solid) theories. However, conventional beam models face a limit in capturing cross-sectional deformation, which is quite usual but significant for the mechanical properties of a thin-walled structure. Therefore, refined beam models must be developed, taking out-of-plane warping and in-plane distortion into consideration. The issue is that an efficient beam theory needs a general procedure for identifying a complete set of sectional deformation modes [1], which are hierarchically capable of forming a reduced model, and presenting the physical interpretation in a clear way. For this reason, the development of advanced beam theories is still appealing.

During the last few decades, many refined beam models have been proposed. Some main contributions are outlined in the review article by Carrera et al. [2]. For the sake of completeness, a brief review of refined beam theories is given here. First, particular attention should be paid to the work of Vlasov [3], who introduced warping functions in modeling thin-walled beams. The Saint–Venant solution is also useful in developing advanced beam theory. For example, Yoon et al. [4] proposed a finite element formulation for nonlinear torsional analysis of 3D beams with arbitrary composite cross-sections, based on the Saint–Venant solution. In addition, the proper generalized decomposition method is useful in reducing the numerical complexity of reproducing three-dimensional behaviors of thin-walled structures; recent progress can be seen in the work of Sibileau et al. [5]. Asymptotic methods are powerful tools for describing the three-dimensional displacement field of beam models, and have evolved into the well-known variational asymptotic method for thin-walled structures (see Ghorashi [6]). In comparison, the Carrera unified formulation is valuable for defining the displacement field by exploiting arbitrary expansions of unknown variables. In this regard, Carrera et al. [7–9] have made sustained efforts and contributed significantly to development of a theory able to consider various structural problems with no need for ad hoc assumptions.

In practical applications, some structural behaviors are observed and fused in the refinement of thin-walled beam models. For example, the shear lag effect is proven to play a certain role in the performances of box bridges, and has been taken into consideration in the definition of shear warping functions by Cambronero-Barrientos et al. [10] and Yu et al. [11]. In addition, shear correction factors have also been introduced to enhance beam models by accounting for shear deformation effects by Lim and Kim [12], and Akgöz and Civalek [13]. The resulting secondary effects of shear deformation on tall buildings have been studied especially by Lacidogna [14]. In manufacturing, thin-walled metal structures have been shown to exhibit time-varying deformation, which was studied by Tuysuz and Altintas [15] in developing an updated model for reduced-order workpiece dynamic parameters. Furthermore, some new concepts, including reliability, have been introduced into the prediction of dynamic behaviors of thin-walled structures [16]. Meanwhile, experiments have always been one of the main methods for studying thin-walled structures. For example, the deformation in coupled bending and torsional vibrations of non-uniform thin-walled beams has been validated and studied in experiments by Zhou et al. [17].

Recently, some higher-order theories have focused on the identification of a complete set of physically meaningful cross-section deformation modes for thin-walled structures. Among them, generalized beam theory (GBT) is one of the most recent contributions. GBT originates from the work of Schardt [18,19], and has been extended into almost every field of structural analysis of thin-walled beams by Davies et al. [20], Silvestre et al. [21], and Camotim et al. [22]. By applying a piece-wise description of cross-sections and performing cross-section analyses, GBT is able to handle arbitrary prismatic cross-sections [23] and provide a set of deformation modes hierarchically organized into several families. Following the development of GBT, Vieira et al. [24,25] have established a criterion for uncoupling the beam governing equations to derive a set of uncoupled deformation modes representing higher-order effects. One might say that these theories are powerful enough to handle almost any prismatic cross-sections, and any structural analyses, with optimal precision. However, the issue lies in the fact that both of them are based on the solution of the nonlinear eigenvalue problem associated with the government of differential equations in the process of defining deformation modes, which is quite demanding for the nonprofessional. In this sense, a more practicable approach, with fewer features but acceptable accuracy, is more suitable in some cases [26].

Towards this end, a new, more user-friendly procedure for identifying higher-order deformation modes in thin-walled structures with a doubly symmetric cross-section is proposed. The displacement field is considered through a linear combination of a set of linear independent deformation modes, defined over the cross-section, and their amplitudes, only dependent on the longitudinal axis, which has naturally separated the variable dependencies of the cross-section and beam axis dimensions. To present these deformation modes, a modal analysis of the thin-walled structure is carried out employing two-dimensional plate/shell elements, with the shell-like deformation shapes extracted and decomposed into in-plane and out-of-plane components. Then, considering the doubly symmetric cross-section, a novel procedure is implemented by uncoupling classical and higher-order deformation modes first and by defining the shape function for each new mode then. The whole procedure only involves elementary calculation of matrix and vectors, being quite simple but effective enough. The new set of deformation modes are finally adopted in the formulation of the one-dimensional higher-order model for thin-walled structures.

2. One-Dimensional Formulation

The thin-walled cross-section is constituted by a set of rectilinear walls that are symmetrically distributed about two axes vertical to each other. The cross-section may be open or closed, but must be sufficiently thin to suit the Kirchhoff hypothesis. On this basis, a brief review of deriving the one-dimensional formulation is presented in this section.

2.1. Displacement Fields

The displacement of a point on the mid-surface of the cross-section is defined with the axial u, tangential v and normal w components, which are prescribed to be positive along the axial direction of the local coordinate system (n, s, z) adopted for each wall. In addition, a global coordinate system (x, y, z) is set with its origin located in the centroid of the cross-section at one end of the structure. The two coordinate systems are shown in Figure 1.

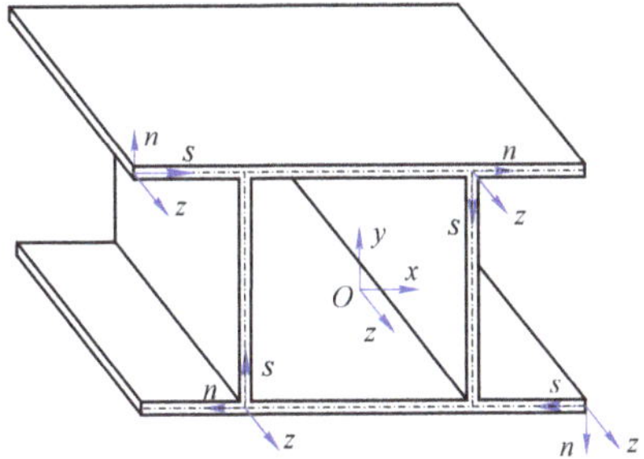

Figure 1. The global (x, y, z) and local (s, n, z) coordinate systems of the thin-walled structure with a doubly symmetric cross-section.

The displacement of an arbitrary point on the structure is described with three components of U, V and W in the global coordinate system. By considering both the membrane and flexural behaviors of the plate, the displacement field, $\mathbf{D} = [U, V, W]$, is obtained as

$$
\begin{aligned}
U(n, s, z) &= u(s, z) - n w_{,s}(s, z), \\
V(n, s, z) &= v(s, z) - n w_{,s}(s, z), \\
W(n, s, z) &= w(s, z),
\end{aligned}
\tag{1}
$$

where a subscript comma denotes differentiation with respect to the following variable. Here, the components, v and w, are in-plane (sn plane) displacements and u is the out-of-plane displacement. By separating the variable dependencies on the s (over the cross-section) and z (along the longitudinal axis) dimensions, these components on the mid-surface are approximated through a set of independent basis functions defined along the coordinate s as

$$u(s,z) = \sum_{k=1}^{N_1} \phi_k(s)\chi_k(z),$$

$$v(s,z) = \sum_{h=1}^{N_2} \psi_h(s)\chi_{N_1+h}(z), \tag{2}$$

$$w(s,z) = \sum_{h=1}^{N_2} \omega_h(s)\chi_{N_1+h}(z),$$

where the subscripts k and h are the numbers of out-of-plane and in-plane deformation modes, respectively; ϕ, ψ, ω are the shape functions of sectional deformation modes on the axial, tangential and normal directions, respectively; and χ is the amplitude function varying along the longitudinal axis, which is also referred to as the generalized displacement. It should be noted that one in-plane mode corresponds to two shape functions ψ and ω, while out-of-plane modes one-to-one match shape functions ϕ.

Substitute Equation (2) into Equation (1), and the three-dimensional displacement $\mathbf{D}$ can be written in a one-dimensional way with a transformation matrix $\mathbf{H}$ as

$$\mathbf{D}(n,s,z) = \mathbf{Hx} = \begin{bmatrix} \boldsymbol{\varphi} & -n\boldsymbol{\omega}\frac{\partial}{\partial z} \\ 0 & \boldsymbol{\psi} - n\boldsymbol{\omega}_{,s} \\ 0 & \boldsymbol{\omega} \end{bmatrix} \left\{ \begin{array}{c} \mathbf{x}_O \\ \mathbf{x}_I \end{array} \right\}, \tag{3}$$

where the generalized displacement vector $\mathbf{x}$, spanned by the set of amplitude functions χ, is separated into two column submatrices, $\mathbf{x}_O$ and $\mathbf{x}_I$; the two submatrices correspond to the out-of-plane and in-plane deformation modes, respectively; and $\boldsymbol{\varphi}$, $\boldsymbol{\psi}$ and $\boldsymbol{\omega}$ are the shape function vectors constituted by the set of ϕ, ψ and ω, respectively.

2.2. Strain and Stress Fields

The strain and stress fields are obtained under the small displacement hypothesis, which are further written in the Kirchhoff's formulation as

$$\varepsilon(n,s,z) = \left\{ \begin{array}{c} \varepsilon_{zz}(n,s,z) \\ \varepsilon_{ss}(n,s,z) \\ \gamma_{sz}(n,s,z) \end{array} \right\} = \mathbf{CD}, \tag{4}$$

$$\sigma(n,s,z) = \left\{ \begin{array}{c} \sigma_{zz}(n,s,z) \\ \sigma_{ss}(n,s,z) \\ \tau_{sz}(n,s,z) \end{array} \right\} = \mathbf{E}\varepsilon, \tag{5}$$

where the compatibility operator $\mathbf{C}$ and the constitutive matrix $\mathbf{E}$ for the plane stress condition are respectively given by

$$\mathbf{C} = \begin{bmatrix} \frac{\partial}{\partial z} & 0 & 0 \\ 0 & \frac{\partial}{\partial s} & 0 \\ \frac{\partial}{\partial s} & \frac{\partial}{\partial z} & 0 \end{bmatrix},$$

$$\mathbf{E} = \begin{bmatrix} \frac{E}{1-v^2} & \frac{Ev}{1-v^2} & 0 \\ \frac{Ev}{1-v^2} & \frac{E}{1-v^2} & 0 \\ 0 & 0 & \frac{E}{2(1+v)} \end{bmatrix}. \tag{6}$$

Here, E and v are the material Young's modulus and Poisson's ratio, respectively.

Substituting Equations (3) and (6) into Equations (4) and (5) yields

$$\varepsilon(n,s,z) = \begin{bmatrix} \boldsymbol{\varphi}\frac{\partial}{\partial z} & -n\boldsymbol{\omega}\frac{\partial^2}{\partial z^2} \\ 0 & \boldsymbol{\psi}_{,s} - n\boldsymbol{\omega}_{,ss} \\ \boldsymbol{\varphi}_{,s} & (-n\boldsymbol{\omega}_{,s} + \boldsymbol{\psi} - n\boldsymbol{\omega}_{,s})\frac{\partial}{\partial z} \end{bmatrix} \left\{ \begin{array}{c} \mathbf{x}_O \\ \mathbf{x}_I \end{array} \right\}, \tag{7}$$

$$\sigma(n,s,z) = \frac{E}{1+v}\left[\begin{array}{cc} \frac{1}{1-v}\varphi\frac{\partial}{\partial z} & -\frac{1}{1-v}n\omega\frac{\partial^2}{\partial z^2} + \frac{v}{1-v}(\psi_{,s} - n\omega_{,ss}) \\ \frac{v}{1-v}\varphi\frac{\partial}{\partial z} & -\frac{v}{1-v}n\omega\frac{\partial^2}{\partial z^2} + \frac{1}{1-v}(\psi_{,s} - n\omega_{,ss}) \\ \frac{1}{2}\varphi_{,s} & \frac{1}{2}(-n\omega_{,s} + \psi - n\omega_{,s})\frac{\partial}{\partial z} \end{array}\right]\left\{\begin{array}{c} x_O \\ x_I \end{array}\right\}. \tag{8}$$

2.3. Beam Governing Equations

The beam energy components are essential for the application of Hamilton's principle, including the strain energy U_{st}, the kinetic energy T_{kn} and the potential energy U_{pt}. By definition, the former two are respectively given by

$$U_{st} = \frac{1}{2}\iiint_V \varepsilon^T \sigma \, dV, \tag{9}$$

$$T_{kn} = \frac{1}{2}\iiint_V \rho \frac{\partial D^T}{\partial t}\frac{\partial D}{\partial t}\, dV, \tag{10}$$

where V is the beam volume and ρ is the material density. The beam is subjected to distributed loads, being defined with the load vector $\mathbf{p} = [p, q, r]^T$. Here p, q and r represent the force densities in the axial, tangential and normal directions, respectively. The potential energy U_{pt} can then be given by

$$U_{pt} = -\int_L \int_A D^T \mathbf{p}\, dA dz. \tag{11}$$

Hamilton's principle states

$$\delta\int_{t_1}^{t_2} (T_{kn} - U_{st} - U_{pt})\, dt = 0, \tag{12}$$

where t_1 and t_2 are the start time and the end time, respectively.

Substituting Equations (3)–(5) and Equations (9)–(11) into Equation (12) yields

$$\int_L \int_A \delta x^T H^T \rho H \frac{\partial^2 x}{\partial t^2}\, dA dz + \int_L \int_A \delta x^T H^T c^T EcHx\, dA dz - \int_L \int_A \delta x^T H^T \mathbf{p}\, dA dz = 0, \tag{13}$$

where A and L are the cross-section area and the beam length, respectively. By applying the condition

$$\delta x\big|_{t=t_1} = \delta x\big|_{t=t_2} = 0, \tag{14}$$

Equation (13) becomes

$$\begin{aligned} &\int_L \int_A \rho\varphi^T\varphi\frac{\partial^2 x_O}{\partial t^2}\, dA dz + \int_L \int_A E^*\varphi^T\varphi x''_O\, dA dz + \int_L \int_A G\varphi_{,s}{}^T\varphi_{,s}x_O\, dA dz + \int_L \int_A (E^*v\varphi^T\psi_{,s} + G\varphi_{,s}{}^T\psi)x'_I\, dA dz \\ &= \int_L \int_A \varphi^T\mathbf{p}\, dA dz \end{aligned}, \tag{15}$$

$$\begin{aligned} &\int_L \int_A \rho n^2\omega^T\omega\frac{\partial^2 x'_I}{\partial t^2}\, dA dz + \int_L \int_A \rho\left(\psi^T\psi + n^2\omega_{,s}{}^T\omega_{,s} + \omega^T\omega\right)\frac{\partial^2 x_I}{\partial t^2}\, dA dz + \int_L \int_A E^*n^2\omega^T\omega x_I^{(4)}\, dA dz \\ &+ \int_L \int_A \left[E^*vn^2\omega^T\omega_{,ss} + E^*vn^2\omega_{,ss}{}^T\omega + G\left(4n^2\omega_{,s}{}^T\omega_{,s} + \psi^T\psi\right)\right]x''_I\, dA dz \\ &+ \int_L \int_A E^*\left(\psi_{,s}{}^T\psi_{,s} + n^2\omega_{,ss}{}^T\omega_{,ss}\right)x_I\, dA dz + \int_L \int_A \left(E^*v\psi_{,s}{}^T\varphi + G\psi^T\varphi_{,s}\right)x'_O\, dA dz \\ &= \int_L \int_A \left(\psi^T q + \omega^T r\right)\, dA dz \end{aligned}, \tag{16}$$

where the superscripts $'$ and $''$ denote the first and second derivatives with respect to the variable z, respectively; $E^* = E/(1-v^2)$ and $G = E/2(1+v)$. Equations (15) and (16) are the governing differential equations of thin-walled structures with a doubly symmetric cross-section.

For the ease of computing, the governing equations are usually interpolated along the longitudinal axis to form one-dimensional finite elements. In view of the two-order partial differential operator in the governing equations, quadratic Lagrange functions are recommended for the interpolation. A special study can be seen in Zhang et al. [26].

3. Higher-Order Deformation Modes

The procedure to identify sectional deformation modes is begun with the modal analysis based on the shell/plate theory. In this process, the shell-like deformation of the cross-section is presented. The cross-section may be open or closed, with or without branches. Without loss of generality, a thin-walled cross-section possessing all the features above, as shown in Figure 1, is chosen as an example to illustrate the proposed approach.

3.1. Shell-Like Deformation

Figure 2 shows a thin-walled structure with a doubly symmetric cross-section. For the convenience of modeling, the structure is fixed at one end with the other end free. Associated geometry and material parameters are set as: section height $h = 0.3$ m, section width $b_1 = 0.3$ m, flange width $b_2 = 0.15$ m, axial length $L = 1.5$ m, wall thickness $\tau = 0.01$ m, Young's modulus $E = 2 \times 10^{11}$ Pa, Poisson's ratio $\nu = 0.3$ and material density $\rho = 7850$ kg/m^3. It should be pointed out that the choice of these parameters is arbitrary; however, there should be noticeable cross-section deformation.

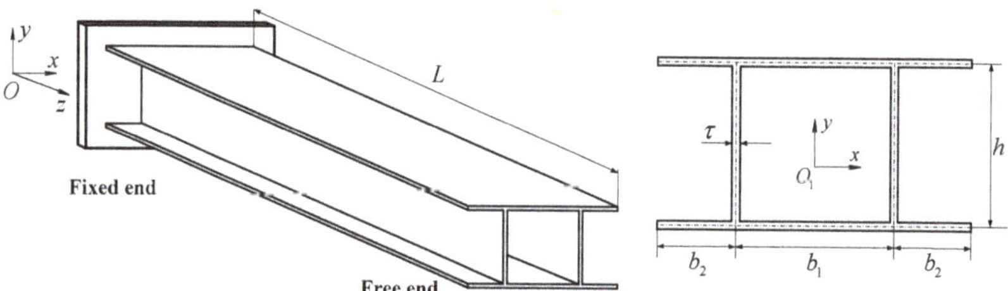

Figure 2. A cantilevered thin-walled structure with a branched, doubly symmetric cross-section.

Since the walls are thin enough, the shell/plate theory is applicable in presenting the shell-like deformation of the cross-section. The shell/plate theory assumes that a mid-surface plane can be used to represent a three-dimensional plate/shell in a two-dimensional form. Currently, the theory has been developed for various shell/plate elements that have been employed in commercial finite element software, such as ANSYS, ABAQUS, ADINA and MSC. ANSYS Shell 181 element is available to be used to model the thin-walled structure in Figure 2. A total of 1080 quadrilateral elements are employed, with 30 elements evenly distributed in the longitudinal direction and 36 over the cross-section. By applying the modal analysis function, the first 12 modal shapes are obtained as the object modes. It should be noted that the number of object modes is related to the number of sectional deformation modes to be identified, which can affect the accuracy of the final one-dimensional model.

Figure 3 presents the deformed contours of the thin-walled structure that have been projected onto the global xy plane. The results show that sectional deformations are dominant for almost every mode shape and that they have become non-negligible factors for the performances of thin-walled structures. In fact, the phenomenon has also been observed in some experiments. For example, noticeable cross-section deformations due to axial and transversal loadings are recreated and exhibited by Debski et al. [27] (see Figure 4) and Ciesielczyk and Studziński [28] (see Figure 5), respectively. In this sense, the conventional beam theory is no longer suitable for these cases. At the same time, it also explains why it is important to study the higher-order deformation of thin-walled structures. Moreover, the sectional deformation modes have been presented, and the next challenge is how to retrieve them.

To obtain these deformation modes, the nodal displacements of the free end cross-section are extracted and decomposed into in-plane (distortion in the xy plane) and out-of-plane (warping vertical to the xy plane, see Carpinteri et al. [29]) components. Figures 6 and 7 exhibit the two deformation mode families, respectively. It should be noted that the out-of-plane family members are not definite before a numerical analysis since not all deformation modes possess out-of-plane components. For example, the z-direction displacements of modes 5, 6, 9 and 11 are almost zero or less than 1/100,000 of the ones along the x- or y-axis. In these cases, the relevant deformation will be passed over. Therefore, only eight out-of-plane deformation modes are obtained from the first 12 modal shapes.

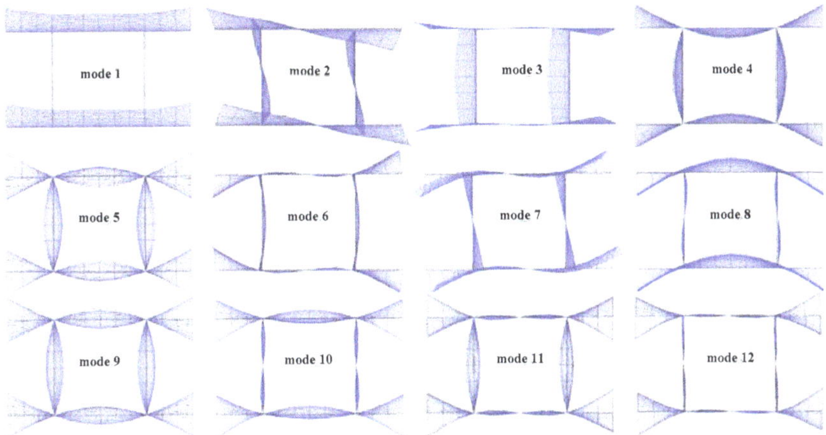

Figure 3. Axial projections of the first 12 modal shapes of the thin-walled structure with a branched, doubly symmetric cross-section.

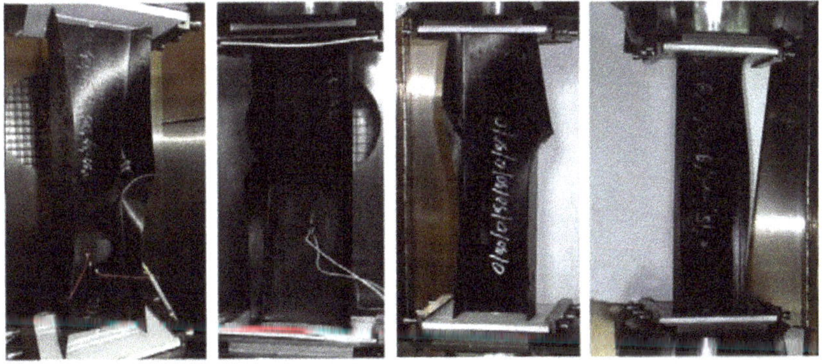

Figure 4. Deformation of the thin-walled structure with a channel section under axial loading.

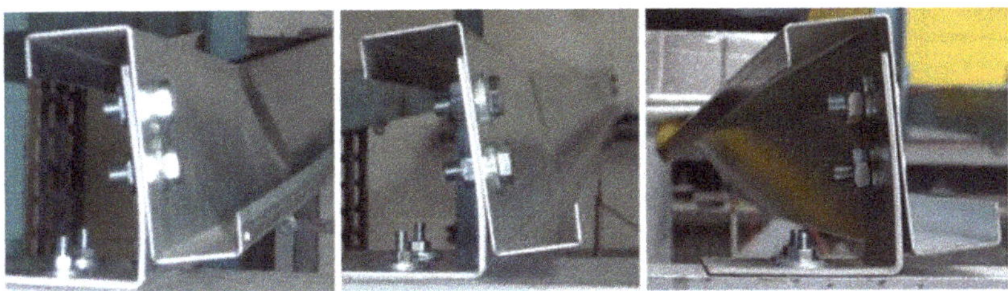

Figure 5. Deformation of the Z-section thin-walled structure under transversal loading.

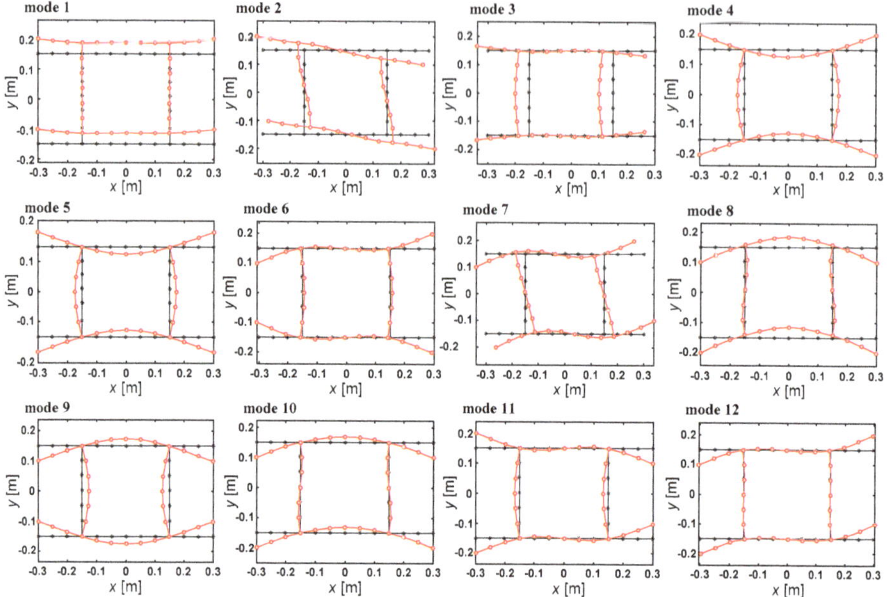

Figure 6. In-plane deformation mode family retrieved from the first 12 modal shapes of the thin-walled structure with a branched, doubly symmetric cross-section.

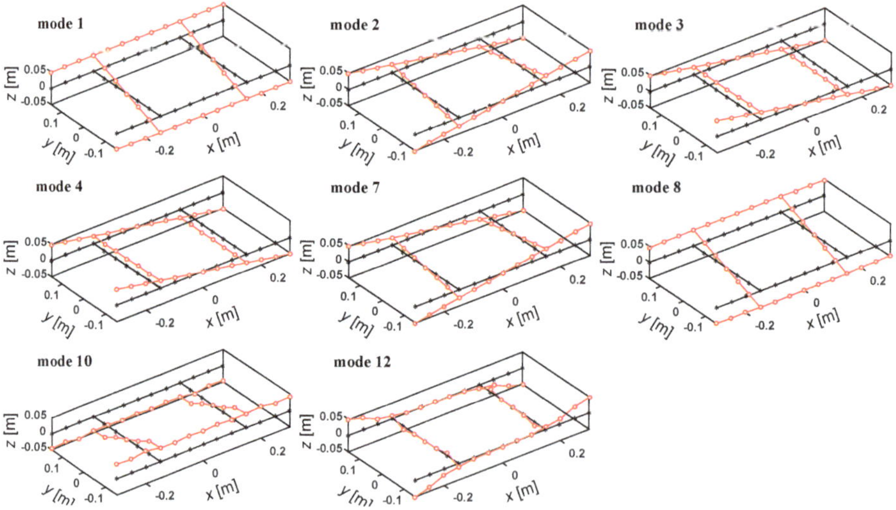

Figure 7. Out-of-plane deformation mode family retrieved from the first 12 modal shapes of the thin-walled structure with a branched, doubly symmetric cross-section.

Obviously, those deformed cross-section shapes in Figures 6 and 7 are the superposition of rigid movements of the whole cross-section (classical Timoshenko modes including three translations and three rotations) and sectional elastic deformations (higher-order deformation modes). Since the classical modes have been obtained, the next procedure is pivotal to retrieve the higher-order deformation modes from the modal shapes by removing the components of classical modes.

3.2. Benchmark Points

The retrieval of higher-order deformation modes is also mandatory from the view of obtaining a set of sectional deformation modes in hierarchy to form a reduced model. The implementation presupposes the uncoupling of the classical and higher-order modes in the sectional deformation shapes shown in Figures 6 and 7.

For the implementation with a computer, a set of benchmark points are defined on the cross-section to indicate deformation patterns in a digitized way. In other words, each deformation pattern corresponds to a kind of cross-section deformation shapes described with a group of displacement signs stemming from the benchmark points. Therefore, the deformation pattern can indicate whether one sectional deformation mode participates in an object modal shape or not. In Figure 8, 12 benchmark points are set on the corner nodes (points 1, 3, 6 and 9), free end nodes (points 4, 7, 10 and 12) and intermediate nodes (2, 5, 8 and 11). The displacement component of one benchmark point may be one of the three cases: positive (marked as "+"), negative (marked as "-") or null (marked as "0"). The in-plane deformation modes are expressed with the normal and tangential displacements (two degrees of freedom) of the benchmark points while axial displacements for the out-of-plane modes. In this sense, the 12 benchmark points can identify a total of 3 (number of displacement signs) × 12 (number of benchmark points) × 2 (degrees of freedom) = 72 in-plane modes. Similarly, half of out-of-plane modes can be distinguished since half of degrees of freedom are set for them.

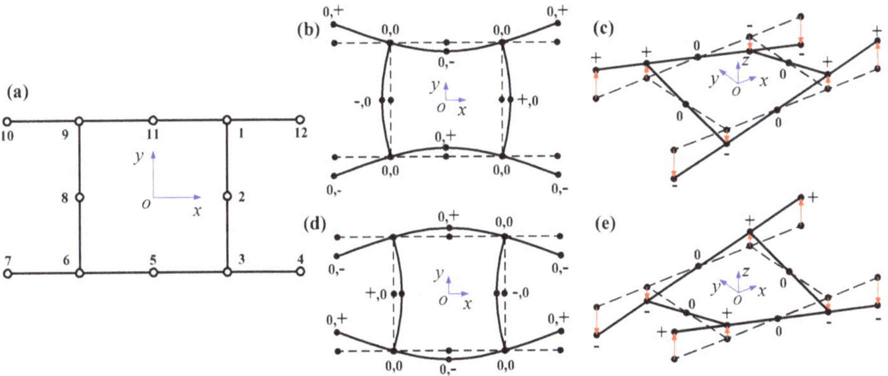

Figure 8. Benchmark points on the doubly symmetric cross-section: (**a**) 12 benchmark points; (**b**) one in-plane mode and (**c**) one out-of-plane mode indicated with benchmark points; (**d**) equivalent form of the in-plane mode shown in (**b**); (**e**) equivalent form of the out-of-plane mode shown in (**c**).

Figure 8b,c show one in-plane and one out-of-plane deformation modes indicated with the deformation signs of benchmark points, respectively. The problem is that two deformation modes, such as the two shown in Figure 8b,d or the two shown in Figure 8c,d, may have opposite displacement components. Essentially, the two forms describe the same deformation mode from the view of the energy method. That is to say, the two forms of deformation modes are equivalent in mechanics. Accordingly, the set of 12 benchmark points can distinguish only half of the sectional deformation modes, namely 36 in-plane modes and 18 out-of-plane ones. In fact, these modes are enough to form a reduced one-dimensional model with a qualified accuracy in most structural analyses.

Moreover, it should be noted that the number of benchmark points needed is related not only to the number of sectional deformation modes to be identified but also to the cross-section configuration. Generally, the corner points and free end points on the cross-section are basic, and a certain number of intermediate points are optional. For example, eight benchmark points may be employed in the cross-section shown in Figure 9a, which can determine 24 in-plane deformation modes and 12 out-of-plane ones. However, the number of benchmark points can be reduced to 4, as shown in

Figure 9b, when less sectional deformation modes are needed. In addition, for a cross-section where the lengths of walls vary a lot, some amount of intermediate nodes may be essential to separate the walls into several segments with approximately equal lengths. This measurement is supposed to contribute to the accurate description of the sectional deformation. For example, the I-section in Figure 9c employs two intermediate benchmark points on the web wall, while no intermediate points are introduced in the flange walls of the dual-cell cross-section in Figure 9d. The difference stems from the experience that relatively longer walls usually deform more complexly, where more interpolation nodes are needed to capture the deformation features.

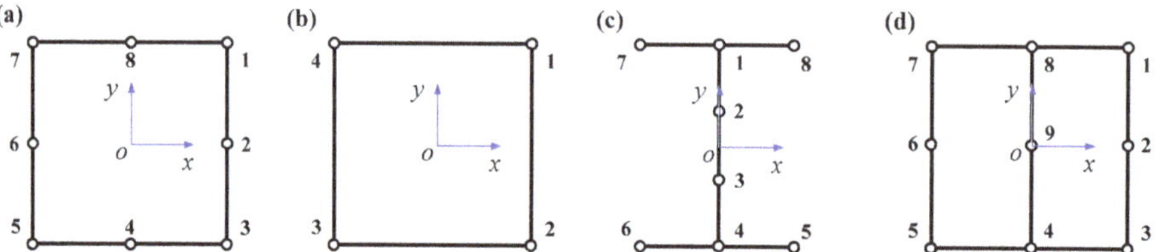

Figure 9. Benchmark points for different thin-walled cross-sections: (**a**) eight benchmark points and (**b**) four benchmark points for a rectangular cross-section; (**c**) eight benchmark points for the I-section; and (**d**) nine benchmark points for a dual-cell cross-section.

3.3. Identification of Deformation Modes

As shown in Figures 6 and 7, the modal shapes consist of rigid movements of the whole cross-section and the sectional elastic deformation. A novel approach is needed to uncouple them by removing the components of rigid movements of the whole cross-section.

Here, classical Timoshenko modes [30], including three translations and three rotations, are adopted to describe the rigid movements of the cross-section, which are shown in Figure 10. The first three indicate the out-of-plane modes, being numbered as modes I, II and III corresponding to the rotation about z-axis, the translations along y- and x-axis, respectively, while the latter three embody the in-plane ones, which are given the numbers as modes i, ii and iii to represent the extension along z-axis, the rotations about x- and y-axis, respectively. In addition, 12 benchmark points are employed to distinguish these modes for the identification with a computer.

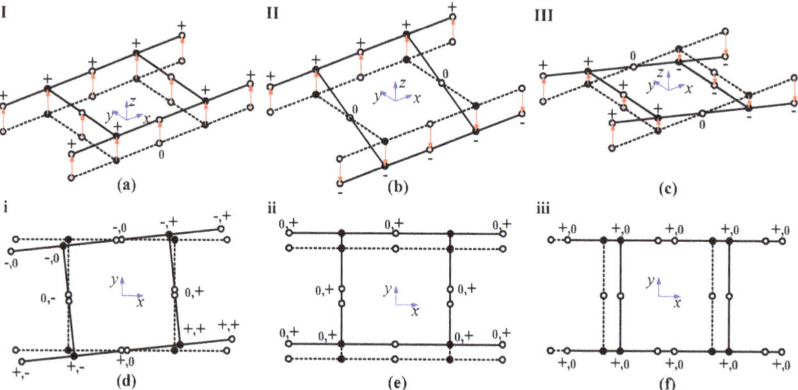

Figure 10. Classical modes indicated with benchmark points, numbered as modes I, II, III for the out-of-plane ones and modes i, ii, iii for the in-plane ones: (**a**) rotation about z-axis; (**b**) translation along y-axis; (**c**) translation along x-axis; (**d**) extension along z-axis; (**e**) rotation about x-axis and (**f**) rotation about y-axis.

To remove the components of rigid movements of the cross-section, the key is to determine their participations in the sectional deformation shape. Here, a criterion is proposed to uncouple deformation modes in this subsection.

3.3.1. Uncoupling Deformation Modes with a Novel Criterion

The uncoupling is first implemented within each category of sectional deformations. The concept means a group of deformation shapes which contain the components of the same classical Timoshenko mode. In consideration of the doubly symmetric cross-section, the concept can be further simplified to be the deformation patterns indicated with the displacement component signs of the four corner benchmark points (the solid circles in Figure 10). Actually, there are six categories of sectional deformation modes in accordance with the six classical Timoshenko modes. We might as well classify them into categories 1–6 for convenience. In other words, focusing on the four corner benchmark points, if all their nodal displacements along z-, y- or x-axis are with the same values, the corresponding deformation shapes are supposed to belong to category 1, category 5 or category 6, respectively; if the nodal displacements are opposite about x- or y-axis, the corresponding deformation shapes are supposed to belong to category 2 or category 3, respectively; if the nodal displacement components along x- and y-axis are central symmetric about the cross-section centroid, the corresponding deformation shape is supposed to belong to category 4.

Hence, the modal shapes shown in Figures 6 and 7 can be separated into several categories. Among the in-plane family, modes 1, 8 and 10 belong to category 5, modes 2, 7 and 12 belong to category 4, and modes 3, 6 and 11 belong to category 6. In particular, the deformation of modes 4, 5 and 9 are zero on the corner benchmark points, implying no components of classical modes. Among the out-of-plane family shown in Figure 7, modes 1, 8 and 10 belong to category 2, and modes 3 and 4 belong to category 3.

Then, within each category, the participations of classical modes can be determined and removed from the modal shapes. Since the cross-section is doubly symmetric, the process can be implemented just on a quarter of the cross-section. For the categories above except category 4, the process can be carried out with the solution of the following equation:

$$\frac{\mathbf{\Theta}_k}{\|\mathbf{\Theta}_k\|_\infty} - \gamma \frac{\mathbf{\Theta}_c}{\|\mathbf{\Theta}_c\|_\infty} = \mathbf{\Theta}_i, \ \gamma = \begin{cases} \frac{\mathbf{\Theta}_k(m,1)\|\mathbf{\Theta}_c\|_\infty}{\mathbf{\Theta}_c(m,1)\|\mathbf{\Theta}_k\|_\infty}, \ \text{for } \mathbf{\Theta}_k(m,1) \neq 0, \mathbf{\Theta}_k(m,2) = 0 \\ \frac{\mathbf{\Theta}_k(m,2)\|\mathbf{\Theta}_c\|_\infty}{\mathbf{\Theta}_c(m,2)\|\mathbf{\Theta}_k\|_\infty}, \ \text{for } \mathbf{\Theta}_k(m,1) = 0, \mathbf{\Theta}_k(m,2) \neq 0 \end{cases} \tag{17}$$

where $\|\ \|_\infty$ means the infinite norm; $\mathbf{\Theta}_k$ is the vector constructed by the corresponding nodal displacement components of a deformation shape to be identified; γ is a ratio to guarantee the displacement components of a corner benchmark point to be null after the process; $\mathbf{\Theta}_c$ is the vector consisting of nodal displacements of an classical mode; $\mathbf{\Theta}_i$ is the vector constituted with the nodal displacements of the newly identified deformation mode, and m is the node number of one corner benchmark point.

For category 4, where both the tangential and normal displacement components are nonzero, another formulation needs to be solved to determine the participation of one sectional mode:

$$\frac{\mathbf{\Theta}_k}{\|\mathbf{\Theta}_k\|_\infty} - \gamma \frac{\mathbf{\Theta}_c}{\|\mathbf{\Theta}_c\|_\infty} - \lambda \frac{\mathbf{\Theta}_r}{\|\mathbf{\Theta}_r\|_\infty} = \mathbf{\Theta}_i, \ \gamma = \frac{[\mathbf{\Theta}_k(m,1) + \mathbf{\Theta}_k(m,2)]\|\mathbf{\Theta}_c\|_\infty}{[\mathbf{\Theta}_c(m,1) + \mathbf{\Theta}_c(m,1)]\|\mathbf{\Theta}_k\|_\infty}, \tag{18}$$

where $\mathbf{\Theta}_r$ is the vector constructed by the nodal displacements of Vlasov distortion [30] and λ is the ratio to guarantee removing the corresponding deformation components.

Actually, a novel criterion has just been expressed in Equations (17) and (18) in uncoupling classical modes and sectional deformation modes. With the process above, the components of sectional rigid movements are removed from the modal shapes, with the residual deformation shapes left to define sectional deformation modes. Furthermore, the process is also useful in the identification of new sectional deformation modes from residual deformation shapes sharing the same deformation pattern. The implementation is to be illustrated in the next subsection.

3.3.2. Higher-Order Deformation Modes

The identification of higher-order deformation modes can be classified into three cases, considering the physical interpretation and the clear hierarchy to form a reduced one-dimensional model.

In the first case, the object modal shapes that fall outside of the six categories are directly identified as higher-order deformation modes, which reflect the distortion or warping of the thin-walled cross-section. In modeling a thin-walled structure, these modes play important roles in the improvement of the model accuracy. In this respect, they are classified as the primary deformation modes, next to the classical Timoshenko modes. In the present example, two deformation modes of this kind can be identified, including mode 4 of the in-plane family in Figure 6, and mode 2 of the out-of-plane family in Figure 7.

In the second case, the object modal shapes that fall inside of the six categories are decomposed into the components of classical modes and those of residual deformation shapes, by applying the criterion shown in Equations (17) and (18). The residual deformation shapes can be identified as higher-order deformation modes. Since always working with the classical modes, rather than playing roles independently in a modal shape, they are not vital for the accuracy of a one-dimensional model. Hence, they are classified as the secondary deformation modes, next to the primary ones. For example, within category 5, removing the components of mode iii from the deformation shape of mode 3 yields mode vi, as a secondary mode. Figure 11 demonstrates the whole process, where one in-plane and one out-of-plane secondary deformation modes are identified, respectively.

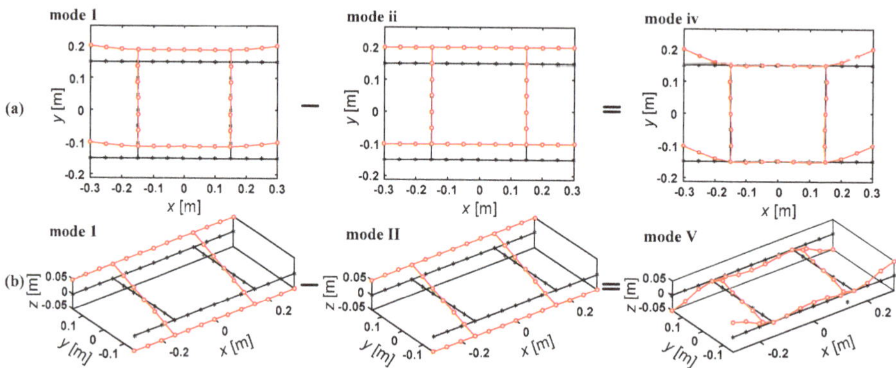

Figure 11. Deriving the secondary deformation modes for the doubly symmetric thin-walled cross-section: (**a**) in-plane mode iv from in-plane mode 1; (**b**) out-of-plane mode V from out-of-plane mode 1.

In the third case, more than one higher-order mode, sharing the same deformation pattern, is identified in the former two cases. These modes are different in physical interpretation, but cannot be distinguished through deformation patterns indicated with the displacement signs of benchmark points. To guarantee their independency, among these residual deformation shapes, the ones derived from a higher order modal will eliminate the components of the one derived from a lower order modal. In this process, the criterion expressed with Equations (17) and (18) should be applied again. The final residual deformation shapes can be identified as new higher-order deformation modes. However, since they are not dominant in any modal shapes, their priority is lower than the secondary modes. Therefore, they are classified as the spare deformation modes. In fact, they can be neglected in a reduced model when the computation efficiency is more significant. Figure 12 has displayed the process, where one in-plane and one out-of-plane spare modes are identified, respectively.

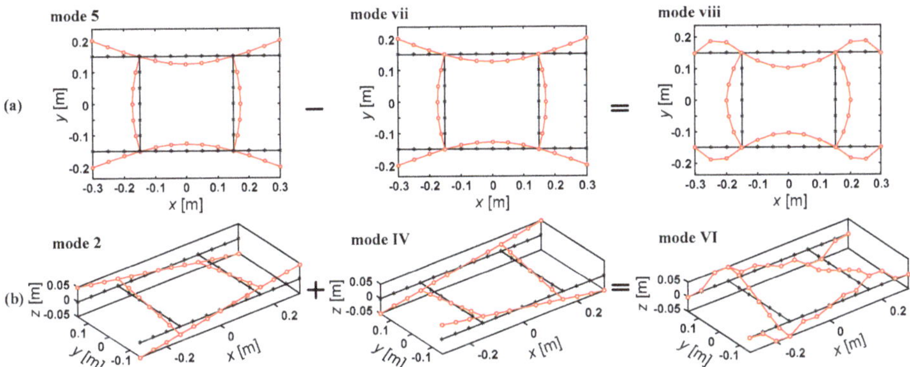

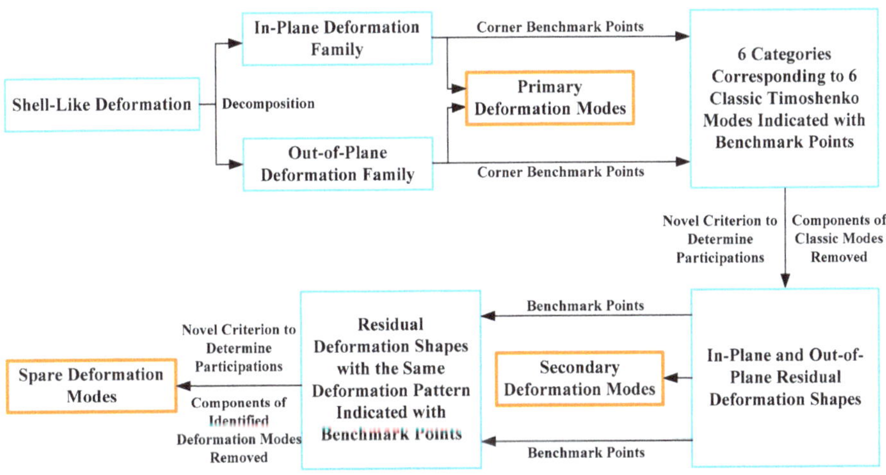

Figure 12. Deriving the spare deformation modes for the doubly symmetric thin-walled cross-section: (**a**) mode viii from mode 5; (**b**) mode xiii from mode 10.

To better exhibit the procedure above, a flow chart is provided in Figure 13, which shows how to uncouple the deformation modes to retrieve primary, secondary and spare deformation modes, respectively. It should be noted that the procedure does not refer to the geometry parameters of the cross-section. However, the double symmetry of the cross-section should be guaranteed since it is essential in the uncoupling of deformation modes by means of the novel criteria. That is to say, the proposed procedure of identifying higher-order deformation modes is general for any doubly symmetric cross-sections.

Figure 13. The flowchart providing a brief view of the process involved in uncoupling higher-order deformation modes of the doubly symmetric thin-walled cross-section.

3.3.3. Shape Functions of Sectional Deformation Modes

To be adopted in the approximation of the displacement field shown in Equation (2), these identified deformation modes need to be described with shape functions in a mathematical way. This process can be carried out by means of the curve fitting technique or the interpolation method. Since the shape functions should be continuous at end nodes of each cross-section wall, the interpolation method is more advisable in this case for the determinacy of passing through one point. Furthermore, the piecewise interpolation is deemed to enhance the capability of capturing the sectional deformation with lower-order polynomials.

As shown in Figure 14, the interpolation is conducted in the first quadrant, actually. Then, the piecewise shape functions for the other three quadrants are completed by applying the symmetry condition. Even in the first quadrant, the interpolation has been separated into three pieces corresponding to the three wall segments. The derived shape function is also related to the number of interpolation nodes to be applied. Generally, the most frequently used interpolation polynomials, namely the linear, quadratic and cubic ones, need two, three and four nodes to determine the unknown coefficients, respectively. Just as shown in Figure 14, the two in-plane modes are interpolated with cubic polynomials, involving four nodes for each segment, while the two out-of-plane modes apply quadratic and linear polynomials, respectively. Naturally, the latter two employ fewer interpolation nodes. It should be pointed out that the two end nodes must be taken into account in the interpolation though a variable number of nodes may be needed for different polynomials. The purpose is to ensure the continuity of the shape function over the whole cross-section.

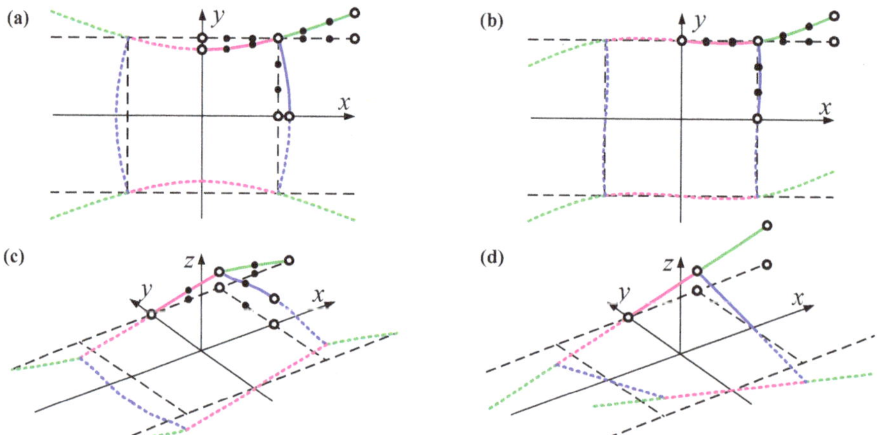

Figure 14. Interpolation of the shape functions of the doubly symmetric thin-walled cross-section: (**a**) one in-plane mode in symmetry and (**b**) one in-plane mode in anti-symmetry described with cubic functions, respectively; (**c**) one out-of-plane mode in symmetry described with quadratic functions; (**d**) one out-of-plane mode in anti-symmetry described with linear functions.

Although polynomials of any order may be acceptable in describing a sectional mode, some conventional experience still deserves attention. For example, the tangential component of a sectional deformation mode is usually interpolated with a linear polynomial since it usually varies moderately along the cross-section mid-surface, while the normal component needs a cubic function to make its two-order partial derivative with respect to s keep nonzero in the governing equations. As for the axial component, it is always approximated with linear functions in view of the relatively small values compared with the normal ones. However, quadratic or cubic polynomials may also be used in some cases where an in-depth cross-section analysis is in need. For example, the identified deformation modes of the presented cross-section are described with cubic polynomials for the normal and axial components and with linear ones for the tangential component, as shown in Figure 15.

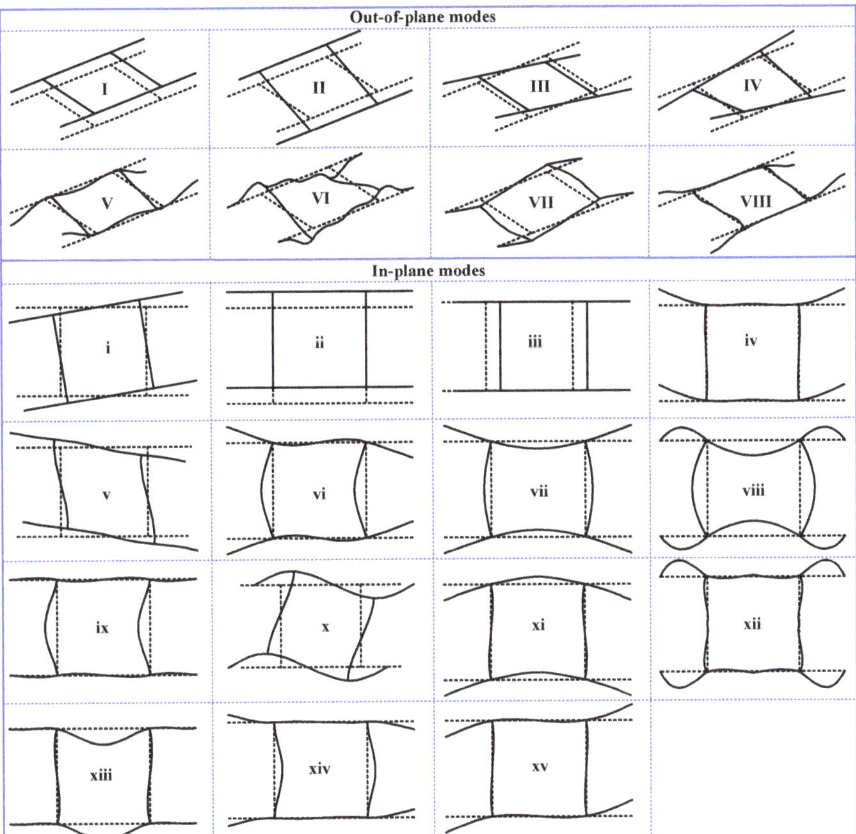

Figure 15. The shape functions of the identified out-of-plane and in-plane higher-order deformation modes of the doubly symmetric thin-walled cross-section.

4. Applications and Illustrative Examples

The application of the proposed approach to the presented doubly symmetric cross-section leads to a set of 15 sectional deformation modes, being exhibited in Figure 15. They can be divided into classical modes and higher-order deformation modes, and the latter ones can be further classified into primary, secondary and spare ones according to the way they are identified. By substituting them into the governing equations, a new one-dimensional higher-order model can be obtained for the analysis of thin-walled structures. In this section, numerical studies are carried out to validate the versatility of the new model. Furthermore, for the ease, the thin-walled cross-section in Figure 1 is employed in numerical studies.

4.1. Convengence of the Finite Element

A quadratic finite element has been developed based on the governing equations, and its convergence is checked in this part. By applying the element in the free vibration analysis of the structure in Figure 2, the relative errors of the first 15 natural frequencies have been demonstrated in Figure 16, varying with the number of employed elements. It should be noted that the structure is meshed equally along the length, and that the converged frequencies are obtained with 100 proposed finite elements, which are considered to become stable.

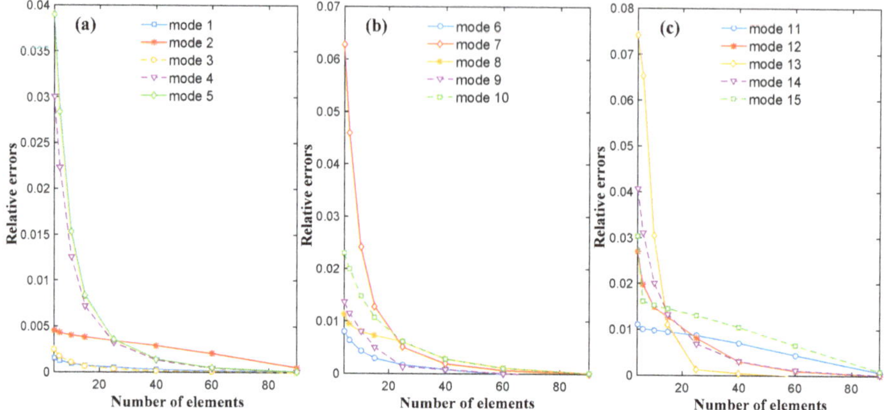

Figure 16. Convergence of the first 15 natural frequencies of the thin-walled structure varying along with the number of employed proposed elements: (**a**) the 1st-5th modes; (**b**) the 6th-10th modes and (**c**) the 11th-15th modes.

The presentation reveals that the frequencies of the first 15 modes converge with different rates but finally achieve similar relative errors smaller than 0.1%, when 90 proposed elements are employed. As a consequence, we just employ no fewer than 90 elements in the numerical examples of thin-walled structures.

4.2. Case Study 1: A Thin-Walled Structure Fixed at One End

In order to demonstrate the validity and the accuracy of the proposed element, the thin-walled structure with one end fixed and the other end free is considered as the first numerical example. Accordingly, 90 proposed elements are employed to mesh it along the beam axis with an equal length. The results are compared with those of the ANSYS shell model, which consists of 1440 Shell 181 4-node shell elements, distributed as 30 along the length and 48 over the cross-section. Table 1 presents the results about the natural frequencies of the first 15 modes. It should be pointed out that the relative errors are calculated based on the assumption that the results derived from ANSYS shell theory are accurate enough.

Table 1. Natural frequencies of the first 15 modes of the cantilevered thin-walled structure.

Mode	Present Model [Hz]	ANSYS Shell [Hz]	Relative Errors [%]
1st	145.63	140.38	3.74
2nd	178.26	171.89	3.71
3rd	181.35	175.63	3.26
4th	183.62	185.45	−0.99
5th	207.44	215.37	−3.68
6th	256.86	262.87	−2.29
7th	257.87	263.73	−2.22
8th	262.62	263.86	−0.47
9th	263.46	270.28	−2.52
10th	274.70	277.07	−0.86
11th	279.30	278.06	0.45
12th	297.71	309.49	−3.81
13th	310.89	324.74	−4.26
14th	312.69	324.75	−3.71
15th	332.30	346.36	−4.06

The results in Table 1 indicate that the natural frequencies obtained from the proposed model are very close to those from ANSYS shell theory, with relative errors smaller than 4.5%. The results might be not as accurate as those of two- or three-dimensional models, but one should bear in mind that a one-dimensional theory has achieved this with fewer than 1/10 even 1/100 degrees of freedom. It is a great improvement on the computation efficiency. In addition, not any one-dimensional model can predict the first 15 natural frequencies of a thin-walled structure with a similar accuracy. In this sense, the proposed model possesses the advantage of giving balanced consideration both on the precision and the efficiency.

In addition, the longitudinal analysis has also been carried out to prove the hierarchic capability of the identified deformation modes. Since in-plane modes have been specially researched by Zhang et al. [28], this study is focused on the out-of-plane ones. As shown in Figure 17, the amplitudes of these modes (χ_i in Equation (2)), also known as generalized displacements, fluctuate along the beam axis. Plainly, each modal shape consists of the components of several deformation modes. The most typical case is that a classical mode plays the predominant role with a secondary mode being auxiliary, such as the first, the third, the fourth, the seventh, the ninth and the tenth modes among the first 12 modes. The second case is that the participations of higher-order modes can almost be neglected compared with those of classical modes, such as the fifth, the sixth and the eighth modes. The third case is that the primary modes are dominant with the spare modes playing supplementary roles, such as the second and the eleventh modes. The results confirm that the identified deformation modes possess the hierarchic capability, being able to obtain a reduced model.

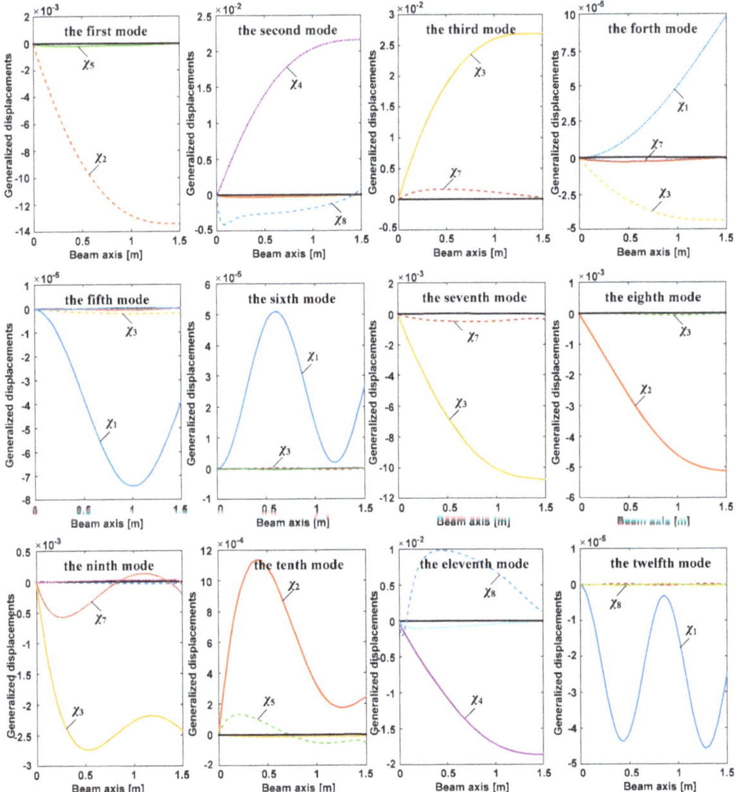

Figure 17. Longitudinal variations of generalized displacements of the thin-walled structure in free vibration analyses.

Moreover, the proposed model may also have the potential of reproducing three-dimensional behaviors of the thin-walled structure. To check this, the free vibration shapes are analyzed and presented in Figure 18. The results are compared with those of a three-dimensional model based on ANSYS Shell 181. The comparison reconfirms that the proposed model agrees well with ANSYS shell theory. Meanwhile, it is also proved that the presented model can accurately reproduce three-dimensional deformations of the first 12 natural modes, with the higher-order deformation modes identified from the first 12 shell-like deformation shapes.

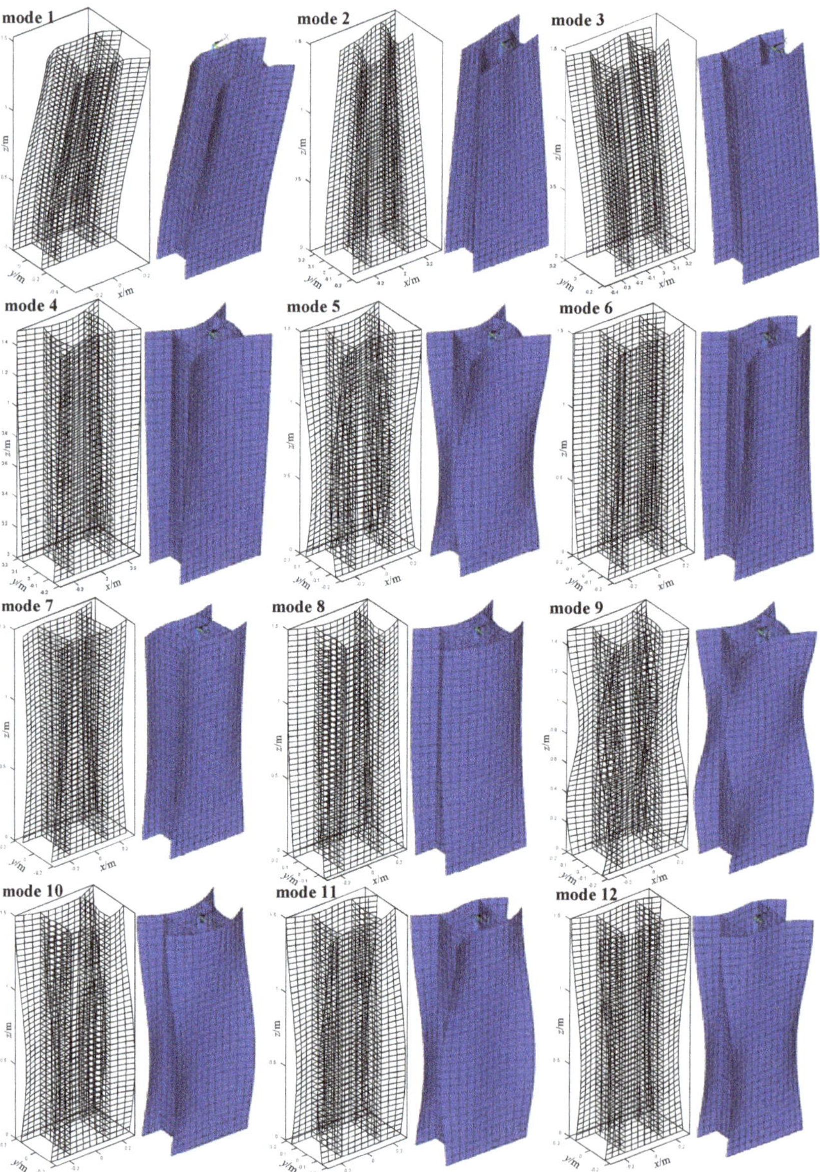

Figure 18. Comparison of free vibration shapes of the cantilevered thin-walled structure between ANSYS shell model (**right**) and proposed model (**left**) concerning the first 12 modes.

Actually, with the higher-order deformation modes identified from the first 12 shell-like deformation shapes, the proposed one-dimensional model can also predict higher order mode shapes. In order to validate the idea, the free vibration shapes of the 13th-15th are also studied and presented in Figure 19. The results support the dedication that the proposed model can accurately calculate more modal shapes with fewer sectional deformation modes. It is also a clear proof of the hierarchic capability of the set of the deformation modes.

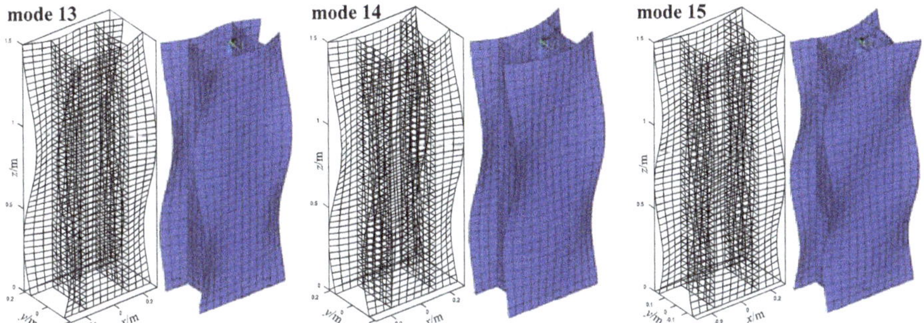

Figure 19. Comparison of free vibration shapes of the cantilevered thin-walled structure between ANSYS shell model (**right**) and proposed model (**left**) concerning the 13th–15th modes.

4.3. Case Study 2: A Thin-Walled Structure Fixed at Two Ends

As a next example, the thin-walled structure in Figure 2, being fixed at two ends, is chosen for the free vibration analysis. The example stems from the idea of testing the proposed model with different boundary conditions. One should bear in mind that the set of sectional deformation modes are derived through the cross-section analysis implemented on a cantilevered thin-walled structure; then, whether they are applicable to a structure with different boundary conditions will be very significant for the practicability of the derived model. In fact, they should be equally effective if the set of sectional deformation modes are positively the natural characteristic of the thin-walled cross-section.

Table 2 presents the information about the first 10 modes, consisting of the values of the natural frequencies, obtained with the proposed model and ANSYS shell theory, and the relative errors. Similarly, the data of the proposed model are calculated with 90 quadratic finite elements equally distributed along the axial direction. As a comparison, the ANSYS shell model is discretized into 1120 Shell 181 4-node elements, distributed as 30 elements along the length, and 48 over the cross-section.

Table 2. Natural frequencies of the first 10 modes of the fixed-fixed thin-walled structure.

Mode	Present Model (Hz)	ANSYS Shell (Hz)	Relative Errors (%)
1st	192.92	199.38	−3.24
2nd	236.09	247.04	−4.43
3rd	251.2	262.52	−4.31
4th	261.05	264.38	−1.26
5th	272.95	285.35	−4.35
6th	290.23	303.77	−4.46
7th	290.85	303.95	−4.31
8th	309.11	319.79	−3.34
9th	321.77	335.58	−4.12
10th	351.7	367.43	−4.28

The results in Table 2 show that the natural frequencies obtained from the proposed model agree well with those from ANSYS shell theory, with relative errors smaller than 4.5%. The accuracy is similar to that of the cantilevered structure, and even the values of some modes are more accurate than those of the cantilevered structure, such as the first mode. Thus, it can be seen as a preliminary proof of the supposition above.

Furthermore, the capability of reproducing three-dimensional behaviors of the model is checked again, focusing on the modal shapes. Figure 20 provides the comparison concerning the 1st to 9th modal shapes of the fixed-fixed thin-walled structure. The results reconfirm the good agreements with ANSYS shell theory. Hence, it is rational to declare that the proposed model with the same set of sectional deformation modes can accurately reproduce three-dimensional behaviors of thin-walled structures with different boundary conditions. It also embodies the applicability and the generality of the proposed model.

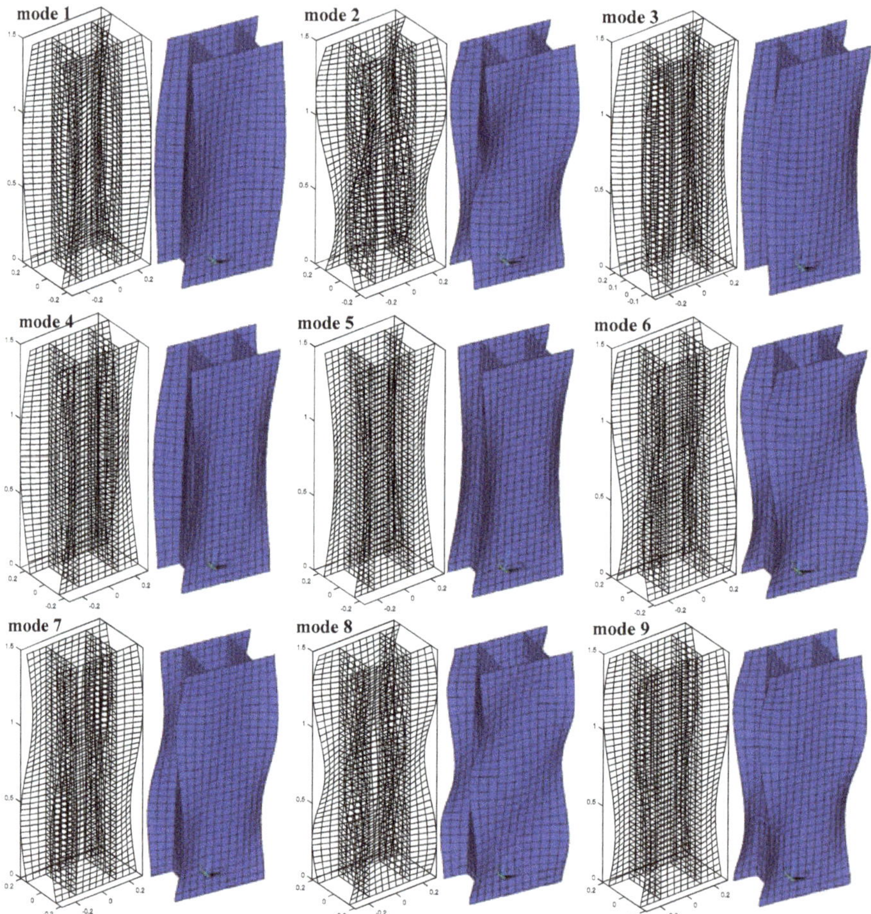

Figure 20. Comparison of free vibration shapes of the fixed-fixed thin-walled structure between ANSYS shell model (**right**) and proposed model (**left**) concerning the first nine modes.

5. Conclusions

The paper focuses on the identification of sectional deformation modes of the doubly symmetric cross-section. The plate/shell theory is utilized to present the shell-like deformation shapes of a thin-walled structure, which are further uncoupled by applying a proposed criterion, with residual deformation shapes left. By introducing benchmark points, these shapes are classified into several deformation patterns, within which the higher-order deformation modes are derived by removing the components of identified ones. By considering the doubly symmetric cross-section, these modes are approximated with shape functions. The identified modes are finally used to formulate the governing equations of the thin-walled structure. The proposed model is examined in numerical examples, and the results validate the accuracy, the efficiency and the practicability of the new model in reproducing three-dimensional behaviors of thin-walled structures. Actually, the authors plan to develop a similar procedure for thin-walled cross-sections with curved walls, by improving the method of approximating the shape functions. In addition, the approach may be also applicative for arbitrary cross-sections, and related study is currently under development.

Author Contributions: Methodology, L.Z.; Project Administration, A.J.; Supervision, A.J.; Visualization, L.P.; Writing—Original Draft, L.Z.; Writing—Review and Editing, W.Z.

Funding: This research was funded by the National Natural Science Foundation of China (Grant No. 51805144), the Foundation of the State Key Laboratory of Mineral Processing (Grant No. BGRIMM-KJSKL-2016-04), the Natural Science Foundation of Jiangsu Province (Grant No. BK20170300), and the Fundamental Research Funds for the Central Universities (Grant No. 2016B14914).

Conflicts of Interest: The authors declare no conflict of interest.

References

1. Peres, N.; Gonçalves, R.; Camotim, D. GBT-based cross-section deformation modes for curved thin-walled members with circular axis. *Thin-Walled Struct.* **2018**, *127*, 769–780. [CrossRef]
2. Carrera, E.; Pagani, A.; Petrolo, M.; Zappino, E. Recent developments on refined theories for beams with applications. *Mech. Eng. Rev.* **2015**, *2*, 1400298. [CrossRef]
3. Vlasov, V.Z. *Thin-Walled Elastic Beams*; Israel Program for Scientific Translations: Jerusalem, Israel, 1961.
4. Yoon, K.; Kim, D.N.; Lee, P.S. Nonlinear torsional analysis of 3D composite beams using the extended St. Venant solution. *Struct. Eng. Mech.* **2017**, *62*, 33–42. [CrossRef]
5. Sibileau, A.; García-González, A.; Auricchio, F.; Morganti, S.; Díez, P. Explicit parametric solutions of lattice structures with proper generalized decomposition (PGD). *Comput. Mech.* **2018**, *62*, 871–891. [CrossRef]
6. Ghorashi, M. Nonlinear static and stability analysis of composite beams by the variational asymptotic method. *Int. J. Eng. Sci.* **2018**, *128*, 127–150. [CrossRef]
7. Carrera, E.; Pagani, A.; Banerjee, J.R. Linearized buckling analysis of isotropic and composite beam-columns by Carrera Unified Formulation and dynamic stiffness method. *Mech. Adv. Mater. Struct.* **2016**, *23*, 1092–1103. [CrossRef]
8. Petrolo, M.; Nagaraj, M.H.; Kaleel, I.; Carrera, E. A global-local approach for the elastoplastic analysis of compact and thin-walled structures via refined models. *Comput. Struct.* **2018**, *206*, 54–65. [CrossRef]
9. Pagani, A.; Carrera, E. Large-deflection and post-buckling analyses of laminated composite beams by Carrera Unified Formulation. *Compos. Struct.* **2017**, *170*, 40–52.
10. Cambronero-Barrientos, F.; Díaz-del-Valle, J.; Martínez-Martínez, J.A. Beam element for thin-walled beams with torsion, distortion, and shear lag. *Eng. Struct.* **2017**, *143*, 571–588. [CrossRef]
11. Yu, J.; Hu, S.W.; Xu, Y.C.; Fan, B. Coupled mechanism on interfacial slip and shear lag for twin-cell composite box beam under even load. *J. Mech.* **2018**, *34*, 601–616. [CrossRef]
12. Lim, T.K.; Kim, J.H. Thermo-elastic effects on shear correction factors for functionally graded beam. *Compos. Part B-Eng.* **2017**, *123*, 262–270. [CrossRef]
13. Akgöz, B.; Civalek, Ö. Effects of thermal and shear deformation on vibration response of functionally graded thick composite microbeams. *Compos. Part B-Eng.* **2017**, *129*, 77–87. [CrossRef]
14. Lacidogna, G. Tall buildings: Secondary effects on the structural behaviour. *Proc. Inst. Civ. Eng.-Struct. Build.* **2017**, *170*, 391–405. [CrossRef]
15. Tuysuz, O.; Altintas, Y. Time-domain modeling of varying dynamic characteristics in thin-wall machining using perturbation and reduced-order substructuring methods. *J. Manuf. Sci. Eng.-Trans. ASME* **2018**, *140*, 011015. [CrossRef]
16. de Lacalle, L.N.L.; Viadero, F.; Hernández, J.M. Applications of dynamic measurements to structural reliability updating. *Probab. Eng. Mech.* **1996**, *11*, 97–105. [CrossRef]

17. Zhou, J.; Wen, S.; Li, F.; Zhu, Y. Coupled bending and torsional vibrations of non-uniform thin-walled beams by the transfer differential transform method and experiments. *Thin-Walled Struct.* **2018**, *127*, 373–388. [CrossRef]

18. Schardt, R. Eine erweiterung der technischen biegetheorie zur berechnung prismatischer faltwerke. *Der Stahlbau* **1966**, *35*, 161–171.

19. Schardt, R. Lateral torsional and distortional buckling of channel-and hat-sections. *J. Constr. Steel Res.* **1994**, *31*, 243–265. [CrossRef]

20. Davies, J.M.; Leach, P.; Heinz, D. Second-order generalised beam theory. *J. Constr. Steel Res.* **1994**, *31*, 221–241. [CrossRef]

21. Silvestre, N.; Camotim, D.; Silva, N.F. Generalized beam theory revisited: From the kinematical assumptions to the deformation mode determination. *Int. J. Struct. Stab. Dyn.* **2011**, *11*, 969–997. [CrossRef]

22. Martins, A.D.; Camotim, D.; Gonçalves, R.; Dinis, P.B. Enhanced geometrically nonlinear generalized beam theory formulation: Derivation, numerical implementation, and illustration. *J. Eng. Mech.* **2018**, *144*, 04018036. [CrossRef]

23. de Miranda, S.; Madeo, A.; Melchionda, D.; Patruno, L.; Ruggerini, A.W. A corotational based geometrically nonlinear Generalized Beam Theory: Buckling FE analysis. *Int. J. Solids Struct.* **2017**, *121*, 212–227. [CrossRef]

24. Vieira, R.F.; Virtuoso, F.B.E.; Pereira, E.B.R. Buckling of thin-walled structures through a higher order beam model. *Comput. Struct.* **2017**, *180*, 104–116. [CrossRef]

25. Vieira, R.F.; Virtuoso, F.B.; Pereira, E.B.R. A higher order model for thin-walled structures with deformable cross-sections. *Int. J. Solids Struct.* **2014**, *51*, 575–598. [CrossRef]

26. Zhang, L.; Zhu, W.; Ji, A.; Peng, L. A simplified approach to identify sectional deformation modes of thin-walled beams with prismatic cross-sections. *Appl. Sci.* **2018**, *8*, 1847. [CrossRef]

27. Debski, H.; Teter, A.; Kubiak, T.; Samborski, S. Local buckling, post-buckling and collapse of thin-walled channel section composite columns subjected to quasi-static compression. *Compos. Struct.* **2016**, *136*, 593–601. [CrossRef]

28. Ciesielczyk, K.; Studziński, R. Experimental and numerical investigation of stabilization of thin-walled Z-beams by sandwich panels. *J. Constr. Steel Res.* **2017**, *133*, 77–83.

29. Carpinteri, A.; Lacidogna, G.; Nitti, G. Open and closed shear-walls in high-rise structural systems: Static and dynamic analysis. *Curved Layer. Struct.* **2016**, *3*, 154–171. [CrossRef]

30. Zhu, Z.; Zhang, L.; Shen, G.; Cao, G. A one-dimensional higher-order theory with cubic distortional modes for static and dynamic analyses of thin-walled structures with rectangular hollow sections. *Acta Mech.* **2016**, *227*, 2451–2475. [CrossRef]

 *symmetry*

Article

Robust Adaptive Full-Order TSM Control Based on Neural Network

Qianlei Cao [3], Chongzhen Cao [1,*], Fengqin Wang [2], Dan Liu [1] and Hui Sun [1]

[1] College of Transportation, Shandong University of Science and Technology, Qingdao 266590, China; danaliu@yeah.net (D.L.); sunhuiluck@163.com (H.S.)

[2] College of Mechanical and Electronic Engineering, Shandong University of Science and Technology, Qingdao 266590, China; skdwangfengqin@126.com

[3] Qingdao Topscomm Communication Co., Ltd., Qingdao 266024, China; qianleicao@foxmail.com

* Correspondence: caochongzhen@126.com; Tel.: +86-136-8542-3639

Received: 13 November 2018; Accepted: 4 December 2018; Published: 6 December 2018

Abstract: Existing full-order terminal sliding mode (FOTSM) control methods often require a priori knowledge of the system model. To tackle this problem, two novel neural-network-based FOTSM control methods were proposed. The first one was model based but did not require knowledge of the uncertainties' bounds. The second one was model free and did not require knowledge of the system model. Finite-time convergence of the two schemes was verified by theoretical analysis and simulation cases. Meanwhile, the designed methods avoided singularity as well as chattering.

Keywords: finite-time control; terminal sliding mode; chattering; neural network

---

1. Introduction

Terminal sliding mode (TSM) control is a sound feedback control method to ensure that nonlinear systems have (i) finite-time convergence and (ii) robustness to system uncertainties [1–3]. Thus, TSM-based control strategies have enjoyed great popularity in many control fields during the past decades, such as robotic manipulators [4–6], spacecraft attitude [7–9], linear motors [10,11], and automatic train operation [12].

However, conventional TSM control is often confronted with singularity and chattering problems. Singularity arises from the existence of negative fractional powers in the control. It makes the control become infinitely large around the equilibrium point and, further, may damage the actuator device. Chattering results from the existence of discontinuous terms in the control. It leads to high-frequency oscillations across the terminal sliding mode and may undermine system stability [13]. Therefore, both of these problems are undesirable in practice.

To tackle the aforementioned problems, many methods have been designed, ranging from discontinuous [14–16] to continuous control [17–19]. A two-phase TSM controller was constructed in [14] to sequentially reach a set of recursive manifolds. Following this scheme, the occurrence of singularity could be avoided. In [15], a nonsingular TSM manifold was designed to avoid singularity during the reaching period. The manifold was, to some extent, equivalent to conventional TSM when it was reached. In [16], a saturation function was utilized to erase the singularity and the stability was analyzed with the help of Landau symbols. Besides the discontinuous methods, continuous schemes have aroused much attention for their natural advantage in eliminating chattering. A novel TSM accompanied with continuous controllers was presented in [17] and the states of a system with dynamic uncertainties could be driven to the neighborhood of origin. In [18], the continuous controllers were extended to systems troubled by mismatched disturbance, and in [19], a continuous TSM controller was constructed by exactly observing uncertainties and was applied to electronic throttle systems.

However, the aforementioned methods adopted the reduced-order terminal sliding mode (ROTSM). Despite the improvement in terms of alleviating chattering, these ROTSM-based methods encountered the loss of finite-time convergence to the equilibrium point. Recently, a novel TSM control concept was proposed in [20]. In the concept, the designed TSM was of full order. That is to say, the order of the TSM was the same as that of the system model. It has been proven that this full-order terminal sliding mode (FOTSM) has many attractive features, such as avoidance of singularity and chattering. This concept was then extended to the trajectory tracking control of robotic manipulators [21–23]. Two-link robotic manipulators can exhibit nonsingular and chattering-free behaviors by using the FOTSM control. However, the FOTSM strategy proposed in [20] requires a priori knowledge of the system model, which may be difficult to obtain in some industrial applications. Despite that this restriction of system knowledge was, to some extent, removed in [21–23], it was only applied to a second-order nonlinear system. To the best of the authors' knowledge, there has not been much work which provides the FOTSM control scheme for high-order nonlinear systems, the dynamic models of which are not known a priori.

This study investigated the robust adaptive FOTSM control problem. Two adaptive FOTSM control schemes were proposed by using neural networks. The first one utilized neural networks to approximate the derivatives of system uncertainties. Then, a model-based controller was designed using the form of an integrator. The second one utilized neural networks to estimate the system model. Then, a model-free controller was designed only based on the system state. By using the FOTSM control technique, the designed schemes both made the states converge to the equilibrium point within a period of time. Since the controllers adopted the form of an integrator or a low-pass filter, the chattering phenomenon was eliminated. Moreover, the proposed schemes avoided singularity. Lyapunov stability derivation and computer simulation both validated the proposed schemes.

2. Problem Formulation

The considered nonlinear system [20] is as follows:

$$
\begin{cases}
\dot{x}_1 = x_2 \\
\dot{x}_2 = x_3 \\
\cdots \\
\dot{x}_{n-1} = x_n \\
\dot{x}_n = f(x,t) + d(x,t) + b(x,t)u
\end{cases}
\tag{1}
$$

in which $x = [x_1, x_2, \ldots, x_n]^\mathrm{T} \in R^n$ stands for the system state vector, $f(x,t)$ and $b(x,t)$ stand for smooth functions, $d(x,t)$ stands for the unknown and bounded uncertainties, and $u \in R$ stands for the control input.

Assumption 1. *The time derivative of $d(x,t)$ (i.e., $\dot{d}(x,t)$) is continuous and bounded.*

Definition 1. *A FOTSM manifold can be defined by the following equation [20]:*

$$
s = \dot{x}_n + \beta_n x_n{}^{\alpha_n} + \cdots + \beta_1 x_1{}^{\alpha_1}
\tag{2}
$$

where β_i fulfills that $p^n + \beta_n p^{n-1} + \cdots + \beta_2 p^1 + \beta_1$ is Hurwitz, and α_i $(i = 1, 2, \ldots, n)$ is designed as

$$
\begin{cases}
\alpha_1 = \alpha, \; n = 1 \\
\alpha_{i-1} = \dfrac{\alpha_i \alpha_{i+1}}{2\alpha_{i+1} - \alpha_i}, \; i = 2, \ldots, n \quad \forall n \geq 2
\end{cases}
\tag{3}
$$

where $\alpha_n = \alpha$, $\alpha_{n+1} = 1$, $\alpha \in (1 - \varepsilon, \, 1)$, and $\varepsilon \in (0, 1)$.

Remark 1. *It has been discussed in* [20] *that once the FOTSM s* $= 0$ *is established,* $x = 0$ *will also be established within finite time.*

Thus, the control objective can be summarized as: design a suitable FOTSM controller for the system described by Equation (1), the parameters of which are not known a priori. Then, the system's finite-time stability can be ensured; meanwhile, the singularity and chattering phenomena are erased.

3. Model-Based FOTSM Controller

For the system model where the information of $f(x, t)$ and $b(x, t)$ is exactly obtained but $d(x, t)$ is unknown, a robust FOTSM control scheme was proposed in [20]. However, the scheme requires a priori knowledge of the upper bounds of $d(x, t)$ and $\dot{d}(x, t)$, which may not be easy to obtain in some situations. Hereinafter, a FOTSM control method is designed even when the upper bounds of $d(x, t)$ and $\dot{d}(x, t)$ are unknown.

The radial basis function (RBF) neural network proves to be a useful technique to approximate nonlinear functions [24]. Since $\dot{d}(x, t)$ is unknown, we define $\rho_1 = \dot{d}(x, t)$ and utilize the RBF neural network to approximate it. Thus, we have

$$\rho_1 = W^{*T}\phi(y) + \varepsilon \tag{4}$$

in which $W^* = [w_1, w_2, \ldots, w_l]^T \in R^l$ stands for the ideal weight matrix, $y = [x_1, x_2, \ldots, x_n, \dot{x}_n]^T$, and the values of $\dot{x}_n$ can be obtained by taking the difference of x_n in consecutive sample times, $|\varepsilon| \leq \varepsilon_N$, and $\phi(y) = [\phi_1(y), \phi_2(y), \ldots, \phi_l(y)]^T \in R^l$ is

$$\phi_i(y) = \exp\left[\frac{-(y - v_i)^T(y - v_i)}{2\omega_i^2}\right], \quad i = 1, 2, \ldots, l \tag{5}$$

in which $v_i = [\mu_{i1}, \mu_{i2}, \ldots, \mu_{in}]^T$ and ω_i denote the kernel unit's parameters, and $0 < \phi_i(y) < 1$.

According to the RBF network and FOTSM manifold defined by Equation (2), design the following controller:

$$u = b^{-1}(x, t)(u_1 + u_2) \tag{6}$$

$$u_1 = -f(x, t) - \beta_n x_n^{\alpha_n} - \cdots - \beta_1 x_1^{\alpha_1} \tag{7}$$

$$\dot{u}_2 = -\hat{W}^T\phi(y) - (k + \eta)\mathrm{sgn}(s) \tag{8}$$

where $\hat{W}$ is used to estimate W^*, $\eta > 0$, $k > \varepsilon_N$, and $\mathrm{sgn}()$ denotes the signum function.

Then, the adaptive law for $\hat{W}$ is adopted as

$$\dot{\hat{W}} = \frac{1}{\gamma_1}s\phi(y) \tag{9}$$

where $\gamma_1 > 0$.

To sum up, the design procedures of the proposed model-based controller can be formulated by the block diagram of Figure 1.

Theorem 1. *Consider the nonlinear system given by Equation (1). If the controllers are designed as Equations (2) and (6)–(9), then the system state is bounded.*

Proof. Design a Lyapunov function candidate as

$$V_1 = \frac{1}{2}s^2 + \frac{1}{2}\gamma_1\tilde{W}^T\tilde{W} \tag{10}$$

where $\widetilde{\boldsymbol{W}} = \boldsymbol{W}^* - \hat{\boldsymbol{W}}$.

Differentiating V_1 with respect to time yields:

$$\dot{V}_1 = s\dot{s} + \gamma_1 \widetilde{\boldsymbol{W}}^{\mathrm{T}} \dot{\widetilde{\boldsymbol{W}}}. \tag{11}$$

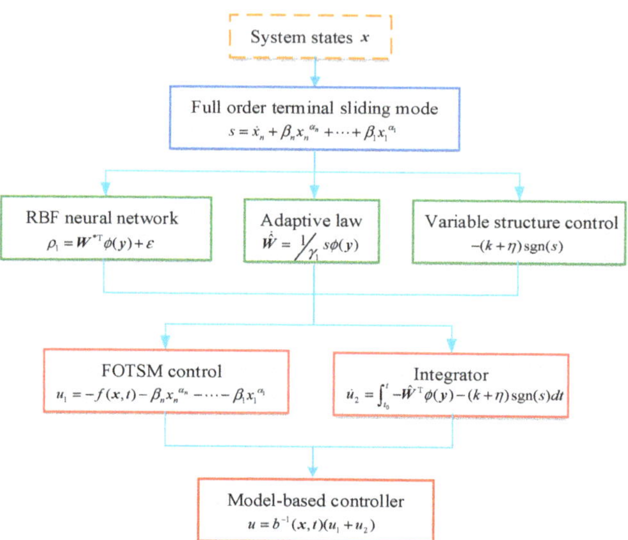

Figure 1. Design procedures of the proposed model-based controller.

Substituting Equation (1) into Equation (2), we have

$$s = \beta_n x_n^{\alpha_n} + \cdots + \beta_1 x_1^{\alpha_1} + f(\boldsymbol{x},t) + b(\boldsymbol{x},t)u + d(\boldsymbol{x},t). \tag{12}$$

From Equations (6) and (7), it can be obtained that

$$\begin{aligned} s &= f(\boldsymbol{x},t) + d(\boldsymbol{x},t) + u_1 + u_2 + \beta_n x_n^{\alpha_n} + \cdots + \beta_1 x_1^{\alpha_1} \\ &= d(\boldsymbol{x},t) + u_2 \end{aligned} \tag{13}$$

Then, the time derivative of the FOTSM manifold is

$$\dot{s} = \dot{d}(\boldsymbol{x},t) + \dot{u}_2 = \rho_1 + \dot{u}_2. \tag{14}$$

According to Equations (4) and (8), rewrite the above equation as

$$\begin{aligned} \dot{s} &= -\hat{\boldsymbol{W}}^{\mathrm{T}}\phi(\boldsymbol{y}) + \boldsymbol{W}^{*\mathrm{T}}\phi(\boldsymbol{y}) + \varepsilon - (k+\eta)\mathrm{sgn}(s) \\ &= \widetilde{\boldsymbol{W}}^{\mathrm{T}}\phi(\boldsymbol{y}) + \varepsilon - (k+\eta)\mathrm{sgn}(s) \end{aligned} \tag{15}$$

Substituting Equations (9) and (15) into Equation (11) yields

$$\begin{aligned} \dot{V}_1 &= s\widetilde{\boldsymbol{W}}^{\mathrm{T}}\phi(\boldsymbol{y}) + s\varepsilon - (k+\eta)|s| + \gamma_1\widetilde{\boldsymbol{W}}^{\mathrm{T}}\dot{\widetilde{\boldsymbol{W}}} \\ &= -(k+\eta)|s| + s\varepsilon + \widetilde{\boldsymbol{W}}^{\mathrm{T}}(s\phi(\boldsymbol{y}) + \gamma_1\dot{\widetilde{\boldsymbol{W}}}) \\ &\leq -(\eta + k - |\varepsilon|)|s| \end{aligned} \tag{16}$$

Since $|\varepsilon| \leq \varepsilon_N < k$ and $\eta > 0$, it can be obtained that

$$\dot{V}_1 \leq -\eta|s| \leq 0. \tag{17}$$

According to Equations (10) and (17), s and $\widetilde{W}$ are bounded. From Definition 1 and the TSM control theory [20,25], the boundedness of the system state can also be ensured.

Since $0 < \phi_i(y) < 1$, and $\|\phi(y)\| \leq \sqrt{l}$, then there is [24]

$$\|\widetilde{W}^{\mathrm{T}}\phi(y)\|_{\mathrm{F}} \leq \|\widetilde{W}^{\mathrm{T}}\|_{\mathrm{F}}\|\phi(y)\|. \tag{18}$$

Thus, $\|\widetilde{W}^{\mathrm{T}}\phi(y)\|_{\mathrm{F}}$ is bounded. $\square$

Theorem 2. *For the nonlinear system expressed by Equation (1) and the controllers given by Equations (2) and (6)–(9), if k in Equation (9) is designed so that $k > \|\widetilde{W}^{\mathrm{T}}\phi(y)\|_{\mathrm{F}} + \varepsilon_N$, then the system is finite-time stable.*

Proof. Design a Lyapunov function candidate as

$$V_2 = \frac{1}{2}s^2. \tag{19}$$

Using Equations (6)–(8) and (12)–(15), we have

$$\begin{aligned}
\dot{V}_2 &= s\widetilde{W}^{\mathrm{T}}\phi(y) - (k+\eta)|s| + s\varepsilon \\
&\leq -\eta|s| - (k - \|\widetilde{W}^{\mathrm{T}}\phi(y)\| - |\varepsilon|)|s|
\end{aligned} \tag{20}$$

Since $k > \|\widetilde{W}^{\mathrm{T}}\phi(y)\|_{\mathrm{F}} + \varepsilon_N$ and $\eta > 0$, it can be obtained that

$$\dot{V}_2 < -\eta|s| < 0 \quad \text{for} \quad |s| \neq 0. \tag{21}$$

Therefore, $s = 0$ can be established [20,25]. Then from Remark 1, the system is finite-time stable. $\square$

Remark 2. *Equation (8) adopts the form of an integrator. Although the signum function is employed in it, the real control signal u is continuous due to the effect of the integrator. Therefore, the chattering can be eliminated. Moreover, the designed controller avoids using the derivative of s, which prevents the singularity from occurring.*

Remark 3. *It is noted that in some nonlinear systems, an integrator is more difficult to implement than a low-pass filter. Thus, in the next section, we will design a FOTSM controller using the form of low-pass filter.*

4. Model-Free FOTSM Controller

For the system in which neither $f(x,t)$ nor $b(x,t)$ is known, based on the RBF neural network, an adaptive FOTSM controller is further designed in this section by only using the system state; that is, it is like a model-free controller. To start, we rewrite Equation (1) as

$$\begin{cases}
\dot{x}_1 = x_2 \\
\dot{x}_2 = x_3 \\
\cdots \\
\dot{x}_{n-1} = x_n \\
c(x,t)\dot{x}_n = c(x,t)f(x,t) + u + c(x,t)d(x,t)
\end{cases} \tag{22}$$

where $c(x,t) = 1/b(x,t)$.

The following assumption is given on the nonlinear system described by Equation (22).

Assumption 2. *$c(x,t)$ is Lipschitz continuous and positive; that is, $0 < c_1 \le c(x,t) \le c_2$, where $0 < c_1 \le c_2$. Moreover, $\dot{c}(x,t)$ is also Lipschitz continuous.*

Remark 4. *The above assumption is reasonable and does not lose generality, for many mechanical engineering systems fulfill this assumption, such as the robotic manipulator [6,15] and the spacecraft [7–9].*

The controller is designed as

$$\dot{u} + \chi u = \chi \tau \tag{23}$$

in which $\chi > 0$, τ stands for the virtual control input which will be determined later.

From Equations (22) and (23), one can figure out that

$$c\ddot{x}_n + (\dot{c} + \chi c)\dot{x}_n - c\dot{f} - (\dot{c} + \chi c)f \\ -c\dot{d} - (\dot{c} + \chi c)d = \chi \tau \tag{24}$$

According to Equations (2) and (24), it can be obtained that

$$\begin{aligned} c\dot{s} &= c\ddot{x}_n + \alpha_n \beta_n c x_n{}^{\alpha_n - 1}\dot{x}_n + \cdots \\ &\quad + \alpha_1 \beta_1 c x_1{}^{\alpha_1 - 1}\dot{x}_1 \\ &= \chi \tau + c\dot{f} + c\dot{d} + (\dot{c} + c\chi)f \\ &\quad -(\dot{c} + c\chi)\dot{x}_n + (\dot{c} + c\chi)d \\ &\quad + \alpha_n \beta_n c x_n{}^{\alpha_n - 1}\dot{x}_n + \cdots + \alpha_1 \beta_1 c x_1{}^{\alpha_1 - 1}\dot{x}_1 \\ &= \chi \tau + \rho_2 \end{aligned} \tag{25}$$

where

$$\begin{aligned} \rho_2 = {} & c\dot{f} + (\dot{c} + c\chi)f + (\dot{c} + c\chi)d + c\dot{d} \\ & -(\dot{c} + c\chi)\dot{x}_n + \alpha_n \beta_n c x_n{}^{\alpha_n - 1}\dot{x}_n + \cdots \\ & + \alpha_1 \beta_1 c x_1{}^{\alpha_1 - 1}\dot{x}_1 \end{aligned} \tag{26}$$

Similarly, we utilize the RBF network to estimate ρ_2; that is,

$$\rho_2 = W^{*\mathrm{T}}\phi(y) + \varepsilon \tag{27}$$

where $y = [x_1, x_2, \ldots, x_n, \dot{x}_n]^{\mathrm{T}}$, $|\varepsilon| \le \varepsilon_N$, and the other parameters are defined in Equations (4) and (5). Then, the virtual control input τ can be designed as

$$\tau = -(k + \eta)\mathrm{sgn}(s) - \hat{W}^{\mathrm{T}}\phi(y)/\chi \tag{28}$$

where $k > \varepsilon_N/\chi$ and $\eta > 0$, $\hat{W}$ is updated as

$$\dot{\hat{W}} = \frac{1}{\gamma_2} s\phi(y) \tag{29}$$

where $\gamma_2 > 0$.

To sum up, the design procedures of the proposed model-free controller can be formulated by the block diagram of Figure 2.

Theorem 3. *Consider the nonlinear system expressed by Equation (22). If the controllers are given by Equations (2), (23), (28), and (29), then the system state is bounded.*

Proof. Design a Lyapunov function candidate as

$$V_3 = \frac{1}{2}cs^2 + \frac{1}{2}\gamma_2\widetilde{W}^\mathrm{T}\widetilde{W}. \tag{30}$$

Since $c > 0$, it can be obtained that $V_3 \geq 0$.

It follows that

$$\dot{V}_3 = sc\dot{s} + \frac{1}{2}\dot{c}s^2 + \gamma_2\widetilde{W}^\mathrm{T}\dot{\widetilde{W}}. \tag{31}$$

Substituting Equations (9) and (25)–(28) into Equation (31) yields

$$
\begin{aligned}
\dot{V}_3 &= s\widetilde{W}^\mathrm{T}\phi(y) + s\varepsilon + \gamma_2\widetilde{W}^\mathrm{T}\dot{\widetilde{W}} - \chi(k+\eta)|s| \\
&= s\widetilde{W}^\mathrm{T}\phi(y) + s\varepsilon - \widetilde{W}^\mathrm{T}s\phi(y) - \chi(k+\eta)|s| \\
&\leq -(k - |\varepsilon|)|s| - \eta|s|
\end{aligned}
\tag{32}
$$

Since $k > \varepsilon_N/\chi \geq |\varepsilon|/\chi$ and $\chi,\ \eta > 0$, it can be obtained that

$$\dot{V}_3 \leq -\chi\eta|s| \leq 0. \tag{33}$$

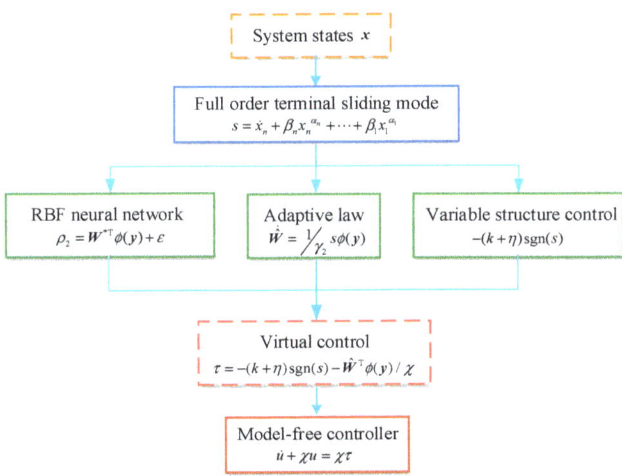

Figure 2. Design procedures of the proposed model-free controller.

Thus, as in Section 3, s and $\widetilde{W}$ are bounded and the boundedness of the system state is also ensured. $\square$

Theorem 4. *Consider the nonlinear system expressed by Equation (22) and the controllers given by Equations (23), (28), and (29). If k is designed so that $k > (\|\widetilde{W}^\mathrm{T}\phi(y)\|_\mathrm{F} + \varepsilon_N)/\chi$, then the system state vector x will converge to zero in finite time.*

Proof. Design a Lyapunov function candidate as

$$V_4 = \frac{1}{2}cs^2. \tag{34}$$

Using Equations (22)–(29), we have

$$\begin{aligned}
\dot{V}_4 \ &= sc\dot{s} + \tfrac{1}{2}\dot{c}s^2 \\
&= s\tilde{W}^{\mathrm{T}}\phi(y) + s\varepsilon - \chi(k+\eta)|s| \\
&\leq -(k - \|\tilde{W}^{\mathrm{T}}\phi(y)\| - |\varepsilon|)|s| - \eta|s|
\end{aligned} \tag{35}$$

Since $k > (\|\tilde{W}^{\mathrm{T}}\phi(y)\|_{\mathrm{F}} + \varepsilon_N)/\chi$ and $\chi,\ \eta > 0$, it can be obtained that

$$\dot{V}_4 < -\chi\eta|s| < 0 \quad for\ |s| \neq 0. \tag{36}$$

Therefore, $s = 0$ can be established according to the TSM control theory [20,25]. Then, from Remark 1, the system is finite-time stable. □

Remark 5. *Equation (23) can smooth the control signal. Therefore, the chattering is attenuated.*

Remark 6. *If the powers are designed as $\alpha_1 = \alpha_2 = \cdots = \alpha_n = 1$, Equation (2) will become a full-order linear sliding mode. In that case, the designed controller still enables the avoidance of chattering while making the system state asymptotically converge.*

Remark 7. *The proposed control concept can be used for many mechanical engineering systems, such as the multiple joint robotic manipulator shown in Figure 3. Its dynamic model can be described as a high-order nonlinear system. In the presence of dynamic uncertainties, the proposed control methods can be used to drive the joints to track the desired trajectories even when the dynamic model is not obtained. The RBF neural network can online estimate the dynamic model to enhance the control performance. Also, the tracking control is finite-time convergent, nonsingular, and chattering free. Thus, the proposed control methods can extend the application range of FOTSM to engineering systems.*

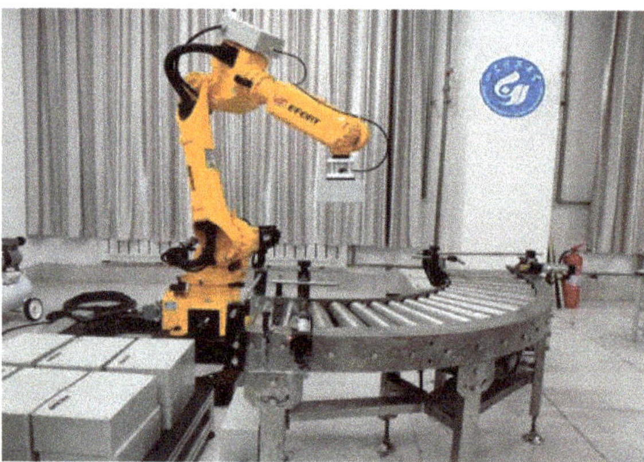

Figure 3. Multiple joint robotic manipulator.

5. Simulation Studies

This section shows how the designed controllers were applied to numerical system models to demonstrate their effectiveness. The numerical models and the simulation procedures exactly followed the literature [20], but more uncertainties were introduced into the model parameters. All simulations were implemented on a laptop computer with an Intel Core i5-8250U CPU. Figure 4 displays the simulation block diagram.

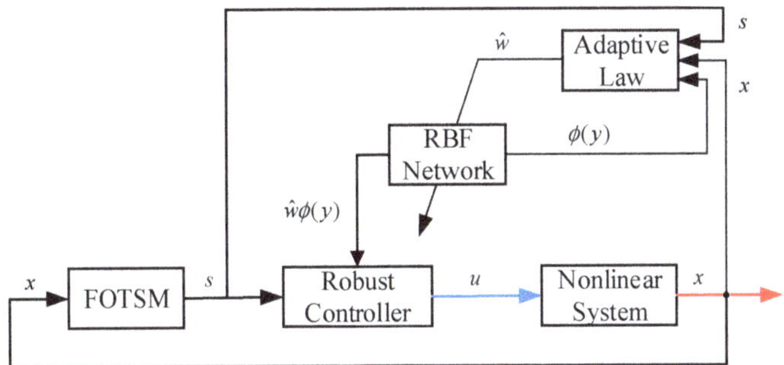

Figure 4. Simulation block diagram of the designed controllers.

In the above block diagram, the control input u was constructed based on the FOTSM and RBF neural network and then applied to the nonlinear system. All the effort aimed at making the system state x converge to zero, which stands for that, in some mechanical engineering systems, the tracking error converges to zero. Similar to [20], the simulation studies were formulated with second- and third-order systems, respectively.

5.1. Control of a Second-Order System

Here, we applied the designed schemes to a second-order system to verify their effectiveness. Two simulation cases are presented: the first case was a test of the model-based FOTSM controller, and the second case was a test of the model-free FOTSM controller.

The system model [20] was

$$\begin{cases} \dot{x}_1 = x_2 \\ \dot{x}_2 = f(x,t) + d(x,t) + u \end{cases}$$

where $f(x,t) = x_2{}^3$, $d(x,t) = 0.1\sin(20t)$, and $x_1(0) = 1$, $x_2(0) = 0$.

Case 1. The test of the model-based FOTSM controller for the second-order nonlinear system.

In this case, the parameters are given in Table 1.

Table 1. Parameters of the full-order terminal sliding mode (FOTSM) manifold and the proposed controller.

Parameters	Values
α_1, α_2	$\alpha_1 = 9/23$, $\alpha_2 = 9/16$
β_1, β_2	$\beta_1 = 10$, $\beta_2 = 7$
k, η, γ_1	$k + \eta = 3$, $\gamma_1 = 50$

The results are given in Figures 5 and 6. Figure 5 shows the response of the system state that is marked as a red arrow in Figure 4. Figure 6 shows the control input that is marked as a blue arrow in Figure 4. From the response of the system state and the controller, one can see that finite-time stability was ensured within 1.8 s by the model-based FOTSM controller. When the system state arrived at the origin, it stayed there even in the presence of uncertainties and disturbance. This demonstrates the robustness of the designed controller. Meanwhile, singularity and chattering problems did not occur, which shows the superior performance of the FOTSM control to overcome the two problems.

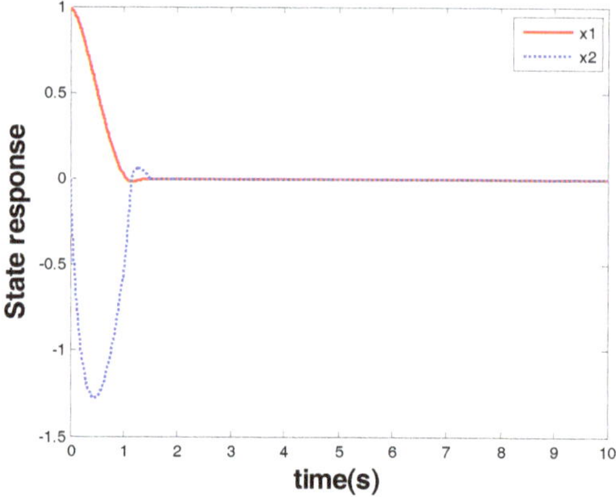

Figure 5. System state vector $x = [x_1, x_2]^T$ of Case 1.

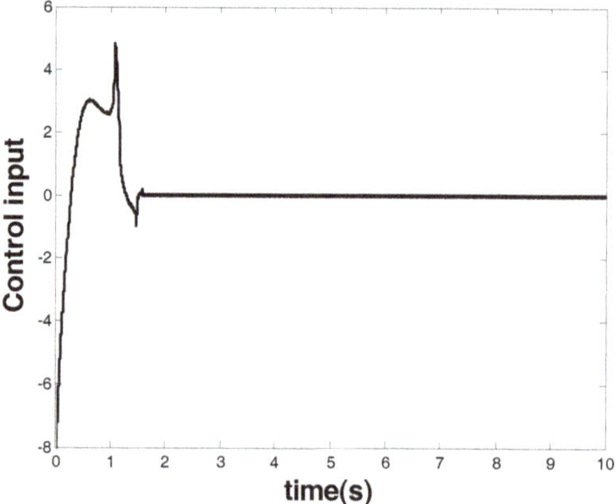

Figure 6. Controller u of Case 1.

Case 2. The test of the model-free FOTSM controller for the second-order nonlinear system.

In this case, the parameters are given in Table 2.

Table 2. Parameters of the FOTSM manifold and the proposed controller.

Parameters	Values
$\alpha_1,\ \alpha_2$	$\alpha_1 = 9/23,\ \alpha_2 = 9/16$
$\beta_1,\ \beta_2$	$\beta_1 = 10,\ \beta_2 = 7$
$k,\ \eta,\ \gamma_2, \chi$	$k + \eta = 40,\ \gamma_2 = 50,\ \chi = 5$

The results are given in Figures 7 and 8. Figure 7 shows the response of the system state that is marked as a red arrow in Figure 4. Figure 8 shows the control input that is marked as a blue arrow in Figure 4. From the response of the system state and the controller, one can see that, similar to the model-based controller in last case, finite-time stability was also ensured by the model-free FOTSM controller. The advantage of the proposed controller over the existing robust FOTSM control methods was also verified, since the existing methods require a priori knowledge of the system model, while the proposed scheme does not.

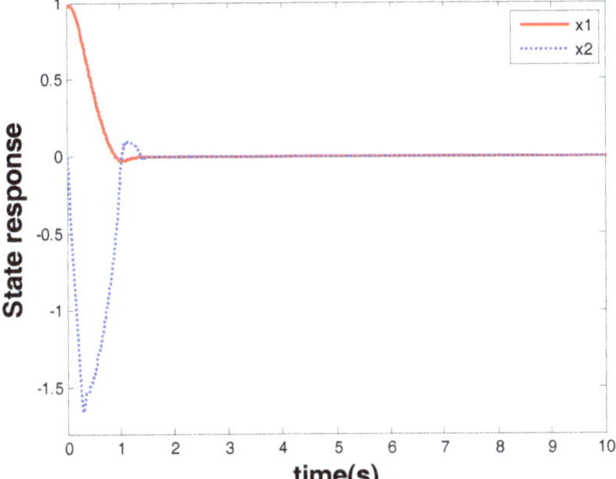

Figure 7. System state vector $x = [x_1, x_2]^{\text{T}}$ of Case 2.

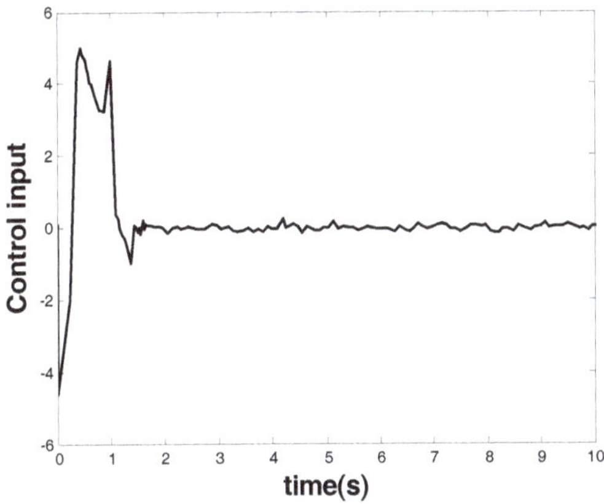

Figure 8. Controller u of Case 2.

5.2. Control of a Third-Order System

Here, we applied the designed schemes to a third-order system to verify their effectiveness. Similar to Section 5.1, two tests are presented: one is for the model-based FOTSM controller, and the other is for the model-free FOTSM controller.

The system model [20] was

$$\begin{cases} \dot{x}_1 = x_2 \\ \dot{x}_2 = x_3 \\ \dot{x}_3 = f(x, t) + d(x, t) + u \end{cases}$$

where $f(x, t) = 0.2x_2^3$, $d(x, t) = 0.1\sin(20t)$, and the initial states of x_1, x_2 and x_3 are $1, -1, 0$.

Case 3. The test of the model-based FOTSM controller for the third-order nonlinear system.

The parameters are given in Table 3.

Table 3. Parameters of the FOTSM manifold and the proposed controller.

Parameters	Values
$\alpha_1,\ \alpha_2,\ \alpha_3$	$\alpha_1 = 4/7,\ \alpha_2 = 2/3,\ \alpha_3 = 4/5$
$\beta_1,\ \beta_2,\ \beta_3$	$\beta_1 = 80,\ \beta_2 = 66,\ \beta_3 = 15$
$k,\ \eta,\ \gamma_1$	$k + \eta = 3,\ \gamma_1 = 50$

The results are given in Figures 9 and 10. Figure 9 shows the response of the system state that is marked as a red arrow in Figure 4. Figure 10 shows the control input that is marked as a blue arrow in Figure 4. From the response of the system state and the controller, one can see that finite-time stability was ensured together with a chattering-free and nonsingular control performance. The convergence time was within 2.1 s, and the controller had robust performance against uncertainties.

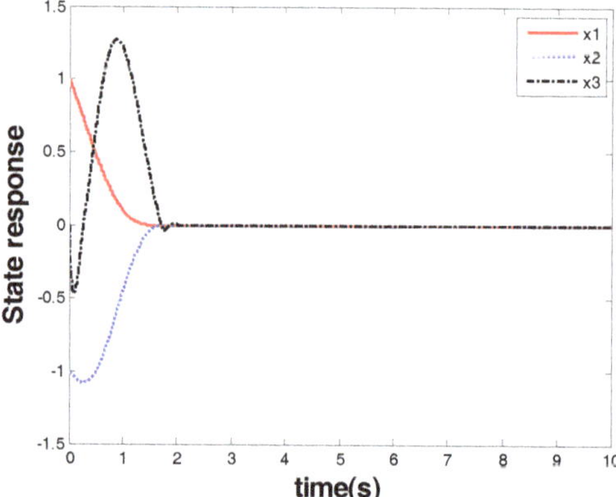

Figure 9. System state vector $x = [x_1,\ x_2,\ x_3]^{\mathrm{T}}$ of Case 3.

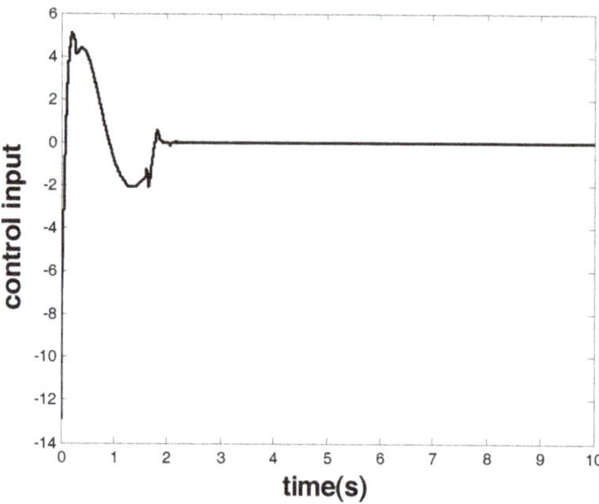

Figure 10. Controller u of Case 3.

Case 4. The test of the model-free FOTSM controller for the third-order nonlinear system.

In this case, the parameters k, η, χ were selected as $k + \eta = 50$ and $\chi = 5$. The other parameters were chosen as in Case 3.

The results are given by Figures 11 and 12. Figure 11 shows the response of the system state that is marked as a red arrow in Figure 4. Figure 12 shows the control input that is marked as a blue arrow in Figure 4. From the figures, one can see that the proposed model-free FOTSM controller also enabled the occurrence of finite-time convergence and achieved the attenuation of chattering and singularity. Therefore, this case verified the effectiveness of the model-free FOTSM controller for the third-order system.

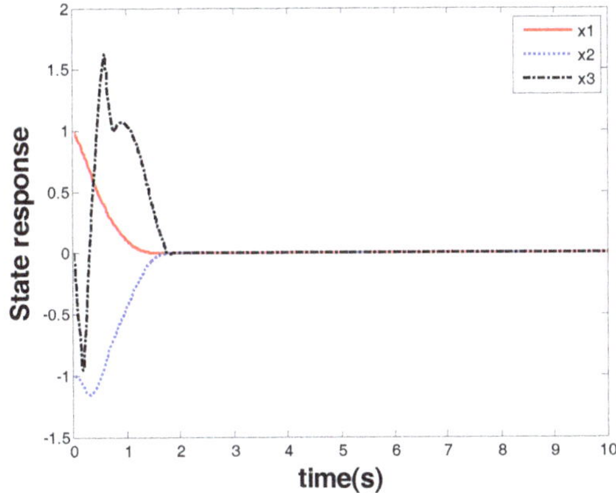

Figure 11. System state vector $x = [x_1, x_2, x_3]^{\mathrm{T}}$ of Case 4.

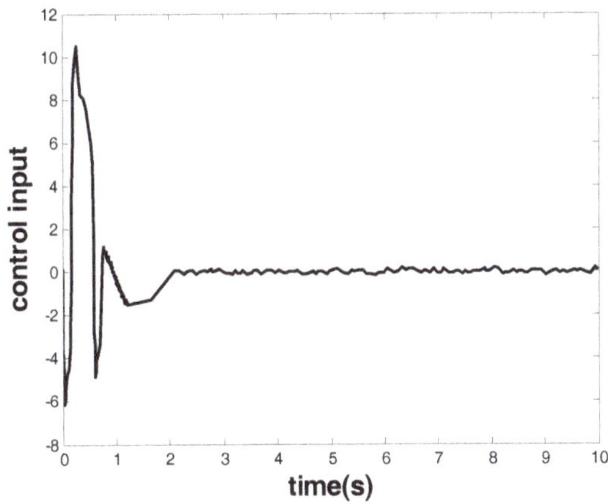

Figure 12. Controller u of Case 4.

6. Conclusions

The FOTSM control problem was investigated in this paper for high-order systems with modeling uncertainties. Two novel neural-network-based FOTSM control schemes were developed, which could achieve finite-time stability of high-order nonlinear systems. The designed schemes could essentially overcome the chattering and singularity problems.

The originality lies in that, unlike the existing FOTSM control methods, the designed control schemes need less information of the system model and the bounds of uncertainties' derivatives. Thus, they can meet the needs of more application scenarios, especially when the information of the system model is difficult to obtain.

The designed controllers are for general high-order nonlinear systems. Thus, many industrial systems described by high-order nonlinear systems, such as n-link robotic manipulators, can utilize the proposed control concept in this paper to achieve finite-time convergent, nonsingular, and chattering-free feedback control performance.

Author Contributions: Conceptualization, Q.C. and C.C.; Data curation, Q.C., F.W., and D.L.; Formal analysis, Q.C. and F.W.; Investigation, Q.C. and D.L.; Methodology, Q.C. and C.C.; Project administration, C.C.; Resources, F.W.; Software, Q.C. and H.S.; Supervision, H.S.; Writing—riginal draft, Q.C.; Writing—review & editing, C.C.

Funding: This research was funded by Project of Shandong Province Higher Educational Science and Technology Program under grant J16LB05.

Conflicts of Interest: The authors declare no conflict of interest. The funders had no role in the design of the study; in the collection, analyses, or interpretation of data; in the writing of the manuscript, or in the decision to publish the results.

References

1. Tao, C.-W.; Taur, J.S.; Chan, M.-L. Adaptive fuzzy terminal sliding mode controller for linear systems with mismatched time-varying uncertainties. *IEEE Trans. Syst. Man Cybern. Part B Cybern.* **2004**, *34*, 255–262. [CrossRef]
2. Li, S.; Du, H.; Yu, X. Discrete-time terminal sliding mode control systems based on Euler's discretization. *IEEE Trans. Autom. Control* **2014**, *59*, 546–552. [CrossRef]
3. Behera, A.K.; Bandyopadhyay, B. Steady-state behaviour of discretized terminal sliding mode. *Automatica* **2015**, *54*, 176–181. [CrossRef]
4. Cao, Q.; Li, S.; Zhao, D. Adaptive motion/force control of constrained manipulators using a new fast terminal sliding mode. *Int. J. Comput. Appl. Technol.* **2014**, *49*, 150–156. [CrossRef]
5. Cao, Q.; Li, S.; Zhao, D. A framework of terminal sliding mode for force/position control of constrained manipulators. *Int. J. Model. Identif. Control* **2015**, *24*, 332–341. [CrossRef]
6. Van, M.; Ge, S.S.; Ren, H. Finite time fault tolerant control for robot manipulators using time delay estimation and continuous nonsingular fast terminal sliding mode control. *IEEE Trans. Cybern.* **2017**, *47*, 1681–1693. [CrossRef]
7. Zou, A.-M.; Kumar, K.D.; Hou, Z.-G.; Liu, X. Finite-time attitude tracking control for spacecraft using terminal sliding mode and Chebyshev neural network. *IEEE Trans. Syst. Man Cybern. Part B Cybern.* **2011**, *41*, 950–963. [CrossRef]
8. Song, Z.; Li, H.; Sun, K. Finite-time control for nonlinear spacecraft attitude based on terminal sliding mode technique. *ISA Trans.* **2014**, *53*, 117–124. [CrossRef]
9. Zhao, L.; Jia, Y. Finite-time attitude tracking control for a rigid spacecraft using time-varying terminal sliding mode techniques. *Int. J. Control* **2015**, *88*, 1150–1162. [CrossRef]
10. Du, H.; Chen, X.; Wen, G.; Yu, X.; Lu, J. Discrete-time fast terminal sliding mode control for permanent magnet linear motor. *IEEE Trans. Ind. Electron.* **2018**, *65*, 9916–9927. [CrossRef]
11. Sun, G.; Ma, Z.; Yu, J. Discrete-time fractional order terminal sliding mode tracking control for linear motor. *IEEE Trans. Ind. Electron.* **2018**, *65*, 3386–3394. [CrossRef]
12. Yao, X.; Park, J.H.; Dong, H.; Guo, L.; Lin, X. Robust adaptive nonsingular terminal sliding mode control for automatic train operation. *IEEE Trans. Syst. Man Cybern. Syst.* **2018**. [CrossRef]
13. Slotine, J.J.E.; Li, E. *Applied Nonlinear Control*; Prantice-Hall: Englewood Cliffs, NJ, USA, 1991.
14. Wu, Y.; Yu, X.; Man, Z. Terminal sliding mode control design for uncertain dynamic systems. *Syst. Control Lett.* **1998**, *34*, 281–287. [CrossRef]
15. Feng, Y.; Yu, X.; Man, Z. Non-singular terminal sliding mode control of rigid manipulators. *Automatica* **2002**, *38*, 2159–2167. [CrossRef]
16. Feng, Y.; Yu, X.; Han, F. On nonsingular terminal sliding-mode control of nonlinear systems. *Automatica* **2013**, *49*, 1715–1722. [CrossRef]
17. Yu, S.; Yu, X.; Shirinzadeh, B.; Man, Z. Continuous finite-time control for robotic manipulators with terminal sliding mode. *Automatica* **2005**, *41*, 1957–1964. [CrossRef]
18. Yang, J.; Li, S.; Su, J.; Yu, X. Continuous nonsingular terminal sliding mode control for systems with mismatched disturbances. *Automatica* **2013**, *49*, 2287–2291. [CrossRef]

19. Wang, H.; Shi, L.; Man, Z.; Zheng, J.; Li, S.; Yu, M.; Jiang, C.; Kong, H.; Cao, Z. Continuous fast nonsingular terminal sliding mode control of automotive electronic throttle systems using finite-time exact observer. *IEEE Trans. Ind. Electron.* **2018**, *65*, 7160–7172. [CrossRef]

20. Feng, Y.; Han, F.; Yu, X. Chattering free full-order sliding-mode control. *Automatica* **2014**, *50*, 1310–1314. [CrossRef]

21. Zhao, D.; Cao, Q.; Li, S.; Zhu, Q. Adaptive full-order sliding mode control of rigid robotic manipulators. In Proceedings of the 2015 34th Chinese Control Conference (CCC), Hangzhou, China, 28–30 July 2015; pp. 657–662. [CrossRef]

22. Cao, Q.; Li, S.; Zhao, D.; Cao, Q.; Lu, S. Full-order sliding mode control of robotic manipulators including actuator dynamics. *J. Syst. Sci. Math. Sci.* **2015**, *35*, 848–859.

23. Cao, Q.; Li, S.; Zhao, D. Full-order multi-input/multi-output terminal sliding mode control for robotic manipulators. *Int. J. Model. Identif. Control* **2016**, *25*, 17–27. [CrossRef]

24. Wang, L.; Chai, T.; Zhai, L. Neural-network-based terminal sliding-mode control of robotic manipulators including actuator dynamics. *IEEE Trans. Ind. Electron.* **2009**, *56*, 3296–3304. [CrossRef]

25. Man, Z.; Paplinski, A.P.; Wu, H.R. A robust MIMO terminal sliding mode control scheme for rigid robotic manipulators. *IEEE Trans. Autom. Control* **1994**, *39*, 2464–2469. [CrossRef]

Article

Research on an Adaptive Variational Mode Decomposition with Double Thresholds for Feature Extraction

Wu Deng [1,2,3,4,5,6], Hailong Liu [1], Shengjie Zhang [1], Haodong Liu [1], Huimin Zhao [1,3,6,*] and Jinzhao Wu [5]

1   Software Institute, Dalian Jiaotong University, Dalian 116028, China; dw7689@djtu.edu.cn (W.D.);
    zhangyazi2015@yeah.net (H.L.); fukaifang@yeah.net (S.Z.); lunwen1209@yeah.net (H.L.)
2   Co-Innovation Center of Shandong Colleges and Universities: Future Intelligent Computing,
    Yantai 264005, China
3   Traction Power State Key Laboratory of Southwest Jiaotong University, Chengdu 610031, China
4   Guangxi Key Lab of Multi-Source Information Mining & Security, Guangxi Normal University,
    Guilin 541004, China
5   Guangxi Key Laboratory of Hybrid Computation and IC Design Analysis, Guangxi University for
    Nationalities, Nanning 530006, China; ziyaarnavut2011@hotmail.com
6   Liaoning Key Laboratory of Welding and Reliability of Rail Transportation Equipment,
    Dalian Jiaotong University, Dalian 116028, China
*   Correspondence: hm_zhao1977@163.com; Tel.: +86-411-8410-5386

Received: 25 October 2018; Accepted: 14 November 2018; Published: 1 December 2018

Abstract: A motor bearing system is a nonlinear dynamics system with nonlinear support stiffness. It is an asymmetry system, which plays an extremely important role in rotating machinery. In this paper, a center frequency method of double thresholds is proposed to improve the variational mode decomposition (VMD) method, then an adaptive VMD (called DTCFVMD) method is obtained to extract the fault feature. In the DTCFVMD method, a center frequency method of double thresholds is a symmetry method, which is used to determine the decomposed mode number of VMD according to the power spectrum of the signal. The proposed DTCFVMD method is used to decompose the nonlinear and non-stationary vibration signals of motor bearing in order to obtain a series of intrinsic mode functions (IMFs) under different scales. Then, the Hilbert transform is used to analyze the envelope of each mode component and calculate the power spectrum of each mode component. Finally, the power spectrum is used to extract the fault feature frequency for determining the fault type of the motor bearing. To test and verify the effectiveness of the DTCFVMD method, the actual fault vibration signal of the motor bearing is selected in here. The experimental results show that the center frequency method of double thresholds can effectively determine the mode number of the VMD method, and the proposed DTCFVMD method can accurately extract the clear time frequency characteristics of each mode component, and obtain the fault characteristics of characteristics; frequency, rotating frequency, and frequency doubling and so on.

Keywords: variational mode decomposition; signal analysis; time-frequency analysis; center frequency method of double thresholds; Hilbert transform

---

1. Introduction

Time-frequency analysis method has been the most commonly applied method in the field of fault diagnosis of rotating machinery [1–3]. Vibration signals of rotating machinery usually contain a lot of information about the equipment health states. During equipment operation, when the equipment health state takes a turn for the worse, the impact and impact pulses of the equipment will increase [4–6]. Therefore, it is very significant to deeply analyze vibration signals for health monitoring and fault prediction of rotating machinery.

Due to the complex structure and changeable working conditions of rotating machinery, the vibration signals often take on the characteristics of multi-component and non-stationary. The time-frequency analysis method is widely applied to process the vibration signals of rotating machinery in order to extract fault features by using signal decomposition and filtering [7–16]. There exists a lot of time-frequency analysis methods, such as the Gabor transform, wavelet transform, Hilbert-Huang transform, empirical mode decomposition (EMD), local mean decomposition (LMD), empirical wavelet transform (EWT), and so on [17–32]. These time-frequency analysis methods can extract fault features from vibration signals, but they have their own problems in vibration signal analysis. The wavelet transform needs to select the wavelet basis function in advance, thus it is a non-adaptive signal analysis method. The applications of EMD and LMD are restricted because of the existence of mode mixing. Although the EEMD and ensemble LMD methods alleviate the mode mixing problem to a certain extent, the computation complexity is sharply increased and the added white noises cannot be removed. variational mode decomposition (VMD) is an adaptive and non-recursive signal decomposition method developed by Dragomiretskiy and Zosso in 2014 [33]. The VMD method is used to the transform mode decomposition problem into a variational solution problem. Additionally, the alternating direction multiplier method is used to optimize the VMD. The signal is decomposed into a k discrete number of sub-signals; near the corresponding center frequency, each component is considered to be tight. The constrained variational optimization problem is used to evaluate the bandwidth of the component. In the optimization, it can obtain a set of mode ensemble with the limited band feature. It is essentially a set of adaptive Wiener filter bank and can separate the modes with different center frequencies. Therefore, the VMD method is widely used to extract the principal modes for fault diagnosis in bearings' signals. Wang et al. [34] studied the equivalent filtering characteristics of VMD, and proposed a rub-impact fault detection method for realizing rotor-stator fault diagnosis. Yi et al. [35] proposed a novel method of fault feature extraction based on the combination of VMD and the particle swarm optimization (PSO) algorithm. Liu et al. [36] proposed a novel signal denoising method that combines variational mode decomposition and detrended fluctuation analysis. Zhang et al. [37] proposed a novel hybrid fault diagnosis approach by using variational mode decomposition and majoriation-minization based total variation denoising for the denoising and non-stationary feature extraction. Liu et al. [38] proposed an improved variational mode decomposition for time-frequency analysis of fault diagnosis of a turbine rotor. Henao et al. [39] reviewed diagnostic techniques for electrical machines. An et al. [40] proposed a gear fault diagnosis method based on variational mode decomposition and envelope analysis according to the modulation characteristics of the gear vibration signal arising from faults therein. Mahgoun et al. [41] proposed a gear fault detection method by using the VMD method at a variable rotating speed. Abdoos et al. [42] proposed a new hybrid algorithm based on VMD, S-transform, and support vector machines for power quality disturbances' detection in electrical power systems. Li et al. [43] proposed an independence-oriented VMD method for identifying the fault feature of the wheel set bearing of a high-speed locomotive. From the application of the VMD method, it can see that the VMD method can better extract the multiple features from a vibration signal. However, the VMD method needs to give the mode number in advance. If the given mode number is unreasonable, the VMD method will cause the loss of important modes. Now, the VMD method is widely used to analyze the signal, and the central frequency observation method is used to determine the appropriate mode number. In this method, a number of values of the mode number is preset to repeatedly run VMD several times, and then the results of all trials are observed to determine the appropriate value of the mode number. This method requires manual judgments and experience to determine the appropriate value of the mode number, thus it will greatly limit the adaptability of VMD method. In order to overcome this drawback, some researchers have deeply studied and improved the VMD method. Immovilli and Cocconcelli [44] investigated shaft radial load effect on bearing fault signatures' detection. Wang et al. [45] proposed a complex variational mode decomposition method for the analysis of complex-valued data. Zhang et al. [46] proposed an improved VMD method with adaptive parameter selection to analyze vibration signals. Mohanty et al. [47] proposed a novel fault identification method based on combining variational mode decomposition, the Hurst exponent, and correlation coefficient. Choi et al. [48] proposed an improvement of variational mode decomposition that can effectively handle problems caused by missing values. Wang et al. [49] proposed a quasi-bivariate VMD method for extracting features with various scales. In addition, the other feature extraction methods are proposed in recent years [50–52].

These improved VMD methods can better determine the mode number and extract the multiple features from a vibration signal. However, some limitations remain regarding the adaptability of the VMD method. Therefore, it is necessary to continue to deeply study the appropriate mode number in the VMD. In this paper, a method called VMD with the center frequency method of double thresholds (DTCFVMD) is proposed to determine the appropriate mode number. Then, the DTCFVMD method is introduced into signal analysis to identify the bearing fault type. Finally, the actual fault vibration signal of the motor bearing is selected to test and verify the effectiveness of the DTCFVMD method.

2. Basic Methods

2.1. Variational Mode Decomposition (VMD) Method

The VMD method is an adaptive signal decomposition method by Dragomiretskiy and Zosso [33]. It is used to decompose a real signal into a set of sub-signals. Near the corresponding center frequency, each component is considered to be tight. The constrained variational optimization problem is used to evaluate the bandwidth of the component. Finally, in the optimization, the VMD method can obtain a solution for the constrained variational problem.

In the VMD method, the intrinsic mode function (IMF) is defined as an amplitude modulation-amplitude frequency (AM-FM) signal. Its expression is described as follows:

$$u_k(t) = A_k(t) \cos(\varphi_k(t)) \tag{1}$$

where, $\varphi_k(t)$ is a non-decreasing function, $\varphi_k'(t) \geq 0$, $A_k(t)$ is the envelopment; the change of the envelopment, $A_k(t)$, and the instantaneous frequency, $w_k(t) = \varphi_k'(t)$, is much slower than the phase, $\varphi_k(t)$. In other words, in sufficiently long intervals, $[t - \delta, t + \delta]$, $\delta \approx 2\pi / \varphi_k'(t)$, $u_k(t)$ can be regarded as a pure harmonic signal with the amplitude, $A_k(t)$, and frequency, $w_k(t)$.

To evaluate the bandwidth of the mode, for each mode function, $u_k(t)$, the Hilbert transform is used to obtain the unilateral spectrum of the signal, $(\delta(t) + j/\pi t) * u_k(t)$. $\delta(t)$ is the distribution function of Dirichlet, * represents the convolution. An exponential term, $e^{jw_k t}$, is added to adjust the estimated center frequency of each mode function of the corresponding analytic signal, and the spectrum of each mode is transferred to the base band, $[(\delta(t) + j/\pi t) * u_k(t)]e^{jw_k t}$. w_k represents the center frequency. Then, the Gauss smoothing of the signal is demodulated to obtain the constraint variational problem.

$$\min_{\{u_k\}, \{w_k\}} \left\{ \sum_k ||\partial_t [(\delta(t) + \frac{j}{\pi t}) * u_k(t)] e^{-jw_k t}||_2^2 \right\} \tag{2}$$

subject to:

$$\sum_k u_k = f \tag{3}$$

To transform the constraint problem into an unconstrained problem, Lagrangian multipliers, λ, and quadratic penalty term, α, are applied to render the unconstrained variational problem. The Lagrangian multipliers, λ, can strictly enforce constraints. The quadratic penalty, α, can ensure the reconstruction accuracy. The augmented Lagrangian is described as:

$$L(\{u_k\}, \{w_k\}, \lambda) = \alpha \|\sum_k \partial_t[(\delta(t) + \frac{j}{\pi t}) * u_k(t)]e^{-jw_k t}\|_2^2 + \|f(t) - \sum_k u_k(t)\|_2^2 + <\lambda(t), f(t) - \sum_k u_k(t)> \tag{4}$$

In VMD, the alternating direction method of the multiplicative operator is used to solve the above variational problem. The mode function, u_k^{n+1}, center frequency, w_k^{n+1}, and λ are alternated and updated to seek to expand the saddle point of the Lagrange expression. The value problem of the mode function, u_k^{n+1}, is described as:

$$u_k^{n+1} = \underset{u_k \in X}{\operatorname{argmin}} \left\{ \alpha \|\sum_k \partial_t[(\delta(t) + \frac{j}{\pi t}) * u_k(t)]e^{-jw_k t}\|_2^2 + \|f(t) - \sum_k u_k(t) + \frac{\lambda(t)}{2}\|_2^2 \right\} \tag{5}$$

The equidistance transformation of Parseval/Plancherel Fourier is used to transform the formula (4) into the frequency domain.

$$\hat{u}_k^{n+1} = \underset{\hat{u}_k, u_k \in X}{\operatorname{argmin}} \left\{ \alpha \|j(w)[(1 + sng(w + w_k))\hat{u}_k(w + w_k)\|_2^2 + \|\hat{f}(w) - \sum_i \hat{u}_i(w) + \frac{\hat{\lambda}(t)}{2}\|_2^2 \right\} \tag{6}$$

The w in the first item is replaced by $w - w_k$ to obtain the following expression:

$$\hat{u}_k^{n+1} = \underset{\hat{u}_k, u_k \in X}{\operatorname{argmin}} \{\alpha \|j(w - w_k)[(1 + sng(w))\hat{u}_k(w)\|_2^2 + \|\hat{f}(w) - \sum_i \hat{u}_i(w) + \frac{\hat{\lambda}(t)}{2}\|_2^2 \tag{7}$$

In the approximation terms of reconstruction, the conjugate symmetry of real signals is used to transform the expression (7) into the form of a non-negative frequency integral.

$$\hat{u}_k^{n+1} = \underset{\hat{u}_k, u_k \in X}{\operatorname{argmin}} \{ \int_0^\infty 4\alpha(w - w_k)^2 |\hat{u}_k(w)|^2 + 2|\hat{f}(w) - \sum_i \hat{u}_i(w) + \frac{\hat{\lambda}(w)}{2}|^2 dw \} \tag{8}$$

By conversion, the solution of quadratic optimization problem can be obtained as follows:

$$\hat{u}_k^{n+1}(w) = \left((\hat{f}(w) - \sum_{i=k} \hat{u}_i(w) + \frac{\hat{\lambda}(w)}{2} \right) \frac{1}{1 + 2\alpha(w - w_k)^2} \tag{9}$$

where, $\hat{u}_k^{n+1}(w)$ is equivalent to the wiener filtering of the current residualwiener filtering equivalent to the current residual, $\hat{f}(w) - \sum_{i=k} \hat{u}_i(w)$. The full spectrum of the real mode can be completed symmetrically by Hermite symmetry, and the $\hat{u}_k(w)$ is the real part, $u_k(t)$, of the inverse Fourier transform.

The center frequency, w_k, is the first item of the right side of the expression (5) and the expression (6). Therefore, the updated expression, w_k, is described as follows:

$$w_k^{n+1} = \underset{w_k}{\operatorname{argmin}} \{\|\sum_k \partial_t[(\delta(t) + \frac{j}{\pi t}) * u_k(t)]e^{-jw_k t}\|_2^2 \} \tag{10}$$

It is like as the mode function, u_k, and the center frequency, w_k, in the Fourier domain is optimized as follows:

$$w_k^{n+1} = \underset{w_k}{\operatorname{argmin}} \left\{ \int_0^\infty (w - w_k)^2 |\hat{u}_k(w)|^2 dw \right\} \tag{11}$$

The solution of the above quadratic optimization problem is given:

$$w_k^{n+1} = \frac{\int_0^\infty w|\hat{u}_k(w)|_2 dw}{\int_0^\infty |\hat{u}_k(w)|_2 dw} \tag{12}$$

The algorithm process of VMD is described as the follows:

Step 1. Initialize the parameters of VMD, including $\{\hat{u}_k^1\}$, $\{w_k^1\}$, and n.

Step 2. Update the u_k and w_k according to the expression (9) and expression (12).

Step 3. Update the λ.

$$\hat{\lambda}^{n+1}(w) \leftarrow \hat{\lambda}^n(w) + \tau(\hat{f}(w) - \sum_k \hat{u}_k^{n+1}(w)) \tag{13}$$

Step 4. Set the error to $\varepsilon > 0$. If $\frac{\sum_k ||\hat{u}_k^{n+1} - \hat{u}_k^n||_2^2}{||\hat{u}_k^n||_2^2} < \varepsilon$ is met, the iteration is terminated. Otherwise, return to Step 2.

2.2. Empirical Mode Decomposition and Ensemble Empirical Mode Decomposition

The EMD method is an adaptive decomposition method. The EMD method can decompose non-linear data into a series of amplitude modulated-frequency modulated (AM-FM) components. The EMD decomposes the original signal $(S(x))$ into a number of IMFs:

$$S(t) = \sum_{i=1}^n c_i(t) + r_n(t) \tag{14}$$

where $r_n(t)$ represents residual error function, and IMF components, $c_1, c_2, c_3, \ldots, c_n$, contain different elements, respectively, from a low to high frequency of signals.

The EEMD method is an improved EMD method [53,54]. It can solve the mode mixing problem in the VMD method by adding Gaussian white noise. The decomposition principle of the EEMD method is that the additional white noise is uniformly distributed in the whole time-frequency space, and the time-frequency space is composed of different scale components separated by the filter group. When the signal is added with a uniformly distributed white noise, the signal regions under different scales are automatically mapped to the appropriate scales.

3. An Adaptive VMD with the Center Frequency Method of Double Thresholds

Compared with the EMD and EEMD methods, VMD is a new signal decomposition method. This method has a better theoretical basis. Its essence is multiple adaptive Wiener filter banks, which take on better noise robustness. In the field of mode decomposition, the VMD can successfully separate two pure harmonic signals with similar frequencies and effectively suppress the mode mixing. However, the VMD decomposition needs to set the number of the decomposition in advance. How to effectively determine the number of the decomposition will directly affect the final decomposition results. If the presupposed value of it is greater than the effective mode number in the signal, it will generate the excessive decomposition phenomenon, decomposing the useless false components to interfere with the analysis of the effective components in the original signal. If the presupposed value is less than the effective mode number in the signal, it will generate the undecomposed phenomenon, which cannot completely decompose the effective component. Therefore, it is a very important role to determine the mode number of the VMD method. The research of the VMD method has just started, and it is a research hotspot topic to determine the mode number for the VMD method at present. There are some methods, which are used to determine the mode number, such as the central frequency method, kurtosis method, and optimization algorithms, and so on. The central frequency value of the corresponding IMF component can be obtained by using the VMD decomposition. The central frequency method can directly determine the mode number by using the central frequency value obtained by VMD decomposition, so the central frequency method has been widely used. However, the central frequency method depends on the subjective factors when the mode number is selected. Therefore, on the basis of the central frequency method, the dual threshold of the central frequency method is proposed for the adaptive selection mode number in the VMD decomposition. The central frequency method of double thresholds is proposed to adaptively select the mode number in the VMD decomposition.

3.1. The Center Frequency Method

3.1.1. The Basic Principle and Implementation Steps

Each IMF component obtained by VMD decomposition corresponds to a central frequency, cf, and the center frequency of the decomposed IMF component under each scale is distributed from low frequency to high frequency. Set the maximum value, k_{max}, of the decomposition scale, k, and the signal is decomposed under the decomposition scale, $k = 1, 2, 3, \cdots, k_{max}$. If the difference between the maximum value of the central frequency of the IMF component under the scale, k, and the maximum value of the central frequency of the IMF component under the scale, $k + 1$, is very small, the IMF component under the scale, k, is determined to firstly obtain the maximum value of the central frequency. According to the central frequency method, the number, k, of the VMD decomposition is determined.

The implementation steps of the central frequency method are described as follows:

Step 1. Determine the maximum value, k_{max}, of the decomposition scale.

Step 2. The VMD method is used to decompose the original signal into a series of IMF components under different scales $(IMF_k(k = 1, 2, 3, \cdots, k_{max}))$. The central frequency values of each IMF component are calculated to obtain the set, $\Delta cf_{IMF_k}(k = 1, 2, 3, \cdots, k_{max})$.

Step 3. Calculate the maximum value of the center frequency of the set, cf_{IMF_k}, under different scales, as $\Delta cf\_max_{IMF_k}(k - 1, 2, 3, \cdots, k_{max})$.

Step 4. Calculate the maximum values of the center frequencies of the set, $\Delta cf\_max_{IMF_k}$, as Δcf_max_{IMF}.

Step 5. Calculate the difference between the maximum center frequency value of all modes and the maximum center frequency value of the mode under scale, k, as $\Delta cf\_max_{IMF_k} = \Delta cf\_max_{IFM} - \Delta cf\_max_{IFM_k}$. If there is $\Delta cf\_max_k << \Delta cf\_max_{IMF}$, the k is the mode decomposition number of the VMD method.

When the central frequency method is used to determine the mode number, k, it is necessary to determine the value, k, according to the $\Delta cf\_max_k << cf\_max_{IFM_k}$. In the process, the subjective factors are influenced by human factors. The result will be that the optimal value cannot be obtained to some extent.

3.1.2. Experimental Analysis

The experiment data comes from the Bearing Data Center of Case Western Reserve University [55]. The deep groove ball bearing of 6205-2RS 6 JEM SKF (SKF company, Goteborg, Sweden) was used in this experiment. The experimental platform is shown in Figure 1. The motor with 1.5 Kw was connected to a dynamometer. The accelerometers at the 12 o'clock position on the drive end and fan end of motor housing were used to collect vibration data. The electro-discharge machining method was used to make the bearing's inner race fault and outer race fault. The bearing information bearing is described in Table 1. The vibration signals were obtained under no-load (0 hp) at a rotating speed of 1797 r/min. The fault diameter was 0.1778 mm. The inner race fault and outer race fault vibration signals are selected in here. The vibration signals were sampled at 12,800 Hz. Each vibration signal duration was 10 s. The vibration signals were divided into sample segmentations. Each sample covered 2048 data points.

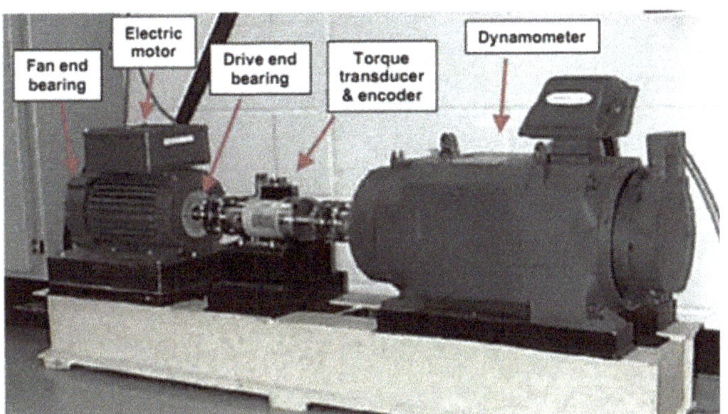

Figure 1. The experimental platform.

Table 1. The bearing information of the 6205-2RS JEM SKF deep groove ball bearing.

Inside Diameter	Outside Diameter	Thickness	Ball Diameter	Pitch Diameter	Roller Number	Rotating Speed
25 mm	52 mm	15 mm	8.182 mm	44.2 mm	9	1797 r/min

The theoretical calculation values of the fault characteristic frequency were calculated according to the calculation method, as shown in Table 2.

Table 2. Fault characteristic frequencies.

Inner Race	Outer Race	Rolling Element	Switching Frequency
162.2 (Hz)	107.3 (Hz)	141.1 (Hz)	29.2 (Hz)

The inner race and outer race faults of the roller bearing are similar in Figures 2 and 3.

The time domain waveform and frequency domain waveform of the inner race fault vibration signal are shown in Figures 4 and 5.

Figure 2. The inner race fault of the roller bearing.

Figure 3. The outer race fault of the roller bearing.

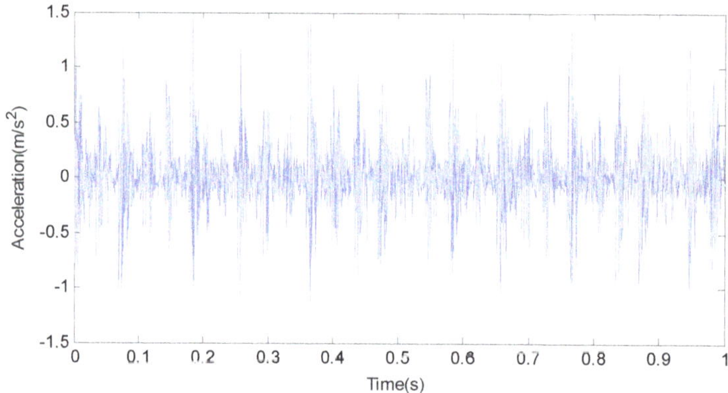

Figure 4. The time domain waveform of the inner race fault vibration signal.

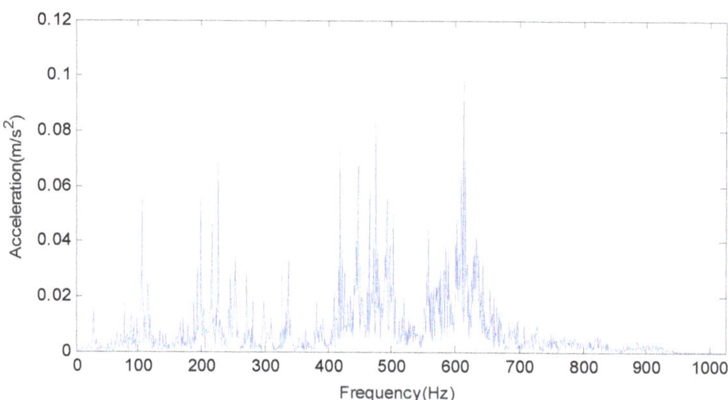

Figure 5. The frequency domain waveform of the inner race fault vibration signal.

Firstly, the VMD method was used to decompose the inner race fault vibration signal to obtain the central frequency of each IMF component under different scales, as shown in Figure 6.

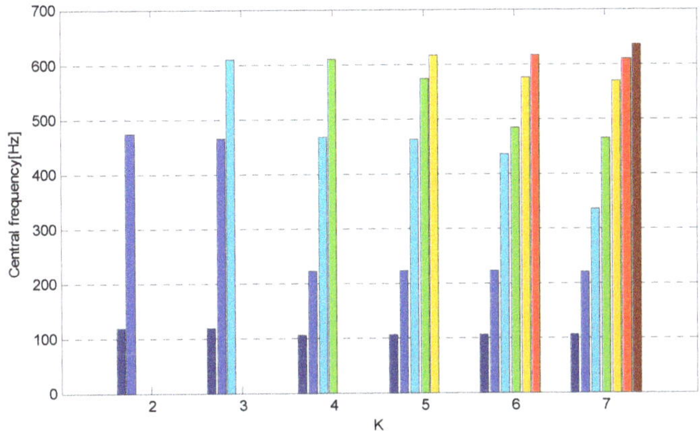

Figure 6. The central frequency of each IMF component under different scales.

As can be seen from Figure 6, when there is $k = 7$, the center frequency of the seventh IMF component obtains the maximum value of all IMF components. When there is $k \geq 3$, the difference between the maximum value of the central frequency of the IMF component and the maximum value of all IMF components is very small under different scales. According to the center frequency method, the value of k is determined as $k = 3$. When there is $k \geq 4$, the new frequency value appears in the mode. The analysis considers that the corresponding mode of the new frequency value is the effective mode. Therefore, when there is $k = 3$, the effective mode cannot be completely decomposed. From the experimental results, it can be found that there is an undecomposed phenomenon for determining the value of k by using the central frequency method.

When the value of k is 5, the time domain waveform and frequency domain waveform of the inner race fault vibration signal are shown in Figures 7 and 8.

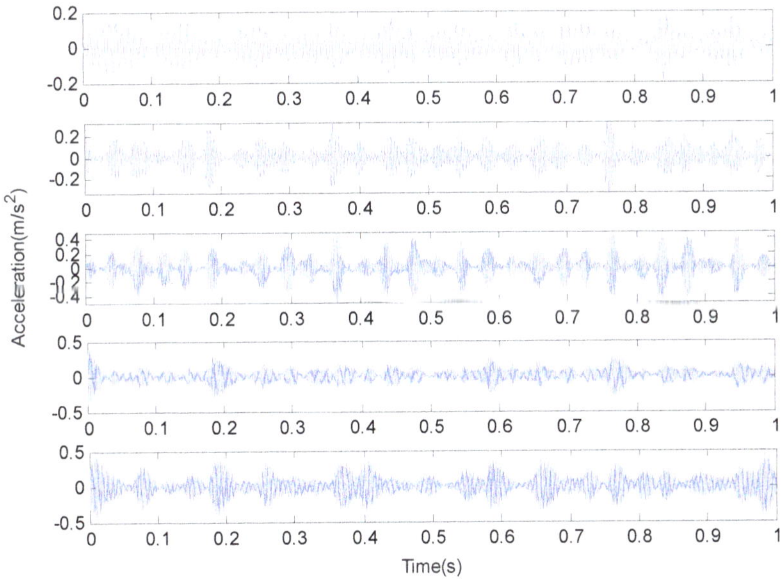

Figure 7. The time domain waveform of each IMF for the inner race at $k = 5$.

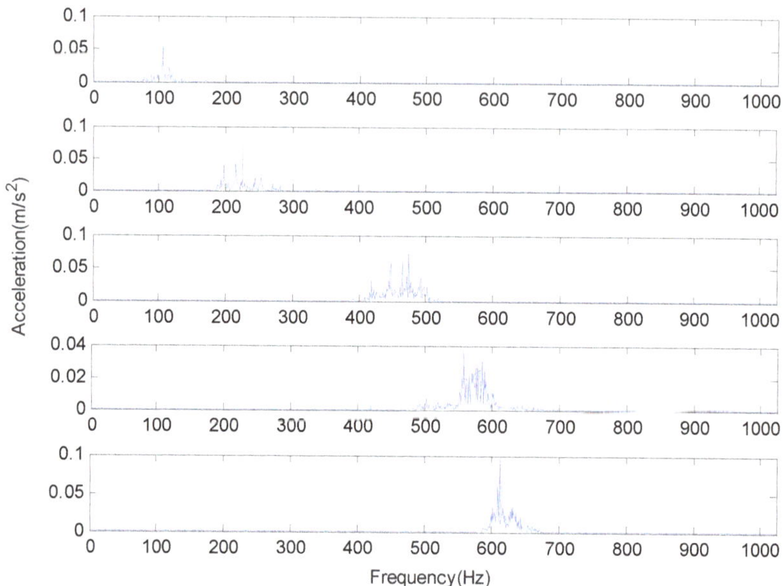

Figure 8. The frequency domain waveform of each IMF for the inner race at $k = 5$.

3.2. The Center Frequency Method of Double Thresholds

3.2.1. The Idea of the Center Frequency Method of Double Thresholds

For the shortcomings of the center frequency method, a center frequency method of double thresholds is proposed to determine the value of the decomposition scale, k, in this paper. The center frequency method of double thresholds is a symmetry method. Based on the center frequency, the double thresholds of T_1 and T_2 are set. The threshold, T_1, is used to measure the difference between the maximum value of the central frequency of the corresponding modes under different scales and the maximum value of center frequencies of all modes. It is known by the central frequency method that the center frequency of the obtained IMF component under a certain scale is distributed from low frequency to high frequency. If the difference is greater than T_1, it indicates that the maximum IMF component of the central frequency in the original signal is not decomposed. The threshold, T_2, is used to measure the difference between the value of the central frequency of the Lth IMF component and the value of the central frequency of the $(L-1)$th IMF component under the same decomposition scale, k ($k = L$). The VMD can effectively decompose the two pure harmonic signals with similar frequencies, and the IMF components with similar frequencies are retained according to the threshold, T_2. Set the maximum value, $k_{\max}$, of the decomposition scale, k. The signal is decomposed under decomposition scale, $k = 1, 2, 3, \cdots, k_{\max}$, to obtain the value and the maximum value of the center frequency of the corresponding modes, and the maximum value of the center frequency of all modes under different scales. If the difference between the maximum value of the central frequency of the IMF component under a certain scale, k, and the maximum value of the central frequency of the IMF component under a certain scale, $k + 1$, is less than T_1, the difference between the value of the center frequency of the Lth IMF component and the value of the center frequency of the $(L-1)$th IMF component under the decomposition scale, k, is calculated ($k = L$). If the difference value is less than the threshold, T_2, the minimum value of k meeting the above conditions is regarded as the optimal mode number of the VMD decomposition. The expressions of T_1 and T_2 are described as follows:

$$T_1 = a * cf\_\max_{IMF_k} \tag{15}$$

$$T_2 = b * cf\_\max_{IMF_k} \tag{16}$$

where there is $k = 1, 2, 3, \cdots, k_{\max}$.

In a practical application, according to the different application backgrounds, the values of a and b are determined according to the experiences. Therefore, according to the practical application in this paper, there are $a = 0.08$ and $b = 0.15$.

3.2.2. The Flow and Steps of the Center Frequency Method of Double Thresholds

The flow of the center frequency method of double thresholds is shown in Figure 9.

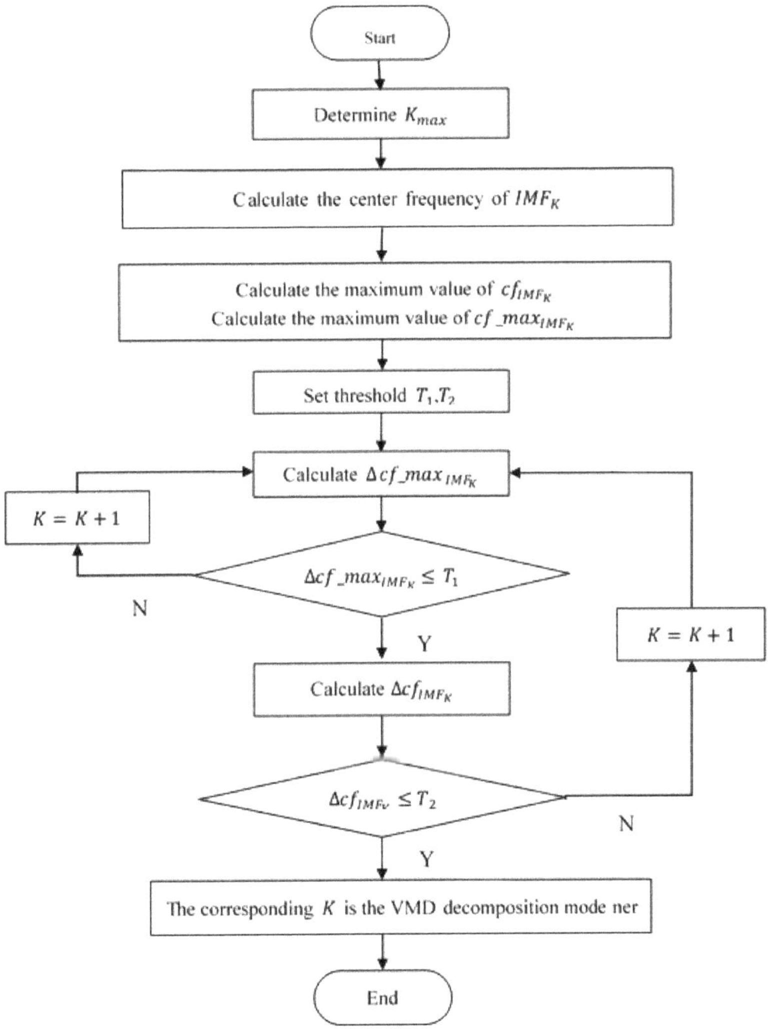

Figure 9. The flow of the center frequency method of double thresholds.

The implementation steps of the center frequency method of double thresholds are as follows:

Step 1. Determine the maximum value, k_{max}, of the decomposition scale.

Step 2. The VMD method is used to decompose the original signal into a series of IMF components under different scales ($IMF_k(k = 1, 2, 3, \cdots, k_{max})$). The central frequency values of each IMF component is calculated to obtain the set, $cf_{IMF_k}(k = 1, 2, 3, \cdots, k_{max})$.

Step 3. Calculate the maximum value of the center frequency of the set, cf_{IMF_k}, under different scales, as $cf\_max_{IMF_k}(k = 1, 2, 3, \cdots, k_{max})$.

Step 4. Calculate the maximum value of the center frequency of the set, $cf\_max_{IMF_k}$, as cf_max_{IMF}.

Step 5. Set thresholds, $a = 8\%$ and $b = 15\%$, for the expressions (15) and (16).

Step 6. Calculate the difference between the maximum center frequency value of all modes, as $\Delta cf\_max_{IMF_k} = cf\_max_{IFM} - cf\_max_{IFM_k}(k = 1, 2, 3, \cdots, k_{max})$.

Step 7. If there is $\Delta cf\_max_k \leq T_1$, Step 8 is executed. Otherwise, execute $k = k + 1$ and go to Step 6.

Step 8. Calculate the difference value of the central frequency of the Lth IMF component and the value of the central frequency of the $(L - 1)$th IMF component of the set, cf_{IMF} ($k = L$), and is calculated as $\Delta cf_{IMF} = cf_{IMF_{k_L}} - cf_{IMF_{k_{L-1}}}$. If there is $\Delta cf_{IMF} \leq T_2$, the corresponding k ($k = 1, 2, 3, \cdots, k_{max}$) of Δcf_{IMF} is the optimal mode number of the VMD method. Otherwise, execute $k = k + 1$ and go to Step 6.

3.3. Effectiveness Analysis of the Center Frequency Method of Double Thresholds

The center frequency method of double thresholds is used to determine the number of VMD decomposition modes. When there is $k = 2$, the $\Delta cf\_max_{IMF_k} \leq T_1$ is not be met. When there is $k = 3, 4$, the $\Delta cf\_max_{IMF_k} \leq T_2$ is not be met. When there is $k = 5$, the condition of judgment is met at the same time. Therefore, $k = 5$ is the optimal mode number of VMD decomposition. Compared with the central frequency method, the difference of the maximum value of the central frequency of the IMF component in the determined modes by two methods is very small, which shows that the component of the maximum value of the central frequency in the original signal has been decomposed. When there is $k = 5$, the central frequency value of two additional IMF components did not appear in $k = 3$. Therefore, the $k = 5$ is better than $k = 3$ according to the effective mode.

The time domain waveform and frequency domain waveform of the outer race fault vibration signal are shown in Figures 10 and 11.

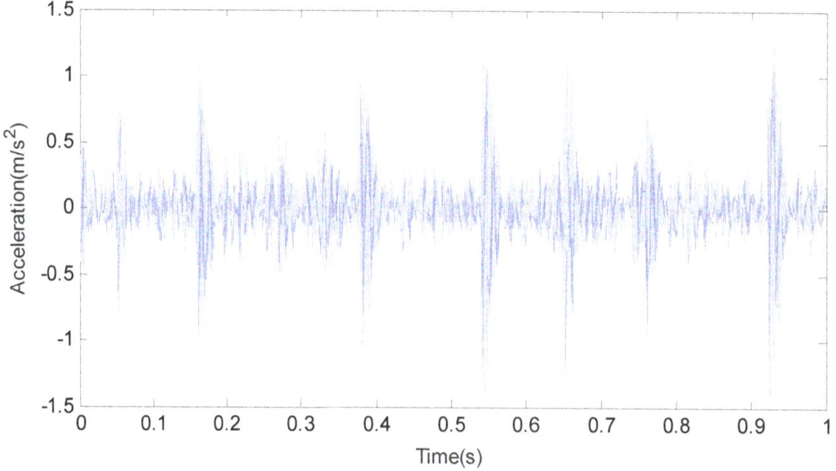

Figure 10. The time domain waveform of the outer race fault vibration signal.

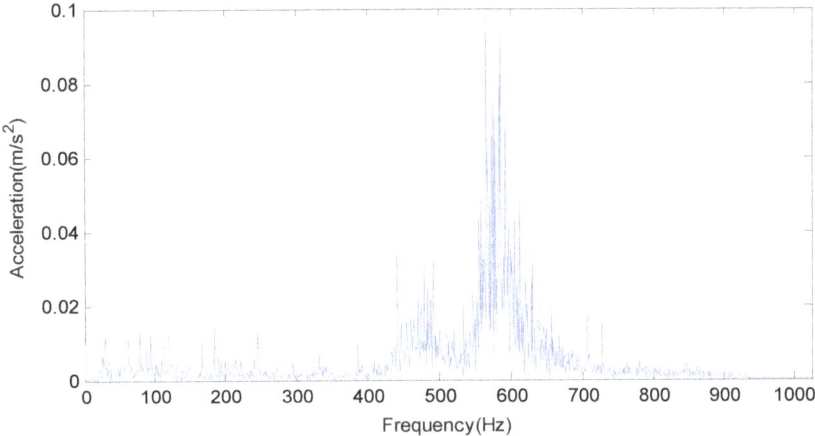

Figure 11. The frequency domain waveform of the outer race fault vibration signal.

The outer race fault vibration signal is decomposed by VMD, and the central frequency value of each IMF component under different scales are shown in Figure 12.

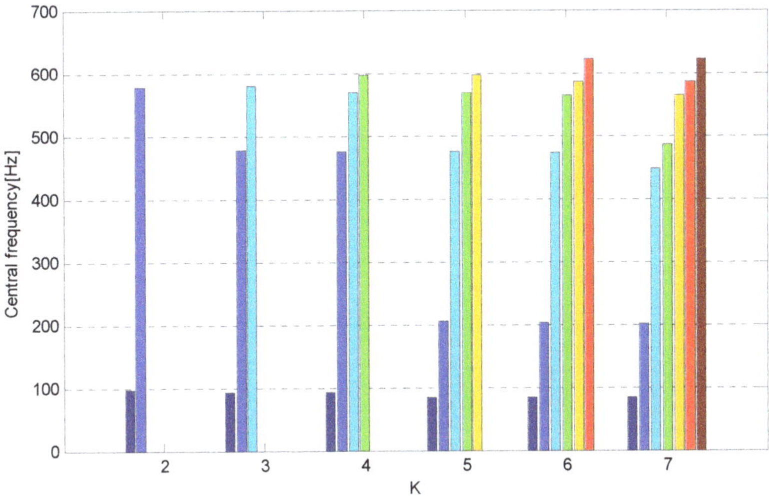

Figure 12. The central frequency values of each IMF component under different scales.

As can be seen from Figure 12, when there is $k = 7$, the center frequency value of the seventh IMF component obtains the maximum value of all IMF components. When there is $k \geq 2$, the difference between the maximum value of the central frequency of the IMF component and the maximum value of all IMF components is very small under different scales. According to the center frequency method, the value of k is determined as $k = 2$. When there is $k \geq 3$, the new frequency value appears in the mode. The analysis considers that the corresponding mode of the new frequency value is the effective mode. Therefore, when there is $k = 2$, the effective mode cannot be completely decomposed. The center frequency method of double thresholds is used to determine the number of VMD decomposition modes. When there is $k = 2, 3$, the $\Delta cf\_\max_{IMF_k} \leq T_2$ is not be met. When there is $k = 4$, the condition of judgment is met at the same time. Therefore, $k = 4$ is the optimal mode number of VMD decomposition. Compared with the central frequency method, the difference of the maximum value of the central frequency of the IMF component in the determined modes by the two methods is very small, which shows that the component of the maximum value of the central frequency in the original signal has been decomposed. When there is $k = 4$, the central frequency value of two additional IMF components did not appear in $k = 2$. Therefore, the $k = 4$ is better than $k = 2$ according to the effective mode.

When the value of k is 4, the time domain waveform and frequency domain waveform of the outer race fault vibration signal are shown in Figures 13 and 14.

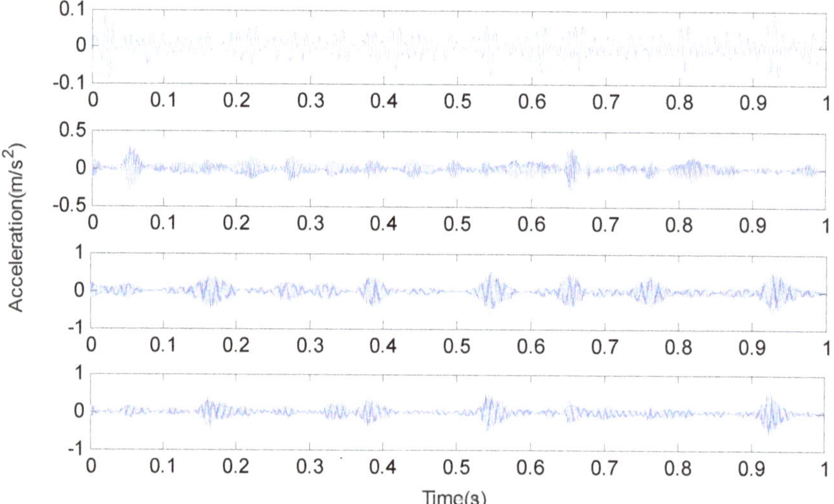

Figure 13. The time domain waveform of each IMF for the outer race at $k = 4$.

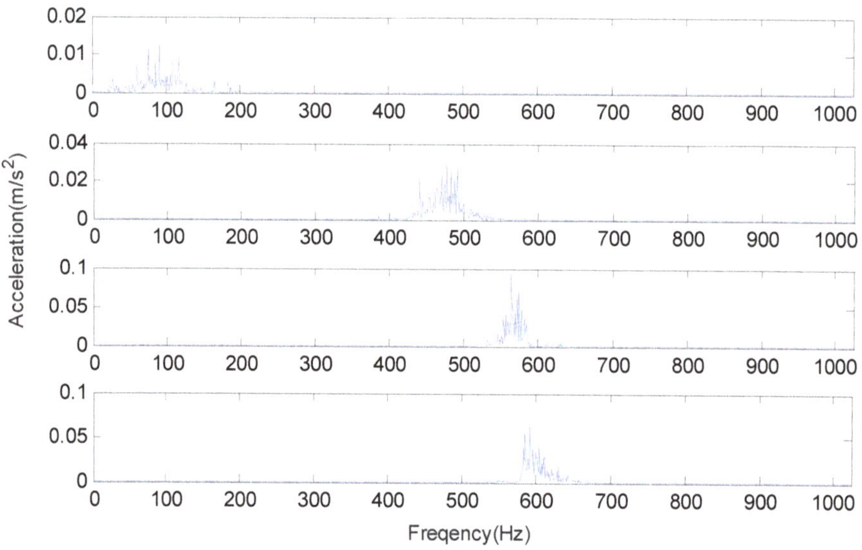

Figure 14. The frequency domain waveform of each IMF for the inner race at $k = 4$.

From the above experimental results and analysis, it can be found that the mode number of VMD with the central frequency method of double thresholds is superior to the mode number of VMD with the central frequency method. In the VMD method, the central frequency method of double thresholds can effectively solve the problem that the person determines the value of k according to $\Delta cf\_max_k \ll \Delta cf\_max_{IMF_k}$. At the same time, the central frequency method of double thresholds makes up for the defect that the central frequency method cannot keep the IMF components with similar frequency values. Therefore, the central frequency method of double thresholds can determine the value of k in the maximum extent, which is the optimal mode number.

4. Feature Extraction Method Based on the DTCFVMD and Hilbert Transform

4.1. Feature Extraction Method

When the rolling bearings have a fault, its vibration signal contains a lot of running state information, which are non-stationary and multicomponent modulation signals. Due to the weak modulation source, early fault signals and noise interference from surrounding equipment and the environment, it is difficult to extract and identify the fault feature frequency. The VMD method is an adaptive and non-recursive signal decomposition method. It adopts a completely non-recursive method to realize the frequency domain decomposition of a signal, which has a strong decomposition ability, good anti-noise performance, and fast processing speed. Therefore, the proposed DTCFVMD method is used to decompose the nonlinear and non-stationary vibration signals of bearings to obtain a series of intrinsic mode functions under different scales. Then, the Hilbert transform is used to analyze the envelope of each mode component and calculate the power spectrum of each mode component, which is used to extract the fault feature frequency for determining the fault type of the motor bearing. The feature extraction model based on the DTCFVMD and Hilbert transform is shown in Figure 15.

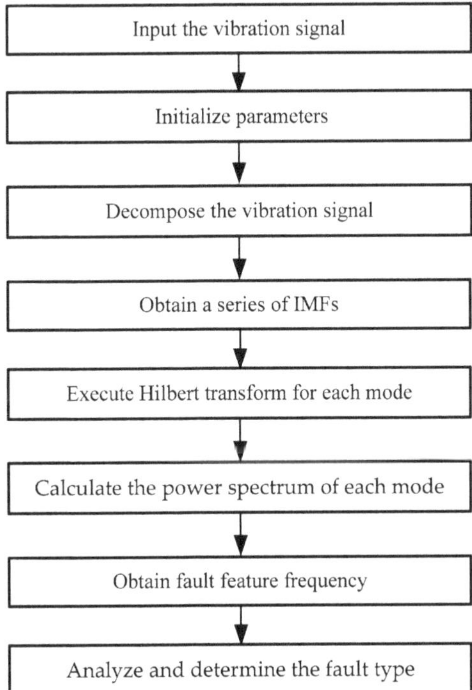

Figure 15. The feature extraction model based on the DTCFVMD and Hilbert transform.

4.2. Steps of Feature Extraction

The steps of the feature extraction model based on the DTCFVMD and Hilbert transform are described as follows:

Step 1. Preprocess the collected vibration signals of the motor rolling bearing, including denoising, filtering, and so on.

Step 2. Propose a center frequency method of double thresholds to improve the VMD method for obtaining the adaptive VMD(DTCFVMD) method.

Step 3. The preprocessed vibration signal is decomposed by using the proposed DTCFVMD method to obtain a series of IMFs.

Step 4. Analyze the envelope of each mode component by using the Hilbert transform.

Step 5. Calculate the power spectrum of each mode component to extract the fault feature frequency.

Step 6. The fault feature frequency is compared with theoretical inherent fault frequency to accurately determine the fault type of the motor rolling bearing.

5. Verification and Results Analysis

5.1. The Effectiveness Verification

The power spectrum represents the relationship between signal power and frequency; that is, the distribution of the signal power in the frequency domain. The Hilbert transform is used to solve the envelope of each IMF component, then the power spectrum of the envelope is solved. The maximum frequency value of the power spectral density is marked in order to express the fault characteristic frequency, rotating frequency, and frequency multiplication by power spectrum analysis.

According to the center frequency method of double thresholds, the optimal mode number of VMD decomposition for the inner race vibration signal and outer race vibration signal are 5 and 4, respectively. In order to verify the validity of the optimal mode, the signal power spectrum is used to extract the fault characteristic and estimate the effective frequency.

For the inner race fault vibration signal, the power spectrum of each IMF component under $k = 4$, $k = 5$, and $k = 6$ is shown in Figure 16 and Table 3.

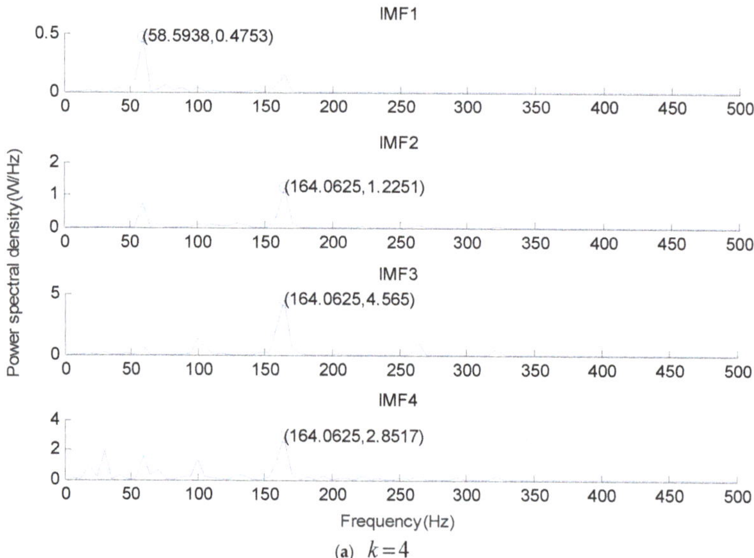

Figure 16. *Cont.*

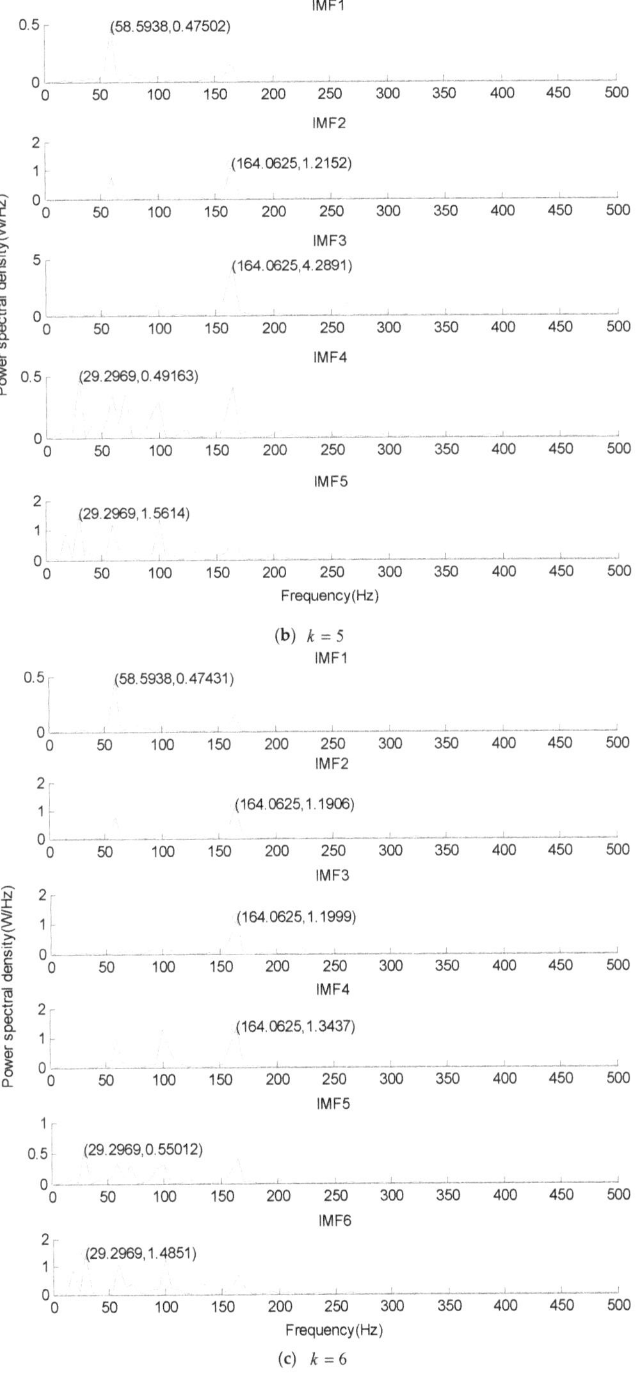

(**b**) *k* = 5

(**c**) *k* = 6

Figure 16. The power spectrum of each IMF component under (**a**) *k* = 4, (**b**) *k* = 5, and (**c**) *k* = 6.

Table 3. Frequency values of power spectrum of each IMF under $k = 4$, $k = 5$, and $k = 6$.

	IMF1 (Hz)	IMF2 (Hz)	IMF3 (Hz)	IMF4 (Hz)	IMF5 (Hz)	IMF6 (Hz)
$k = 4$	58.6	164.1	164.1	164.1		
$k = 5$	58.6	164.1	164.1	29.3	29.3	
$k = 6$	58.6	164.1	164.1	164.1	29.3	29.3

As can be seen from Figure 16 and Table 3, 29.3 Hz represents the rotating frequency, 58.6 Hz represents the frequency doubling of the rotating frequency, and 164.1 Hz represents the fault characteristic frequency of the inner race. When there is $k = 5$, the rotating frequency can be obtained from the frequency spectrum, but the rotating frequency cannot be obtained from the frequency spectrum when there is $k = 4$. When there are $k = 5$ and $k = 6$, the rotating frequency, the frequency doubling of the rotating frequency, and the fault characteristic frequency of the inner race can be obtained from the frequency spectrum. When there is $k = 6$, it has one more fault characteristic frequency at 164.1 Hz than $k = 5$. However, the frequency value repeats, so $k = 5$ is the optimal mode number of the VMD decomposition.

For the outer race vibration signal, the power spectrum of each IMF component under $k = 4$, $k = 5$, and $k = 6$ is shown in Figure 17 and Table 4.

As can be seen from Figure 17 and Table 4, 29.3 Hz represents the rotating frequency, 87.9 Hz represents the frequency tripling of the rotating frequency, 105.5 Hz represents the fault characteristic frequency of the outer race, and 46.9 Hz represents the difference between the fault characteristic frequency and the frequency doubling of the rotating frequency of the outer race. When there is $k = 4$, the rotating frequency can be obtained from the frequency spectrum, but the rotating frequency cannot be obtained from the frequency spectrum when there is $k = 3$. When there are $k = 4$ and $k = 5$, the frequency tripling of the rotating frequency, the difference between the fault characteristic frequency, and the frequency doubling of the rotating frequency of the outer race can be obtained from the frequency spectrum. When there is $k = 5$, it is not able to obtain the rotating frequency. Therefore, $k = 4$ is the optimal mode number of the VMD decomposition.

The frequency value of each IMF component extracted from the power spectrum shows that the value of k determined by the central frequency method of double thresholds is the optimal mode number of the VMD method. Therefore, it is proved that the proposed central frequency method of double thresholds is effective.

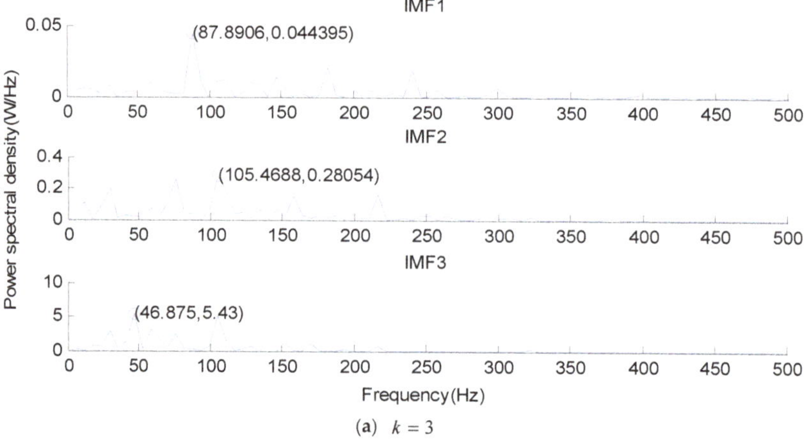

Figure 17. *Cont.*

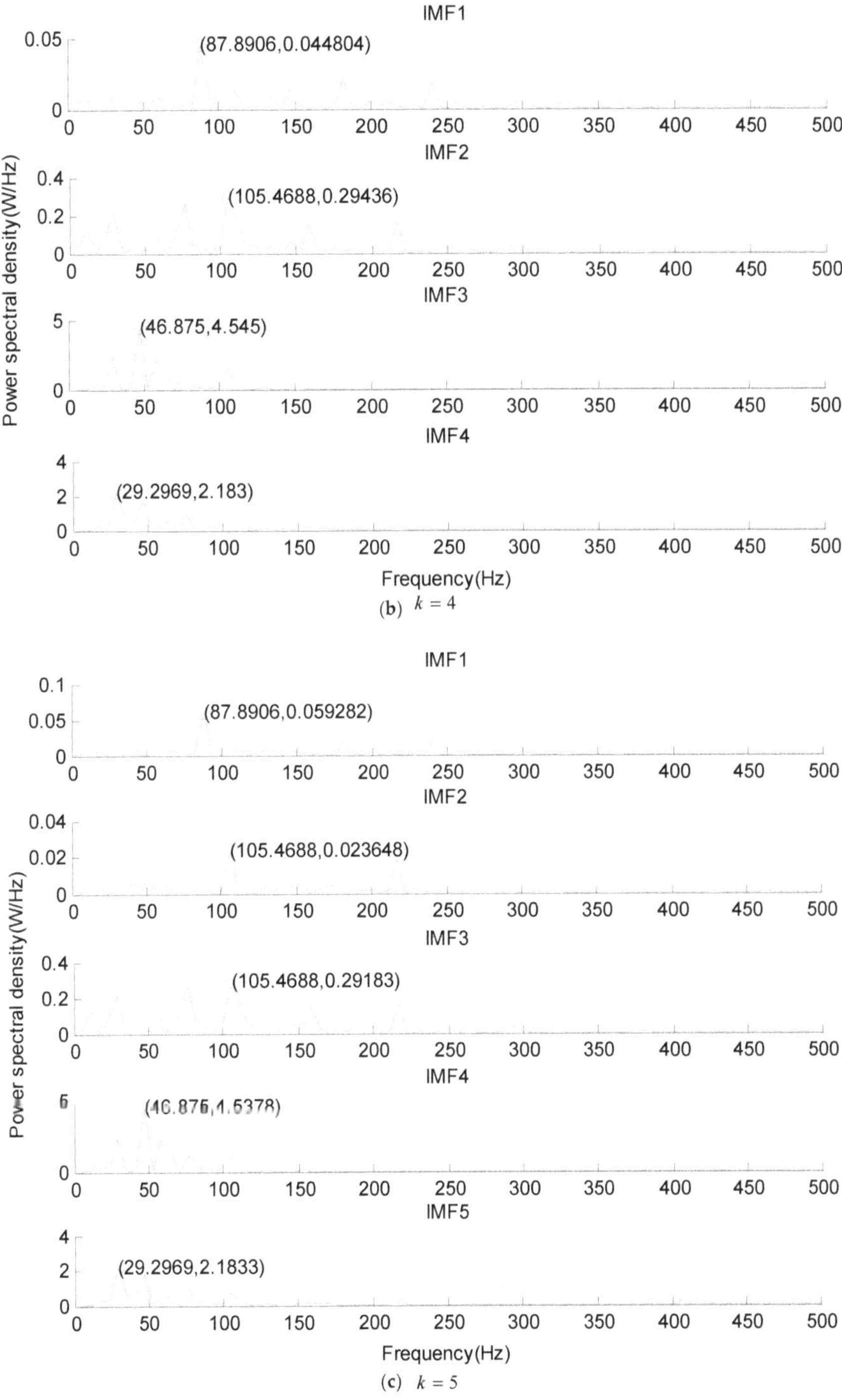

Figure 17. The power spectrum of each IMF component under (**a**) $k = 3$, (**b**) $k = 4$, and (**c**) $k = 5$.

Table 4. Frequency values of power spectrum of each IMF component under $k = 3$, $k = 4$, and $k = 5$.

	IMF1 (Hz)	IMF2 (Hz)	IMF3 (Hz)	IMF4 (Hz)	IMF5 (Hz)
$k = 3$	87.9	105.5	46.9		
$k = 4$	87.9	105.5	46.9	29.3	
$k = 5$	87.9	105.5	105.5	46.9	46.9

5.2. Comparison and Analysis

To verify the effectiveness of VMD with the center frequency method of double thresholds, the VMD with the center frequency method of double thresholds (DTCFVMD) is compared with the EMD and EEMD methods. The DTCFVMD, EMD, and EEMD methods are used to decompose the vibration signals of the inner ring and outer ring to obtain a series of IMFs. Then, the Hilbert transform is used to obtain the power spectrum of the decomposed IMF components. The effectiveness of DTCFVMD is explained by analyzing the effective frequency components extracted from the power spectrum.

The frequency values extracted by the power spectrum of each IMF component of the inner race vibration signal by using the DTCFVMD ($k = 5$), EMD, and EEMD methods are shown in Table 5.

Table 5. Frequency values extracted by the power spectrum of each IMF component of the inner race vibration signal.

	M1 (Hz)	M2 (Hz)	M3 (Hz)	M4 (Hz)	M5 (Hz)	M6 (Hz)	M7 (Hz)	M8 (Hz)	M9 (Hz)	M10 (Hz)	M11 (Hz)	M12 (Hz)
EMD	164.1	164.1	164.1	29.3	23.4	41.0	5.9	5.9	5.9	5.9	5.9	
EEMD	164.1	164.1	164.1	58.6	35.2	29.3	5.9	11.7	11.7	5.9	5.9	5.9
DTCFVMD	58.6	164.1	164.1	29.3	29.3							

In Table 5, 29.3 Hz represents the rotating frequency, 58.6 Hz represents the frequency doubling of the rotating frequency, and 164.1 Hz represents the fault characteristic frequency of the inner race. As can be seen from Table 5, the number of effective frequencies of the power spectrum in the IMF component by using the EEMD method is more than that of the EMD method. Therefore, the decomposition effect of the EEMD method is better than that of the EMD method. Compared with the DTCFVMD method, the number of effective frequencies of IMF components by using the EEMD method is less than that of the EMD method and the EEMD method, and each mode can extract the effective frequency value. However, there exists some invalid frequency values in the modes of the EMD decomposition and the EEMD decomposition. In general, the decomposition effect of the DTCFVMD method is better than the decomposition effect of the EMD method and the EEMD method.

The frequency values extracted by the power spectrum of each IMF component of the outer race vibration signal by using the DTCFVMD ($k = 4$), EMD, and EEMD methods are shown in Table 6.

Table 6. Frequency values extracted by the power spectrum of each IMF component of the outer race vibration signal.

	M1 (Hz)	M2 (Hz)	M3 (Hz)	M4 (Hz)	M5 (Hz)	M6 (Hz)	M7 (Hz)	M8 (Hz)	M9 (Hz)	M10 (Hz)	M11 (Hz)	M12 (Hz)
EMD	105.5	105.5	111.3	87.9	17.6	23.4	11.7	11.7	11.7	5.9		
EEMD	105.5	105.5	105.5	105.5	87.9	29.3	52.7	5.9	5.9	5.9	5.9	5.9
DTCFVMD	87.9	105.5	46.9	29.3								

In Table 6, 29.3 Hz represents the rotating frequency, 87.9 Hz represents the frequency tripling of the rotating frequency, 105.5 Hz represents the fault characteristic frequency of the outer race, 52.7 Hz represents the fault characteristic frequency of 1/2, 11.7 Hz represents the fault characteristic frequency of 1/9, and 46.9 Hz represents the difference between the fault characteristic frequency and the frequency doubling of the rotating frequency of the outer race. As can be seen from Table 6, the number of effective frequencies of the power spectrum in the IMF component by using the EEMD method is more than that of the EMD method, and the EEMD method can obtain the rotating frequency. Therefore, the decomposition effect of the EEMD method is better than that of the EMD method. Compared with the DTCFVMD method, the number of effective frequencies of IMF components by using the EEMD method is less than that of the EMD method and the EEMD method, and each mode can extract the effective frequency value. However, there exists some invalid frequency values in the modes of the EMD decomposition and the EEMD decomposition. In general, the decomposition effect of the DTCFVMD method is better than the decomposition effect of the EMD method and the EEMD method.

6. Conclusions

Time-frequency analysis methods are the most commonly applied methods. They are very significant to deeply analyze vibration signals for monitoring the health state of rotating machinery. The VMD is an adaptive and non-recursive signal decomposition method, which can decompose a real valued input signal into an ensemble of sub-signals or modes with specific sparsity properties. However, it has the disadvantage of the experienced selection mode number. Therefore, a center frequency method of double thresholds was proposed to determine the decomposed mode number of the VMD method. It can adaptively obtain the optimal mode number that matches the analyzed signal, and decompose the vibration signal into a series of IMFs under different scales. The Hilbert transform method was used to calculate the power spectrum of each mode component to determine the fault type. The actual fault vibration signal of the motor bearing was selected to test and verify the effectiveness of the proposed DTCFVMD method. The experiment results demonstrate that the proposed DTCFVMD method can determine the optimal value of k by using the central frequency method of double thresholds for the inner race and outer race of the motor bearing. The proposed DTCFVMD method was compared with the VMD method, EMD, and EEMD methods for verifying the effectiveness of vibration analysis and feature extraction. Therefore, the proposed DTCFVMD method is an effective method to analyze the vibration signal. It can provide a new idea to improve the VMD method for analyzing the vibration analysis and extracting fault features of rotating machinery.

Due to the ineffectiveness of the proposed DTCFVMD method for analyzing the vibration signal of the rolling element of motor bearings, and the equipment and working conditions becoming more complex, the VMD method is needed to be further deeply studied. In future work, the new methods are proposed to analyze the vibration signal of different equipment, conditions, and industries.

Author Contributions: The methodology, H.Z.; validation, S.Z. and H.L. (Hailong Liu); writing—original draft preparation, W.D.; writing—review and editing, H.L. (Haodong Liu); funding acquisition, H.Z. and J.W.

Funding: This research was funded by the National Natural Science Foundation of China under Grant 51605068, Grant 51475065, Grant 61771087,Grant 51875072, Grant 11461006 and Grant 61772006, the Innovative Talents Promotion Plan of Liaoning Colleges and Universities under Grant LR2017058, Open Project Program of the Traction Power State Key Laboratory of Southwest Jiaotong University under Grant TPL1705 and Grant TPL1803, the Research Fund of the Guangxi Key Lab of Multi-source Information Mining and Security under Grant MIMS17-03, Shandong Social Science Planning Research Special under Grant 18CHLJ18, High Education Science and Technology Planning Program of Shandong Provincial Education Department under Grant J18KA340, the Science and Technology Program of Guangxi under Grant AB17129012, the Science and Technology Major Project of Guangxi under Grant AA17204096, the Special Fund for Scientific and Technological Bases and Talents of Guangxi under Grant 2016AD05050, the Special Fund for Bagui Scholars of Guangxi and Liaoning BaiQianWan Talents Program.

Acknowledgments: The authors would like to thank all the reviewers for their constructive comments. The program for the initialization, study, training, and simulation of the proposed algorithm in this article was written with the tool-box of MATLAB 2016b produced by the Math-Works, Inc.

Conflicts of Interest: The authors declare no conflict of interest.

References

1. Yu, J.; He, Y.J. Planetary gearbox fault diagnosis based on data-driven valued characteristic multigranulation model with incomplete diagnostic information. *J. Sound Vib.* **2018**, *429*, 63–77. [CrossRef]
2. Deng, W.; Yao, R.; Zhao, H.M.; Yang, X.H.; Li, G.Y. A novel intelligent diagnosis method using optimal LS-SVM with improved PSO algorithm. *Soft Comput.* **2017**, 1–18. [CrossRef]
3. Zhu, D.C.; Zhang, Y.X.; Liu, S.Y.; Zhu, Q. Adaptive combined HOEO based fault feature extraction method for rolling element bearing under variable speed condition. *J. Mech. Sci. Technol.* **2018**, *32*, 4589–4599. [CrossRef]
4. Zhou, J.C.; Du, Z.X.; Liao, Y.H.; Tang, A.H. An optimization design of vehicle axle system based on multi-objective cooperative optimization algorithm. *J. Chin. Inst. Eng.* **2018**. [CrossRef]
5. Wang, L.P.; Chen, B.Y.; Chen, C.; Chen, Z.; Liu, G. Application of linear mean-square estimation in ocean engineering. *China Ocean Eng.* **2016**, *30*, 149–160. [CrossRef]
6. Liu, Y.Q.; Yi, X.K.; Zhai, Z.G.; Gu, J.X. Feature extraction based on information gain and sequential pattern for English question classification. *IET Softw.* **2018**. [CrossRef]
7. Zhang, S.F.; Shen, W.; Li, D.S.; Zhang, X.W.; Chen, B.Y. Nondestructive ultrasonic testing in rod structure with a novel numerical Laplace based wavelet finite element method. *Latin Am. J. Solids Struct.* **2018**, *15*, 1–17. [CrossRef]
8. Zhou, J.; Fang, X.Y.; Tao, L. A sparse analysis window for discrete Gabor transform. *Circuits Syst. Signal Process.* **2017**, *36*, 4161–4180. [CrossRef]
9. Luo, J.; Chen, H.L.; Zhang, Q.; Xu, Y.T.; Huang, H.; Zhao, X.H. An improved grasshopper optimization algorithm with application to financial stress prediction. *Appl. Math. Model.* **2018**, *64*, 654–668. [CrossRef]
10. Lu, S.L.; He, Q.B.; Yuan, T.; Kong, F.R. Online fault diagnosis of motor bearing via stochastic–resonance-based adaptive filter in an embedded system. *IEEE Trans. Syst. Man Cybern. Syst.* **2017**, *47*, 1111–1122. [CrossRef]
11. Shen, L.; Chen, H.L.; Yu, Z.; Kang, W.C.; Zhang, B.; Li, H.; Yang, B.; Liu, D. Evolving support vector machines using fruit fly optimization for medical data classification. *Knowl.-Based Syst.* **2016**, *96*, 61–75. [CrossRef]
12. Deng, W.; Zhao, H.M.; Zou, L.; Li, G.Y.; Yang, X.H.; Wu, D.Q. A novel collaborative optimization algorithm in solving complex optimization problems. *Soft Comput.* **2017**, *21*, 4387–4398. [CrossRef]
13. Zhang, Q.; Jin, B.; Wang, X.; Lei, S.; Shi, Z.; Zhao, J.; Liu, Q.; Peng, R. The mono (catecholamine) derivatives as iron chelators: Synthesis, solution thermodynamic stability and antioxidant properties research. *R. Soc. Open Sci.* **2018**, *5*, 171492. [CrossRef] [PubMed]
14. Xu, J.F.; Ren, Z.R.; Li, Y.; Skjetne, R.; Halse, K.H. Dynamic simulation and control of anactive roll reduction system using freeflooding tanks with vacuum pumps. *J. Offshore Mech. Arctic Eng.* **2018**, *140*, 061302. [CrossRef]
15. Guo, S.K.; Chen, R.; Wei, M.M.; Li, H.; Liu, Y.Q. Ensemble data reduction techniques and Multi-RSMOTE via fuzzy integral for bug report classification. *IEEE Access* **2018**, *6*, 5934–45950. [CrossRef]
16. Zhang, Q.; Chen, H.L.; Luo, J.; Xu, Y.T.; Wu, C.; Li, C. Chaos enhanced bacterial foraging optimization for global optimization. *IEEE Access* **2018**. [CrossRef]
17. Kang, L.; Du, H.L.; Zhang, H.; Ma, W.L. Systematic research on the application of steel slag resources under the background of big data. *Complexity* **2018**, *2018*. [CrossRef]
18. Wu, D.Q.; Huo, J.Z.; Zhang, G.F.; Zhang, W.H. Minimization of logistics cost and carbon emissions based on quantum particle swarm optimization. *Sustainability* **2018**, *10*, 3791. [CrossRef]
19. Immovilli, F.; Bianchini, C.; Lorenzani, E.; Bellini, A.; Fornasiero, E. Evaluation of combined reference frame transformation for interturn fault detection in permanent-magnet multiphase machines. *IEEE Trans. Ind. Electron.* **2015**, *62*, 1912–1920. [CrossRef]
20. Sun, F.R.; Yao, Y.D.; Li, X.F. The heat and mass transfer characteristics of superheated steam coupled with non-condensing gases in horizontal wells with multi-point injection technique. *Energy* **2018**, *143*, 995–1005. [CrossRef]
21. Fu, H.; Li, Z.; Liu, Z.; Wang, Z. Research on big data digging of hot topics about recycled water use on micro-blog based on particle swarm optimization. *Sustainability* **2018**, *10*, 2488. [CrossRef]
22. Guo, S.K.; Chen, R.; Li, H.; Gao, J.; Liu, Y.Q. Crowdsourced Web application testing under real-time constraints. *Int. J. Softw. Eng. Knowl. Eng.* **2018**, *28*, 751–779. [CrossRef]

23. Ren, Z.R.; Jiang, Z.Y.; Skjetne, R.; Zhen, G. Development and application of a simulator for offshore windturbine blades installation. *Ocean Eng.* **2018**, *166*, 380–395. [CrossRef]

24. Jiang, S.; Lian, M.; Lu, C.; Gu, Q.; Ruan, S.; Xie, X. Ensemble prediction algorithm of anomaly monitoring based on big data analysis platform of open-pit mine slope. *Complexity* **2018**, *2018*. [CrossRef]

25. Sun, F.R.; Yao, Y.D.; Chen, M.Q.; Li, X.F.; Zhao, L.; Meng, Y.; Sun, Z.; Zhang, T.; Feng, D. Performance analysis of superheated steam injection for heavy oil recovery and modeling of wellbore heat efficiency. *Energy* **2017**, *125*, 795–804. [CrossRef]

26. Peng, Y.; Lu, B.L. Discriminative extreme learning machine with supervised sparsity preserving for image classification. *Neurocomputing* **2017**, *261*, 242–252. [CrossRef]

27. Mandic, D.P.; Rehman, N.U.; Wu, Z.H. Empirical mode decomposition-based time-frequency analysis of multivariate signals. *IEEE Signal Process. Mag.* **2013**, *30*, 74–86. [CrossRef]

28. Wang, L.; Xu, X.; Liu, G.; Chen, B.; Chen, Z. A new method to estimate wave height of specified return period. *J. Oceanol. Limnol.* **2017**, *35*, 1002–1009. [CrossRef]

29. Guo, S.K.; Liu, Y.Q.; Chen, R.; Sun, X.; Wang, X.X. Using an improved SMOTE algorithm to deal imbalanced activity classes in smart homes. *Neural Process. Lett.* **2018**. [CrossRef]

30. Yang, A.M.; Yang, X.L.; Chang, J.C.; Bai, B.; Kong, F.B.; Ran, Q.B. Research on a fusion scheme of cellular network and wireless sensor networks for cyber physical social systems. *IEEE Access* **2018**, *6*, 18786–18794. [CrossRef]

31. Deng, W.; Zhao, H.M.; Yang, X.H.; Xiong, J.X.; Sun, M.; Li, B. Study on an improved adaptive PSO algorithm for solving multi-objective gate assignment. *Appl. Soft Comput.* **2017**, *59*, 288–302. [CrossRef]

32. Gilles, J. Empirical wavelet transform. *IEEE Trans. Signal Process.* **2013**, *61*, 3999–4010. [CrossRef]

33. Dragomiretskiy, K.; Zosso, D. Variational mode decomposition. *IEEE Trans. Signal Process.* **2014**, *62*, 531–544. [CrossRef]

34. Wang, Y.X.; Markert, R.; Xiang, J.W. Research on variational mode decomposition and its application in detecting rub-impact fault of the rotor system. *Mech. Syst. Signal Process.* **2015**, *60–61*, 243–251. [CrossRef]

35. Yi, C.C.; Lv, Y.; Dang, Z. A fault diagnosis scheme for rolling bearing based on particle swarm optimization in variational mode decomposition. *Shock Vib.* **2016**, *2016*, 9372691. [CrossRef]

36. Liu, Y.Y.; Yang, G.L.; Li, M. Variational mode decomposition denoising combined the detrended fluctuation analysis. *Signal Process.* **2016**, *125*, 349–364. [CrossRef]

37. Zhang, S.F.; Wang, Y.X.; He, S.L. Bearing fault diagnosis based on variational mode decomposition and total variation denoising. *Meas. Sci. Technol.* **2016**, *27*, 075101. [CrossRef]

38. Liu, S.K.; Tang, G.J.; Wang, X.L. Time-frequency analysis based on improved variational mode decomposition and teager energy operator for rotor system fault diagnosis. *Math. Probl. Eng.* **2016**, *2016*, 1713046. [CrossRef]

39. Henao, H.; Capolino, G.A.; Fernandez-Cabanas, M.; Filippetti, F.; Bruzzese, C.; Strangas, E.; Pusca, R.; Estima, J.; Riera-Guasp, M.; Hedayati-Kia, S. Trends in fault diagnosis for electrical machines: A review of diagnostic techniques. *IEEE Ind. Electron. Mag.* **2014**, *8*, 31–42. [CrossRef]

40. An, X.L.; Pan, L.P.; Zhang, F. Analysis of hydropower unit vibration signals based on variational mode decomposition. *J. Vib. Control* **2017**, *23*, 1938–1953. [CrossRef]

41. Mahgoun, H.; Chaari, F.; Felkaoui, A. Detection of gear faults using variational mode decomposition (VMD). *Mech. Ind.* **2016**, *17*, 207. [CrossRef]

42. Abdoos, A.A.; Mianaei, P.K.; Ghadikolaei, M.R. Combined VMD-SVM based feature selection method for classification of power quality events. *Appl. Soft Comput.* **2016**, *38*, 637–646. [CrossRef]

43. Li, Z.P.; Chen, J.L.; Zi, Y.Y.; Pan, J. Independence-oriented VMD to identify fault feature for wheel set bearing fault diagnosis of high speed locomotive. *Mech. Syst. Signal Process.* **2017**, *85*, 512–529. [CrossRef]

44. Immovilli, F.; Cocconcelli, M. Experimental investigation of shaft radial load effect on bearing fault signatures detection. *IEEE Trans. Ind. Appl.* **2017**, *53*, 2721–2729. [CrossRef]

45. Wang, Y.X.; Liu, F.Y.; Jiang, Z.S. Complex variational mode decomposition for signal processing applications. *Mech. Syst. Signal Process.* **2017**, *86*, 75–85. [CrossRef]

46. Zhang, X.; Miao, Q.; Zhang, H.; Wang, L. A parameter-adaptive VMD method based on grasshopper optimization algorithm to analyze vibration signals from rotating machinery. *Mech. Syst. Signal Process.* **2018**, *108*, 58–72. [CrossRef]

47. Mohanty, S.; Gupta, K.K.; Raju, K.S. Hurst based vibro-acoustic feature extraction of bearing using EMD and VMD. *Measurement* **2018**, *117*, 200–220. [CrossRef]

48. Choi, G.B.; Oh, H.S.; Kim, D.H. Enhancement of variational mode decomposition with missing values. *Signal Process.* **2018**, *142*, 75–86. [CrossRef]
49. Wang, W.K.; Pan, C.; Wang, J.J. Quasi-bivariate variational mode decomposition as a tool of scale analysis in wall-bounded turbulence. *Exp. Fluids* **2018**, *59*, 1. [CrossRef]
50. Zhao, H.M.; Yao, R.; Xu, L.; Yuan, Y.; Li, G.Y.; Deng, W. Study on a novel fault damage degree identification method using high-order differential mathematical morphology gradient spectrum entropy. *Entropy* **2018**, *20*, 682. [CrossRef]
51. Duan, C.; Huo, J.; Li, F.; Yang, M.; Xi, H. Ultrafast room-temperature synthesis of hierarchically porous metal–organic frameworks by a versatile cooperative template strategy. *J. Mater. Sci.* **2018**, *53*, 16276–16287. [CrossRef]
52. Deng, W.; Zhang, S.J.; Zhao, H.M.; Yang, X.H. A novel fault diagnosis method based on integrating empirical wavelet transform and fuzzy entropy for motor bearing. *IEEE Access* **2018**, *6*, 35042–35056. [CrossRef]
53. Park, S.; Kim, S.; Choi, J.H. Gear fault diagnosis using transmission error and ensemble empirical mode decomposition. *Mech. Syst. Signal Process.* **2018**, *108*, 262–275. [CrossRef]
54. Zhao, H.M.; Sun, M.; Deng, W.; Yang, X.H. A new feature extraction method based on EEMD and multi-scale fuzzy entropy for motor bearing. *Entropy* **2017**, *19*, 14. [CrossRef]
55. Bearing Data Center. Available online: http://csegroups.case.edu/bearingdatacenter/home (accessed on 6 July 2016).

Article

On the Existence of Self-Excited Vibration in Thin Spur Gears: A Theoretical Model for the Estimation of Damping by the Energy Method

Yanrong Wang [1,2], Hang Ye [1,2], Long Yang [3] and Aimei Tian [4,*]

[1]   School of Energy and Power Engineering, Beihang University, Beijing 100191, China; yrwang@buaa.edu.cn (Y.W.); yeahhang155@gmail.com (H.Y.)
[2]   Collaborative Innovation Center for Advanced Aero-Engine, Beijing 100191, China
[3]   4DPower Tech. Co., Ltd., Beijing 100191, China; leon.young50@gmail.com
[4]   School of Astronautics, Beihang University, Beijing 100191, China
*   Correspondence: amtian@buaa.edu.cn

Received: 26 October 2018; Accepted: 22 November 2018; Published: 22 November 2018

Abstract: The gear is a cyclic symmetric structure, and each tooth is subjected to a periodic mesh force. These mesh forces have the same phase difference tooth by tooth, which can excite gear vibrations. The mechanism of additional axial force caused by gear bending is shown and examined, which can significantly affect the stability of a self-excited thin spur gears vibration. A mechanical model based on energy balance is then developed to predict the contribution of additional axial force, leading to the proposed numerical integration method for vibration stability analysis. By analyzing the change in the system energy, the occurrence of the self-excited vibration is validated. A numerical simulation is carried out to verify the theoretical analysis. The impacts of modal damping, contact ratio, and the number of nodal diameters on the stability boundaries of the self-excited vibration are revealed. The results prove that the backward traveling wave of the driven gear as well as the forward traveling wave of the driving gear encounter self-excited vibration in the absence of sufficient damping. The model can be used to predict the stability of the gear self-excited vibration.

Keywords: traveling waves; vibration stability; self-excited vibration; thin spur gear

---

1. Introduction

Gears are commonly used in machines, automobiles, wind turbines, and aerospace industries as a transmission component. The status of the operational gear significantly influences the performances of the machines. In order to meet specific operational requirements, some gears are designed to be thin. As a consequence, the bending stiffness and the associated flexural natural frequencies are relatively low, which could easily lead to some kind of dynamic phenomena, such as transverse vibration.

Recent research on fracture failure of aero-engine gears showed that the characteristics of the gear crack were found to be typical fatigue fractures caused by transverse vibrations. However, the operational conditions showed that there was no resonance during the operation, which indicates that there must be some other reasons that could cause the fracture. Gears' self-excited vibration was suspected to be the main reason, but this needs to be verified.

In most investigations on the vibration of gears, gears are usually simplified as thin, moderate, thick annular plate, or thick-walled cylinders [1]. Many different theories of plate vibration can be applied when studying the vibration of circular or annular plates with variable thickness. These theories include the classical thin plate theory, the shear deformable plate theory (for the thicker plates), and even three-dimensional (3D) elasticity theory [2]. The study of plate vibration can be traced back to the 19th century, and many studies have contributed considerably to this research area.

Leissa [3] and Houser [4] summarized the theoretical studies of plate vibration and gear dynamics in detail. Some of the complex effects—such as the distributed masses at the outer edge, the outer reinforcing ring, the shear deformation, the rotary inertia, and initial tension on the free vibration of the spinning disk—were studied by Sinha [5,6], Cote [7], and Suzuki [8]. Chen [9] also theoretically studied the vibration and stability of rotating polar orthotropic sandwich plates.

In recent years, the finite element method (FEM) and experimental tests have been increasingly used to analyze gear vibration characteristics including resonance, modal shape identification, and the effect of the thickness on the natural frequencies [10–13]. In addition, the traveling wave vibrations occur more easily in the running gears due to a high meshing frequency, and the traveling wave vibration has been widely investigated in the field of rotating disks [14–17]. Many researchers focused on gear engagement dynamics using FEM [18,19], lumped parameters models [20–24] and experimental tests [18,25–27]. FEM can capture accurately nonlinear dynamics but requires a well-defined geometry and a long calculation time. Lumped parameter models have high computational efficiency but need some results from FEM models [20]. These works have contributed to the establishment of the traveling wave vibration theory. A number of papers can be found on the traveling wave vibration of a rotating gear [28–30]; however, few papers have dealt with gear self-excited vibration [31], even though this phenomenon was the main reason for the aero-engine gear crack failures.

The purpose of this paper is to meet the actual demand to verify gear self-excited vibration observed in the field and to analyze the stability, if it does exist. We study the transverse self-excited vibration of a thin spur rotating gear using a numerical method, and the conditions that could cause transverse self-excited vibration are predicted. The self-excited force is calculated based on the analysis of the traveling wave vibration. Through numerical simulation, the impacts of modal damping, contact ratio, and the number of nodal diameters on the stability boundaries are studied. The work is organized as follows: Theoretical derivation of the proposed method, including the principles of self-excited vibration, the source of spur gear self-excitation, and self-excited vibration prediction, are shown in Section 2. Convergence analysis and stability boundary analysis is performed on a thin spur gear in Section 3, followed by conclusions in Section 4.

2. Theoretical Analysis

For a pair of precise meshing spur gears, there is no axial force when they are running, so only torsional vibration can occur. Therefore, there must be an axial force that could excite the transverse vibration and results in fatigue failure. Actually, the gears used in industry cannot be precisely meshed as expected. The deflections of axes can create an angle between the contact line of the gears and the rotation axis. As a result, an axial force component is contained in the mesh force vector, and it can become a self-excited force of the system. This kind of axial force can excite transverse vibrations of the gear. Once the vibration status changes, the axial force changes accordingly, and subsequently the changed axial force will, in turn, continue to affect the transverse vibration. This consecutive interaction will be enhanced when the feedback is positive, and the gear system will constitute a self-excited vibration system.

Figure 1 depicts a schematic diagram of the mechanism of gear self-excited vibration. Through the axial force and the damping force applied to the transverse traveling wave vibration, the change in the system energy can be determined. If the excitation work is more than the work consumed by the damping, the system energy increases, which indicates that self-excited gear vibration will occur. This is the theory upon which the analysis of this paper is based.

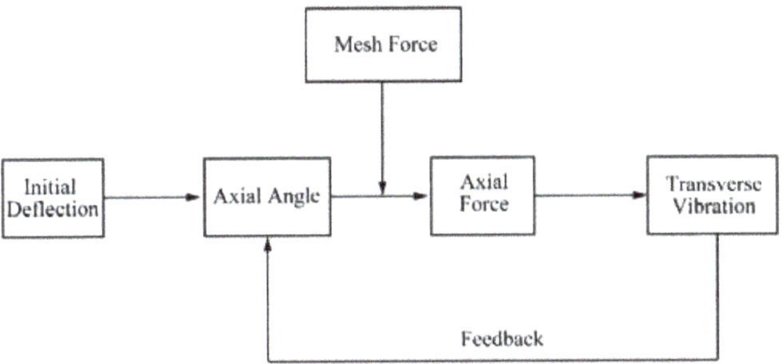

Figure 1. The mechanism of the gear self-excited vibration.

2.1. Traveling Wave Vibration of the Rotating Gear

A cylindrical coordinate (r, θ, z) fixed on the rotating gear is defined, and in order to describe the theoretical analysis as well as the subsequent numerical results more clearly, the phenomenon of the traveling wave vibration of the thin rotating gear is discussed first.

The transverse vibration of a thin spur gear has the same expressions as the circular plate, since it is usually simplified as an annular plate. According to the plate vibration theory, displacement of the circular plate transverse vibration could be expressed as:

$$w(r,\theta,t) = A(r) \cos m\theta \cdot \cos \omega t \tag{1}$$

where w is the vibration displacement in the z direction, r is the radial coordinate, θ is the circumferential coordinates, t is time, A is the vibration amplitude that depends on the radial coordinate r, m is the number of nodal diameters, and ω is the vibration frequency.

Equation (1) is the vibration displacement of a stationary gear or the displacement in the rotary coordinate fixed on the rotating gear. It can also be decomposed into two components:

$$w(r,\theta,t) = \frac{1}{2}A(r)[\cos(m\theta - \omega t) + \cos(m\theta + \omega t)] \tag{2}$$

Since the gear is rotating, the rotation speeds of the nodal diameters are different in the inertia coordinate for these two components; therefore, the transverse vibration of a rotating gear is a kind of traveling wave vibration. The first component of the right item of Equation (2) represents a traveling wave rotating in the same direction of the gear, which is called the forward traveling wave (FTW), whereas the second component represents a traveling wave rotating in the opposite direction of the gear, called the backward traveling wave (BTW).

2.2. Mechanical Model of the Gear Self-Excited Force

According to the gear self-excited vibration mechanism, a mechanical model is proposed to describe the self-excited force of the spur gear. Figure 2a is the schematic diagram of an engaged spur gear pair and the defined rotary coordinates. If there is an initial axial deflection when the gears are running, the normal direction of the surface of the action is no longer perpendicular to the rotation axis, and there occurs transverse deformation along the tangential direction at the mesh point, as shown in Figure 2b. Thus, the mesh force will generate an axial component—the gear self-excited force $F_z(t)$.

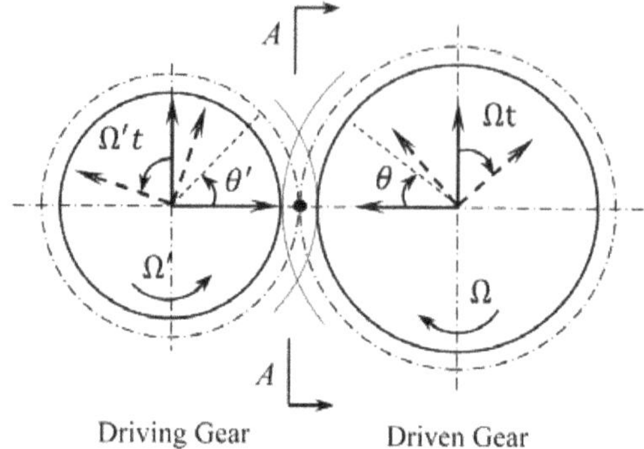

Driving Gear   Driven Gear

(**a**) The engaged gear pair.

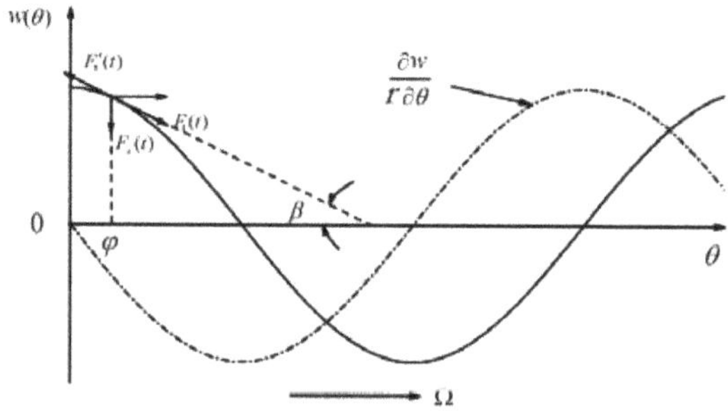

(**b**) The self-excited force.

Figure 2. Schematic diagram of the engaged gear pair and the self-excited force.

If the transverse vibration displacement at the gear rim develops along the circumferential direction, the displacement is approximately along a cosine curve, shown as the solid line in Figure 2b when looking from the A-A direction. According to the relationship between the forces shown in Figure 2b, and taking the direction of $w(\theta)$ as the positive direction, the self-excited force can be expressed as:

$$F_z(t) = -F_t(t)\sin\beta = F_t(t)\frac{\partial w}{r\partial\theta} \tag{3}$$

where β is the axial angle that represents the angle between the normal direction and the axial direction of the meshing force.

Based on the mechanism of the gear meshing, the engagement process of the gears with contact ratios from 1–2 consists of two parts: The single tooth mesh process and the double teeth mesh process. If the gear torque load is T_0, take the pitch diameter d_0 instead of the mesh point diameter d_1, assuming that the height of the gear tooth is small enough compared to the gear diameter. Accordingly, the mesh force $F_t(t)$ for the engaged gear can be expressed as:

$$F_t(t) = \begin{cases} 4T_0/d_0 & \text{Single tooth mesh process} \\ 2T_0/d_0 & \text{Double teeth mesh process} \end{cases} \tag{4}$$

Since the engagement of the operating gear is a continuous process, the single tooth and the double tooth mesh processes are converting to each other continually. So, the mesh force $F_t(t)$ is not only time-dependent when observed from the whole gear, but is also periodic. Thus, two different mesh points should be considered.

Let $F_0 = 2T_0/d_0$, and after the coordinate transformation $t' = 2\pi nzt/60$, the expressions of the mesh force at mesh point 1 and mesh point 2 are shown in Figure 3, where n is the rotational speed, z is the number of gear teeth, and ε is the gear contact ratio.

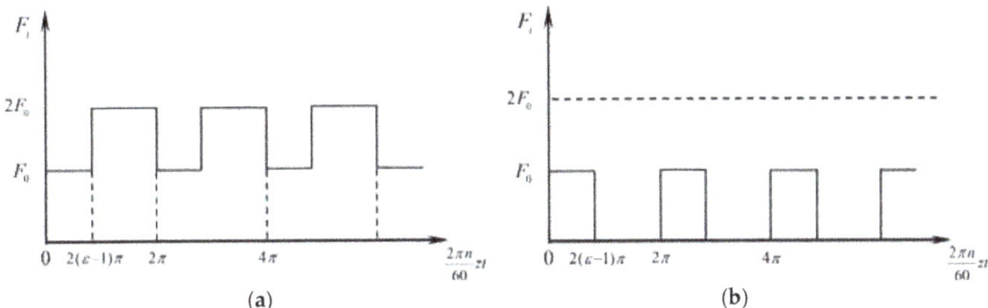

Figure 3. Expressions of the mesh force: (**a**) Mesh point 1, and (**b**) mesh point 2. The abscissa indicates time, and the ordinate indicates the meshing force.

Figure 3 shows that both forces are periodic functions with a period time of 2π. To facilitate further analysis, Fourier expansion of these two mesh forces in one period is completed, and the harmonic terms of the mesh forces are:

$$\begin{pmatrix} F_{1t}(t) \\ F_{2t}(t) \end{pmatrix} = \begin{pmatrix} 3-\varepsilon \\ \varepsilon-1 \end{pmatrix} F_0 + \begin{pmatrix} 1 \\ 1 \end{pmatrix} \cdot \frac{2F_0}{\pi} \sum_{k=1}^{\infty} \frac{1}{k} \sin k\varepsilon\pi \cos[kz\Omega t - k\varepsilon\pi] \tag{5}$$

where k is excitation order.

Substitute Equation (5) into Equation (3), then the expression of the self-excited force can be obtained.

Although the mechanical model of the self-excited force proposed here is simple, it meets all of the requirements of the mechanical principles. Thus, the mechanical model is sufficient to describe the overall behavior of gear.

2.3. The Work of the Self-Excited Force on the Vibration

Take the backward traveling wave of the driven gear as an example. The position angle θ of the mesh point varies with time in the rotary coordinate, and the relationship between θ and time t ($\theta = 0$ when $t = 0$) is:

$$\theta = -\Omega t \tag{6}$$

where Ω is the gear angular speed.

In addition, mesh points 1 and 2 are on the two adjacent teeth, and the phase φ between these two points is shown in Figure 4. In an actual situation, the phase φ changes as the engagement process continues. However, in order to simplify the analysis, let $\varphi = 1.5\beta' = 3\pi/2z$. Then, the expressions of the backward traveling wave at mesh point 1 and mesh point 2 are:

$$w_{b,1} = \frac{1}{2}A_0 \cos(\omega - m\Omega)t \tag{7}$$

$$w_{b,2} = \frac{1}{2}A_0 \cos[(\omega - m\Omega)t + \frac{3\pi}{2z}] \tag{8}$$

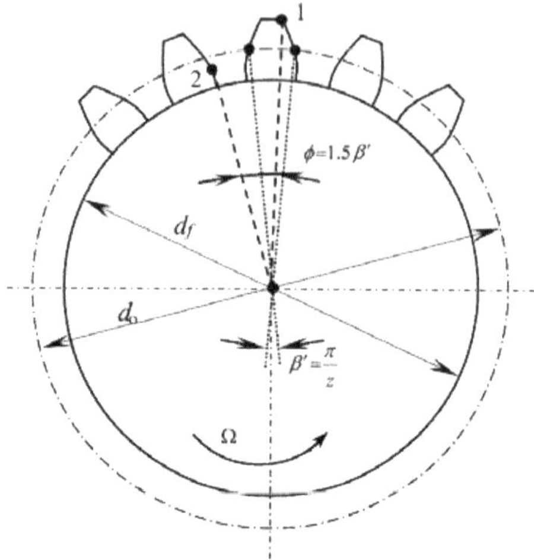

Figure 4. Schematic diagram of the phase between the two mesh points.

Therefore, the work done by the additional force to the backward traveling wave during one vibration cycle can be determined by the following equation:

$$W_F = \int_0^T F_t(t) \frac{\partial w_b}{r \partial \theta} \cdot \frac{\partial w_b}{\partial t} \cdot dt \tag{9}$$

The time of one vibration cycle for the backward traveling wave is $T = 2\pi/\omega$. If we substitute Equation (5) into Equation (9), then the excitation work done to the BTW is:

$$W_F = W_{1cbF} + W_{2cbF} \tag{10}$$

where

$$W_{1cbF} = W_{1cbF1} + W_{1cbF2} + W_{1cbF3} + \ldots \tag{11}$$

$$W_{2cbF} = W_{2cbF1} + W_{2cbF2} + W_{2cbF3} + \ldots \tag{12}$$

And

$$
\begin{aligned}
W_{1cbF1} &= \int_0^T (3-\varepsilon) F_0 \tfrac{1}{2r} A_0 m \sin(\omega - m\Omega)t \cdot [\tfrac{1}{2} A_0 (\omega - m\Omega) \sin(\omega - m\Omega)t] dt \\
&= \tfrac{1}{8r}(3-\varepsilon) F_0 A_0^2 m (\omega - m\Omega)\left[T - \tfrac{1}{2(\omega - m\Omega)} \sin 2(\omega - m\Omega)T \right]
\end{aligned} \tag{13}
$$

$$
\begin{aligned}
W_{1cbFk} &= \int_0^T \tfrac{2F_0}{\pi} \cdot \tfrac{1}{k} \sin k\varepsilon\pi \cdot \cos(kz\Omega t - k\varepsilon\pi) \cdot \tfrac{1}{4r} A_0^2 m (\omega - m\Omega) \sin^2(\omega - m\Omega)t \cdot dt \\
&= \tfrac{1}{4rk\pi} F_0 A_0^2 m (\omega - m\Omega) \sin k\varepsilon\pi \cdot \Big\{ \tfrac{1}{kz\Omega}[\sin(kz\Omega T - k\varepsilon\pi) + \sin k\varepsilon\pi] \\
&\quad - \tfrac{1}{2} \cdot \Big\langle \tfrac{1}{kz\Omega - 2(\omega - m\Omega)}[\sin(kz\Omega T - 2(\omega - m\Omega)T - k\varepsilon\pi) + \sin k\varepsilon\pi] \\
&\quad + \tfrac{1}{kz\Omega + 2(\omega - m\Omega)}[\sin(kz\Omega T + 2(\omega - m\Omega)T - k\varepsilon\pi) + \sin k\varepsilon\pi] \Big\rangle \Big\} \\
&\quad k = 2,3\ldots
\end{aligned} \tag{14}
$$

$$
\begin{aligned}
W_{2cbF1} &= \int_0^T (\varepsilon - 1) F_0 \tfrac{1}{2r} A_0 m \sin[(\omega - m\Omega)t + \tfrac{3\pi}{2z}] \cdot \{ \tfrac{1}{2} A_0 (\omega - m\Omega) \sin[(\omega - m\Omega)t + \tfrac{3\pi}{2z}] \} dt \\
&= \tfrac{1}{8r}(\varepsilon - 1) F_0 A_0^2 m (\omega - m\Omega) \Big\{ \Big\langle T - \tfrac{1}{2(\omega - m\Omega)} \cdot [\sin(2(\omega - m\Omega)T + \tfrac{3\pi}{z}) - \sin \tfrac{3\pi}{z}] \Big\rangle \Big\}
\end{aligned} \tag{15}
$$

$$
\begin{aligned}
W_{2cbFk} \ =\ & \int_0^T \frac{2F_0}{r\pi}\cdot\frac{1}{k}\sin k\varepsilon\pi\cdot\cos(kz\Omega t - k\varepsilon\pi)\cdot\frac{1}{4}A_0^2 m(\omega - m\Omega)\sin^2\left[(\omega - m\Omega)t + \frac{3\pi}{z}\right]\cdot dt \\[4pt]
=\ & \frac{1}{4rk\pi}F_0 A_0^2 m(\omega - m\Omega)\sin k\varepsilon\pi\cdot\Bigg\{\frac{1}{kz\Omega}\left[\sin(kz\Omega T - k\varepsilon\pi) + \sin k\varepsilon\pi\right] \\[4pt]
& -\frac{1}{2}\cdot\Bigg\langle\frac{1}{kz\Omega-2(\omega-m\Omega)}\left[\sin(kz\Omega T - 2(\omega - m\Omega)T - k\varepsilon\pi - \tfrac{3\pi}{z}) + \sin(k\varepsilon\pi + \tfrac{3\pi}{z})\right] \\[4pt]
& +\frac{1}{kz\Omega+2(\omega-m\Omega)}\left[\sin(kz\Omega T + 2(\omega - m\Omega)T - k\varepsilon\pi + \tfrac{3\pi}{z}) + \sin(k\varepsilon\pi - \tfrac{3\pi}{z})\right]\Bigg\rangle\Bigg\}
\end{aligned}
\tag{16}
$$

$$
k = 2,3\ldots
$$

The excitation work for the forward traveling wave vibration can be quickly obtained by simply replacing w_b with w_f in Equation (9) and the other related equations. Moreover, if the object of the analysis is the driving gear, the term $F_t(t)$ must be replaced with $-F_t(t)$. That is, the relationship of the excitation work between the driving gear and the driven gear is:

$$
W_{zbF} = -W_{cbF}
\tag{17}
$$

2.4. The Work of the Damping Force on the Vibration

The damping of an engaged rotating gear is quite complex and hard to calculate. However, because the damping always dissipates the energy, the work of the damping force could be obtained by taking advantage of the loss factor η, which represents the energy loss of the system caused by the damping effect [32]. The definition is:

$$
\eta = \frac{\Delta U}{2\pi U_{\max}}
\tag{18}
$$

where ΔU is the energy dissipated of the system during one damping cycle (vibration cycle) and $U_{\max}$ is the initial system energy, which is approximately equal to the maximum system kinetic energy when the damping is relatively low, which is $U_{\max} = T_{\max}$.

For an engaged gear, the maximum kinetic energy is the integration of the kinetic energy throughout the whole gear web, which is:

$$
T_{\max} = \frac{1}{2}\int_{r_1}^{r_2}\int_0^{2\pi}\rho h\left(\frac{\partial w_b}{\partial t}\right)^2_{\max}\cdot r\,dr\,d\theta
\tag{19}
$$

where $A(r) = A_0\cdot R(r)$. According to Equations (18) and (19), and the relationship $\eta = 2\zeta$ for the situation of low damping [32], the damping work during one vibration cycle can be determined as:

$$
W_D = 4\pi\zeta U_{\max} = \frac{\rho h\zeta\omega^2\pi^2}{2}A_0^2\int_{r_1}^{r_2} R^2(r)\,dr
\tag{20}
$$

where the term $\int_{r_1}^{r_2} R^2(r)\,dr$ can be calculated from the deformation results of the FEM modal analysis:

$$
\int_{r_1}^{r_2} R^2(r)\,dr = \sum_{i=2}^{num}\frac{1}{2}\left(R_{(i)}^2 r_i + R_{(i-1)}^2 r_{i-1}\right)\cdot|r_i - r_{i-1}|
\tag{21}
$$

where *num* is the number of the numerical integration data, and r_i and $R_{(i)}$ are the values of the radius and normalized deformation at point i, respectively.

2.5. Theoretical Prediction of the Self-Excited Vibration

From the perspective of the system energy, if the excitation work applied to the traveling waves is larger than the damping work, the total energy of the system will increase over time, indicating that self-excited vibration will occur. Thus the condition of the gear self-excited vibration occurrence is:

$$
W_F > W_D
\tag{22}
$$

The mesh force amplitude F_0 when the self-excited vibration occurs can then be determined by solving the equation:

$$W_F = W_D \tag{23}$$

Moreover, once the mesh force amplitude F_0 is obtained, the critical transmission power, which also represents the minimum transmission power that can cause self-excited vibration, or the maximum transmission power for safe operation, can then be predicted according to the definition of the transmission power.

$$P_0 = \frac{F_0 d_0 n}{2 \times 9550} \text{KW} \tag{24}$$

Therefore, if the actual operation transmission power P is larger than the critical transmission power P_0, then self-excited vibration will occur and the system will be unstable.

3. Results and Discussion

According to the theoretical analysis of self-excited vibration in thin spur gears, the calculation program of the proposed method is developed by using Fortran language. The block diagram of proposed analysis method is shown in Figure 5. And the interface of the program is shown in Figure 6.

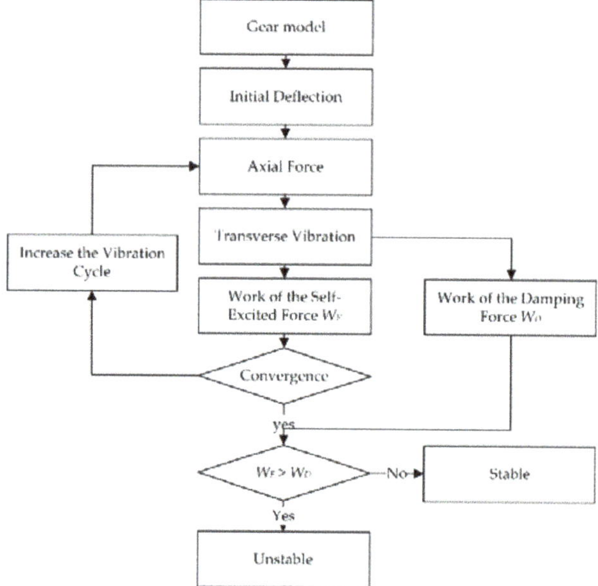

Figure 5. The block diagram of proposed analysis method.

The program shown in this paper is applied to a model of a virtual thin spur gear. The parameters of the thin spur gear used in this numerical simulation are listed in Table 1, with radius of the rim $r_1 = 95$ mm, radius of the hole $r_2 = 25$ mm, thickness $h = 4$ mm, and the number of teeth $z = 50$. The finite element model is shown in Figure 7. The gear material properties are: Young modulus $E = 212$ GPa, density $\rho = 7.85 \times 10^3$ kg/m^3, and Poisson's ratio $\mu = 0.284$. Several different parameters were also chosen for the numerical simulation in order to demonstrate the effects of the nodal diameter and damping ratio on the self-excited vibration stability. The maximum simulation speed was 10,000 rpm, and the results for each speed were calculated.

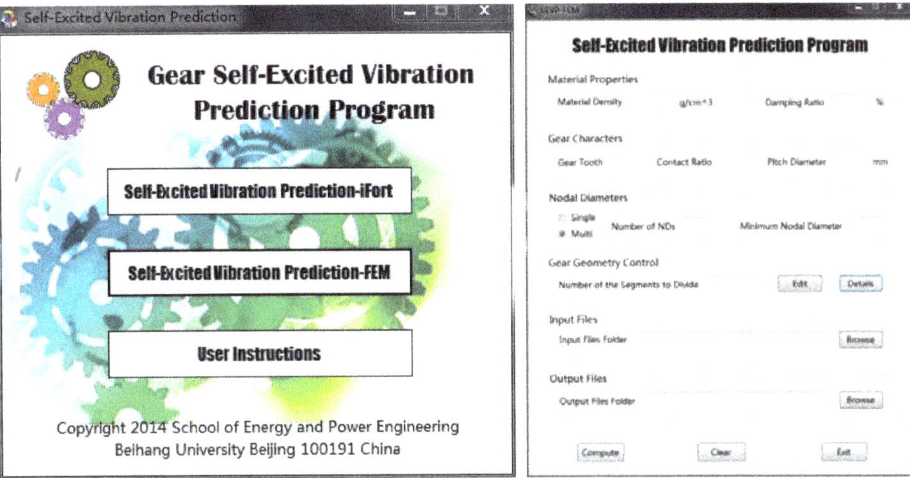

Figure 6. The interface of the program.

Table 1. Gear parameters.

Parameter	Value
r_1	25 mm
r_2	95 mm
h	4 mm
z	50

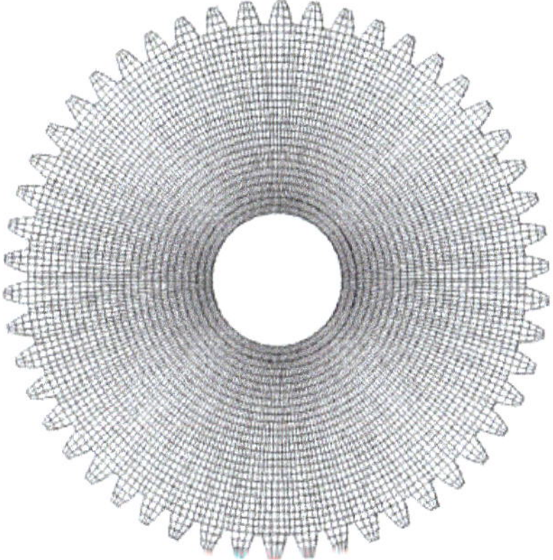

Figure 7. Finite element model (FEM) of the gear.

3.1. Convergence Analyses

For high-speed gears, the meshing period is smaller than the low nodal diameter' vibration period. For the convenience of calculation, the initial moment of the vibration period during the calculation is defined as the initial moment of the single-tooth engagement period, or the initial moment of the double-tooth engagement. A vibration cycle consists of a number of complete engagement cycles and an incomplete engagement cycles. In the program, the incomplete engagement cycle is treated as the truncation error. Therefore, the greater vibration cycle considered during the calculation, the smaller the truncation error. Figure 8 shows the impact of the calculation cycle on the results. When the vibration cycle is 1, the numerical calculation results have a certain degree of fluctuation. When the vibration cycle is large enough (more than 100), the work done by the excitation force becomes stabilized and approaches the limit value. Therefore, in the calculation below, the vibration cycle is selected as 100.

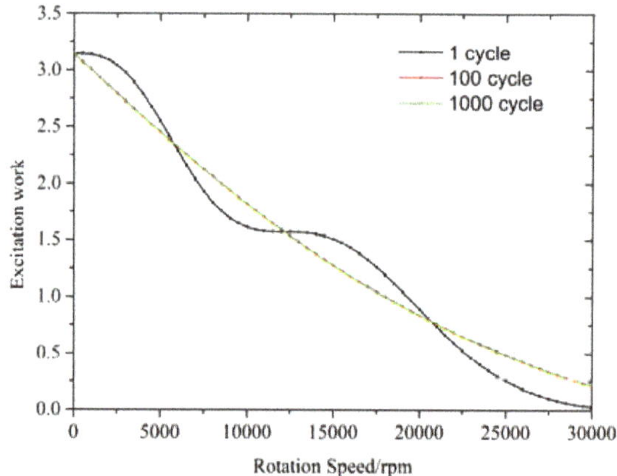

Figure 8. The impact of the calculation cycle on the results.

3.2. Stability Boundary Analysis

Figure 9 provides an example of the excitation work done to the BTW and FTW vibrations of the driven gear with two nodal diameters. The most obvious difference between the BTW and FTW is that the excitation work done to the BTW is positive, whereas the excitation work done to the FTW is negative. This means the self-excited force promotes the vibration of the BTW and decreases the vibration of the FTW. In this circumstance, the vibration energy of the BTW increases over time if there is insufficient damping, which indicates that self-excited vibration may occur. On the other hand, self-excited vibration cannot occur for the FTW vibration because of the negative excitation work, which means that the self-excited force consumes the vibration energy, behaving like damping. For the driving gear, since the excitation work done to the traveling waves is opposite to that of the driven gear, self-excited vibration will occur in the FTW vibration rather than the BTW vibration.

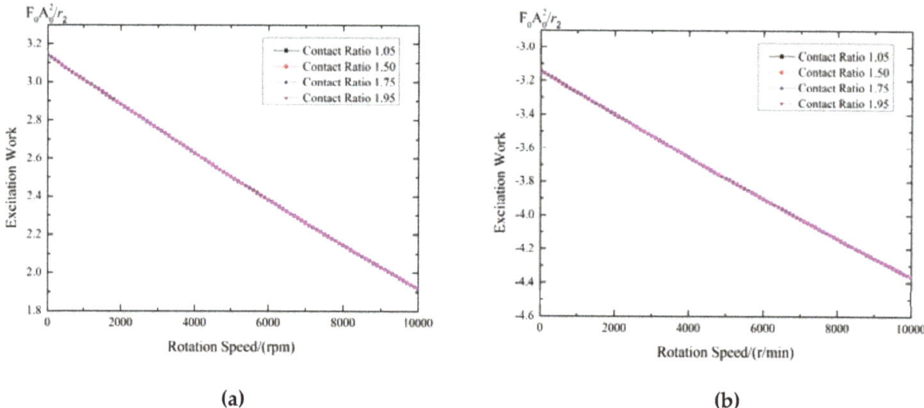

(a)   **(b)**

Figure 9. Excitation work of the vibration with two nodal diameters: (**a**) Backward traveling wave (BTW) and (**b**) forward traveling wave (FTW).

The vibrations with three, four, and five nodal diameters have the same conclusions, and the figures will not be superfluously presented here. The rest of the discussion in this paper is based on the results of the backward traveling wave vibration of the driven gear as an example.

Figure 10 depicts the amplitude of the mesh force and the transmission power for the vibration with three nodal diameters when self-excited vibration occurs, which are called the critical mesh force and the critical transmission power, respectively. For each of the vibration parameters, e.g., nodal diameter and contact ratio, the entire image region is divided into two parts by the value line: An unstable area and a stable area. That means if the operating point is within the unstable area, then self-excited vibration will occur, and the gear will become unstable. Therefore, the possibility of self-excited vibration occurrence for thin spur gears indeed exists, and the size of the stable area determines the stability of the gear's self-excited vibration. Figure 10 shows that the stable area is not affected by the contact ratio.

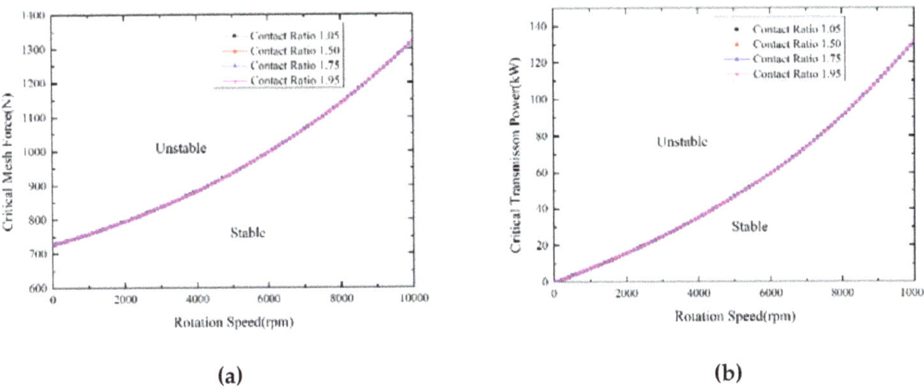

(a) **(b)**

Figure 10. (**a**) Critical mesh force and (**b**) critical transmission power for the vibration with three nodal diameters.

In order to analyze the influences of the number of nodal diameters and the damping ratio on the stability of the gear system, some additional comparisons between different nodal diameters, as well as damping ratios, are completed.

Figure 11 shows the critical mesh force and critical transmission power for vibrations with different nodal diameters. From the figure, we conclude that the number of nodal diameters considerably affects both the critical mesh force and critical transmission power. The more the number of nodal diameters, the larger the stable area. This is because a vibration with more nodal diameters always has a higher vibration frequency. As a result, the maximum vibration velocity, as well as the maximum kinetic energy of the system, are greater than those vibrations of the smaller nodal diameters, resulting in larger damping work for the same damping ratio. Thus, through the discussions above, we conclude that when the other conditions are constant, the vibration with fewer nodal diameters more easily creates self-excited vibration and becomes unstable.

Figure 12 shows the critical mesh force and critical transmission power for vibrations with different damping ratios. The results show that the damping ratio influences both the critical mesh force and critical transmission power. In fact, according to Equation (20), the capacity of the gear stability linearly increased along with the increase of the damping ratio. Therefore, increasing the damping ratio is one of the most effective methods to improve the stability and avoid self-excited vibration for the selected gear operating condition.

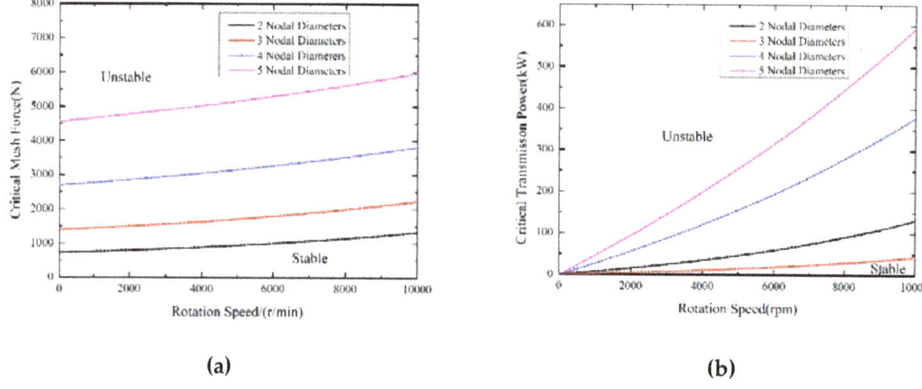

(a)   **(b)**

Figure 11. Influences of the nodal diameters: (**a**) Critical mesh force and (**b**) critical transmission power.

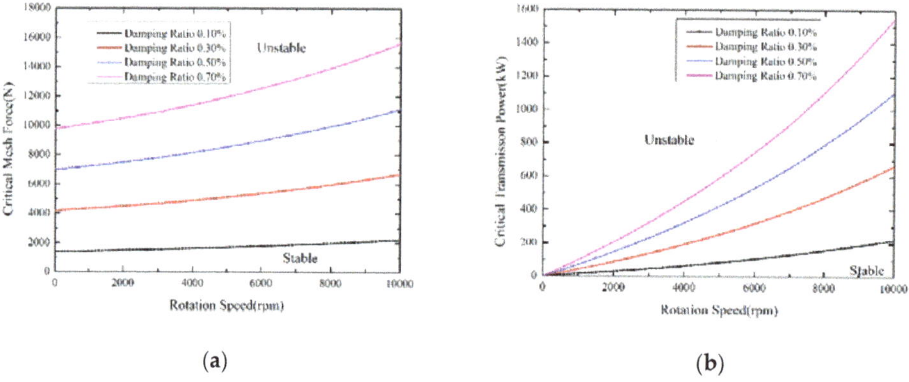

(a)   **(b)**

Figure 12. Influences of the damping ratios, $m = 3$, $\varepsilon = 1.75$ for (**a**) critical mesh force and (**b**) critical transmission power.

The results for the driving gear are similar to those of the driven gear discussed above, and are not present in this paper.

4. Conclusions

This paper addresses the self-excited vibration of a thin spur gear caused by the initial transverse vibration. Based on the theory of self-excited vibration, a method is proposed to predict the condition of the self-excited vibration occurrence and the capacity of the self-excited vibration stability for a thin spur gear. The self-exciting force and the associated excitation and damping work are analyzed theoretically, and a numerical simulation is performed to verify the theoretical analysis. The operating conditions when the self-excited vibration occurs are predicted, and the influences of the nodal diameter, contact ratio, and damping ratio on the gear stability are also discussed. The results from the numerical simulation prove that the proposed method is capable of predicting the stability of the gear system and several main conclusions from the simulation are drawn as follows:

(1) The additional axial force caused by the initial transverse vibration is the self-exciting force of the gear system, which promotes backward traveling wave vibration of the driven gear and forward traveling wave vibration of the driving gear. That means self-excited vibration may occur under these cases.

(2) The effect of the contact ratio on the excitation work and the stability can be neglected. The contact ratio is not a key parameter that affects the self-excited vibration of the gear.

(3) The nodal diameter significantly influences both the critical mesh force and critical transmission power, which represent the capacity for stability, indicating that the vibration with a smaller nodal diameter would more easily encounter self-excited vibration.

(4) The damping ratio also has a significant impact on system stability. Increasing the damping ratio is the most effective method for improving the stability and avoiding self-excited vibration for a selected operating condition for the gear system.

As the method proposed in this paper can predict self-excited vibration in spur gears, it is useful for troubleshooting the transmission box, especially for those with thin gears. The method can also be applied in industry for the analysis of gear crack failures with non-resonance conditions.

Author Contributions: Conceptualization and Methodology, Y.W. Software development and data analysis, L.Y. and H.Y. Writing—Original Draft Preparation, H.Y. Review and Editing, A.T.

Funding: This work was supported by the National Nature Science Foundation of China (No. 51475022).

Conflicts of Interest: The authors declare no conflict of interest.

Notation

$A(r)$	vibration deformation of gear web		U_{max}	initial system energy
A_0	vibration amplitude at the gear rim		w	vibration displacement
BTW	backward traveling wave		w_f	displacement of FTW
d_0	pitch diameter		w_b	displacement of BTW
d_1	diameter of mesh point		W_{1cbF}	excitation work of BTW at mesh point 1
$F_t(t)$	mesh force		W_{2cbF}	excitation work of BTW at mesh point 2
F_0	amplitude of the mesh force		W_{cbF}	excitation work of BTW for driven gear
$F_z(t)$	self-excited force of driven gear		W_{zbF}	excitation work of BTW for driving gear
FTW	forward traveling wave		W_F	excitation work
h	thickness of the gear web		W_D	damping work
k	excitation order		β	axial angle
m	number of nodal diameters		β'	central angle of one gear tooth
n	rotational speed (r/min)		ΔU	energy dissipated in one damping cycle
P_0	transmission power		ε	contact ratio
r	radius of the gear		ϕ	phase angle between adjacent teeth
$R(r)$	normalized deformation function		φ	circumferential angle of the mesh point
r_1	radius of the inner hole		η	loss factor
r_2	radius of the gear rim		Ω	angular speed of the gear
T_{max}	maximum kinetic of the gear		θ	circumferential coordinates
T_0	torque		ζ	damping ratio

References

1. Byrtus, M.; Zeman, V. On modeling and vibration of gear drives influenced by nonlinear couplings. *Mech. Mach. Theory* **2011**, *46*, 375–397. [CrossRef]
2. Hosseini-Hashemi, S.; Es'Haghi, M.; Taher, H.R.D.; Fadaie, M. Exact closed-form frequency equations for thick circular plates using a third-order shear deformation theory. *J. Sound Vib.* **2010**, *329*, 3382–3396. [CrossRef]
3. Leissa, A.W. *Vibration of Plates*; Report Number: NASA SP-160; NASA: Washington, DC, USA, 1969; pp. 7–36.
4. Nevzat Özgüven, H.; Houser, D.R. Mathematical models used in gear dynamics—a review. *J. Sound Vib.* **1988**, *121*, 383–411. [CrossRef]
5. Sinha, S.K. Free vibrations of a thick spinning annular disk with distributed masses at the outer edge. *J. Sound Vib.* **1988**, *122*, 217–231. [CrossRef]
6. Sinha, S.K. On free vibrations of a thin spinning disk stiffened with an outer reinforcing ring. *J. Vib. Acoust. Stress Reliab.* **1988**, *110*, 507–514. [CrossRef]
7. Cote, A.; Atalla, N.; Nicolas, J. Effects of shear deformation and rotary inertia on the free vibration of a rotating annular plate. *J. Vib. Acoust.* **1997**, *119*, 641–643. [CrossRef]

8. Suzuki, K.; Yoshida, H.; Zheng, X. Influence of initial tension on out-plane vibrations of a rotating circular plate: Advanced in dynamics and design of continuous systems. *JSME Int. J. Ser. C-Mech. Syst. Mach. Elem. Manuf.* **2002**, *45*, 54–59. [CrossRef]

9. Chen, Y.R.; Chen, L.W. Vibration and stability of rotating polar orthotropic sandwich annular plates with a viscoelastic core layer. *Compos. Struct.* **2007**, *78*, 45–57. [CrossRef]

10. Drago, R.J.; Brown, F.W. The analytical and experimental evaluation of resonant response in high-speed, lightweight, highly loaded gearing. *J. Mech. Des.* **1981**, *103*, 346–356. [CrossRef]

11. Li, S. Experimental investigation and FEM analysis of resonance frequency behavior of three dimensional, thin-walled spur gears with a power-circulating test rig. *Mech. Mach. Theory* **2008**, *43*, 934–963. [CrossRef]

12. Bogacz, R.; Noga, S. Free transverse vibration analysis of a toothed gear. *J. Arch. Appl. Mech.* **2012**, *82*, 1159–1168. [CrossRef]

13. Qin, H.; Lu, M.; She, Y.; Wang, S.; Li, X. Modeling and solving for transverse vibration of gear with variational thickness. *J. Cent. South Univ.* **2013**, *20*, 2124–2133. [CrossRef]

14. Honda, Y.; Matsuhisa, H.; Sato, S. Modal response of a disk to a moving concentrated harmonic force. *J. Sound Vib.* **1985**, *102*, 457–472. [CrossRef]

15. Ouyang, H.; Mottershead, J.E. Unstable travelling waves in the friction-induced vibration of discs. *J. Sound Vib.* **2001**, *248*, 768–779. [CrossRef]

16. Lee, C.W.; Kim, M.E. Separation and identification of travelling wave modes in rotating disk via directional spectral analysis. *J. Sound Vib.* **1995**, *187*, 851–864. [CrossRef]

17. Tian, J.; Hutton, S.G. Traveling-wave modal identification based on forced or self-excited resonance for rotating discs. *J. Vib. Control* **2001**, *7*, 3–18. [CrossRef]

18. Parker, R.G.; Vijayakar, S.M.; Imajo, T. Non-linear dynamic response of a spur gear pair: Modelling and experimental comparisons. *J. Sound Vib.* **2000**, *237*, 435–455. [CrossRef]

19. Tamminana, V.K.; Kahraman, A.; Vijayakar, S. A study of the relationship between the dynamic factors and the dynamic transmission error of spur gear pairs. *J. Mech. Des.* **2006**, *129*, 75–84. [CrossRef]

20. Oliveri, L.; Rosso, C.; Zucca, S. Influence of actual static transmission error and contact ratio on gear engagement dynamics. In Proceedings of the 35th IMAC, A Conference and Exposition on Structural Dynamics, Garden Grove, CA, USA, 30 January–2 February 2016.

21. Liu, Z.X.; Liu, Z.S.; Zhao, J.M.; Zhang, G.H. Study on interactions between tooth backlash and journal bearing clearance nonlinearity in spur gear pair system. *Mech. Mach. Theory* **2017**, *107*, 229–245. [CrossRef]

22. Wei, J.; Bai, P.X.; Qin, D.T.; Lim, T.C.; Yang, P.W.; Zhang, H. Study on vibration characteristics of fan shaft of geared turbofan Engine with sudden imbalance caused by blade off. *J. Vib. Acoust.* **2018**, *140*, 041010. [CrossRef]

23. Liu, F.H.; Jiang, H.J.; Liu, S.N.; Yu, X.H. Dynamic behavior analysis of spur gears with constant & variable excitations considering sliding friction influence. *J. Mech. Sci. Technol.* **2016**, *30*, 5363–5370. [CrossRef]

24. Rosso, C.; Bonisoli, E. An unified framework for studying gear dynamics through model reduction techniques. In Proceedings of the 34th IMAC, A Conference and Exposition on Structural Dynamics, Orlando, FL, USA, 25–28 January 2016.

25. Hotait, M.A.; Kahraman, A. Experiments on the relationship between the dynamic transmission error and the dynamic stress factor of spur gear pairs. *Mech. Mach. Theory* **2013**, *70*, 116–128. [CrossRef]

26. Moss, J.; Kahraman, A.; Wink, C. An experimental study of influence of lubrication methods on efficiency and contact fatigue life of spur gears. *J. Tribol.* **2018**, *140*, 051103. [CrossRef]

27. Talbot, D.; Sun, A.; Kahraman, A. Impact of tooth indexing errors on dynamic factors of spur gears: Experiments and model simulations. *J. Mech. Des.* **2016**, *138*, 093302. [CrossRef]

28. Costantino, C.; Paola, F.; Gabriele, M.; Giovanni, D.C. Numerical investigation on traveling wave vibration of bevel gears. In Proceedings of the ASME 2007 International Design Engineering Technical Conferences Computers and Information in Engineering Conference, Las Vegas, NV, USA, 4–7 September 2007.

29. Bucher, I. Estimating the ratio between travelling and standing vibration waves under nonstationary conditions. *J. Sound Vib.* **2004**, *270*, 341–359. [CrossRef]

30. Talbert, P.B.; Gockel, R.R. Modulation of gear tooth loading due to traveling wave vibration. In Proceedings of the ASME 2003 International Design Engineering Technical Conferences and Computers and Information in Engineering Conference, Chicago, IL, USA, 2–6 September 2003.

31. Yang, L.; Wang, Y.R. Prediction on self-excited vibration of thin spur gear based on energy method. *J. Aerospace Power* **2016**, *31*, 241–248. (In Chinese) [CrossRef]

32. De Silva, C.W. *Vibration Damping, Control, and Design*, 3rd ed.; CRC Press: Boca Raton, FL, USA, 2007; ISBN 978-142-005-321-0.

MDPI

St. Alban-Anlage 66

4052 Basel

Switzerland

Tel. +41 61 683 77 34

Fax +41 61 302 89 18

www.mdpi.com

Symmetry Editorial Office

E-mail: symmetry@mdpi.com

www.mdpi.com/journal/symmetry